Studies in Big Data

Volume 31

Series editor

Janusz Kacprzyk, Polish Academy of Sciences, Warsaw, Poland
e-mail: kacprzyk@ibspan.waw.pl

About this Series

The series "Studies in Big Data" (SBD) publishes new developments and advances in the various areas of Big Data- quickly and with a high quality. The intent is to cover the theory, research, development, and applications of Big Data, as embedded in the fields of engineering, computer science, physics, economics and life sciences. The books of the series refer to the analysis and understanding of large, complex, and/or distributed data sets generated from recent digital sources coming from sensors or other physical instruments as well as simulations, crowd sourcing, social networks or other internet transactions, such as emails or video click streams and other. The series contains monographs, lecture notes and edited volumes in Big Data spanning the areas of computational intelligence incl. neural networks, evolutionary computation, soft computing, fuzzy systems, as well as artificial intelligence, data mining, modern statistics and Operations research, as well as self-organizing systems. Of particular value to both the contributors and the readership are the short publication timeframe and the world-wide distribution, which enable both wide and rapid dissemination of research output.

More information about this series at http://www.springer.com/series/11970

Sergio Flesca · Sergio Greco
Elio Masciari · Domenico Saccà
Editors

A Comprehensive Guide Through the Italian Database Research Over the Last 25 Years

Springer

Editors
Sergio Flesca
University of Calabria
Rende
Italy

Elio Masciari
ICAR-CNR
Rende
Italy

Sergio Greco
University of Calabria
Rende
Italy

Domenico Saccà
University of Calabria
Rende
Italy

ISSN 2197-6503 ISSN 2197-6511 (electronic)
Studies in Big Data
ISBN 978-3-319-87187-5 ISBN 978-3-319-61893-7 (eBook)
DOI 10.1007/978-3-319-61893-7

Softcover reprint of the hardcover 1st edition 2017

Printed on acid-free paper

This Springer imprint is published by Springer Nature
The registered company is Springer International Publishing AG
The registered company address is: Gewerbestrasse 11, 6330 Cham, Switzerland

Preface

Database research played a crucial role for the growth of the computer science field since its early stage. The research on this topic has greatly evolved during the last 25 years, and many research directions have been explored that had a relevant impact on several industrial applications: data mining and knowledge discovery, information extraction, semi-structured data and Web information systems to cite a few can be considered today as mature research fields with numerous algorithms and studies to manage heterogeneous data. Indeed, all these application scenarios have their roots in database field. Moreover, the advances in information technology lead to a new data management paradigm, referred to as Big Data, that will drive the research and the industry in the future.

The Italian Database Community has been active in all those research fields, and many well-reputed Italian scientists provided interesting original contributions.

This book sets out to give the reader a comprehensive view of the evolution of the database research field from classical relational database management systems to the current Big Data metaphor. The individual contributions of this book will illustrate the point of view of several research groups in Italy that had in the last 25 year quite active collaboration with leading research groups spread all around the world.

This book will be of great interest for readers willing to catch the evolution of the research on databases from a different perspective yet in an unified proposal. It is composed of five parts and a total of twenty-nine chapters.

Part I focuses on **Big Data** by illustrating several aspects ranging from data pre-elaboration to data warehousing. It consists of seven chapters. Part II analyses issues posed by **Incomplete/Inconsistent Data and Uncertain Reasoning** dealing with ontologies, schema mappings and uncertain queries. It consists of five chapters. Part III illustrates results related to **Data Modelling and Querying**. It consists of four chapters. Part IV analyses issues posed by **Knowledge Discovery and Data Mining** from relational data, social networks and complex data. It consists of eight chapters. Finally, Part V deals with **Security, Privacy and Health Systems** issues. It consists of five chapters.

We would like to thank all the authors who submitted their contributions for publication in this book. We also thank Thomas Ditzinger and Janusz Kacprzyk of Springer for their continuous support.

Rende, Italy
June 2017

Sergio Flesca
Sergio Greco
Elio Masciari
Domenico Saccà

Contents

Part I
Big Data

The Visual Side of the Data

Marco Angelini, Tiziana Catarci, Massimo Mecella and Giuseppe Santucci

Abstract In the last decades, visually querying data and visualizing information have been investigated in order to allow users to get insights and extract knowledge from data. Nowadays, these functionalities should be adapted to big data, including streaming ones. In this chapter, we will review the main approaches to visual queries and provide an historical overview of information visualization.

Keywords Big data · Visual query languages · Information visualization

1 Introduction

A Visual Query System (VQS) can be defined as a system that uses a visual representation for both the domain of interest and requests related to such a domain. The first graphical query language which was referred to as Query-By-Example was introduced in the mid-1970s (QBE, [76]). A wide range of implementations were built using the QBE concepts and there are several tools using this paradigm today.

Since the purpose of any VQS is to provide access to the information contained in a database, the main users' tasks are understanding the database content, focusing on meaningful items, finding query patterns, and reasoning on the query result. These tasks require specific techniques to be effectively accomplished, and such techniques

M. Angelini · T. Catarci · M. Mecella (✉) · G. Santucci
Dipartimento di Ingegneria Informatica Automatica e Gestionale Antonio Ruberti,
Sapienza Università di Roma, via Ariosto 25, 00185 Roma, Italy
e-mail: mecella@diag.uniroma1.it

M. Angelini
e-mail: angelini@diag.uniroma1.it

T. Catarci
e-mail: catarci@diag.uniroma1.it

G. Santucci
e-mail: santucci@diag.uniroma1.it

S. Flesca et al. (eds.), *A Comprehensive Guide Through the Italian Database Research Over the Last 25 Years*, Studies in Big Data 31, DOI 10.1007/978-3-319-61893-7_1

involve typical activities such as pointing, browsing, filtering, and zooming; all activities that nicely fit with visual representations and direct manipulation mechanisms. For instance, if the result of an information request can be organized as a visual display, or a sequence of visual displays, the information throughput is immensely superior to the one that can be achieved using only textual support, and the users can directly point at the information they are looking for, without any need to be trained in the complex syntax of query languages. Alternatively, users can navigate in the information space, following visible paths that will lead them to the target items. Again, thanks to the visual support, users are able to easily understand how to formulate queries and they are likely to achieve the task more rapidly and less prone to errors than with traditional textual interaction modes.

In modern VQSs, *information visualization* (a.k.a. infovis) mechanisms are used for displaying the query results and making sense of them. Indeed, infovis relies on basic features that the human perceptual system inherently assimilates very quickly: color, size, shape, proximity, and motion. These features can be used by the designers of information systems to increase the data density of the displayed information. Because the users perceive such features in a natural way, and because each feature can be used to represent different attributes of data, good visualizations allows for not only perceiving results in a simple way but also to perceive more information at one time. In this way visualizations can reduce the search for data by grouping or visually relating information. While visualizations compact information into a small space, they can also allow hierarchical search by using overviews to locate areas for more detailed search. In fact, they also allow zooming in or popping up details on demand. Infovis, through aggregation and abstraction, enables users to recognize gaps in the data, discover outliers or errors, pinpoint minimum and maximum values, identify clusters, compare objects, visually draw some conclusions, discover trends and patterns.

The term "big data" refers to structured or unstructured data sets that are hard to store and process using common software tools (e.g., relational databases), regardless of the computing power or the physical storage at hand. Typically, *volume*, *velocity*, and *variety* are used to characterize the key properties of big data. They are the so-called three V's of big data.

Table 1 Main characteristics of UNECE data sources. In a spectrum, human-sourced are the less structured data and machine-generated are the more structured data. Process-based data have mixed characteristics of human-sourced and machine-generated data

Source	Structure	Human influence
Human-sourced	Loosely structured	Direct
Process-mediated	Structured	Indirect (e.g., data entry activities)
Machine-generated	Well structured	None

According to a *classification* proposed by *UNECE (United Nations Economic Commission for Europe)*,[1] there are three main types of data sources that can be viewed as big data: *human-sourced* (e.g., blog comments), *process-mediated* (e.g., banking records), and *machine-generated* (e.g., sensor measurements, including streaming data), cf. Table 1 for a summary.

In this chapter, we will first review the main approaches to visual queries, and later on we will consider an historical view on infovis, by specifically addressing recent trends as visualization of big data, including streaming ones.

2 Classifying Visual Query Systems

A general overview and classification of VQSs can be found in [16]. According to the visual representation adopted for the database and the queries, VQSs are categorized into form-based, diagram-based, icon-based, and a combination of these.

A form is a generalization of a table, and it is possible to represent relationships among cells, subset or the overall set, allowing a three level answer. There are VQSs in which it is possible to manipulate both the intensional and extensional parts of the database, focusing on different parts of the dataset. Diagrams are frequently used in VQSs, which generally use some visual components (e.g., shapes, colors, arrows) that are univocally mapped into a concept. In an icon-based system there is a mapping between a real concept and an icon that hides the schema of the data. It is possible to query the database by combining icons according to spatial concepts. The main problem in designing an iconic system is to define a non-ambiguous mapping. While different attempts are made to find a common mapping, still there are no universal standards.

Another possible categorization is made by considering the top-down strategies that are used to understand the reality of interest: iterative refinement, selective hierarchical zoom, or user-system dialogue. A different approach relies on browsing, which enables getting more knowledge by exploring the neighbourhood concepts. Browsing can take various forms, including extensional browsing, intensional browsing, or mixed browsing. An alternative approach is schema simplification, which brings the schema close to the query. This can be realized through transformations of concepts of the original schema in a user view.

The Visual Query Languages (VQLs) are also classified according to the query formulation strategy. In a schema navigation strategy, the user starts from a concept and can reach the other concepts of interest. There can be different paths to navigate the schema. The first possibility is to use an arbitrary path to explore the schema, reach the concepts of interest and apply condition(s) on them. It is also possible to select one concept from the database and then navigate the schema by a hierarchical view built using the chosen concept as a root. Moreover, users can choose the starting

[1] cf. `http://www1.unece.org/stat/platform/display/bigdata/Classification+of+Types+of+Big+Data`, accessed February 2017.

Table 2 Classification of approaches

Approach	Visual representation	Strategy for understanding the reality	Strategy for query formulation
QBD* [5]	Diagram	Browsing	Sub-queries
MURAL [63]	Diagram	Browsing	Sub-queries
QBB [54]	Icon	Browsing	n/a
QBI [49]	Icon	Filtering	Select-project
Flow [50]	Hybrid	Filtering	Design a flow
Kaleidoquery [51]	Hybrid	Filtering	Design a flow
VISUAL [10]	Hybrid	n/a	Hierarchical

concept and then build their own relationships. A second possible strategy in the query formulation process is by using sub-queries. This can be accomplished using the following two approaches: by composition of concepts, usually an iconic based language in which several icons are combined to produce the final query, or reusing previously defined queries. Another strategy for query formulation is by matching, which can be done by example or by pattern. In a matching-by-example strategy, users can provide an example of a query and the system generalizes the example and builds the query. In a pattern-matching strategy, the system searches in the database for a pattern specified by the user. The last strategy for query formulation is by using range selection. In this strategy, it is possible to specify a range on different data-set through graphical widgets.

A summary view of the most prominent approaches is reported in Table 2, whereas Fig. 1 report some historical screenshots of them.

2.1 Comparing VQLs

A VQL should provide different kinds of interaction because there is no unique paradigm that leads to the best results. An empirical experiment [9] about the ease of use of two different query languages shows that there can be some advantages as well as disadvantages, in both iconic and diagram-based approaches. In the experiment comparing QBD and QBI systems, different strategies are used for the query formulation (navigation vs. composition), recursive algorithms [59] vs stored queries, as well as some different visual formalisms (diagrams vs. icons), which are basic aspects of a VQL. Thus, the results can be extended to larger classes of VQSs. The experiment focuses on discovering which relation occurs between the query language type and both the query class and the experience of the user. In particular, the queries were classified according to the semantic distance of the path involved in the query and the overall number of the cycles in the query, where the notion of path derives from the

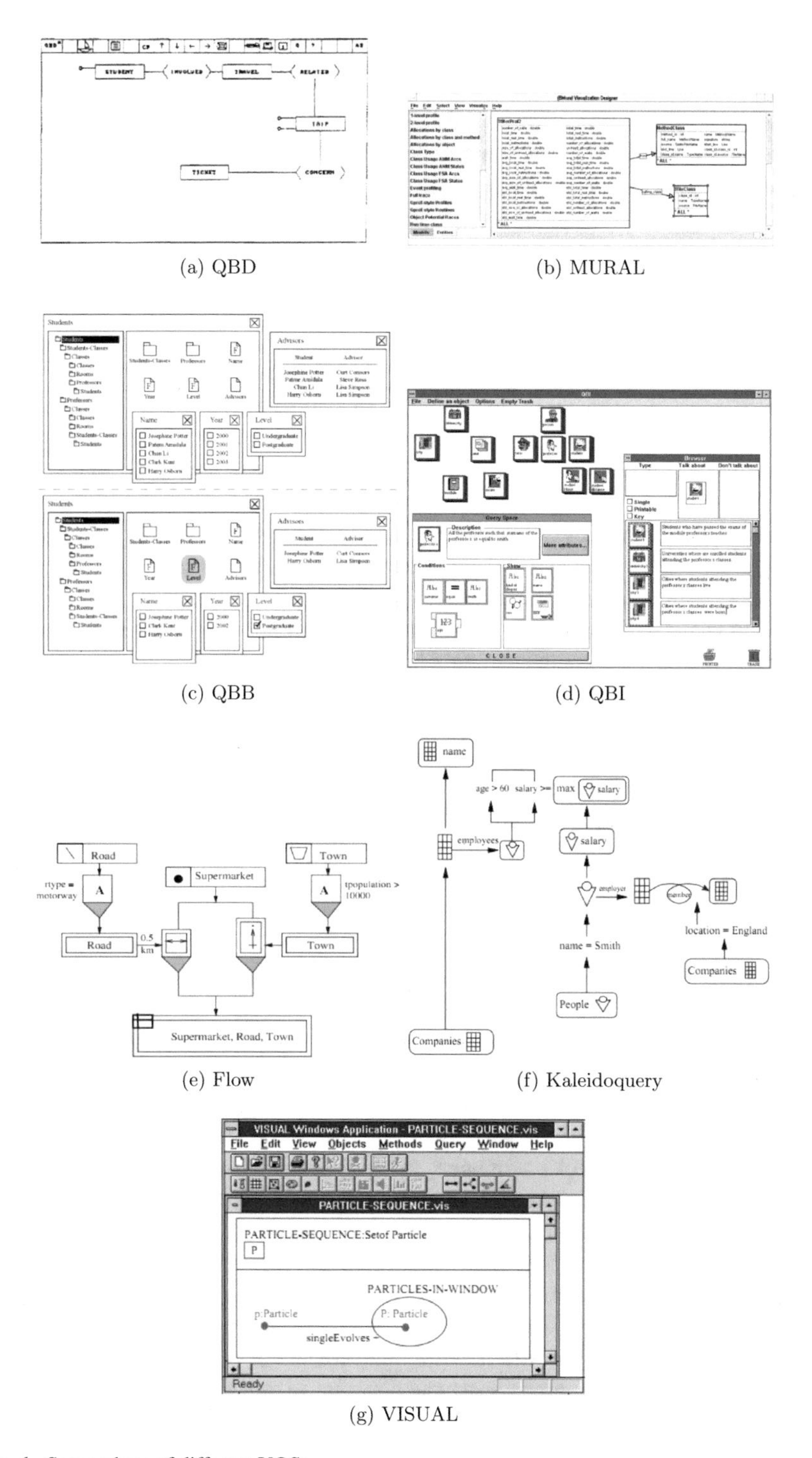

(a) QBD (b) MURAL

(c) QBB (d) QBI

(e) Flow (f) Kaleidoquery

(g) VISUAL

Fig. 1 Screenshots of different VQSs

Graph Model described in QBI. The main result is that both accuracy and response time seem to be highly sensitive to the semantic distance of the query path, whilst QBD shows independence for both criteria. In addition, QBD is less accurate and requires more time when there are cycles in the query, and QBI seems not to be affected by the presence of cycles.

3 Visual Querying of Semantic and Big/streaming Data

VQSs mainly deal with traditional databases, i.e., databases containing alphanumeric data. However, in recent years the application realms of databases have increased greatly in terms of both the number and variety of data types. As a consequence, specialized systems have been proposed for accessing such new kinds of databases, containing non-conventional data, such as images, videos, temporal series, maps, etc. Furthermore, the idea of information repository has been deeply influenced by the beginning of the Web age. Different visual systems have been proposed to cope with the need for extracting information residing on the Web. In particular, providing users with visual representations and intuitive user interfaces can significantly aid the understanding of the domains and knowledge represented by ontologies and linked data.

As ontologies grow in size and complexity, the demand for comprehensive visualization and sophisticated interaction also increases. Ontology visualization and ontology visual querying are not new topics and a number of approaches have become available in recent years, e.g., [17, 45, 62], with some being already well-established, although more work is needed to provide the users with powerful querying and navigational aids and comprehensive visualization techniques.

Another important class of VQSs are those ones specifically targeted for data stream processing that, despite its growing importance, has not been covered sufficiently in the literature. As in classical VQLs, they provide a language, consisting of a set of visual constructs, to express, in a visual format, queries on data transmitted in a continuous and unbounded fashion (i.e., data streams). It is important to note that this class of languages can be seen as an extension of the generic visual query languages, due to the fact that they can query in a visual manner using the same criteria of classical VQLs both data streams and classical relational databases. They are oriented to a wide spectrum of users, even those ones who have some knowledge about concepts related to data streaming but with no skills in developing code. The importance of data streams is continuously increasing. At the same time, all the existing VQLs working on classical databases are not good anymore for interacting with this huge and potentially infinite amount of data. VQLs for data streams are being developed to address these challenges. Basically, they use the same approach of the classical VQLs extending it with the new data stream operators. However, querying a data stream usually requires constructs different than a relational language. Due to the lack of a standard proposal, a large number of academic and commercial data stream query languages and their corresponding Data Stream Management Systems

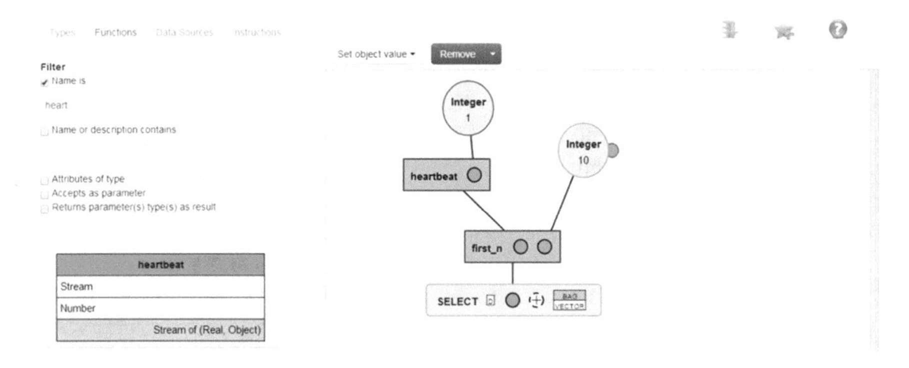

Fig. 2 The visual query editor of the SmartVortex Visual Query System

(DSMSs) have been developed. The main functionalities of these languages rely on some relational query language, typically SQL, which is extended to provide features dealing with characteristics present in streams, mainly temporal aspects, i.e., the fact that events change over time (e.g. window and filtering operators). Some examples of data stream query languages are SARI-SQL, EP-SPARQL, CQL, Espers EPL , SiddhiQL and StreamSQL. Although there are a lot of data stream query languages, there are only few VQLs able to also interact with data streams. Only in recent years, the first VQLs for data streams have appeared, either in the form of research prototypes or real commercial systems. In particular, StreamSQL, along with its graphical counterpart EventFlow, is used within the StreamBase Complex Event Processing system, a commercialization of the Aurora project [1]. To date, this is the only existing commercial system.

As for research prototypes, the SmartVortex Visual Query System [11], has a powerful Visual Query Editor (VQE) containing the canvas and four tabs with all the necessary visual constructs to build a query. The VQE can be seen in Fig. 2. The VQL is built on top of a Federated Data Stream Management System and generates its object-relational and functional query language. Therefore, the four tabs of the VQE contain all the types and functions of the textual language along with the instructions to develop the query. Since the system is intended for use in industrial applications, the last tab is specific to this context, containing all the machines/sensors managed by the system (i.e., the input streams). The four tabs along with an example of visual query are shown in Fig. 2.

4 Data Visualization

After the analysis of visual querying of data, we revert our attention to (big) data visualization. Data visualization has its root in ancient times before the computer and automation was present, with the first example of data visualization identifiable in the Minard's map (1869) of disastrous Napoleonic campaign; this static drawing was able to summarize mulch multi-dimensional information, like the decreasing trend in soldiers number during the campaign, correlated with geographic elements (bridges, rivers), temperature trends and major battles. With the emergence of computers, the need for visualizing data eventually arose. From a visualization point of view, the need to visualize huge amount of data resulted almost immediately, in problems related to low computational power and lacking of visualization primitives and theoretical concepts. Nonetheless, this complexity represented the starting point for researching and developing novel solutions. In what follows, we will make an overview of this evolution in (big) data visualization.

4.1 1980–1993: From Scientific Visualization to Abstract Visualization

In the second half of the decade 1980–1990, the need for a computer visual representation of a dataset arises. Previous efforts in graphical representation of a dataset were intertwined more with computer graphics work and no theoretical foundations, apart for very common charts mutuated from the maths and statistic domains, were used. One of the reasons was the low graphical capabilities of the computers, not allowing complex visual representations. Interestingly enough, the problem of having to cope with large data sources ("big" for the time) was present even at that time, e.g., [22] citing large data sources in various field, like military intelligence, weather forecast, astronomy, geophysics, medical scanners, with their magnitudo as big as gigabytes of data per day. [22] is one of the first to propose the visualization, targeted at the scientific computing, as a solution for allowing the interpretation and comprehension of a big data set. The scientific community, in fact, was the first to experiment with visual representations of scientific data sets, ranging from volume rendering to iso-surfaces visualization to tensors and flows visualizations. First attempts were focused on precise mapping of the data into a visual representation and can be found in [27] relatively to the simulation of a tornadic storm (see Fig. 3a) and the representation of a neuron (see Fig. 3b).

As stated above, the low computational power allowed most of the time for very approximated graphical representations, that were in most cases completely static or at a non interactive frame-rate ($<$ 10 frames per second). Nonetheless, the search for more abstract and powerful ways of visually representing information can be traced in the two works [36, 38]: both of them present a 2D data visualization; the first solution copes with the dimensionality of the data, representing each tuple as a segment that intersect a series of vertical axes, each of them representing a dimension of the dataset. In this way multidimensional data can be compactly represented on a large number of dimensions (order of tens of dimensions) allowing fast recognition of possible correlations/anti-correlations and distribution of the dataset on the different dimensions (see Fig. 4a).

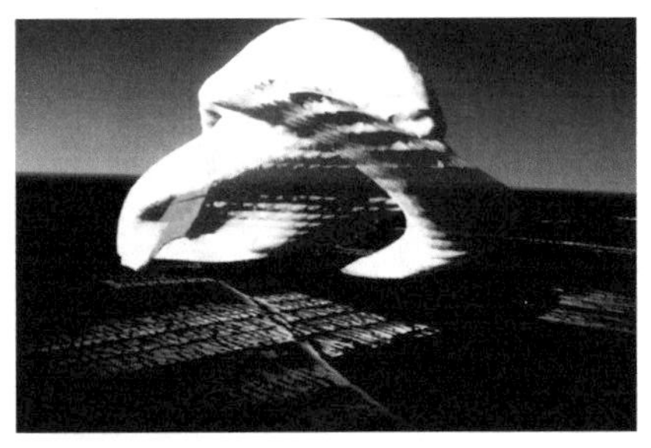

(a) Volume rendering of a storm

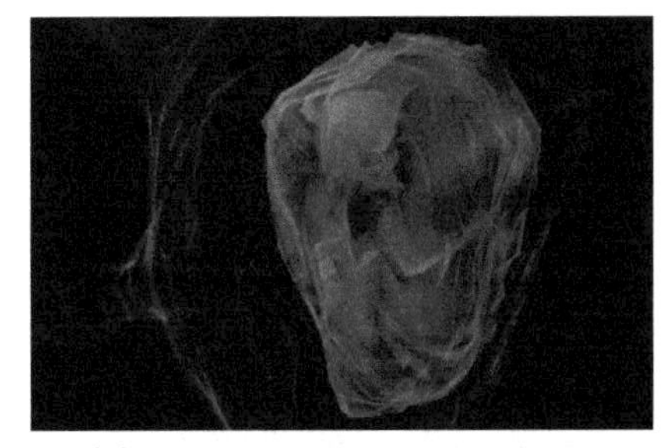

(b) Visual rendering of a neuron

Fig. 3 **a** - on the *left*, a volume rendering representing a storm derived by scientific data. **b** - on the *right*, yielding clues into the causes of Alzheimer's disease, this neuron's surface membrane was rendered using polygons to connect the slices. Surfaces of interior structures, whose distribution in the cell is also being studied, can be rendered similar

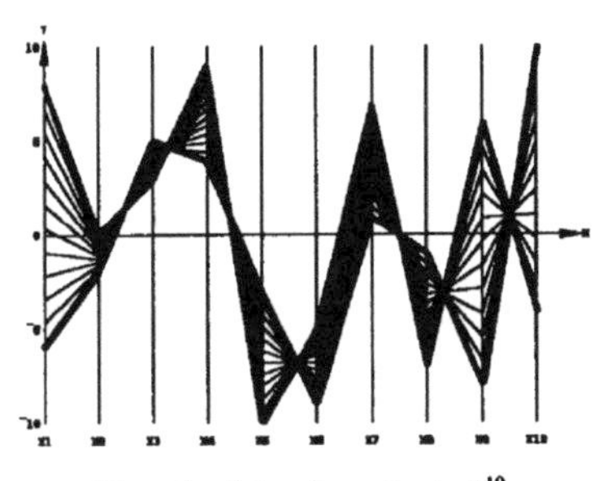

(a) Parallel coordinates

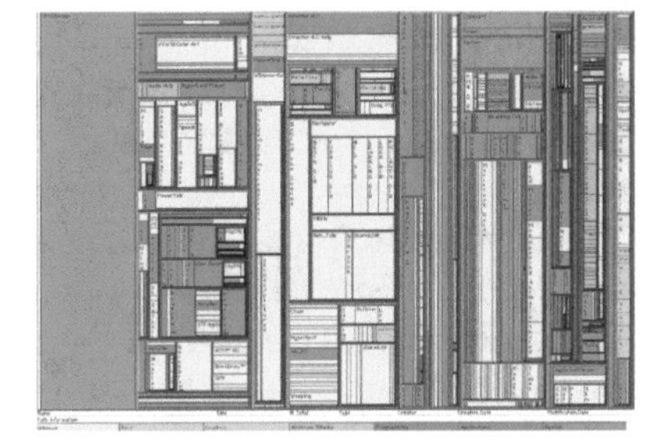

(b) Treemap

Fig. 4 **a** - parallel coordinates, **b** - treemap

The second solution instead provides a compact visual representation for hierarchical datasets, representing some common dimensions (e.g. cardinality of elements) as rectangles, and reiterating the same behaviour on the lower levels of the hierarchy. The final representation is like a tree observed from the top, therefore the name of treemap (see Fig. 4b).

Since these seminal works, more effort is spent towards the abstract representation of collected data. In 1993, the work in [42] proposes a first example of pixel-based representation of query results from large databases, with tens of thousands of data items. This approach first reduces the cardinality of the data to be displayed and then maps them to the pixel space using several distance functions (see Fig. 5a). The layout starts from the center for the exact matching results and then continues following a spiral. The more a data item is in the center, the more it fits to the query. The color scale can map again the fitness of the result with respect to the query (top-left) or can be mapped on the distribution of a single parameter of the query, allowing to judge data item relevance on a single predicate.

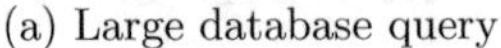
(a) Large database query

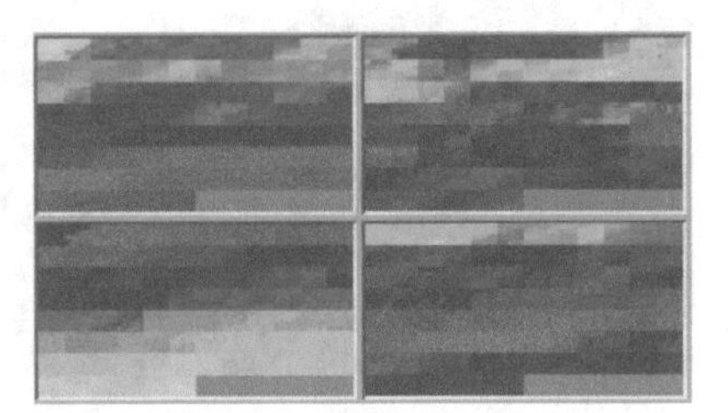
(b) Recursive patterns

Fig. 5 **a** - large database query, **b** - recursive patterns

A generalization of the pixel-based approach is presented in [7, 39, 41]; particularly interesting is the work in [40], where the strategy used to improve the final visualization is organizing the pixels in small groups and then arranging the groups to form some global pattern. This process can then be reiterated to an arbitrary level, as shown in Fig. 5b for the IBM stock trends in the year 1987–1995 (about 50.000 values).

This approach has a potential limitation in the number of pixels of the screen. Since in general these techniques only use one pixel per data value, they allow for visualizing the largest amounts of data which are possible with respect to display resolution (different depending on the times, nowadays 1 million data values for a full-HD display). In order to overcome this limitation, a technique is to use interactive environments that allow the user to focus on single portions of the dataset. An example is [68], where a solution to estimate properties of the atmosphere of Venus by comparing measured and simulated data is proposed. As it is visible in Fig. 6, the user is presented with multiple simple displays simultaneously (center), where each display shows a different selection of the data, augmented with many options to simplify browsing through the data (left).

4.2 1993–2006: The Information Visualization Era - Comprehend the Data

Starting from 1995, the research in pure information visualization, independently from the scientific domain, got a proper form with the creation of the 1st symposium and then conference in Information Visualization (InfoVis). In this scenario, an acceleration in abstract dataset visualization had been achieved. Particular emphasis is given to visual ways of representing high cardinality and high dimensionality of a dataset. An example is presented in [20], instantiated on the e-mails traffic of members of a hierarchical organization. The root of the hierarchy represents the company CEO and the first level of the hierarchy shows the vice presidents within the company

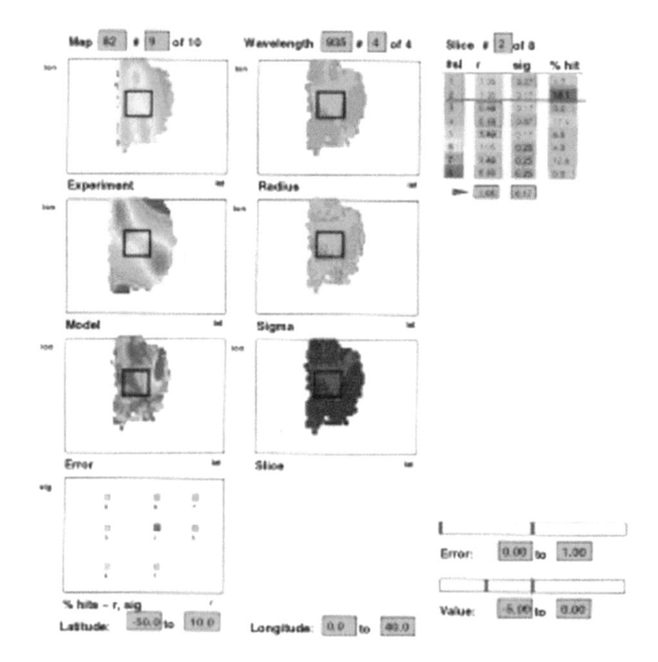

Fig. 6 Multiple selections

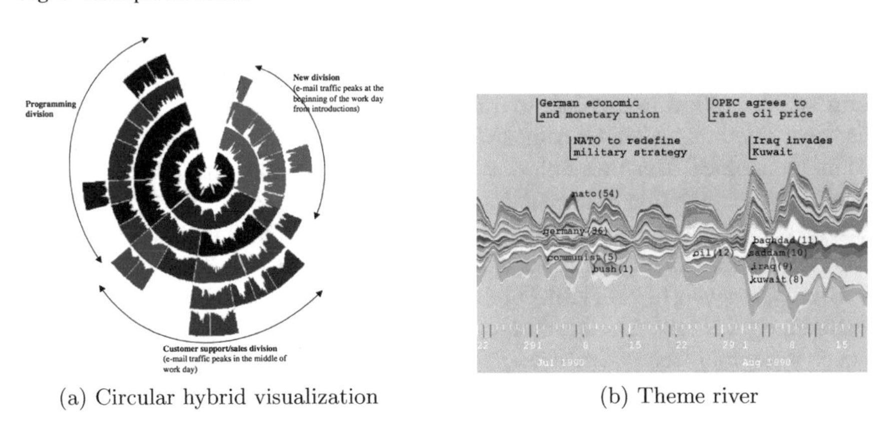

(a) Circular hybrid visualization (b) Theme river

Fig. 7 **a** - circular hybrid visualization encoding a company employee hierarchy and the e-mail traffic of each employee within the period of one day; **b** - theme river

and so on. The histograms within each node in the hierarchy represent the amount of electronic mails received by each person in the organization throughout the period of one day (see Fig. 7a). Other fields of application include software visualization [23], time-series [71], cyber-security [66].

Another technique is theme rivers, first introduced in [31]; it is an abstraction that allows for aggregating collected data in a timeline and highlights variations and

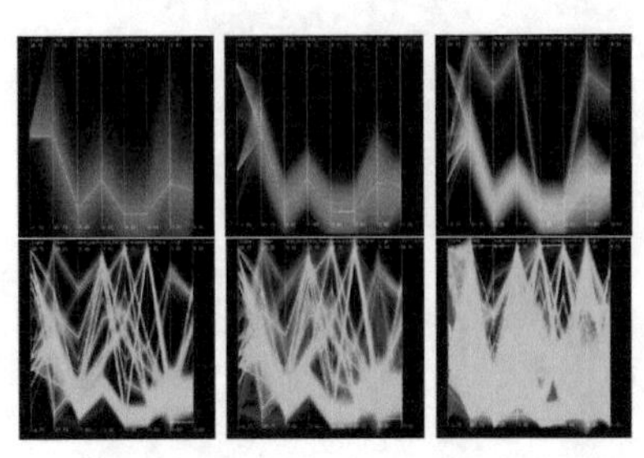

(a) Hierarchical Parallel Coordinates

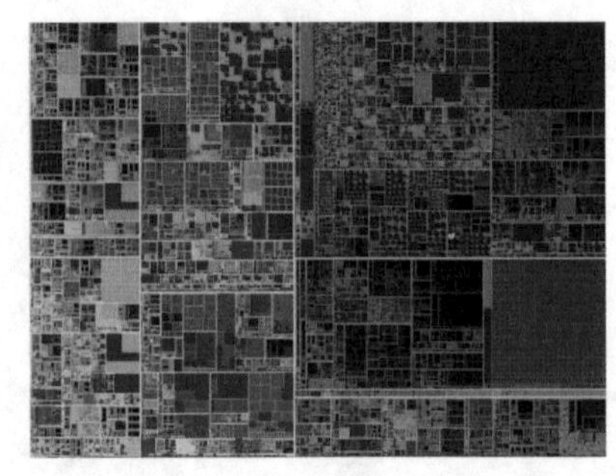

(b) Scalable Treemap

Fig. 8 **a** - hierarchical parallel coordinates; **b** - this treemap gives an overview of 970.000 files of a file system containing 1 million files (smaller files are smaller than one pixel and not counted), on a 1600x1200 display. The size of each rectangle is determined by the file size; color represents file type. Deeply nested directories appear darker

zones of interest; the data, grouped into categories, are visualized as a set of continuous stacked flows. Figure 7b shows its application to explore the 1990 Associated Press (AP) newswire data from the TREC5 distribution disks, a set of over 100.000 documents, highlighting the trending topics and visualizing the most discussed and their time-extension.

In the same years, different solutions were proposed to offer the capability of exploration of a dataset. The techniques rely heavily on aggregating the data in some form and then allowing the user to traverse the data towards the details, following the "overview first, details on demand" strategy [61]. A first attempt can be found in [4] for hierarchical data, then followed by a general adoption as a standard technique for allowing exploration of large datasets; [65] applies the same idea on kernel lock activities (lock requests are shown in blue; time holding a lock is shown in yellow) collected from a 125 million cycle simulation of the Argus parallel graphics library.

Given the raising of both visual power of computers (introduction of GPUs, better performant primitives and creation of visual toolkits), previously introduced techniques were improved and evolved in both scopes and functionalities. An example is the work in [28] that adapts parallel coordinates to hierarchical datasets in order to allow their exploration (see Fig. 8a). The work in [26] refines and applies the treemap technique to 970.000 files of a file system (see Fig. 8b), where the size of each rectangle represents the file size and the color the file type.

In 1999 the works [21, 43] proposed an overall approach and design solutions on how to manage the visualization process of new grown large data, i.e., hundreds of Gigabytes for simulation models. An approach that is followed is to use different visual representations of the same dataset, with the goal of highlighting particular properties that can be visible only with a specific representation. The work in [64] proposes well-known visual representations of a coffee store chain dataset, organized in an analysis environment that allows user interaction and customization (see Fig. 9).

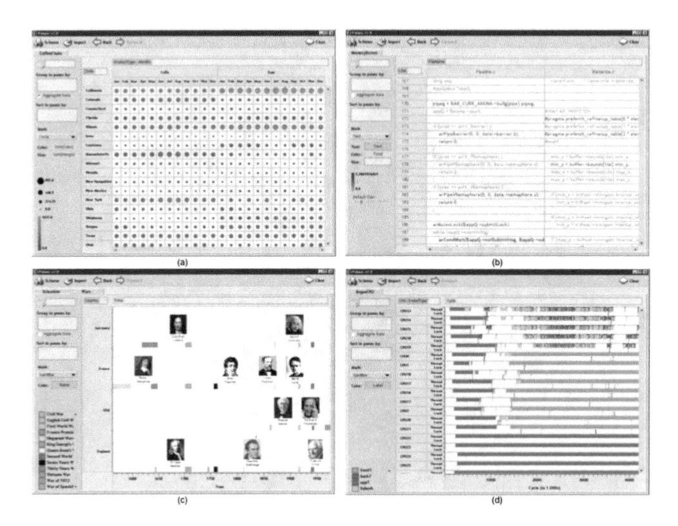

Fig. 9 Graphical table, source code display, Gantt, thread execution

While the exploration of the dataset is well-supported, the need for changing visual representation each time is detrimental to the user task execution.

The concept of steering a visualization opposed to post-processing a visualization (the classic way of generating a visualization) after obtaining the data, together with the introduction of dimensional reduction of the dataset, begins to appear in the literature in order to support the representation of even larger datasets.

The raise in data cardinality led to developing solutions for reducing the visual clutter/congestion resulting from the rendering of a sheer amount of elements on the screen [13] explicitly using formal quality metrics [14]. The work in [72] presents an interactive technique that curves graph edges away from a person's focus of attention without changing the node positions. This opens up sufficient space to disambiguate node and edge relationships and to see underlying information while still preserving node layout (see Fig. 10). [8] proposes a solution that computes frequency or density information from the data set, and uses such information in Parallel Coordinate visualizations to filter out the information to be presented to the user, thus reducing visual clutter and allowing the analyst to observe relevant patterns in the data.

The improvement in screen resolution led also to the born of multi-coordinated view environment, where different visualizations are simultaneously rendered and the user can interact with one of them and see the effect propagated on to the others. Examples can be found in [12, 33, 69]. [18] proposes a highly coordinated and interactive urban visualization tool that provides intuitive understanding of the urban data, combining buildings and city blocks into legible clusters, thus providing continuous levels of abstraction while preserving the user's mental model of the city (see Fig. 11a). [70] expands this approach by providing design strategies for constructing coordinated multiple view interfaces for cross-filtered visual analysis of

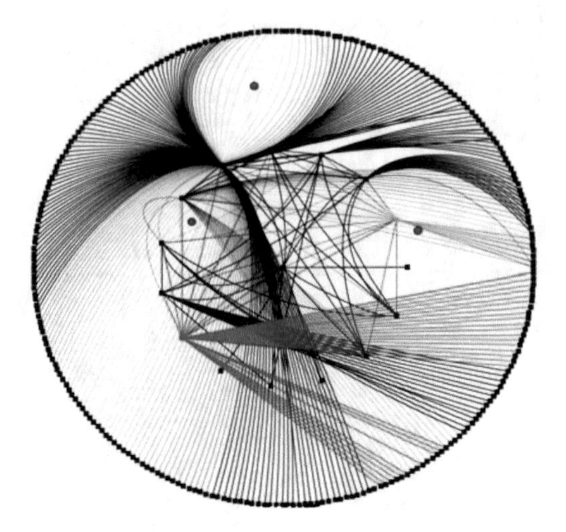

Fig. 10 Application of three EdgeLenses with selected edges excluded from the visualization

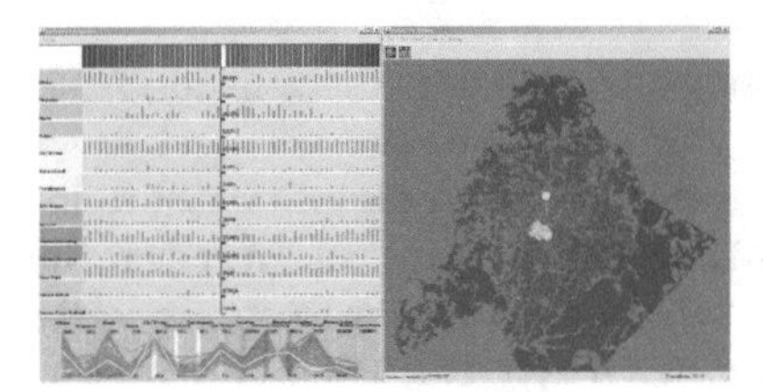

(a) UrbanVis multi-coordinated views environment

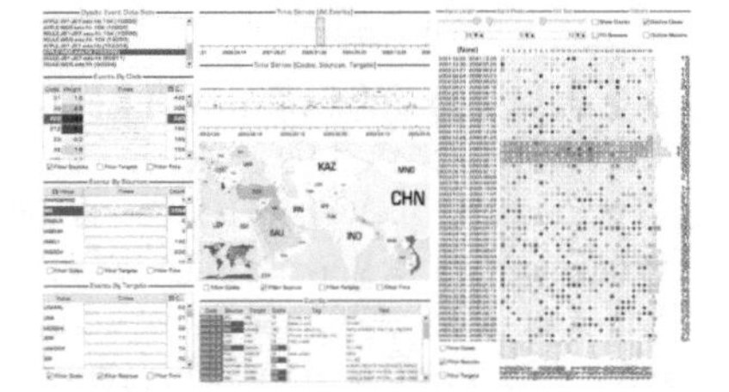

(b) Cross-filtered visualization system

Fig. 11 **a** - UrbanVis overview: the data view on the *left* shows demographic data of the areas around the focus point (focus in the middle). The model view on the *right* shows the clustered building models. The color gradient indicates the distance from the focus point, and provides a visual link between the two different data views (matrix view and parallel coordinates) and the model view. The data shown are census data for the city of Charlotte in Mecklenburg county, North Carolina. **b** - Cross-filtered visualization of geographic and temporal patterns in 150.000+ citations of political activity in international events reported by Agence France-Presse from May 1991 to January 2007. Cross-filtering on events related to Iraq reveals a spike in conflictual events in early 2003. Further cross-filtering with military engagement as the chosen event type reveals the United States military as a frequent target actor

multidimensional data sets, and evaluate them on different datasets (see Fig. 11b for France Press).

Coordinated multiple visualizations are still evolving, not necessarily following the common grid layouts but possibly experimenting with more immersive layouts; as an example, [46] proposes a layout where the entire data set is patched together from smaller visualizations; there is one VisBrick for each cluster in each group of interdependent dimensions. Whereas the total impression of all VisBricks together

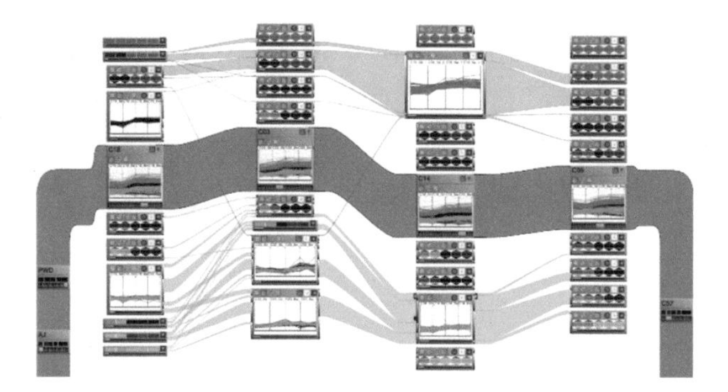

Fig. 12 Four different groups of dimensions with different numbers of clusters per group. The *gray* arch connects the overviews of the groups. The arches show how the data are distributed over the clusters in that group, thus summarizing the characteristics of a dimension group. The clusters themselves are shown in stacked VisBricks *above* and *below* the arch depending on whether their average data values are higher or lower than the overall average for the group. *Colored ribbons* indicate how data items are distributed across clusters of multiple dimension groups

gives a comprehensive high-level overview of the different groups of data, each VisBrick independently shows the details of the group of data it represents (see Fig. 12). A similar approach applied to comparative analysis of multidimensional data is presented in [35, 47].

At the beginning of 2000s, the advent of data streams and the raise in velocity of update of new data sources led to the study of progressive/incremental/out-of-core solutions in order to manage the timely update of visual feedback, in particular for scientific visualizations, see [15, 24]). The work in [73] presented an adaptive visualization technique based on data stratification to ingest stream information adaptively when influx rate exceeds processing rate, and an incremental visualization technique based on data fusion to project new information directly onto a visualization subspace spanned by the singular vectors of the previously processed neighboring data.

The growth of streaming data also led to the proposal of novel visual techniques to convey at the same time representations of aggregated values elaborated from the real trend of the stream; [34] presents visual sedimentation, a novel design metaphor for directly visualizing data streams inspired by the physical process of sedimentation. Data are visually depicted as falling objects using a force model to land on a surface, aggregating them into strata over time. The approach is applicable to different visualization paradigms, cf. Figure 13a.

Another example is in [2], where data coming from an Intrusion Detection System (IDS) are visualized using a rain-falling metaphor allowing for identifying temporal trends and actual and past alerts, cf. Figure 13b.

Social data played an important role in this, presenting challenging scenarios for visualization and exploration [30]. A first attempt of visualization of social networks is in [32], presenting node-link network layouts as customized techniques for exploring connectivity in large graph structures, supporting visual search and analysis, and automatically identifying and visualizing community structures (see

(a) The Visual Sedimentation metaphor applied to a bar chart (left), a pie chart (center), and a bubble chart (right).

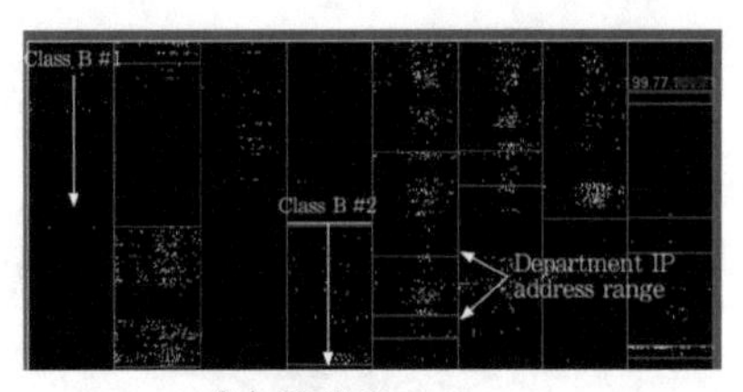

(b) IDS Rainstorm

Fig. 13 **a** - visual sedimentation; **b** - IDS RainStorm main view: the 8 vertical axes represent the 2.5 Class B IP addresses. The *thicker horizontal lines* between these axes show where each Class B starts. The other *horizontal lines* show the start and end of each department. Those addresses not in a department are either unallocated or reserved for special use by OIT and other departments. This screenshot shows the real alarms generated during an entire day

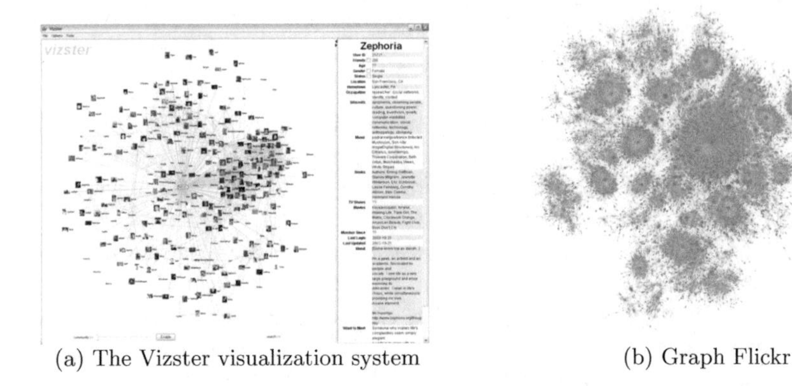

(a) The Vizster visualization system

(b) Graph Flickr

Fig. 14 **a** - the Vizster visualization system: the *left* side presents a network display with controls for community analysis and keyword search. The *right* side consists of a panel displaying a selected member's profile information. Words in the profile panel that occur in more than one profile will highlight on mouse-over; clicking these words will initiate searches for those terms. **b** - graph Flickr edge filtered with edges colored by degree

Fig. 14a). [37] proposes novel methods for specifically visualizing power-law graphs typical in social networks/sociology, cd. Figure 14b for Flickr.

4.3 2006-Present: The Visual Analytics Era - Let the Data Speak

The improved interactivity of the visualization solutions and the tight coupling with algorithmic solution based on data mining and machine learning led in 2006 to a further specialization of Information Visualization in the form of Visual Analytics.

Fig. 15 Anonymized outgoing traffic connections from University of Konstanz gateway on November 29th, 2005 showing all 197427 IP prefixes

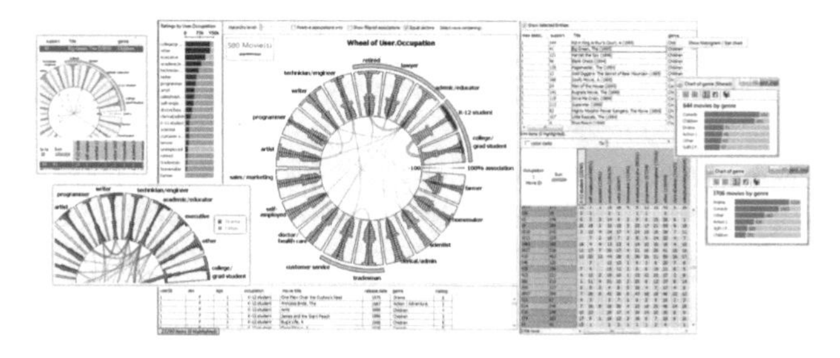

Fig. 16 Five levels of abstraction to explore the user-genre table (1 million dataset from MovieLens) and the underlying information: a bar chart of the column categories (genres), the wheel view showing sectors for the items selected in the bar chart; detail view for items selected in the wheel view; contingency table with cells in active parts colored in *dark gray* and the categorical data summarized in the cell highlighted in *red*

Visual Analytics put the user at the center of the analysis work-flow, extracting from the visualization of complex analytics new knowledge, that is reinserted as input of the analytic engine in order to move further the process of data exploration.

Visual Analytics found a broad application for big data analysis [44], including analysis of multi-variate and high-dimensional data [25, 47, 52, 58], data streams [57], multi-dimensional-scaling [53]. In [48] the authors propose a visual analytics solution for exploration of large network traffic; the traffic data are represented in an interactive treemap visualization; a signature-based detection analytic engine allows for monitoring and predicting anomalous behaviours (see Fig. 15).

In [3], Contingency Wheel++, a new scalable visual analytics solution for large categorical data is presented; the solution offers an interactive exploration environment, implementing a multi-level overview+detail interface that enables exploration of individual data items that are aggregated in the visualization or in the table using coordinated views (see Fig. 16).

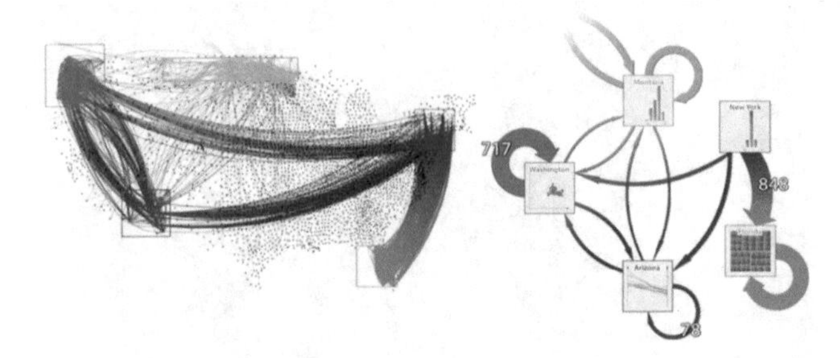

Fig. 17 Multivariate network exploration using selections of interest, detail view (*left*) and high-level infographic-style overview (*right*)

In [67], an interesting approach on illustrative visualizations for exploring complex network data (transport, telecommunications), based on the paradigm"from Detail to Overview via Selections and Aggregations" (DOSA) is presented: users are enabled to gain insights through the creation of selections of interest (manually or automatically), and producing high-level, infographic-style overviews simultaneously (see Fig. 17).

Since 2012, in particular, survey and systematization research papers that explicitly summarize existing solutions for big data analysis start to appear: [75] compares state-of-the-art commercial systems; [29] conducts a review study of the visualization methods for Big Data, identifying strong and weak points of various paradigms (treemaps, circle packing, sunburst, circular network diagrams, parallel coordinates and steamgraph). [74] classifies big data domains according to the 5w model (Why the data occured, where the data came from, what the data are, how the data were transferred, who received the data and when the data occured) and then proposes fitting visualizations accordingly.

At the same time, many Visual Analytic solutions still worked in a monolithic way, first computing the data and then visualizing them. Continuous increases in volume of data produced started to ask for different techniques that allowed for a fast visualization of the results in order not to interfere with the cognitive work-flow of the user. This research activity led to the design of progressive techniques applied to Visual Analytics [56], providing new results for both the analytics and visualization aspects. On the former, the work in [19] adds customization and guidelines to the conventional methods based on aggregated data (e.g., dimension reduction, clustering) in order to adapt them to real-time exploration. [6] presents a first generalization of the design for a progressive visual analytics solution, further formalized in [60].

Nowadays, new ways to conduct big data analysis are investigated using augmented and virtual reality; [55] proposes a solution for visualizing large and heterogeneous data in various hybrid-reality environments based on CAVE2 digital lens (high resolution computer display); this environment allows the user visualize and juxtapose multiple interrelated data-sets, reducing window switching and increasing the amount of data that can be viewed at one time (see Fig. 18).

Fig. 18 Students having an immersive meeting in CAVE2. The space can comfortably accommodate up to seven individuals with chairs and tables. CAVE2 can be used to effectively juxtapose 2D, 3D, and abstract data, as well as text documents

5 Concluding Remarks

Through this overview, that covers several years from the beginning of visual query languages to novel Visual Analytic solutions expressly designed for Big Data, it is possible to see the trend that data analysis followed in the last 30 years, from the emergence of databases to the modern solutions for storing and querying huge information repositories. Tracing the evolution of data analysis makes it clear how the paradigm shifted from a scenario in which the user (e.g., scientist, decision-maker) had a very well defined task and used instruments for refining the results obtained by the automated system (VQS), to a modern scenario in which the amount of data and computational power allow the system to automatically extract knowledge from data, present them to the user in a comprehensible way through visualizations, so enabling the user to explore the data in near or real-time and steer the computation again, driving user's interactions by the discovered insights. We can label the first approach as "human-driven" approach, where the visualization supported the user in refining the way in which she specified her task, to a "data-driven" approach, where the data analysis allows the user to form hypotheses and the visualization system permits the user to verify them or trying new ones.

Another interesting point is that the concept of big data already existed since a long time ago. In visualization, the rendering capabilities posed a challenge in visualizing even the simplest simulations back in the 1970, whereas the more the computational power increased, the more the complexity of the problem to cope with increased as well. This led to a virtuous cycle in which more complex data called for novel visualizations, and vice-versa.

With the advent of new solutions for human-computer interaction and sensors that collect more and more granular data, it is clear that the next decades will see even more emphasis on data analysis, in order to identify the important aspects, comprehend their reasons, and leverage this knowledge in anticipating the future.

References

1. D.J. Abadi, D. Carney, U. Çetintemel, M. Cherniack, C. Convey, S. Lee, M. Stonebraker, N. Tatbul, S.B. Zdonik, Aurora: a new model and architecture for data stream management. VLDB J. **12**(2), 120–139 (2003)
2. K. Abdullah, C.P. Lee, G.J. Conti, J.A. Copeland, J.T. Stasko, Ids rainstorm: visualizing ids alarms, in *VizSEC* (2005), p. 1
3. B. Alsallakh, W. Aigner, S. Miksch, M.E. Gröller, Reinventing the contingency wheel: scalable visual analytics of large categorical data. IEEE Trans. Vis. Comput. Graph. **18**(12), 2849–2858 (2012)
4. K. Andrews, H. Heidegger, Information slices: visualising and exploring large hierarchies using cascading, semi-circular discs, in *Proceedings of IEEE Infovis 98 Late Breaking Hot Topics* (1998), pp. 9–11
5. M. Angelaccio, T. Catarci, G. Santucci, Qbd*: a graphical query language with recursion. IEEE Trans. Softw. Eng. **16**, 1150–1163 (1990)
6. M. Angelini, G. Santucci, Modeling incremental visualizations, in *Proceedings of the EuroVis Workshop on Visual Analytics (EuroVA 13)* (2013), pp. 13–17
7. M. Ankerst, D.A. Keim, H-P. Kriegel, Circle segments: a technique for visually exploring large multidimensional data sets, in *Visualization* (1996)
8. A.O. Artero, M.C.F. de Oliveira, H. Levkowitz, Uncovering clusters in crowded parallel coordinates visualizations, in *IEEE Symposium on Information Visualization, 2004, INFOVIS 2004* (IEEE, 2004), pp. 81–88
9. A.N. Badre, T. Catarci, A. Massari, G. Santucci, Comparative ease of use of a diagrammatic versus an iconic query language, in *Interfaces to Databases (IDS-3), Proceedings of the 3rd International Workshop on Interfaces to Databases, Napier University, Edinburgh, 8-10 July 1996* (1996)
10. N.H. Balkir, G. Özsoyoglu, Z.M. Özsoyoglu, A graphical query language: VISUAL and its query processing. IEEE Trans. Knowl. Data Eng. **14**(5), 955–978 (2002)
11. E. Bauleo, S. Carnevale, T. Catarci, S. Kimani, M. Leva, M. Mecella, Design, realization and user evaluation of the smartvortex visual query system for accessing data streams in industrial engineering applications. J. Vis. Lang. Comput. **25**(5), 577–601 (2014)
12. E. Bertini, L. Dell'Aquila, G. Santucci, Springview: cooperation of radviz and parallel coordinates for view optimization and clutter reduction, in *Proceedings - Third International Conference on Coordinated and Multiple Views in Exploratory Visualization, CMV 2005*, vol. 2005 (2005), pp. 22–29
13. E. Bertini, G. Santucci, By chance is not enough: preserving relative density through non uniform sampling. Proc. Intern. Conf. Inf. Vis. **8**, 622–629 (2004)
14. E. Bertini, G. Santucci, Quality metrics for 2d scatterplot graphics: Automatically reducing visual clutter. Lecture Notes in Computer Science (including subseries Lecture Notes in Artificial Intelligence and Lecture Notes in Bioinformatics) **3031**, 77–89 (2004)
15. S.P. Callahan, L. Bavoil, V. Pascucci, C.T. Silva, Progressive volume rendering of large unstructured grids. IEEE Trans. Vis. Comput. Graph. **12**(5), 1345–1352 (2006)
16. T. Catarci, M.F. Costabile, S. Levialdi, C. Batini, Visual query systems for databases: a survey. J. Vis. Lang. Comput. **8**(2), 215–260 (1997)
17. T. Catarci, T. Di Mascio, E. Franconi, G. Santucci, S. Tessaris, An ontology based visual tool for query formulation support. Lect. Notes Comput. Sci. **2889**, 32–33 (2003)
18. R. Chang, G. Wessel, R. Kosara, E. Sauda, W. Ribarsky, Legible cities: focus-dependent multi-resolution visualization of urban relationships. IEEE Trans. Visual Comput. Graph. **13**(6), 1169–1175 (2007)

19. J. Choo, H. Park, Customizing computational methods for visual analytics with big data. IEEE Comput. Graph. Appl. **33**(4), 22–28 (2013)
20. M.C. Chuah, Dynamic aggregation with circular visual designs, in *Proceedings of the IEEE Symposium on Information Visualization, 1998* (IEEE, 1998), pp. 35–43
21. M. Cox, Large data management for interactive visualization design, in *Proceedings of the SIGGRAPH'99 System Designs for Visualizing Large-Scale Scientific Data Course Notes* (1999), pp. 5–29
22. T.A. Defanti, M.D. Brown, Visualization in scientific computing. Adv. Comput. **33**, 247–307 (1991)
23. J.R. Eagan Jr, M.J. Harrold, J.A. Jones, J.T. Stasko, Visually encoding program test information to find faults in software. Technical report, Georgia Institute of Technology (2001)
24. D. Ellsworth, B. Green, P. Moran, Interactive terascale particle visualization, in *Proceedings of the Conference on Visualization'04* (IEEE Computer Society, 2004), pp. 353–360
25. N. Elmqvist, J. Stasko, P. Tsigas, Datameadow: a visual canvas for analysis of large-scale multivariate data. Inf. Vis. **7**(1), 18–33 (2008)
26. J-D. Fekete, C. Plaisant, Interactive information visualization of a million items, in *IEEE Symposium on Information Visualization, 2002, INFOVIS 2002* (IEEE, 2002), pp. 117–124
27. K.A. Frenkel, The art and science of visualizing data. Commun. ACM **31**(2), 111–121 (1988)
28. Y-H. Fua, M.O. Ward, E.A. Rundensteiner, Hierarchical parallel coordinates for exploration of large datasets, in *Proceedings of the Conference on Visualization'99: Celebrating Ten Years* (IEEE Computer Society Press, 1999), pp. 43–50
29. J. Elect. Comput. Eng. Analytical review of data visualization methods in application to big data. **2013**, 22 (2013)
30. D. Guo, Flow mapping and multivariate visualization of large spatial interaction data. IEEE Trans. Vis. Comput. Graph. **15**(6), 1041–1048 (2009)
31. S. Havre, B. Hetzler, L. Nowell, Themeriver: visualizing theme changes over time, in *IEEE Symposium on Information Visualization, 2000, InfoVis 2000* (IEEE, 2000), pp. 115–123
32. J. Heer, D. Boyd, Vizster: visualizing online social networks, in *IEEE Symposium on Information Visualization, 2005, INFOVIS 2005* (IEEE, 2005), pp. 32–39
33. J. Hong, D.H. Jeong, C.D. Shaw, W. Ribarsky, M. Borodovsky, C.G. Song, Gvis: a scalable visualization framework for genomic data, in *EuroVis*, vol. 5 (2005), pp. 191–198
34. S. Huron, R. Vuillemot, J.-D. Fekete, Visual sedimentation. IEEE Trans. Visual Comput. Graph. **19**(12), 2446–2455 (2013)
35. J.-F. Im, M.J. McGuffin, R. Leung, Gplom: the generalized plot matrix for visualizing multidimensional multivariate data. IEEE Trans. Visual Comput. Graph. **19**(12), 2606–2614 (2013)
36. A. Inselberg, B. Dimsdale, Parallel coordinates: a tool for visualizing multi-dimensional geometry, in *Proceedings of the 1st Conference on Visualization'90* (IEEE Computer Society Press, 1990), pp. 361–378
37. Y. Jia, J. Hoberock, M. Garland, J. Hart, On the visualization of social and other scale-free networks. IEEE Trans. Visual Comput. Graph. **14**(6), 1285–1292 (2008)
38. B. Johnson, B. Shneiderman, Tree-maps: a space-filling approach to the visualization of hierarchical information structures, in *Proceedings of the 2nd Conference on Visualization'91* (IEEE Computer Society Press, 1991), pp. 284–291
39. D.A. Keim, Pixel-oriented visualization techniques for exploring very large data bases. J. Comput. Graph. Stat. **5**(1), 58–77 (1996)
40. D.A. Keim, M. Ankerst, H.-P. Kriegel, Recursive pattern: a technique for visualizing very large amounts of data, in *Proceedings of the 6th Conference on Visualization'95* (IEEE Computer Society, 1995), p. 279
41. D.A. Keim, M.C. Hao, J. Ladisch, M. Hsu, U. Dayal, Pixel bar charts: a new technique for visualizing large multi-attribute data sets without aggregation, in *IEEE Symposium on Information Visualization, 2001, INFOVIS 2001* (2001), pp. 113–120
42. D.A. Keim, H.-P. Kriegel, T. Seidl, Visual feedback in querying large databases, in *Proceedings of the 4th Conference on Visualization'93* (IEEE Computer Society, 1993), pp. 158–165

43. D. Kenwright, Automation or interaction: what's best for big data? in *Proceedings of the Visualization'99* (IEEE, 1999), pp. 491–495
44. J. Kohlhammer, D. Keim, M. Pohl, G. Santucci, G. Andrienko, Solving problems with visual analytics. Procedia Comput. Sci- Europ. Future Technol. Conf. Exhib. **7**, 117–120 (2011)
45. D. Lembo, D. Pantaleone, V. Santarelli, D.F. Savo, Easy OWL drawing with the graphol visual ontology language, in *Principles of Knowledge Representation and Reasoning: Proceedings of the Fifteenth International Conference, KR 2016, Cape Town, South Africa, 25–29 April, 2016* (2016), pp. 573–576
46. A. Lex, H.-J. Schulz, M. Streit, C. Partl, D. Schmalstieg, Visbricks: multiform visualization of large, inhomogeneous data. IEEE Trans. Visual Comput. Graph. **17**(12), 2291–2300 (2011)
47. A. Lex, M. Streit, C. Partl, K. Kashofer, D. Schmalstieg, Comparative analysis of multidimensional, quantitative data. IEEE Trans. Visual Comput. Graph. **16**(6), 1027–1035 (2010)
48. F. Mansmann, D.A. Keim, S.C. North, B. Rexroad, D. Sheleheda, Visual analysis of network traffic for resource planning, interactive monitoring, and interpretation of security threats. IEEE Trans. Visual Comput. Graph. **13**(6), 1105–1112 (2007)
49. A. Massari, S. Pavani, L. Saladini, P.K. Chrysanthis, Qbi: query by icons (1995)
50. A.J. Morris, A.I. Abdelmoty, B.A. El-Geresy, A visual query language for large spatial databases, in *Proceedings of the Working Conference on Advanced Visual Interfaces, AVI 2002, Trento, Italy, 22–24 May, 2002* (2002), pp. 359–360
51. N. Murray, N. Paton, C. Goble, Kaleidoquery: a visual query language for object databases (1998), pp. 247–257
52. E.J. Nam, Y. Han, K. Mueller, A. Zelenyuk, D. Imre, Clustersculptor: a visual analytics tool for high-dimensional data, in *IEEE Symposium on Visual Analytics Science and Technology, 2007, VAST 2007* (IEEE, 2007), pp. 75–82
53. F.V. Paulovich, C.T. Silva, L.G. Nonato, Two-phase mapping for projecting massive data sets. IEEE Trans. Visual Comput. Graph. **16**(6), 1281–1290 (2010)
54. S. Polyviou, G. Samaras, P. Evripidou, A relationally complete visual query language for heterogeneous data sources and pervasive querying, in *Proceedings of the 21st International Conference on Data Engineering, ICDE 2005, 5–8 April 2005, Tokyo, Japan* (2005), pp. 471–482
55. K. Reda, A. Febretti, A. Knoll, J. Aurisano, J. Leigh, A. Johnson, M.E. Papka, M. Hereld, Visualizing large, heterogeneous data in hybrid-reality environments. IEEE Comput. Graph. Appl. **33**(4), 38–48 (2013)
56. S. Rinzivillo, D. Pedreschi, M. Nanni, F. Giannotti, N. Andrienko, G. Andrienko, Visually driven analysis of movement data by progressive clustering. Inf. Vis. **7**(3–4), 225–239 (2008)
57. S. Rose, S. Butner, W. Cowley, M. Gregory, J. Walker, Describing story evolution from dynamic information streams, in *IEEE Symposium on Visual Analytics Science and Technology, 2009, VAST 2009* (IEEE, 2009), pp. 99–106
58. S.J. Rysavy, D. Bromley, V. Daggett, Dive: a graph-based visual-analytics framework for big data. IEEE Comput. Graph. Appl. **34**(2), 26–37 (2014)
59. G. Santucci, P.A. Sottile, Query by diagram: a visual environment for querying databases. *Softw. Pract. Exp.*, **23**(3), 317–340 (1993), cited By 12
60. H.-J. Schulz, M. Angelini, G. Santucci, H. Schumann, An enhanced visualization process model for incremental visualization. IEEE Trans. Visual Comput. Graph. **22**(7), 1830–1842 (2016)
61. B. Shneiderman, The eyes have it: a task by data type taxonomy for information visualizations, in *Proceedings of the IEEE Symposium on Visual Languages, 1996* (IEEE, 1996), pp. 336–343
62. A. Soylu, M. Giese, R. Schlatte, E. Jiménez-Ruiz, E. Kharlamov, Ö.L. Özçep, C. Neuenstadt, S. Brandt, Querying industrial stream-temporal data: an ontology-based visual approach[1]. JAISE **9**(1), 77–95 (2017)
63. S.P. Reiss, A visual query language for software visualization (2002), pp. 80–82
64. C. Stolte, D. Tang, P. Hanrahan, Polaris: a system for query, analysis, and visualization of multidimensional relational databases. IEEE Trans. Visual Comput. Graph. **8**(1), 52–65 (2002)

65. C. Stolte, D. Tang, P. Hanrahan, Multiscale visualization using data cubes. IEEE Trans. Visual Comput. Graph. **9**(2), 176–187 (2003)
66. S.T. Teoh, K.L. Ma, S.F. Wu, X. Zhao, Case study: interactive visualization for internet security, in *Proceedings of the Conference on Visualization'02* (IEEE Computer Society, 2002), pp. 505–508
67. S. Van den Elzen, J.J. Van Wijk, Multivariate network exploration and presentation: from detail to overview via selections and aggregations. IEEE Trans. Visual Comput. Graph. **20**(12), 2310–2319 (2014)
68. J.J. van Wijk, H.J. Spoelder, W.-J. Knibbe, K.E. Shahroudi, Interactive exploration and modeling of large data sets: a case study with venus light scattering data, in *Proceedings of the 7th Conference on Visualization'96* (IEEE Computer Society Press, 1996), pp. 433–ff
69. C. Weaver, Building highly-coordinated visualizations in improvise, in *IEEE Symposium on Information Visualization, 2004, INFOVIS 2004* (IEEE, 2004), pp. 159–166
70. C. Weaver, Multidimensional visual analysis using cross-filtered views, in *IEEE Symposium on Visual Analytics Science and Technology, 2008, VAST'08* (IEEE, 2008), pp. 163–170
71. M. Weber, M. Alexa, W. Müller, Visualizing time-series on spirals. Infovis **1**, 7–14 (2001)
72. N. Wong, S. Carpendale, S. Greenberg, Edgelens: an interactive method for managing edge congestion in graphs, in *IEEE Symposium on Information Visualization, 2003, INFOVIS 2003* (IEEE, 2003), pp. 51–58
73. P.C. Wong, H. Foote, D. Adams, W. Cowley, J. Thomas, Dynamic visualization of transient data streams, in *IEEE Symposium on Information Visualization, 2003, INFOVIS 2003* (IEEE, 2003), pp. 97–104
74. J. Zhang, M.L. Huang, 5ws model for big data analysis and visualization, in *2013 IEEE 16th International Conference on Computational Science and Engineering (CSE)* (IEEE, 2013), pp. 1021–1028
75. L. Zhang, A. Stoffel, M. Behrisch, S. Mittelstadt, T. Schreck, R. Pompl, S. Weber, H. Last, D. Keim, Visual analytics for the big data eraa comparative review of state-of-the-art commercial systems, in *2012 IEEE Conference on Visual Analytics Science and Technology (VAST)* (IEEE, 2012), pp. 173–182
76. M.M. Zloof, Query-by-example: a database language. IBM Syst. J. **16**(4), 324–343 (1997)

Digital Libraries: From Digital Resources to Challenges in Scientific Data Sharing and Re-Use

Maristella Agosti, Nicola Ferro and Gianmaria Silvello

1 Introduction

Digital libraries and digital archives are the information management systems for storing, indexing, searching, accessing, curating and preserving digital resources which manage our cultural and scientific knowledge heritage (KH). They act as the main conduits for widespread access and exploitation of KH related digital resources by engaging many different types of users, ranging from generic and leisure to students and professionals.

In this chapter, we describe the evolution of digital libraries and archives over the years, starting from *Online Public Access Catalog (OPAC)*, passing through monolithic and domain specific systems, up to service-oriented and component-based architectures. In particular, we present some specific achievements in the field: the DELOS Reference Model and the DelosDLMS, which provide a conceptual reference and a reference implementation for digital libraries; the FAST annotation service, which defines a formal model for representing and searching annotations over digital resources as well as a RESTful Web service implementation of it; the NESTOR model for digital archives, which introduces an alternative model for representing and managing archival resources in order to enhance interoperability among archives and make access to them faster; and, the CULTURA environment, which favours user engagement over multimedia digital resources.

M. Agosti · N. Ferro · G. Silvello (✉)
Department of Information Engineering, University of Padua, Padua, Italy
e-mail: gianmaria.silvello@unipd.it

M. Agosti
e-mail: maristella.agosti@unipd.it

N. Ferro
e-mail: nicola.ferro@unipd.it

S. Flesca et al. (eds.), *A Comprehensive Guide Through the Italian Database Research Over the Last 25 Years*, Studies in Big Data 31, DOI 10.1007/978-3-319-61893-7_2

Finally, we discuss how digital libraries and archives are a key technology for facing upcoming challenges in data sharing and re-use. Indeed, due to the rapid evolution of the nature of research and scientific publishing which are increasingly data-driven, digital libraries and archives are also progressively addressing the issues of managing scientific data. In this respect, we focus on some key building blocks of this new vision: data citation to foster accessibility to scientific data as well as transparency and verifiability of scientific claims, reproducibility in science as an exemplar showcase of how all these methods are indispensable for addressing fundamental challenges, and keyword-based search over relation/structured data to empower natural language access to scientific data.

2 Evolution of Digital Libraries

The term "digital libraries" corresponds to a very complex notion with several diverse aspects and it cannot be captured by a simple definition. Indeed, the term is used to refer to systems that are very heterogeneous in scope and provide very different functionalities [28]. These systems span from digital object and metadata repositories, reference-linking systems, archives, and content administration systems, to complex systems that integrate advanced services. Furthermore, digital libraries represent the meeting point of many disciplines and research fields – i.e. database management, information retrieval, library and information sciences, document and information systems, the Web, information visualization, artificial intelligence, human-computer interaction, and others [49].

Initially, digital libraries were almost monolithic systems, each one built for a specific kind of information resources – e.g. images or videos – and with very specialized functions developed ad-hoc for those contents. This approach caused a flourishing of systems where the very same functions were developed and re-developed many times from scratch. Moreover, these systems were confined to the realm of traditional libraries, since they were the digital counterpart of the latter, and they had a kind of static view of their role, which was document-centric rather than user-centric.

In the 1980s the most advanced library automation systems were designed to include procedures also able to collect log data that were used to manage the system itself, and especially to monitor the usage of system search facilities by users, where the search facility which was designed for user search and access to catalog data was an OPAC [48]; some OPACs were reachable in a distributed environment: an example of such a system is the DUO OPAC system from the early 1990s [15]. Towards the end of the 1980s/beginning of 1990s it became apparent that a library automation system could not only manage catalog data or metadata describing physical objects, but also digital files representing physical objects. Digital libraries started to be seen as increasingly user-centered systems, where the original content management task is partnered with new communication and cooperation tasks.

In this evolving scenario, the design and development of effective services which foster cooperation among users and the integration of heterogeneous information resources becomes a key factor. Digital libraries are thus no longer perceived as isolated systems but, on the contrary, as systems that need to cooperate with each other to improve the user experience and give personalized services. Nowadays, there are several accepted conceptions of digital libraries:

- *User-centric systems*: Digital libraries as user-centered information infrastructures able to support content management tasks together with tasks devoted to communication and cooperation. Although they are still places where information resources can be stored and made available to end users, recent design and development efforts move in the direction of transforming them into infrastructures able to support the user in different information centric activities.
- *Dynamic interactions*: Digital libraries as dynamic forms of facilitation of communication, collaboration and other forms of interaction among scientists, researchers and the general public.
- *Large capabilities*: Digital libraries as systems able to handle distributed multimedia document collections, sensor data, mobile information, and pervasive computing services.

Digital libraries have contributed to supporting the creation of innovative applications and services to access, share and search our cultural KH. In this context, another key feature we have to consider to understand the world of digital libraries is that they have to take into account several distributed and heterogeneous information sources with different community background and different information objects ranging from full content of digital objects to the metadata describing them. These objects can be exchanged between distributed systems or they can be aggregated and accessed by users with distinct information needs and living in different countries. Indeed, one of the most important contributions of digital libraries is to make available collections of digital resources from different cultural institutions such as *libraries*, *archives* and *museums*, to make them accessible in different languages and to provide advanced services over them. We have to consider that the above mentioned institutions are different from several point-of-views: their internal organization has different peculiarities, the resources they collect and manage have different structure and nature, these resources are described with different means and for different purposes, their users have different information needs and require different methods to access the resources. Thus, digital libraries are heterogeneous systems with peculiarities and functions that range from data representation to data exchange and data management. Furthermore, digital libraries are meaningful parts of a global information network which includes scientific repositories, curated databases and commercial providers. All these aspects need to be taken into account and balanced to support final users with effective and interoperable information systems.

A fundamental role of digital libraries therefore is to provide data models, protocols, applications and services to handle all these resources, all the while preserving their characteristics and addressing the issues related to their differences.

3 Models and Services for Digital Libraries

Digital libraries have shaped the way for accessing our cultural heritage and have become primary knowledge conduits thanks to the development of formal and conceptual models of what digital libraries are, such as the DELOS Reference Model [28] and the *Streams, Structures, Spaces, Scenarios, Societies (5S)* model [45]. These models have then been specialised to specific domains, such as archives [41], and to specific services, such as digital annotations [10]. Finally, thanks to the recent development of semantic technologies, it has been also possible to provide formal mappings between the DELOS Reference Model and the 5 S models to improve interoperability among digital libraries [13]. In the following, we briefly present our main contributions in this context.

3.1 DELOS Reference Model and DelosDLMS

The DELOS Reference Model approaches the problem of modelling the digital library universe by highlighting six domains or main concepts [28], which are at the core of what digital libraries are and what their purpose is:

- *Content*: the data and information that digital libraries handle and make available to their users;
- *User*: the actors (whether human or not) entitled to interact with digital libraries;
- *Functionality*: the services that digital libraries offer to their users;
- *Quality*: the parameters that can be used to characterize and evaluate the content and behaviour of digital libraries;
- *Policy*: a set of rules that govern the interaction between users and digital libraries;
- *Architecture*: a mapping of the functionality and content offered by a digital library onto hardware and software components.

These six domains represent the high level containers that help organize the DELOS Reference Model. For each of these concepts, the fundamental entities and their relationships are clearly defined and discussed. Note that these six domains are not separate, but, on the contrary, are strongly inter-related; the entities within one domain are often related to or influenced by the entities in other domains.

Moreover, the DELOS Reference Model distinguishes between three different "systems" which constitute the digital library universe and rely on the six domains introduced above for their definition:

- *Digital Library (DL)*: an organisation, which might be virtual, that comprehensively collects, manages and preserves for the long term rich digital content, and offers to its user communities specialised functionality on that content, of measurable quality and according to codified policies.
- *Digital Library System (DLS)*: a software system that is based on a defined (possibly distributed) architecture and provides all functionality required by a particular

Digital Library. Users interact with a Digital Library through the corresponding Digital Library System.
- *Digital Library Management System (DLMS)*: a generic software system that provides the appropriate software infrastructure both (i) to produce and administer a Digital Library System incorporating the suite of functionality considered fundamental for Digital Libraries and (ii) to integrate additional software offering more refined, specialised or advanced functionality.

The three systems are at different levels of abstraction and constitute a kind of hierarchy: at the more general level there is the notion of DL, which is what is actually perceived by the end-users and what they interact with; in-between, there is the DLS, which mainly concerns system designers and administrators who have to instantiate and manage it; at the lower level, there is the DLMS, which typically interests system developers who implement the actual components that are used by the upper layers.

3.2 The FAST Annotation Model

The *Flexible Annotation Service Tool (FAST)* [33] covers many of the uses and applications of annotations, since it is able to represent and manage annotations which range from metadata to full content; its flexible and modular architecture makes it suitable for annotating general Web resources as well as digital objects managed by different digital library systems; the annotations themselves can be complex multimedia compound objects, with a varying degree of visibility which ranges from private to shared and public annotations and different access rights. The FAST annotation service has proven its flexibility and adaptability to different applicative contexts in many different ways. It has been integrated into the DelosDLMS [3], the prototype of the next generation digital library system developed by DELOS, and in the CULTURA environment [5, 6].

FAST adopts and implements the formal model for annotations proposed in [10]. Annotations are compound multimedia objects constituted by different *signs of annotation* that materialize the annotation itself. For example, we can have *textual signs*, which contain the textual content of the annotation, *image signs*, if the annotation is made up of images, and so on. In turn, each sign is characterized by one or more *meanings of annotation* that specify the semantics of the sign. Moreover, an annotation is uniquely identified by a handle, which usually takes the form of a pair (namespace, identifier), where the namespace provides logical grouping of the identifiers, it has a scope which defines its visibility, and it can be shared with different groups of users.

Annotations can be linked to digital objects with two main types of links: (1) *annotate link* an annotation annotates a digital object, which can be either a document or another annotation; (2) *relate-to link* an annotation relates to a digital object, which can be either a document or another annotation. The hypertext between annotations and annotated objects can be exploited for providing alternative navigation and browsing capabilities. In addition, it can span and cross the boundaries of the single digital library and also related to Web resources. Most importantly, this hypertext can be exploited to develop advanced search functionalities [9]. Based on the proposed formal model, we developed a fully-fledged search model, mixing exact match and best match queries, paired with an intuitive query language expressed in the *Contextual Query Language (CQL)* syntax [32]. In this way, we can not only search for annotations, by means of a mix of full text and queries based on the structure, but we can also retrieve annotated resources thanks to the hypertext that allows us to pass from the found annotations to the annotated resources.

3.3 NESTOR: A Model for Digital Archives

Digital archives are one of the pillars of our cultural heritage and, thanks to technologies such as the digital libraries, they are increasingly opening up to end-users by focusing on usability, accessibility, and findability of the resources they manage.

Archives represent the trace of the activities of a physical or juridical person in the course of their business which is preserved because of their continued value over time. Archives and archival descriptions (i.e. metadata) are modeled by using a hierarchical structure, which expresses the relationships and dependency links between the records of the archive.

In recent years, archival descriptions have moved on-line and there have been increasing calls for reconsidering their presentation based on user studies. Indeed, from this new point-of-view, the XML standard for digital description of archives such as the *Encoded Archival Description (EAD)* seriously constrains user orientation of archives; but with EAD several important digital archives operations are not possible: (i) the user cannot access a specific item on-the-fly, instead we have to define fixed access points to the archival hierarchy; (ii) the user cannot reconstruct the context of an item without browsing the archival hierarchy; (iii) we cannot present the users with selected items from an archive, instead we have to give them the archive as a whole.

To tackle these issues, we proposed to model archives through the *NEsted SeTs for Object hieRarchies (NESTOR)* [39, 41] – i.e. a set-based data model allowing for the representation of hierarchical relationships between objects through the inclusion property between sets – which opened-up new ways of representing and handling archival resources. The NESTOR model is defined by two set-based data models: The *Nested Set Model (NS-M)* and the *Inverse Set Data Model (INS-M)*. These models are defined in the context of set theory as a collection of subsets, their properties have been formally proved as well as their equivalence to the tree in terms of expressive power [12, 37, 41].

The most intuitive way to understand how these models work is to relate them to the tree. In the NS-M each node of the tree is mapped into a set, where child nodes become *proper subsets* of the set created from the parent node. Every set is subset of at least one set; the set corresponding to the tree root is the only set without any superset and every set in the hierarchy is subset of the root set. The external nodes are sets with no subsets. The tree structure is maintained thanks to the nested organization and the relationships between the sets are expressed by the set inclusion order.

The second data model is the INS-M where each node of the tree is mapped into a set, where each parent node becomes a subset of the sets created from its children. The set created from the tree's root is the only set with no subsets and the root set is a proper subset of all the sets in the hierarchy. The leaves are the sets with no supersets and they are sets containing all the sets created from the nodes composing tree path from a leaf to the root.

NESTOR is particularly well-suited for advancing user-orientation of archives because it allows for exposing archival data as Linked Data on the Web [40], thus augmenting the understandability of these data. Furthermore, NESTOR can be adopted for "socializing the archives" [41] by means of annotations [38] such that available resources can be augmented with user-generated content which then provides alternative access points for searching and browsing resources. Furthermore, NESTOR has been realized by means of three alternative in-memory dictionary-based data structures, which have been proved to be highly competitive with state-of-the-art solutions for accessing XML data by considering pre-processing and query execution time and memory occupation [42, 43, 52].

3.4 User Engagement: The CULTURA Environment

The main aim of the CULTURA environment was to create a *Virtual Research Environment (VRE)* in which users with a range of different backgrounds and expertise can collaboratively explore, interrogate, interact with, and interpret complex and diverse digital cultural heritage collections [6]. The CULTURA environment is a VRE that pushed forward the frontiers of technology in the creation of community and content aware interfaces to digital humanities collections.

The CULTURA environment adopts a service-oriented approach to offer a rich and engaging experience for different user categories, which range from academic and professional users to the general public. The services are conceived and developed to be applicable to a wide variety of document collections [14]. The potential generality of the environment is demonstrated by the fact that the environment supports different use cases; one of those is represented by the IPSA collection, a digital archive of illuminated manuscripts, while the other major archive is the 1641 Depositions, which is a collection of noisy text documents, mainly of a legal nature, dating from the 17th Century.

In both collections, the managed digital objects – "either scanned illuminated manuscripts or legal documents" – are described by appropriate metadata, according to a traditional record-centric approach. The goal of the environment was to exploit an improved user engagement and interaction with the managed artifacts in order to semantically enrich them with a superimposed layer of user-provided information.

This required a move from a traditional record-centric approach to a resource-centric one, opened towards *Linked Open Data (LOD)* and a better sharing of resources.

The history of art provides a fertile ground for research into semantically enriched metadata and LOD; indeed, in history of art the main way to produce new knowledge is to reveal connections between different items (illuminations, pictures, frescos) that can cast new light on an artist, an artistic movement or an art-historical period. The most valuable connections are the unexpected ones linking elements that may seem to have very few features in common [54]. Therefore, it was decided that the central tool for allowing researchers in history of art to discover new knowledge and unveil new links and relationships among resources would be the semantic annotation tool, called FAST-CAT. This software enables semantically-typed links to be superimposed over the managed digital objects, the traditional record-centric metadata, and Web resources in general. Besides being semantically typed, these links can include fully-fledged multimedia content, which allows for rich description and explanation of the link and provides added value to both specialist users and the general public.

Both the FAST annotation model [10] and the CAT model and tool (FAST-CAT) [33] have been applied to the CULTURA environment and they provided adaptive and personalized access to the IPSA historical collection. FAST-CAT has been integrated into the environment in order to provide users with an additional means of interacting with the portal, as well as for providing feedback on CULTURA user model that stores user interests. It is the belief of the authors that FAST-CAT has huge potential as an annotation tool within the digital humanities field. Indeed, it demonstrates the feasibility of transitioning from a traditional digital archive with a record-centric approach, towards a resource-centric one with semantically enriched information provided by actively engaging users via digital annotations [59]. The process of discovering new unexpected connections among cultural heritage artifacts, i.e. a process that can be defined as serendipity, enabled by the FAST annotation model, is especially encouraged by the LOD paradigm where meaningful links between entities allow us to move across diverse and apparently unrelated knowledge domains.

4 Data-Driven Digital Libraries

The role of data is going to become central in DL as well as in the other fields of computer science. In this section we explore the role of DL in the context of reproducibility in science and the relationships between DL and data citation.

4.1 Reproducibility

Computer science is particularly active in reproducibility, as witnessed by the recent *Association for Computing Machinery (ACM)* policy on result and artifact review and badging.[1] For instance, the database community started an effort called "SIGMOD reproducibility" [44] "to assist in building a culture of sharing results, code, and scripts of database research".[2] Since 2015, the *European Conference in IR (ECIR)* [35, 46], allocated a whole paper track on reproducibility and in 2015 the RIGOR workshop at SIGIR was dedicated to this topic [16]. Moreover, in 2016 the "Reproducibility of Data-Oriented Experiments in e-Science" seminar was held in Dagstuhl (Germany) [1] bringing together researchers from different fields of computer science with the goal "to come to a common ground across disciplines, leverage best-of-breed approaches, and provide a unifying vision on reproducibility" [34, 36].

In recent years, the nature of research and scientific publishing has been rapidly evolving and progressively relying on data to sustain claims and provide experimental evidence for scientific breakthroughs [47]. The preservation, management, access, discovery and retrieval of research data are topics of utmost importance as witnessed by the great deal of attention they are receiving from the scientific and publishing communities [24]. Along with the pervasiveness and availability of research data, we are witnessing the growing importance of citing these data. Indeed, data citation is required to make results of research fully available to others, provide suitable means to connect publications with the data they rely upon [61], give credit to data creators, curators and publishers [23], and enabling others to better build on previous results and to ask new questions about data [22].

Even though *Information Retrieval (IR)* has a long tradition in ensuring that the due scientific rigor is guaranteed in producing experimental data, it does not have a similar tradition in managing and taking care of such valuable data [8, 31]. This represents a serious obstacle for tackling the above mentioned challenges. For example, there is a lack of commonly agreed formats for modeling and describing the experimental data as well as almost no metadata (descriptive, administrative, copyright, etc.) for annotating and enriching them. The semantics of the data themselves is often not explicit and it is demanded to the scripts typically used for processing them, which are often not well documented, rely on rigid assumptions on the data format or even on side effects in processing the data. Finally, IR lacks a commonly agreed mechanism for citing and linking data to the papers describing them [57].

There have been early examples of systems to manage IR experimental data, such as EvaluatOR [17] and *Distributed Information Retrieval Evaluation Campaign Tool (DIRECT)*[3] [7, 11], but they have not been designed with reproducibility and/or data citation as goals. More recently, steps toward more fine grained models and systems have been proposed, as for example LOD-DIRECT[4] [56] which uses semantic Web and LOD technologies to model IR evaluation data and make them linkable, or nanopublications for IR evaluation [50].

[1] https://www.acm.org/publications/policies/artifact-review-badging.

[2] http://db-reproducibility.seas.harvard.edu/.

[3] http://direct.dei.unipd.it/.

[4] http://lod-direct.dei.unipd.it/.

The situation is even more severe in the context of keyword-based search over relational databases, which is a key technology for lowering the barriers of access to the huge amounts of data managed by databases. It is an extremely difficult and open challenge [2] since it comes up against the "conflict of impedance" between vague and imprecise user information needs and rigorously structured data, allowing users to express their queries in natural language against a potentially unknown database.

Even if there have been attempts to reproduce state-of-the-art solutions and provide shared benchmarks [29], we still need to move beyond the evaluation of keyword search components in isolation or not related to the actual user needs, and, instead, to consider the whole system, its constituents, and their inter-relations with the ultimate goal of supporting actual user search tasks [20, 21]. Moreover, there is a lack of commonly shared open source platforms implementing state-of-the-art algorithms for keyword-based access to relational data as, for example, Terrier[5] in the information retrieval field, and we just started to move towards providing open source implementations of these algorithms[6] [18].

4.2 Data Citation

The practice of citation is foundational for scientific advancement and the propagation of knowledge and it is one of the basic means on which scholarship and scientific publishing rely. In recent years, the nature of research and scientific publishing has been rapidly evolving and progressively relying on data to sustain claims and provide experimental evidence for scientific breakthroughs [47]. The preservation, management, access, discovery and retrieval of research data are topics of utmost importance as witnessed by the great deal of attention they are receiving from the scientific and publishing communities [24, 25].

Along with the pervasiveness and availability of research data, we are witnessing the growing importance of citing these data. Indeed, data citation is required to make results of research fully available to others, provide suitable means to connect publications with the data they rely upon, give credit to data creators, curators and publishers, and enable others to better build on previous results and to ask new questions about data [58]. Furthermore, data citation plays a central role for providing better transparency and reproducibility in science, a challenge taken up by several fields.

In the traditional context of printed material, the practice of citation has been evolving and adapting over the centuries [24] reaching a stable and reliable state; nevertheless, traditional citation methods and practices cannot be easily applied for citing data. Indeed, citing data poses new and significant challenges, such as the use

[5] http://www.terrier.org/.

[6] https://bitbucket.org/ks-bd-2015-2016/ks-unipd.

of heterogeneous data models and formats requiring different methods to manage, retrieve and access the data; the transience of data calling for versioning and archiving methods and systems; the necessity to cite data at different levels of coarseness requiring methods to identify, select and reference specific subsets of data; and the necessity to automatically generate citations to data because we cannot assume that the people citing the data understand the complexity of a dataset, know how data should be cited in a specific context, or select relevant information to form a complete and correct citation.

As described in [26], from the computational perspective the problem of data citation can be formulated as follows: "Given a dataset D and a query Q, generate an appropriate citation C". Several of the existing approaches to address this problem allow us to reference datasets as a single unit having textual data serving as metadata source, but as pointed out by [51] most data citations "can often not be generated automatically and they are often not machine interpretable". Until now, the problem of how to cite a dataset at different levels of coarseness, to automatically generate citations and to create human- and machine-readable citations has been tackled only by a few working systems [27, 30, 51, 53, 55]. From these experiences, it clearly emerges that data citation is a compound and complex problem and a "one size fits all" system to address it does not exist, yet. As a consequence, within the context of data citation, there are several open issues and research directions we can take into account:

- *Citation identity and containment*: This problem refers to the necessity of uniquely identifying a citation to data and of being able to discriminate between two citations referring to different data or different versions of the same data and between two different citations referring to the same data.
- *Versioning*: One of the main differences between traditional citations and data citations is that data may not be fixed, but may evolve through time; indeed, new data may be added to a dataset, some changes may occur, some mistakes may be fixed or new information may be added. A citation needs to ensure that the data a citation uses is identical to that cited.
- *Provenance*: Provenance information plays a central role because we may need to reconstruct the chain of ownership of a data object or the chain of modifications that occurred to it in order to produce a reliable citation [30]. New solutions have to be provided to integrate data citation with currently employed systems controlling and managing the data workflow.
- *Supporting scientific claims*: Scientific claims are often based on evidence gathered from data. They could be related to a single datum or to multiple data coming from the same source or from different sources. Data citation can be used to support such claims and to provide a means to verify their reliability.

References

1. Report from Dagstuhl seminar 16041: reproducibility of data-oriented experiments in e-science, in *Report from Dagstuhl Seminar 16041: Reproducibility of Data-Oriented Experiments in e-Science* ed. By J. Freire, N. Fuhr, A Rauber. Dagstuhl Reports, Vol. 6(1) (Schloss Dagstuhl–Leibniz-Zentrum fuer Informatik, Germany, 2016)
2. D. Abadi, R. Agrawal, A. Ailamaki, M. Balazinska, P.A. Bernstein, M.J. Carey, S. Chaudhuri, J. Dean, A. Doan, M.J. Franklin, J. Gehrke, L.M. Haas, A.Y. Halevy, J.M. Hellerstein, Y.E. Ioannidis, H.V. Jagadish, D. Kossmann, S. Madden, S. Mehrotra, T. Milo, J.F. Naughton, R. Ramakrishnan, V. Markl, C. Olston, B.C. Ooi, C. Rè, D. Suciu, M. Stonebraker, T. Walter, J. Widom, The Beckman report on database research. ACM SIGMOD Rec. **43**(3), 61–70 (2014)
3. M. Agosti, S. Berretti, G. Brettlecker, A. del Bimbo, N. Ferro, N. Fuhr, D. Keim, C.P. Klas, T. Lidy, D. Milano, M. Norrie, P. Ranaldi, A. Rauber, H.J. Schek, T. Schreck, H. Schuldt, B. Signer, M. Springmann, DelosDLMS – the Integrated DELOS Digital Library Management System, in Thanos et al. [60], pp. 36–45
4. M. Agosti, J. Borbinha, S. Kapidakis, C. Papatheodorou, G. Tsakonas, (eds.), in *Proceedings of 13th European Conference on Research and Advanced Technology for Digital Libraries (ECDL 2009)*. Lecture Notes in Computer Science (LNCS), vol. 5714 (Springer, Heidelberg, Germany, 2009)
5. M. Agosti, O. Conlan, N. Ferro, C. Hampson, G. Munnelly, Interacting with digital cultural heritage collections via annotations: the CULTURA approach, in *Proceedings of 13th ACM Symposium on Document Engineering (DocEng 2013)* ed. By S. Marinai, K. Marriot (ACM Press, New York, USA, 2013), pp. 13–22
6. M. Agosti, O. Conlan, N. Ferro, C. Hampson, G. Munnelly, C. Ponchia, G. Silvello, Enriching digital cultural heritage collections via annotations: the CULTURA approach, in *22nd Italian Symposium on Advanced Database Systems, SEBD 2014*, ed. By S. Greco, A. Picariello (2014), pp. 319–326
7. M. Agosti, E. Di Buccio, N. Ferro, I. Masiero, S. Peruzzo, G. Silvello, DIRECTions: Design and specification of an IR evaluation infrastructure, in *Information Access Evaluation. Multilinguality, Multimodality, and Visual Analytics* ed. By T. Catarci, P. Forner, D. Hiemstra, A. Peñas, G. Santucci, Proceedings of the Third International Conference of the CLEF Initiative (CLEF 2012). Lecture Notes in Computer Science (LNCS), vol. 7488 (Springer, Heidelberg, Germany, 2012)
8. M. Agosti, G.M. Di Nunzio, N. Ferro, The Importance of Scientific Data Curation for Evaluation Campaigns, in Thanos et al. [60], pp. 157–166
9. M. Agosti, N. Ferro, Annotations as context for searching documents, in *Proceedings of 5th International Conference on Conceptions of Library and Information Science – Context: nature, impact and role (CoLIS 5)* ed. By F. Crestani, I. Ruthven. Lecture Notes in Computer Science (LNCS), vol. 3507 (Springer, Heidelberg, Germany, 2005), pp. 155–170
10. M. Agosti, N. Ferro, A formal model of annotations of digital content. ACM Trans. Inf. Syst. (TOIS) **26**(1), 3:1–3:57 (2008)
11. M. Agosti, N. Ferro, Towards an Evaluation Infrastructure for DL Performance Evaluation, in *Evaluation of Digital Libraries: An insight into useful applications and methods*, ed. by G. Tsakonas, C. Papatheodorou (Chandos Publishing, Oxford, 2009), pp. 93–120
12. M. Agosti, N. Ferro, G. Silvello, The NESTOR framework: manage, access and exchange hierarchical data structures, in *Proceedings of 18th Italian Symposium on Advanced Database Systems (SEBD 2010)* ed. By R. Martoglia, S. Bergamaschi, S. Lodi, C. Sartori(Società Editrice Esculapio, Bologna, Italy, 2010), pp. 242–253
13. M. Agosti, N. Ferro, G. Silvello, Digital library interoperability at high level of abstraction. Future Gener. Comput. Syst. (FGCS) **55**, 129–146 (2016)
14. M. Agosti, M. Manfioletti, N. Orio, C. Ponchia, Evaluating the deployment of a collection of images in the CULTURA environment, in *Proceedings of the International Conference on Theory and Practice of Digital Libraries, TPDL 2013*. Lecture Notes in Computer Science, vol. 8092 (Springer, 2013), pp. 180–191

15. M. Agosti, M. Masotti, Design of an OPAC database to permit different subject searching accesses in a multi-disciplines universities library catalogue database, in Belkin et al. [19], pp. 245–255
16. J. Arguello, M. Crane, F. Diaz, J. Lin, A. Trotman, Report on the SIGIR 2015 workshop on reproducibility, inexplicability, and generalizability of results (RIGOR). SIGIR Forum **49**(2), 107–116 (2015)
17. T.G Armstrong, A. Moffat, W. Webber, J. Zobel, EvaluatIR: an online tool for evaluating and comparing IR systems, in *Proceedings of 32nd Annual International ACM SIGIR Conference on Research and Development in Information Retrieval (SIGIR 2009)* ed. By J. Allan, J.A Aslam, M. Sanderson, C. Zhai, J. Zobel (ACM Press, New York, USA, 2009), p. 833
18. A. Badan, L. Benvegnù, M. Biasetton, G. Bonato, A. Brighente, A. Cenzato, P. Ceron, G. Cogato, S. Marchesin, A. Minetto, L. Pellegrina, A. Purpura, R. Simionato, N. Soleti, M. Tessarotto, A. Tonon, F. Vendramin, N. Ferro, Towards open-source shared implementations of keyword-based access systems to relational data, in *Proceedings of 1st EDBT/ICDT Workshop on Keyword-based Access and Ranking at Scale (KARS 2017)*, ed. By N. Ferro, F. Guerra, Z. Ives, G. Silvello, M. Theobald. CEUR Workshop Proceedings, ISSN 1613-0073 (2017), http://CEUR-WS.org
19. N.J. Belkin, P. Ingwersen, A. Mark Pejtersen, E.A. Fox (eds.), in *Proceedings of 15th Annual International ACM SIGIR Conference on Research and Development in Information Retrieval (SIGIR 1992)* (ACM Press, New York, USA, 1992)
20. S. Bergamaschi, N. Ferro, F. Guerra, G. Silvello, A perspective look at keyword-based search over relation data and its evaluation, in *Proceedings of 23rd Italian Symposium on Advanced Database Systems (SEBD 2015)* ed. By P. Atzeni, M. Lenzerini, D. Lembo, R. Torlone (2015)
21. S. Bergamaschi, N. Ferro, F. Guerra, G. Silvello, Keyword-based search over databases: a roadmap for a reference architecture paired with an evaluation framework. LNCS Trans. Comput. Collect. Intell. (TCCI) **9630**, 1–20 (2016)
22. C.L. Borgman, The Conundrum of Sharing Research Data. JASIST **63**(6), 1059–1078 (2012), http://dx.doi.org/10.1002/asi.22634
23. C.L. Borgman, Why are the attribution and citation of scientific data important? in *Report from Developing Data Attribution and Citation Practices and Standards: An International Symposium and Workshop*. National Academy of Sciences' Board on Research Data and Information (National Academies Press, Washington DC, 2012), pp. 1–8
24. C.L. Borgman, *Big Data, Little Data, No Data* (MIT Press, Cambridge, 2015)
25. J. Brase, Y. Socha, S. Callaghan, C.L. Borgman, P.F. Uhlir, B. Carroll, Data Citation: Principles and Practice, *Research Data Management: Practical Strategies for Information Professionals* (Purdue University Press, West Lafayette, 2014), pp. 167–186
26. P. Buneman, S.B. Davidson, J. Frew, Why data citation is a computational problem. Commun. ACM (CACM) **59**(9), 50–57 (2016)
27. P. Buneman, G. Silvello, A rule-based citation system for structured and evolving datasets. IEEE Data Eng. Bull. **33**(3), 33–41 (2010)
28. L. Candela, D. Castelli, N. Ferro, Y. Ioannidis, G. Koutrika, C. Meghini, P. Pagano, S. Ross, D. Soergel, M. Agosti, M. Dobreva, V. Katifori, H. Schuldt, The DELOS Digital Library Reference Model. Foundations for Digital Libraries. ISTI-CNR at Gruppo ALI, Pisa, Italy (2007), http://www.delos.info/files/pdf/ReferenceModel/DELOS_DLReferenceModel_0.98.pdf
29. J. Coffman, A.C. Weaver, An empirical performance evaluation of relational keyword search techniques. IEEE Trans. Knowl. Data Eng. (TKDE) **1**(26), 30–42 (2014)
30. S.B. Davidson, D. Deutsch, M. Tova, G. Silvello, A model for fine-grained data citation, in *8th Biennial Conference on Innovative Data Systems Research (CIDR 2017)* (2017)
31. M. Dussin, N. Ferro, Managing the knowledge creation process of large-scale evaluation campaigns, in Agosti et al. [4], pp. 63–74
32. N. Ferro, Annotation search: the fast way, in Agosti et al. [4], pp. 15–26
33. N. Ferro, The FAST annotation service, in *Proceedings of 17th Italian Symposium on Advanced Database Systems (SEBD 2009)*, ed. By V. De Antonellis, S. Castano, B. Catania, G. Guerrini (Seneca Edizioni, Torino, Italia, 2009), pp. 169–176

34. N. Ferro, Reproducibility challenges in information retrieval evaluation. ACM J. Data Inf. Quality (JDIQ) **8**(2), 8:1–8:4 (2017)
35. N. Ferro, F. Crestani, M.F. Moens, J. Mothe, F. Silvestri, G.M. Di Nunzio, C. Hauff, G. Silvello, Advances in information retrieval, in *Proceedings of 38th European Conference on IR Research (ECIR 2016)*. Lecture Notes in Computer Science (LNCS), vol. 9626 (Springer, Heidelberg, Germany, 2016)
36. N. Ferro, N. Fuhr, K. Järvelin, N. Kando, M. Lippold, J. Zobel, Increasing reproducibility in IR: findings from the dagstuhl seminar on reproducibility of data-oriented experiments in e-science. SIGIR Forum 50(1) (2016)
37. N. Ferro, G. Silvello, The NESTOR framework: how to handle hierarchical data structures, in Agosti et al. [4], pp. 215–226
38. N. Ferro, G. Silvello, FAST and NESTOR: how to exploit annotation hierarchies, in *Digital Libraries. Proceedings of the Sixth Italian Research Conference (IRCDL 2010)*, ed. By M. Agosti, F. Esposito, C. Thanos (Springer, Heidelberg, Germany, 2010), pp. 55–66
39. N. Ferro, G. Silvello, The NESTOR model: properties and applications in the context of digital archives, in *Proceedings 19th Italian Symposium on Advanced Database Systems (SEBD 2011)*, ed. By G. Mecca, S. Greco, (Università della Basilicata, Italy, 2011), pp. 274–285
40. N. Ferro, G. Silvello, Modeling archives by means of OAI-ORE, in *Digital Libraries and Archives - Proceedings of 8th Italian Research Conference (IRCDL 2012)*, ed. By M. Agosti, F. Esposito, S. Ferilli, N. Ferro. Communications in Computer and Information Science (CCIS), vol. 354 (Springer, Heidelberg, Germany, 2013)
41. N. Ferro, G. Silvello, NESTOR: a formal model for digital archives. Inf. Process. Manag. **49**(6), 1206–1240 (2013)
42. N. Ferro, G. Silvello, Descendants, ancestors, children and parent: a set-based approach to efficiently address Xpath primitives. Inf. Process. Manag. **52**(3), 399–429 (2016)
43. N. Ferro, G. Silvello, Fast access to XML data: a set-based approach, in *Proceedings of 24th Italian Symposium on Advanced Database Systems (SEBD 2016)*, ed. By P. Paolini, M.A. Bochicchio, G. Mecca (2016)
44. J. Freire, P. Bonnet, D. Shasha, Computational reproducibility: state-of-the-art, challenges, and database research opportunities, in *Proceedings of the ACM SIGMOD International Conference on Management of Data, SIGMOD 2012* (2012), pp. 593–596, http://doi.acm.org/10.1145/2213836.2213908
45. M.A. Gonçalves, E.A. Fox, L.T. Watson, N.A. Kipp, Streams, structures, spaces, scenarios, societies (5s): a formal model for digital libraries. ACM Trans. Inf. Syst. (TOIS) **22**(2), 270–312 (2004)
46. A. Hanbury, G. Kazai, A. Rauber, N. Fuhr (eds.), Advances in information retrieval, in *Proceedings of 37th European Conference on IR Research (ECIR 2015)*. Lecture Notes in Computer Science (LNCS), vol. 9022 (Springer, Heidelberg, Germany, 2015)
47. T. Hey, S. Tansley, K. Tolle (eds.), *The Fourth Paradigm: Data-Intensive Scientific Discovery* (Microsoft Research, USA, 2009)
48. C.R. Hildreth (ed.), *The Online Catalogue: Developments and Directions* (Library Association, London, 1989)
49. Y.E. Ioannidis, Digital libraries at a crossroads. Int. J. Digi. Libr. **5**(4), 255–265 (2005)
50. A. Lipani, F. Piroi, L. Andersson, A. Hanbury, An information retrieval ontology for information retrieval nanopublications, in *Information Access Evaluation – Multilinguality, Multimodality, and Interaction. Proceedings of the Fifth International Conference of the CLEF Initiative (CLEF 2014)*, ed. By E. Kanoulas, M. Lupu, P. Clough, M. Sanderson, M. Hall, A. Hanbury, E. Toms. Lecture Notes in Computer Science (LNCS), vol. 8685, (Springer, Heidelberg, Germany, 2014), pp. 44–49
51. S. Pröll, A. Rauber, Scalable data citation in dynamic, large databases: model and reference implementation, in *Proceedings of the 2013 IEEE International Conference on Big Data*, ed. By X. Hu, T.L. Young, V. Raghavan, B.W. Wah, R. Baeza-Yates, G. Fox, C. Shahabi, M. Smith, Q. Yang, R. Ghani, W. Fan, R. Lempel, R. Nambiar (IEEE, 2013), pp. 307–312

52. G. Silvello, Structural and content queries on the nested sets model, in *Proceedings of 20th Italian Symposium on Advanced Database Systems (SEBD 2012)*, ed. By N. Ferro, L. Tanca (Edizioni Libreria Progetto, Padova, Italy, 2012), pp. 283—288
53. G. Silvello, A Methodology for Citing Linked Open Data Subsets. D-Lib Mag. 21(1/2) (2015), http://dx.doi.org/10.1045/january2015-silvello
54. G. Silvello, Linked open data framework for serendipity in history of art research, in *Proceedings of 1st AI*IA Workshop on Intelligent Techniques At LIbraries and Archives co-located with XIV Conference of the Italian Association for Artificial Intelligence, IT@LIA@AI*IA 2015*, ed. By S. Ferilli, N. Ferro. CEUR Workshop Proceedings, vol. 1509 (2015), http://CEUR-WS.org
55. G. Silvello, Learning to cite framework: how to automatically construct citations for hierarchical data. J. Am. Soc. Inf. Sci. Technol. (JASIST) in print, 1–28 (2016)
56. G. Silvello, G. Bordea, N. Ferro, P. Buitelaar, T. Bogers, Semantic representation and enrichment of information retrieval experimental data. Int. J. Digit. Libr. (IJDL) in press, 1–28 (2016)
57. G. Silvello, N. Ferro, Data citation is coming. Introduction to the special issue on data citation. Bull. IEEE Technical Comm. Digi. Libr. (IEEE-TCDL) **12**(1), 1–5 (2016)
58. G. Silvello, N. Ferro, Data citation is coming. Introduction to the special issue on data citation. Bull. IEEE Tech. Comm. Digit. Libr. Special Issue on Data Citation 12(1), 1–5 (2016)
59. C.M. Steiner, M. Agosti, M.S. Sweetnam, E.C. Hillemann, N. Orio, C. Ponchia, C. Hampson, G. Munnelly, A. Nussbaumer, D. Albert, O. Conlan, Evaluating a digital humanities research environment: the CULTURA approach. Int. J. Digit. Libr. (IJDL) **15**(1), 53–70 (2014)
60. C. Thanos, F. Borri, L. Candela, Digital libraries: research and development, in *First International DELOS Conference. Revised Selected Papers*. Lecture Notes in Computer Science (LNCS), vol. 4877 (Springer, Heidelberg, Germany, 2007)
61. M. Vernooy-Gerritsen, Enhanced Publications: Linking Publications and Research Data in Digital Repositories (Amsterdam University Press, Amsterdam, 2009)

From Data Integration to Big Data Integration

Sonia Bergamaschi, Domenico Beneventano, Federica Mandreoli, Riccardo Martoglia, Francesco Guerra, Mirko Orsini, Laura Po, Maurizio Vincini, Giovanni Simonini, Song Zhu, Luca Gagliardelli and Luca Magnotta

Abstract The Database Group (DBGroup, www.dbgroup.unimore.it) and Information System Group (ISGroup, www.isgroup.unimore.it) research activities have been mainly devoted to the Data Integration Reserach Area. The DBGroup designed and developed the MOMIS data integration system, giving raise to a successful innov-

S. Bergamaschi (✉) · D. Beneventano · F. Guerra · L. Po · M. Vincini · G. Simonini
Dipartimento di Ingegneria Enzo Ferrari, Università di Modena e Reggio Emilia, Modena, Italy
e-mail: sonia.bergamaschi@unimore.it

D. Beneventano
e-mail: domenico.beneventano@unimore.it

F. Guerra
e-mail: francesco.guerra@unimore.it

L. Po
e-mail: laura.po@unimore.it

M. Vincini
e-mail: maurizio.vincini@unimore.it

G. Simonini
e-mail: giovanni.simonini@unimore.it

M. Orsini · L. Magnotta
Datariver s.r.l., Modena, Italy
e-mail: mirko.orsini@datariver.it

L. Magnotta
e-mail: luca.magnotta@datariver.it

F. Mandreoli · R. Martoglia
FIM, Università di Modena e Reggio, Modena, Italy
e-mail: federica.Mandreoli@unimore.it

R. Martoglia
e-mail: riccardo.martoglia@unimore.it

S. Zhu · L. Gagliardelli · L. Magnotta
ICT School, Università di Modena e Reggio, Modena, Italy
e-mail: song.zhu@unimore.it

L. Gagliardelli
e-mail: luca.gagliardelli@unimore.it

S. Flesca et al. (eds.), *A Comprehensive Guide Through the Italian Database Research Over the Last 25 Years*, Studies in Big Data 31, DOI 10.1007/978-3-319-61893-7_3

ative enterprise DataRiver (www.datariver.it), distributing MOMIS as open source. MOMIS provides an integrated access to structured and semistructured data sources and allows a user to pose a single query and to receive a single unified answer. Description Logics, Automatic Annotation of schemata plus clustering techniques constitute the theoretical framework. In the context of data integration, the ISGroup addressed problems related to the management and querying of heterogeneous data sources in large-scale and dynamic scenarios. The reference architectures are the Peer Data Management Systems and its evolutions toward dataspaces. In these contexts, the ISGroup proposed and evaluated effective and efficient mechanisms for network creation with limited information loss and solutions for mapping management query reformulation and processing and query routing. The main issues of data integration have been faced: automatic annotation, mapping discovery, global query processing, provenance, multidimensional Information integration, keyword search, within European and national projects. With the incoming new requirements of integrating open linked data, textual and multimedia data in a big data scenario, the research has been devoted to the Big Data Integration Research Area. In particular, the most relevant achieved research results are: a scalable entity resolution method, a scalable join operator and a tool, LODEX, for automatically extracting metadata from Linked Open Data (LOD) resources and for visual querying formulation on LOD resources. Moreover, in collaboration with DATARIVER, Data Integration was successfully applied to smart e-health.

1 Data Integration at the DBGroup: State of the Art

Data Integration is the problem of combining data residing at different autonomous sources, and providing the user with a unified view of these data. The DBGroup group has been investigating data integration for more than 20 years and almost all of the research activities have been centered around the MOMIS (Mediator EnvirOnment for Multiple Information Sources) Data Integration System; [5, 20], a most recent description is in [18]. An open source version of the MOMIS system is delivered and maintained by Datariver (see Sect. 4).

The MOMIS system is characterized by a classical wrapper/mediator architecture [60], where the local data sources contain the real data, while a *Global Virtual Schema* (*GVS*) provides a reconciled, integrated, and virtual view of the underlying data sources. MOMIS follows a *Global-As-View* approach: each class of the *GVS* is characterized in terms of a view over the sources; then a query over the *GVS* is rewritten on the data sources by *query unfolding* [36]. In the following we describe the main features and applications of the MOMIS system. For a complete version of this section see http://dbgroup.ing.unimore.it/Momis.

1.1 Automatic Global Schema Generation and Annotation

One of the main features of the MOMIS System is the rapid deployment of data integration projects, by means of a process that detects semantic similarities among the involved local source schemata, automatically generates a *GVS* and the mappings among the *GVS* and the local schemata. The theoretical framework of this data integration process is constituted by Description Logics, Automatic Annotation of schemata plus clustering techniques. In particular, with the Annotation step, schema labels (i.e. class/attribute names) are mapped to concepts of a lexical ontology, such as Wordnet [51], in order to perform *schema matching*, i.e., to discover relationships among the elements of different schemata [28]. However, performance of automatic annotation methods on real world schemata suffers from the abundance of non-dictionary words such as compound nouns, abbreviations and acronyms. In [58] we addressed this problem by introducing a *schema label normalization* method which expands abbreviations/acronyms and annotates compound nouns; the techniques was implemented into the NORMS tool, described in the next section.

1.2 The NORMS (NORMalizer of Schemata) Tool

NORMS is a tool that provides automatic normalization and annotation for schema labels [57]; it was developed in collaboration with Datariver[1] and it can be integrated in the MOMIS framework to improve schema matching. To handle automatically a large number of non-dictionary words, NORMS uses, in addition to lexical ontologies, other resources: (1) complementary schemata (other schemata that have to be integrated with the current schema); (2) online abbreviation dictionary; (3) user-defined dictionaries. The main innovative aspect of NORMS is the *Compound Nouns (CN) Interpretation*: i.e., the task of determining the semantic relationships holding among its constituents. NORMS annotates each CN constituent and then performs automatic CN interpretation by using the set of nine semantic relationships (CAUSE, HAVE, MAKE etc.) defined by Levi in [38]. The performance of NORMS was tested in several data integration projects of different semantic domains; moreover, NORMS was also tested w.r.t. schema and data integration test suites, such as Almalgam [52] and TPC-H.[2] The experimental results have shown the effectiveness of NORMS, which significantly improves annotation, and, consequently, schema matching.

[1]NORMS will be included in the next release of the MOMIS Open Source version, available at http://www.datariver.it/data-integration/momis/.

[2]http://www.tpc.org/tpch.

1.3 *Data Fusion*

Data fusion is the process of fusing multiple records representing the same real-world object into a single, consistent, and clean representation; merging different records into a single representation and at the same time resolving existing data conflicts was a problem addressed by several researchers in the past decades, but with few implemented solutions in integration systems [29]. To perform *Data Fusion*, we assumed that *Entity Resolution*, i.e., the identification of the same object in different sources, has been already performed and thus a shared object identifier (ID) exist among different sources. Multiple records with the same ID are fused by means of the Full Join Merge operator [18]. On 2016, we faced and solved the problem of Entity Resolution in a Big Data Integration scenario (see Sect. 5.1). Moreover, as the increasing number of distributed data sources to be integrated introduces join scalability as a critical issue for the scalability of a Data Integration system, we devised a new scalable join operator (see Sect. 5.2).

1.4 *Data Provenance*

Data Integration Systems deal with information coming from different sources, potentially uncertain or even inconsistent with each other, then the problem of "How to incorporate the notion of data quality (source reliability, accuracy, etc.)" is crucial. A common approach for evaluating the *quality* and *trustworthiness* of the data in systems involving a large number of sources is the analysis of the *provenance of information*. *Provenance* describes where data come from, how it is derived and how it was modified over time.

The notion of *PI-CS* provenance defined in [31] encodes all the *possible different derivations* of a output tuple in the query result, by storing a set of input tuples *for each derivation*. In [12] we extended the definition of *PI-CS* provenance to include *resolution functions* and, then, to distinguish between derivations where the contributing tuples have conflicts or not. We proved the usefulness of data provenance in a Data Integration context in several studies, from the use of data provenance in conflict resolution strategies [4] to the development of provenance-aware semantic search engines based on data integration [6].

1.5 *Integration of Data and Multimedia Sources*

The proliferation of multimedia data, and the consequent need of their integration with traditional information, represents nowadays a critical issue. In [13] we extended the MOMIS System to integrate "traditional" and "multimedia" data sources: multimedia queries can be expressed on the GVV without requiring multimedia processing capabilities at the *mediator* level since they are managed at the *local* level by the multimedia system. However, this solution requires a completely new query processing method. with respect to the one used for traditional data (based on a full join operation). Then, in [1] we discussed the question: "How can a multimedia local source supporting ranking queries be integrated into a mediator system without such capabilities?". We first described a naïve approach to support ranking (*Top-K*) queries which keep substantially unchanged the query processing method; this approach does not guarantee the completeness of results for *Top-K* queries, i.e., less than *K* results might be returned. Then, we discussed two alternative solutions for extending a mediator system so as to support multimedia queries.

1.6 The MOMIS System: Applications and Extensions

The MOMIS system was used to integrate genotypic and phenotypic data, for the development of the CEREALAB database [50]. A specific user's requirement coming from this application domain was data provenance capabilities, as discussed in [7]; moreover, in [11] we shown the publication of CEREALAB into the Linked Open Data cloud; finally, in [10] we applied to CEREALAB a fully automatic and semantic method for searching bibliographic data. We applied the MOMIS system either for building a tourism Information provider [19] and to define semantic mappings amongst product classification schemas [14, 16].

In [9] we extended the MOMIS system with a multi-agent architecture based on the P2P model; while a single peer carries out data integration activities, it exchanges knowledge with other peers by means of specialized agents. In [59] we shown as the tests of the THALIA benchmark [35] are fulfilled by the MOMIS system; In [15] we presented a framework able to query an integrated view of data and to search for eServices related to retrieved data.

1.7 Integrating Multidimensional Information

Collaborative business making is emerging as a possible solution for the difficulties that Small and Medium Enterprises (SMEs) are having in the current difficult economic scenarios. Collaboration, as opposed to competition, provides a competitive advantage to companies and organizations that operate in a joint business structure. When dealing with multiple organizations, managers must access unified strategic information obtained from the knowledge repositories of each individual organization; unfortunately, traditional Business Intelligence (BI) tools are not designed with the aim of collaboration so the task becomes difficult from a managerial, organizational and technological point of view. To deal with this shortcoming, we provided an integration, mapping-based, methodology for heterogeneous Data Warehouses that aims at facilitating business stakeholders' access to unified strategic information obtained from a network of heterogeneous collaborating SMEs [17]. In particular, we proposed a mapping-based integration methodology that is able to generate semantic mappings between dimensions of different DWs, either between dimension categories and between members of such categories. Our work is motivated by data-quality which is a crucial aspect when building an inter-organization DW; in fact, given the use of the DW, incorrect or low-quality data may not only make the DW useless, but its use may also lead to wrong decisions with potential negative impact on the organization. For these reasons, we analyzed data-quality requirements for the generated mappings, such as coherency and consistency, to ensure that the integrated information can be correctly aggregated (or disaggregated). This observation is relevant as the multidimensional data is usually explored along aggregation patterns, drilling-down or rolling-up from a starting analysis point.

1.8 MOMIS Dashboard

The MOMIS Dashboard[3] is an interactive visualization tool (available also as a mobile application) either for integrated data obtained with MOMIS or for data of several commercial DBMSs. It makes easier to compare data and capture useful information. It allows to filter the data and visualize the results through different charts. The available charts are : line, bar, pie, bubble on a Google Maps, tabular views; filters applicable on the charts are: date interval, numeric interval, set of different values, free text. The MOMIS Dashboard allows to define advanced access rules on the data, i.e. different views. In this way, different users can access to the same charts but filtering only their proprietary data set. Another interesting feature is the multilingual support, i.e., it can manage different languages for the same Dashboard. The MOMIS Dashboard is developed in collaboration with Datariver that has already used it in several projects (see (http://www.datariver.it/data-integration/)), such as "Open Data Lavoro", a web application for analysis and monitoring of public open data related to job market, and the "Italian FSHD Registry", which will be described in Sect. 4.1.

2 Data Integration at the ISGroup: State of the Art

Peer data management systems (PDMSs) represent a natural step beyond data integration systems, replacing their single logical schema with an interlinked collection of semantic mappings between peers' individual schemas [34]. Moreover, the synergy between PDMSs and Semantic Web technologies has paved the way for large-scale sharing of semantically rich data, like RDF or ontologies. Because of the lack of common understanding of the vocabulary used by peers, the resulting heterogeneity

[3]For a complete description see http://dbgroup.ing.unimore.it/MomisDashboard.

of data representations opens new challenges as to the efficient and effective retrieval of relevant information.

As opposed to viewing semantic misalignment as a limit for interoperability, the research of ISGROUP leveraged on the presence of semantic approximations between the peers' schemas as a means for giving effective hints along the following directions: (1) effective query processing [42, 47]; (2) network creation [49].

A PDMS underlies a potentially very large network able to handle huge amounts of data. For this reason, query routing, i.e. the process of selecting the most promising peers, is a fundamental issue for providing relevant answers to queries over distributed resources. In this context, we proposed a distributed index mechanism where each peer is provided with a novel kind of index, the Semantic Routing Index (SRI), for routing queries effectively. SRIs are also employed in the query answering phase for reducing the space of reformulations and for ranking answers. The proposed approach for query processing founds on a fuzzy settlement which provides a formal semantics of the approximations originated by the heterogeneity of the schemas in a PDMS.

When processing queries in a PDMS, the semantic path needs to relate the terms used in the query with the terms specified by the node providing the data. Hence it is likely that there will be information loss along long paths in the PDMS because of missing (or incomplete) mappings, leading to the well-known problem of how to boost a network of mappings in a PDMS. Our research investigated this issue and proposed the first efficient approach for the flexible creation and maintenance of the network structure, clustering together semantically related peers. The approach we propose is scalable, incremental, and self-adaptive in the creation and maintenance of clusters.

When it comes to network construction, every time a new peer joins the network, there is the need to make explicit the semantics of its schema by associating each schema's term with the right concept. In this parallel research branch, ISGROUP also proposed effective knowledge-based disambiguation techniques working on a variety of structures both at data and metadata level [39, 45, 46].

All the above techniques have been incorporated in a complete PDMS infrastructure, the SUNRISE (System for Unified Network Routing, Indexing and Semantic Exploration) system, that offers network construction and exploration facilities for XML and RDF data sharing [41, 43]. Then, in the context of the FIRB NeP4B project and in collaboration with ISTI-CNR, SUNRISE was extended to multimedia data that needs the support of content-based, aka similarity, predicates in the query language [37].

Another research area where the main concepts of PDMSs where exploited profitably is inter-business collaborative where companies coordinate themselves to develop common and shared opportunities. In collaboration with the Business Intelligence Group of the University of Bologna we introduced a peer-to-peer data warehousing architecture, Business Intelligence Network (BIN) [32, 48]. The original contributions we gave in this area are a language for the definition of semantic mappings between the schemata of peers, using predicates that are specifically tailored for the multidimensional model, a framework for OLAP query reformulation that relies on the translation of mappings and queries towards the underlying relational schemata, a query reformulation algorithm.

In the context of data integration systems, the dataspace model represent a further step towards the integration of loosely-connected information that support data co-existence and pay-as-you-go principles [33]. Such challenge arises in application scenarios that have emerged recently such as exploratory data analysis and personal information management. A very recent line of research is conducted in collaboration with George Fletcher from Tu/E and aims at investigating theoretical and engineering solutions for query-driven mapping discovery. [30]. The project proposes a shift of perspective in mapping management from the state-of-the-art data-centric approach to a user-centric approach where mappings contribute to users satisfaction in their information seeking activities.

Graph-based data models have recently gained much popularity as powerful means for data representation, especially in the dataspace scenario. The largeness and the heterogeneity of most graph-modeled datasets in several database application areas make flexible query answering capabilities an essential need. ISGROUP proposed an approach [40, 44] for the approximate matching of complex queries on such kind of data, whose query answering model gracefully blends label approximation with structural relaxation, under the primary objective of delivering meaningfully approximated results only.

3 Keyword Search over Relational Databases

Despite the effort that the research community has put in the field in the last fifteen years and the significant number of scientific publications and prototypes developed, keyword search over relational databases is still a hot and open challenge. No research prototypes have transitioned from proof-of-concept implementations into deployed systems, and the current research proposals fail in making users able to effectively and efficiently query relational databases. The problem is intrinsically complex and multifaceted, since the information in a database is spread over a number of tables connected with multiple paths. The simple exploitation of full-text search functionalities natively implemented in the DBMS (e.g., the `match-against` function in MySQL) enables users only to discover the attributes of the database containing the query keywords at run-time, thus providing just partial, and incomplete results to be joined to form complete answers. Nevertheless, the existence of a path joining two generic tuples is not assured even in a full connected database, and it can involve tuples in other tables. Moreover, in case of multiple connections, each path carries out different semantics which cannot match with the intended meaning of the user's query. The computation of all paths largely affect the time required for solving the query and the quality of the answer.

The DBGroup is an active member in this area since 2009 and developed techniques and tools in conjunction with other research groups (in particular with Prof. Velegrakis, University of Trento – IT, Prof. Ferro, University of Padua – IT, Prof.

Trillo, University of Zaragoza – ES), and under the big umbrella of the KEYSTONE IC1302 COST Action (http://www.keystone-cost.eu). The main contributions are related to: (i) the introduction of a principled model for the representing the keyword search problem over structured databases; (ii) the development of techniques and prototypes based on heuristic rules and machine learning techniques; (iii) an analysis of the methodology adopted for evaluating keyword based systems and the definition of a roadmap and a reference architecture for an evaluation framework.

Representing keyword search over structured databases. In our research, we have developed a three-step principle model to represent the process of solving keyword queries. The fundamental steps we consider are first to match the keywords into the database structures, then to discover ways these matched structures can be combined, and finally to select the best matches and combinations such that the identified database structures represent what the user had in mind to discover when formulating the keyword query. The first step is focused on trying to capture the meaning of the keywords in the query as they are understood by the user. In some sense, it provides the *user perspective* of the keyword query and it does so by providing a mapping of the keywords into database terms. We call the mappings as *configurations* and the step is referred to as the *forward analysis step* since it starts from the keywords and moves towards the database. The second step tries to capture the meaning of the keywords as they can be understood from the point of view of the data engineers who designed that database structure. In this step we compute the paths connecting the keywords, thus forming the possible solutions (we refer them as the *interpretations*) to the user query. In some sense, this phase provides the *database perspective* of the keyword query and it does so by providing the relationships among the images of the keywords. This task is referred to as the *backward analysis step* since it starts from the database structures and moves towards the query keywords through their images. The third step provides a ranking of the interpretations produced by the second step. These are the *explanations*, i.e., "fully justified" answers to the keyword queries.

Techniques and prototypes to solve keyword queries. We have developed 3 prototypes implementing the heuristics and machine learning based techniques developed. The **KEYMANTIC** [21, 22] system was focused on finding configurations. It provided a solution based on a bipartite graph matching model where user keywords were matched to database schema elements by using an extension of the Hungarian algorithm. In particular, weights qualifying the matches run-time change on the basis of the partial assignments computed by the algorithm. **KEYRY** [27, 53] extended KEYMANTIC by providing a probabilistic framework, based on a Hidden Markov Model (HMM), to compute the configurations. In our approach, the HMM states are database elements, and the observations are the keywords that can be associated to the states. The application of the Viterbi algorithm allows the computation of sequences of states to be associated to the keyword queries. **QUEST** [25, 26] provides a complete keyword search system for relational database, where the configurations, computed by means of a HMM approach, and the interpretations, computed by means of Steiner Trees, are joined by means of the probabilistic framework provided by the Dempster Shafer theory.

Evaluating keyword search systems. We noticed two main issues that are hampering the design and development of next generation systems for keyword search over structured data: (i) the lack of systemic approaches considering all of the issues of keyword search from the interpretation of the user needs, to the computation, retrieval, ranking and presentation of the results; and (ii) the absence of a shared and complete evaluation methodology measuring user satisfaction, achieved utility and required effort for carrying out informative tasks. In light of this, we proposed in [24] a reference architecture including all the components needed for a keyword search engine for relational databases and we have discussed a roadmap for the definition of a fair and solid evaluation framework.

4 Smart Health Care at DataRiver

DataRiver (www.datariver.it) is an Innovative SME, accredited as a Research Innovation Institution of the Emilia Romagna Region, founded on June 2009 as a Spin-Off of the University of Modena and Reggio Emilia from the initiative of professors and researchers of the DBGroup. The company develops innovative software solutions in the healthcare industry and offers specialized consulting services for Clinical Data Management, Big Data Integration (*BDI*) and Analytics concerned to Clinical Trials, Cancer and Rare Diseases Registries, Mobile Data Capture Apps for the collection and management of patient clinical data from mobile and wearable devices. The healthcare industry is naturally rich with data – clinical, patient, claim, hospital system, financial, pharmacy and, most recently, data from wearable technology. In the healthcare context, data flows through heterogeneous systems and are stored in disparate data sources. BDI techniques provide a complete and synthetic view of all the health information about a patient or a set of selected patients. The Big Data analytics applied to integrated patient's health data allow a widespread monitoring to prevent clinical events and plan personalized care treatments. The Internet of Medical Things (IoMT) is the collection of medical devices and applications that connect to healthcare IT systems through online computer networks, including comprehensive solutions that follow patients on their life places, enabling home care and telemedicine. DBGroup and DataRiver are involved in several Smart Health Care projects for the research and development of innovative solutions, addressing the issues for BDI in the healthcare context and providing state-of-the-art applications for the IoMT.

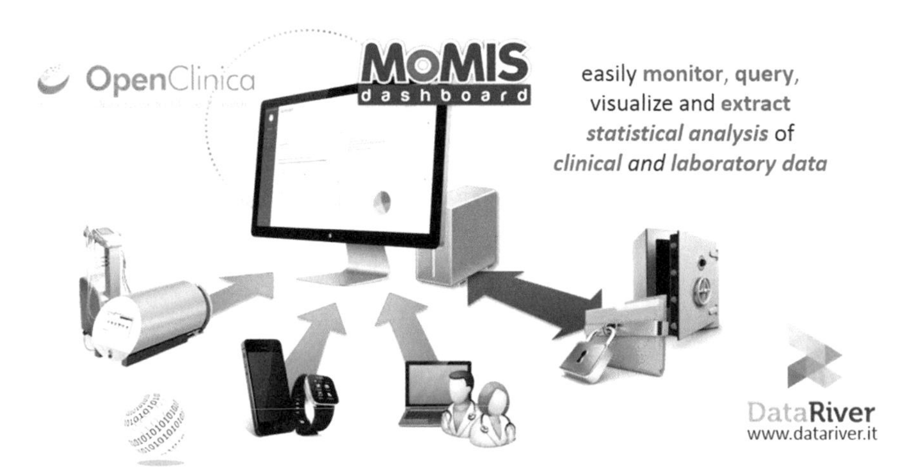

Fig. 1 Enhanced data integration and analytics for Smart Health Care

4.1 The Italian FSHD Registry: An Enhanced Data Integration and Analytics Framework for Smart Health Care

Facioscapulohumeral muscular dystrophy (FSHD) is a common myopathy, that has been associated to the reduction of a string of DNA elements, named D4Z4. BDI and analytics techniques have been applied by DataRiver and DBGroup in the "Italian FSHD Registry" project, to discover new research results by the unified analysis of patient's clinical and genomics data. The Italian FSHD Registry study (www.fshd.it) involves 13 medical research centers of the FSHD network to collect clinical and genomics data of more than 6.000 patients and 2.400 families; the objective is to develop an enhanced data integration and analytic framework to predict the risk of developing a severe form of the disease, and to identify factors that can help to prevent the disease worsening. In the "Italian FSHD Registry" project we successfully used the MOMIS system and the MOMIS Dashboard (see Sect. 1) jointly with OpenClinica (www.openclinica.org), an electronic data capture software for clinical research used to capture cleaner data, ensure compliance and promote patient engagement. The OpenClinica platform have been extended by DataRiver to collect family data and to be fully integrated with the MOMIS system. The MOMIS system provides a unified view of patient's data and biological data coming from different Biobank databases, to run queries both on clinical and genomics parameters and discover new information. The MOMIS Dashboard was used to easily monitor, query, visualize and extract statistical analysis of clinical and laboratory data (Fig. 1).

Fig. 2 Mobile, IoMT platform for remote monitoring and patient empowerment

4.2 *My Smart Age with HIV: A Mobile and IoMT Platform for Remote Monitoring and Empowerment of HIV Patients*

In the My Smart Age with HIV (MySAwH) clinical trial (www.mysmartage.org), an innovative IoMT mobile App has been developed to empower elderly HIV patients via health promotion, assessing reduction in health deficit and improvement in quality of life. DataRiver designed and developed the IoMT framework architecture and MySAwH App to collect patient's data from smartphone and wearable devices, elaborating and analyzing health indicators for the project; the IoMT framework has been designed to expand the traditional healthcare infrastructure and to provide patient monitoring and support outside the hospitals. The exploitation of IoMT, mobile App and wearable technologies, the integration and analysis of all collected patient's data in real time provide the physician with a continuous patient monitoring to measure the response to illness and the life quality improvement. The patient obtains an up to date insight of health condition and a constant support via the direct communication with caregiver (Fig. 2).

5 Big Data Integration at the DBGroup

A huge amount of (semi-)structured data is available on the Web in the form of web tables, marked-up contents, and Linked Open Data. For enterprises, government agencies, and researcher of large scientific projects, this data can be even more valuable if integrated with their proprietary data. With the incoming of these new requirements, the research at the DBGroup has been devoted to the Big Data Integration Research Area. The main topics we are investigating are: scalable entity resolution methods, distributed scalable join methods and summarization and visual querying methods of LOD resources.

5.1 Entity Resolution

As introduced in Sect. 1.3, *Entity Resolution* (ER) is the well-known problem to identify records that refer to the same entity. Generally, to perform ER and vocabulary-based topic detection, traditional techniques are based on schema-alignment among data sources (i.e., deriving a unique homogenous common schema from several heterogeneous ones). Unfortunately, the (semi-) structured data of the Web is usually characterized by high heterogeneity, volume and noise (missing/inconsistent data), making schema-alignment techniques no longer applicable. Therefore, Data Integration techniques dealing with this type of data typically renounce to exploit data source schemata.

At the DBgroup we developed a set of novel techniques [23, 54, 55] to induce loose schema information directly from the data, without exploiting the semantic of the schemas, able to scale to the huge data of the Web. This lose schema information can be employed as a surrogate of the schema-alignment and employed to enhance ER and vocabulary-based topic detection.

For ER, we proposed `Blast` [55] (Blocking with Loosely-Aware Schema Techniques), an approach to reduce the ER complexity with indexing techniques aiming to group similar records in blocks, and limit the comparison to only those records appearing in the same block. For the topic detection, we proposed Whatsit [23] a novel approach that generates signatures of sources that are matched against the signatures of a reference vocabulary. Thus, a description of the topics of the source in terms of this reference vocabulary is generated.

Finally, we developed an open source software[4] for both the approaches and we experimentally evaluated them on real world datasets. The results demonstrate that `Blast` outperforms the state-of-the-art blocking approaches for the big data scenario, and that Whatsit can actually be employed to detect topics of a given data source. Currently, we are developing ER methods for massively parallel and distributed systems (Apache Spark and Apache Flink), and we are performing on the same systems extensive benchmarks to assess their scalability for different scenarios related to big data integration problems.

---

[4] http://stravanni.github.io/blast/.

5.2 Distributed Scalable Join

As we discussed in Sect. 1.3, join scalability is a critical issue for the scalability of a Data Integration system and, then, a key requirement in a Big Data Integration scenario. To fulfill this goal we designed a scalable parallel join engine. In the distributed systems context, the main join paradigms/algorithms are [29]: Map Reduce Joins; Online Joins; Stream Joins.

Map Reduce is one of most popular paradigms for parallel computation. However, the Map Reduce paradigm uses a barrier between the Map and Reduce stages, this hurts performance. It means that the Map step has to be completed before the next step (Reduce) starts. Consequently, data coming by a big data source must be loaded in the system before the join operation can start. In addition, in case of multiple joins, the previous join step have to be completed before the next join can start. Otherwise, the Stream Join merges data in the stream paradigm, and the result of the Join can be used before it is completed. This join paradigm is useful to merge sensor data, where the input data is continuous. However, stream Joins are usually windows based to support infinite streams of data; this requires that input data sources be sorted and the sort is an expensive operation in a Big Data Integration scenario.

For these reasons, we are studying an Online Join algorithm based on a non-blocking join method. Our join engine receives stream inputs from data sources and merges tuples on the fly and, when the result of a tuple is complete, it can be used immediately in the next operation. Our join algorithm is a multi- way equi-join and it works on a cluster of computation nodes. The basis of the algorithm is composed by two steps: the first step is a the distribution step, which distributes tuples on the nodes. We apply a hash function on the join attributes, and the result of the function determines to which computation node a tuple has to be sent. The second step is the local join step, where tuples are merged on the basis of the join condition. Each computation node performs the join locally: if the output tuple is complete, i.e. it have at least one tuple from all data sources, the node emits the result.

5.3 Summarization and Visual Querying Methods for LOD Resources

With more than one thousand of LOD sources available on the Web, we are assisting to an emerging trend in publication and consumption of LOD datasets. However, the pervasive use of external resources together with a deficiency in the definition of the internal structure of a dataset causes many LOD sources are extremely complex to understand. LOD tools lack in producing an high level representation of datasets and in supporting users in the exploration and querying of a source. To overcome the above problems we defined a method to unveil the implicit structure of a LOD dataset by building a *Schema Summary* which contains the main classes and properties used within the datasets, whether they are taken from external vocabularies or not, and is conceivable as an RDFS ontology. This method was implemented in the LODeX tool [2, 3], which extracts statistical indexes for building the Schema Summary, by querying the SPARQL endpoint of a LOD source; LODeX allows users to compose visual queries by selecting objects from the Schema Summary and thus supporting users in exploring and understanding the contents of a LOD source. For a complete description see http://dbgroup.ing.unimore.it/lod.

The great majority of open data is normally published in an unstructured format and is typically accessed only by closed communities. In [56] we proposed a semi-automatic methodology for facilitating resource providers in publishing public data into the LOD cloud, and for helping consumers in efficiently accessing and querying them. The methodology was applied on the research project on Youth Policies of the Emilia-Romagna Region ("Open linked data Osservatorio Giovani della Regione Emilia-Romagna") [8]. The project goals were to identify interesting data sources both from the open data community and from the private repositories of local governments of Emilia Romagna region related to the Youth Policies, to integrate them and, to show up the result of the integration by means of a useful navigator tool; in the end, to publish new information as LOD. We firstly have analyzed the useful open data sources, then using the MOMIS system, we have integrated them with the proprietary data provided by the project partners; the MOMIS Dashboard was then used to provide an easy access to the data. Finally we published the integrated data as LOD. This project has exemplified how a Public Administrations can benefit from the use of Open Data and can effectively extract new and important information by integrating its own datasets with open data sources.

References

1. I. Bartolini, D. Beneventano, S. Bergamaschi, P. Ciaccia, A. Corni, M. Orsini, M. Patella, M.M. Santese, MOMIS goes multimedia: WINDSURF and the case of top-k queries, in *SEBD'15, Gaeta, 14–17 June 2015.* (2015), pp. 200–207
2. F. Benedetti, S. Bergamaschi, L. Po, Lodex: a tool for visual querying linked open data, in *ISWC'15 Posters & Demonstrations Track* (2015)
3. F. Benedetti, S. Bergamaschi, L. Po, Visual querying LOD sources with lodex, in *K-CAP'15, Palisades, NY, USA, 7-10 Oct 2015* (2015), pp. 12:1–12:8
4. D. Beneventano, Provenance based conflict handling strategies, in *DASFAA'12, Busan, South Korea, 15–18 Apr 2012* (2012), pp. 286–297
5. D. Beneventano, S. Bergamaschi, The momis methodology for integrating heterogeneous data sources, in *IFIP 18th World Computer Congress 22–27 Aug 2004 Toulouse, France* (Springer, US, 2004), pp. 19–24
6. D. Beneventano, S. Bergamaschi, Provenance-aware semantic search engines based on data integration systems. IJOCI **4**(2), 1–30 (2014)
7. D. Beneventano, S. Bergamaschi, A.R. Dannaoui, Integration and provenance of cereals genotypic and phenotypic data, in *SEBD'12* (2012), pp. 91–98
8. D. Beneventano, S. Bergamaschi, L. Gagliardelli, L. Po, Driving innovation in youth policies with open data, in *IC3K'15, Revised Selected Papers, Communications in Computer and Information Science* (Springer, 2016)
9. D. Beneventano, S. Bergamaschi, F. Guerra, M. Vincini, The SEWASIE network of mediator agents for semantic search. J. UCS **13**(12), 1936–1969 (2007)

10. D. Beneventano, S. Bergamaschi, R. Martoglia, Exploiting semantics for searching agricultural bibliographic data. J. of Inf. Sci. **42**(6), 748–762 (2016)
11. D. Beneventano, S. Bergamaschi, S. Sorrentino, M. Vincini, F. Benedetti, Semantic annotation of the CEREALAB database by the AGROVOC linked dataset. Ecol. Inf. **26**(2), 119–126 (2015)
12. D. Beneventano, A.R. Dannaoui, A. Sala, On provenance of data fusion queries, in *SEBD'11, 26–29 June 2011* (2011), pp. 84–94
13. D. Beneventano, C. Gennaro, S. Bergamaschi, F. Rabitti, A mediator-based approach for integrating heterogeneous multimedia sources. Multimed. Tools Appl. **62**(2), 427–450 (2013)
14. D. Beneventano, F. Guerra, S. Magnani, M. Vincini, A web service based framework for the semantic mapping amongst product classification schemas. J. Electron. Commer. Res. **5**(2), 114–127 (2004)
15. D. Beneventano, F. Guerra, A. Maurino, M. Palmonari, G. Pasi, A. Sala, Unified semantic search of data and services, in *MTSR'09* (2009), pp. 95–107
16. D. Beneventano, S.E. Haoum, D. Montanari, Mapping of heterogeneous schemata, business structures, and terminologies, in *Workshop at DEXA'07* (2007), pp. 412–418
17. D. Beneventano, M. Olaru, M. Vincini, Analyzing dimension mappings and properties in data warehouse integration, in *OTM'13* (2013), pp. 616–623
18. S. Bergamaschi, D. Beneventano, F. Guerra, M. Orsini, Data integration, in *Handbook of Conceptual Modeling: Theory, Practice and Research Challenges*, ed. By D.W. Embley, B. Thalheim (Springer, 2011)
19. S. Bergamaschi, D. Beneventano, F. Guerra, M. Vincini, Building a tourism information provider with the MOMIS system. J. Inf. Technol. Tour. **7**(3–4), 221–238 (2004)
20. S. Bergamaschi, S. Castano, M. Vincini, Semantic integration of semistructured and structured data sources. SIGMOD Rec. 28(1) (1999)
21. S. Bergamaschi, E. Domnori, F. Guerra, M. Orsini, R. Trillo-Lado, Y. Velegrakis, Keymantic: semantic keyword-based searching in data integration systems. PVLDB 3(2) (2010)
22. S. Bergamaschi, E. Domnori, F. Guerra, R. Trillo-Lado, Y. Velegrakis, Keyword search over relational databases: a metadata approach, in *SIGMOD* (ACM, 2011), pp. 565–576
23. S. Bergamaschi, D. Ferrari, F. Guerra, G. Simonini, Y. Velegrakis, Providing insight into data source topics. J. Data Semant. **5**(4), 211–228 (2016)
24. S. Bergamaschi, N. Ferro, F. Guerra, G. Silvello, Keyword-based search over databases: a roadmap for a reference architecture paired with an evaluation framework. Trans. Comput. Collect. Intell. **21**, 1–20 (2016)
25. S. Bergamaschi, F. Guerra, M. Interlandi, R.T. Lado, Y. Velegrakis, QUEST: a keyword search system for relational data based on semantic and machine learning techniques. PVLDB **6**(12), 1222–1225 (2013)
26. S. Bergamaschi, F. Guerra, M. Interlandi, R.T. Lado, Y. Velegrakis, Combining user and database perspective for solving keyword queries over relational databases. Inf. Syst. **55**, 1–19 (2016)
27. S. Bergamaschi, F. Guerra, S. Rota, Y. Velegrakis, A hidden markov model approach to keyword-based search over relational databases, in *ER*, vol. 6998 (LNCS, Springer, 2011), pp. 411–420
28. S. Bergamaschi, L. Po, S. Sorrentino, Automatic annotation for mapping discovery in integration systems, in *SEBD'08* (2008), pp. 334–341
29. J. Bleiholder, F. Naumann, Data fusion. ACM Comp. Surv. **41**, 1–41 (2008)
30. G.H.L. Fletcher, F. Mandreoli, No users no dataspaces! query-driven dataspace orchestration? in *Proceedings of SEBD* (2016), pp. 150–157
31. B. Glavic, G. Alonso, R.J. Miller, L.M. Haas, Tramp: Understanding the behavior of schema mappings through provenance. PVLDB **3**(1), 1314–1325 (2010)
32. M. Golfarelli, F. Mandreoli, W. Penzo, S. Rizzi, E. Turricchia, Towards OLAP query reformulation in peer-to-peer data warehousing, in *Proceedings of ACM (DOLAP)* (2010), pp. 37–44
33. A.Y. Halevy, M.J. Franklin, D. Maier, Principles of dataspace systems, in *ACM PODS* (2006), pp. 1–9

34. A.Y. Halevy, Z.G. Ives, D. Suciu, I. Tatarinov, Schema mediation for large-scale semantic data sharing. VLDB J. **14**(1), 68–83 (2005)
35. J. Hammer, M. Stonebraker, O. Topsakal, Thalia: test harness for the assessment of legacy information integration, in *ICDE* (2005), pp. 485–486
36. M. Lenzerini, Data integration: a theoretical perspective, in *PODS* (2002), pp. 233–246
37. R. Lenzi, C. Gennaro, F. Mandreoli, R. Martoglia, M. Mordacchini, W. Penzo, S. Sassatelli, A unified multimedia and semantic perspective for data retrieval in the semantic web. Inf. Syst. **36**(2), 174–191 (2011)
38. J.N. Levi, *The Syntax and Semantics of Complex Nominals*(Academic Press, Cambridge, 1978)
39. F. Mandreoli, R. Martoglia, Knowledge-based sense disambiguation (almost) for all structures. Inf. Syst. **36**(2), 406–430 (2011)
40. F. Mandreoli, R. Martoglia, W. Penzo, Approximating expressive queries on graph-modeled data: the gex approach. J. Syst. Softw. **2015**(109), 106–123 (2015)
41. F. Mandreoli, R. Martoglia, W. Penzo, S. Sassatelli, Data-sharing p2p networks with semantic approximation capabilities. IEEE IC **13**(5), 60–70 (2009)
42. F. Mandreoli, R. Martoglia, W. Penzo, S. Sassatelli, G. Villani, Sri@work: efficient and effective routing strategies in a pdms, in *WISE* (2007), pp. 285–297
43. F. Mandreoli, R. Martoglia, W. Penzo, S. Sassatelli, G. Villani, Building a pdms infrastructure for xml data sharing with sunrise, in *EDBT-DATAX* (2008)
44. F. Mandreoli, R. Martoglia, W. Penzo, G. Villani, Flexible query answering on graph-modeled data. Proc. EDBT **2009**, 216–227 (2009)
45. F. Mandreoli, R. Martoglia, E. Ronchetti, Versatile structural disambiguation for semantic-aware applications, in *Proceedings of ACM CIKM* (2005), pp. 209–216
46. F. Mandreoli, R. Martoglia, E. Ronchetti, Strider: a versatile system for structural disambiguation. Proc. EDBT **2006**, 1194–1197 (2006)
47. F. Mandreoli, R. Martoglia, S. Sassatelli, W. Penzo, Sri: exploiting semantic information for effective query routing in a pdms, in *Proceedings of of the ACM CIKM Workshop WIDM* (2006), pp. 19–26
48. F. Mandreoli, W. Penzo, S. Rizzi, M. Golfarelli, E. Turricchia, Olap query reformulation in peer-to-peer data warehousing. Inf. Syst. **37**(5), 393–411 (2012)
49. F. Mandreoli, W. Penzo, S. Sassatelli, S. Lodi, R. Martoglia, Semantic peer, here are the neighbors you want!. Proc. EDBT **2008**, 26–37 (2008)
50. J. Milc, A. Sala, S. Bergamaschi, N. Pecchioni, A genotypic and phenotypic information source: the cerealab database. Database (2011)
51. G.A. Miller, Wordnet: a lexical database for english. C. ACM **38**(11), 39–41 (1995)
52. R.J. Miller, D. Fisla, M. Huang, F. Kymlicka, V. Lee, The amalgam schema and data integration test suite (2001), www.cs.toronto.edu/~miller/amalgam
53. S. Rota, S. Bergamaschi, F. Guerra, The list viterbi training algorithm and its application to keyword search over databases, in *CIKM* (2011), pp. 1601–1606
54. G. Simonini, S. Bergamaschi, Enhancing Entity Resolution Efficiency with Loosely Schema-Aware Techniques (2016), pp. 270–277
55. G. Simonini, S. Bergamaschi, H.V. Jagadish, BLAST: a loosely schema-aware meta-blocking approach for entity resolution. PVLDB **9**(12), 1173–1184 (2016)
56. S. Sorrentino, S. Bergamaschi, E. Fusari, D. Beneventano, Semantic annotation and publication of linked open data. Comput. Sci. Appl. - ICCSA **2013**, 462–474 (2013)
57. S. Sorrentino, S. Bergamaschi, M. Gawinecki, NORMS: an automatic tool to perform schema label normalization, in *ICDE'11* (2011), pp. 1344–1347
58. S. Sorrentino, S. Bergamaschi, M. Gawinecki, L. Po, Schema label normalization for improving schema matching. DKE **69**(12), 1254–1273 (2010)
59. M. Vincini, D. Beneventano, S. Bergamaschi, Semantic integration of heterogeneous data sources in the momis data transformation system. J. UCS - J. Univers. Comput. Sci. **19**(13), 1986–2012 (2013)
60. G. Wiederhold, Intelligent integration of information, in *SIGMOD'93, Washington, D.C., 26–28 May 1993* (ACM Press, 1993), pp. 434–437

Matching Techniques for Data Integration and Exploration: From Databases to Big Data

Silvana Castano, Alfio Ferrara and Stefano Montanelli

Abstract In the last two decades, data matching has been addressed for different purposes and in different application contexts, ranging from data integration, to ontology evolution, to semantic data clouding, until more recent exploratory data analysis over large/big datasets. This paper describes the evolution of research activity on matching techniques for data integration and exploration at the ISLab group of the Università degli Studi di Milano. We analyze the matching techniques according to the structure of target data, the algorithmic pattern of the matching process, and the application focus, and we discuss the results of using our techniques for exploratory analysis of a real dataset composed by all the SEBD proceedings publications in the timeframe 1993–2016.

Keywords Matching techniques · Data integration · Data exploration · Big data

1 Introduction

Techniques for data matching and classification play an essential role in modern information systems architectures, to correctly compare and integrate disparate datasets and information sources as well as to achieve effective data sharing and exploration on the global scale [14]. The data matching problem has been addressed in the literature with different purposes and goals. Initial research was targeted to database schema analysis and classification with focus on development of semi-automated techniques for the data integration process, to build a global, reconciled schema as a unique query interface for the end users [11]. A further set of contributions was devoted to the analysis and classification of semi-structured and semantic web data,

S. Castano · A. Ferrara · S. Montanelli (✉)
Università Degli Studi di Milano,DI - Via Comelico, 39-20135 Milan, Italy
e-mail: stefano.montanelli@unimi.it

S. Castano
e-mail: silvana.castano@unimi.it

A. Ferrara
e-mail: alfio.ferrara@unimi.it

S. Flesca et al. (eds.), *A Comprehensive Guide Through the Italian Database Research Over the Last 25 Years*, Studies in Big Data 31, DOI 10.1007/978-3-319-61893-7_4

with contributions for ontology and instance matching, and for linked data analysis and classification [15]. As a further step ahead, research focused on social data and on "human-in-the-loop" approaches to enforce crowd-collaborative solutions to face the need of performing data classification problems in-the-large, as typically demanded in the era of big data [2].

This paper describes the evolution of research activity on matching techniques for data integration and exploration at the ISLab group of the Università degli Studi di Milano. In the last two decades, data matching has been addressed at ISLab for different purposes and in different application contexts, ranging from data integration, to ontology evolution, to semantic data clouding, until more recent exploratory data analysis over large/big datasets and data streams. To describe the proposed techniques and research contributions by capturing the evolution of their features and purposes, we introduce a multi-dimensional framework with the following main dimensions: (i) *target*, to describe the evolution of the prominent features of the target data of the matching process; (ii) *pattern*, to describe the prominent algorithmic pattern of the matching process; (iii) *focus*, to describe the evolution of the application for which the matching process is executed, ranging from data(base) integration, to ontology mapping/evolution, to semantic data clouding, up to (big) data exploration.

As a real example, we discuss main results of using our techniques for enforcing exploratory analysis of a real dataset composed by all the SEBD proceedings publications in the timeframe 1993–2016.

2 Evolution of Data Matching

A summary overview of the matching techniques evolution along each analysis dimension is shown in Fig. 1. The *application focus* follows the general evolution of the research field, in that it ranges from matching techniques for data(base) integration, to techniques for ontology mapping/evolution, to those for semantic data clouding, up to those for exploratory analysis of large/big datasets. With respect to the target, our matching techniques range from techniques for database schema matching, to techniques for ontology matching, to techniques for linked data and social data matching, up to matching techniques for big data and data streams. Moving from the left to the right side of the figure, we observe a shift from highly structured, semantically rich data, typical of databases, schemas, ontologies, to progressively loosely structured data, with less rich structure, typical of semi-structured data until unstructured data, both textual data from documental repositories and web-based systems and microdata, such as user-generated contents, news, sensor-based data typical of social web and IoT, respectively. With the era of big data, not only the volume and the variability of data format and structure determine the features to be considered in the matching process, but also the fact that streams of data arrive with a continuous flow and that the level of freshness of the data (e.g., the time period of the data plays a role for matching).

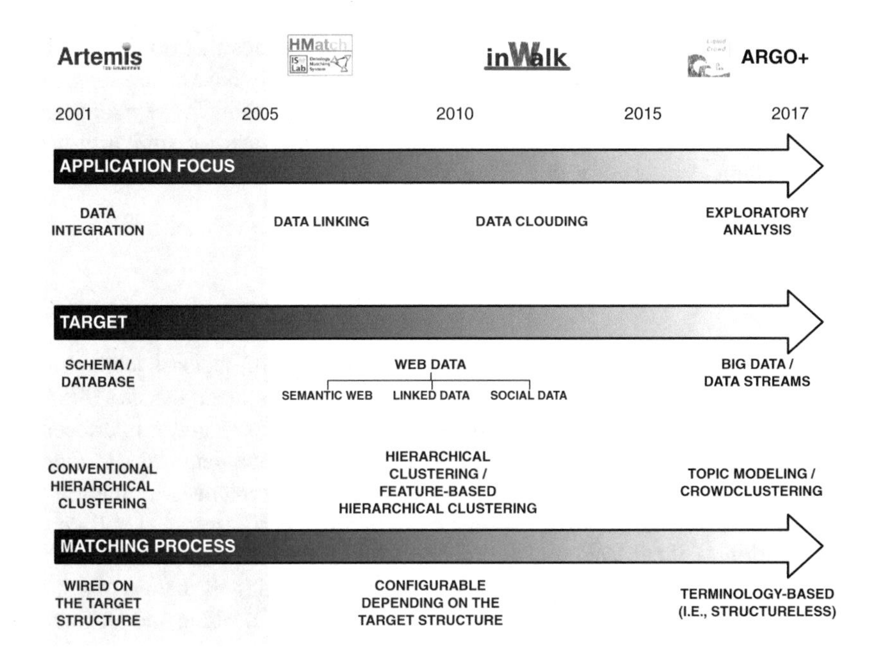

Fig. 1 Evolution of ISLab research on matching techniques

With respect to the pattern dimension, we range from structure-centered techniques, suitable for highly structured data with a well-defined schema structure and rich semantics, originating a matching process pattern whose components are hardly wired on the structure of the target data, by taking into account the way the structure is defined and the role played by matching elements to be compared, as well as their semantics (e.g., the fact that a concept has properties and/or the fact that it is formally defined to be a subclass or a superclass of another concept in a schema is an information to be properly considered for an accurate matching process of concepts of different schemas). By progressively shifting to more unstructured data, we move from a *configurable* matching process pattern, with techniques that can be flexibly configured to the level of structuring and semantics of the target data, to properly consider the available features for matching, up to a *structureless* pattern, that is, a matching process relying solely on sets of terms associated with the target data to be matched, without conveying in the matching process any information about the structure and/or semantics of the data to be matched. To accomodate the needs of data matching for integration and exploration, matching techniques, usually matching on a pairwise basis, have been used in combination with clustering techniques, to exploit results of matching and to aggregate pairs of matching elements into similarity clusters. Moving from structured to unstructured target data and to large/big datasets, we also move from conventional clustering algorithms to more feature-based clustering techniques, capable of highlighting not only the matching value characterizing a given cluster but also the data features that originated those matching results. In addition, additional clustering techniques have been adopted and properly revised/extended, based on topic modeling algorithms for keyword-based clustering of textual data, and on crowdclustering algorithms for similarity-based clustering large datasets of web resources.

3 Matching Techniques for Data Integration

The growing complexity and the overwhelming diffusion of database technologies in the industry stimulated the research on data integration since the late 90s [11, 12]. The main goal of data integration is to provide uniform query interfaces to mediate across heterogeneous data representations in diverse data sources. One of the most successful approaches to the data integration problem is to analyze the heterogeneous schemas of the data sources in order to derive a global and unified representation of them, called global schema. In doing that, data integration systems need to address several kinds of heterogeneities affecting the names of schema elements, the structure of schema elements, as well as the meaning and abstraction level of information represented by schema elements. In this context, we developed the Artemis [4] methodology and tool environment for the analysis and unification of heterogeneous schemas for data integration. The main distinctive contribution of Artemis is the intensive use of schema matching and clustering techniques for the semi-automated construction of the global schema out of the analyzed source schemas.

Artemis is based on an entity-based reference data model to uniformly represent relational, object-oriented, as well as XML data schemas. Each source schema S_i is defined as a set of entities $S_i = \{e_{1i}, e_{2i}, \ldots, e_{ni}\}$, where an entity e_{ji} abstracts the construct used in source schemas to represent real-world objects (e.g., relations, classes) and it is characterized by a name, a set of structural properties, describing real-world object properties (e.g., attributes) and a set of dependency properties, describing facts involving several real-world objects (e.g., foreign keys, relationships).

The algorithmic contribution of Artemis to schema matching is the definition of a matching pattern based on *Name Affinity (NA)* coefficient, *Structural Affinity (SA)*, and *Global Affinity (GA)* coefficients, all measuring the affinity between entities with a strength value in the range [0,1]. The highest the value, the strongest the corresponding affinity between entities. In particular: NA provides a measure of entity affinity based on their name similarity; SA provides a measure of entity affinity proportional to the number of entity properties/relationships with semantic correspondence; GA is a linear combination of NA and SA coefficients to originate a comprehensive affinity value which takes into account both terminological and structural aspects of entity affinity. In both NA and SA, terminological analysis of entity and property names plays a crucial role in the matching pattern of Artemis. One of the most successful contributions introduced by Artemis is a novel methodology for exploiting terminological information in the schema matching process, using a weighted-graph thesaurus where nodes represent terms used as entity and property names and edges represent terminological relationships holding between terms, that is, *SYN* (synonymy), *BT/NT* (broader/narrower terms, or hypernymy/hyponymy), and *RT* (related terms, or positive relation). Such a thesaurus is built in an automated way using WordNet, and it can be enriched by a domain expert with terminological knowledge typical of the considered integration domain. The affinity between two terms n and n' coincides with the maximum strength path between n and n' in the thesaurus, if a path exist, and it is equal to zero otherwise.

The ultimate goal of schema matching in Artemis is to generate the global schema, by grouping together all entities with affinity in the diverse schemas. This activity is performed by exploiting hierarchical agglomerative clustering techniques, in order to create an *affinity tree* where leaves represent source schema entities and internal nodes represent the GA affinity value holding for the all leave entities of the node. Given the affinity tree, a threshold-based mechanism is used to select *candidate clusters* of entities to be unified into a unique global entity in the global schema. Moreover, a global-as-view mapping is defined in form of mapping rules between each global entity and the corresponding local entities in the affinity cluster, to correctly map the entity structure in the global schema on the corresponding entity structure in the source schema and, thus, correctly support global query processing.

4 Matching Techniques for Data Linking and Clouding

The huge multiplicity of web resources, like for example, Web 2.0 and related user-centered services (e.g., news publishing, social networks, microblogging systems), Semantic Web and related knowledge repositories, as well as conventional Web and related documents requires matching and clustering techniques enabling users to move across "multiple webs", to explore resources and pieces of knowledge with different provenance and format in a seamless and comprehensive way. This is the idea of exploratory applications usually based on the linked-data paradigm defined with the aim at retrieving and understanding knowledge about a topic of interest by exploiting information aggregation and learning techniques in a social context (e.g., Google Wonder Wheel and Sig.ma). On the other side, visual and intuitive structures are required to satisfy a target user request by semantically organizing all the retrieved resources according to prominence- and relevance-based ranking criteria. This is the idea of news aggregation applications (e.g., Relevant News, RSS Clusgator System (RCS)) and information mashups based on clustering algorithms and tag/keyword-based search functionalities.

In this context, the fixed schema-matching pattern of Artemis leaves the floor a to light, configurable matching pattern of the H-Match system [5]. Instead of a structured global schema, we proposed the notion of *in*Cloud (*in*formation Cloud) and an approach to web resource clouding for the construction of *in*Clouds spanning over documental, social, and semantic web resources [6]. Starting from a user-specified target entity of interest *e*, namely a keyword-based specification of a topic of interest like a real-world object/person or an event, the goal of the proposed clouding approach is to generate an *in*Cloud structure for enabling smart and intuitive exploration of all the web resources that are prominent for *e*. Prominence captures the importance of a web resource within the *in*Cloud, by distinguishing, also in a visual way "á la tag-cloud", how much prominent resource(s) are with respect to the target entity. Closeness between web resources captures the degree of similarity between different web resources and it is evaluated using matching and clustering techniques.

For *in*Cloud construction, the considered web resources are distinguished in three main categories, that are *tagged resources* coming from bookmarking and social annotation systems, *microdata resources* coming from microblogging systems and news feeds, and *semantic web resources* coming from RDF(S) knowledge repositories and OWL ontologies. Each kind of web resource is differently structured according to a variety of formats, ranging from fast, short, ready-to-consume news/posts to well-structured, formal ontology instances of the Semantic Web. Matching techniques are employed to evaluate the level of closeness (i.e., similarity) between the considered web resources. The choice of the matching techniques to use has to take into account the nature and the different complexity and structure that can characterize the different web resources. For example, when matching is invoked for comparing tagged resources, we need to consider that the closeness evaluation can be only based on term featuring the resources. Resource structure can be additionally exploited when microdata and semantic web resources are matched, respectively. Moreover, also the popularity of terms, namely the number of term occurrences in resource descriptions needs to be taken into account for a more precise closeness evaluation. For this reason, the matching process for web data clouding cannot be hardly wired on the structure of the resources, but rather must be dynamically configurable according to the kind of web resources to match (i.e., according to their level of structuring and the different metadata featuring the resources). As a result, a library of matching techniques has been developed in our H-MATCH tool, where closeness evaluation configuration is dynamically determined according to the kind of web resources to match, using different matching models. Four different *matching models* are defined in the H-Match tool suite, namely surface, shallow, deep, and intensive (see Fig. 2). In particular, the surface model is suitable for matching resources featured only by terminological metadata such as the resource name and/or representative keywords extracted from resource contents. The shallow model is suitable when, not only terms, but also also resource properties are available and can taken into account. The deep and the intensive models are adequate when also semantic relations and property values of resources can be used for matching, respectively.

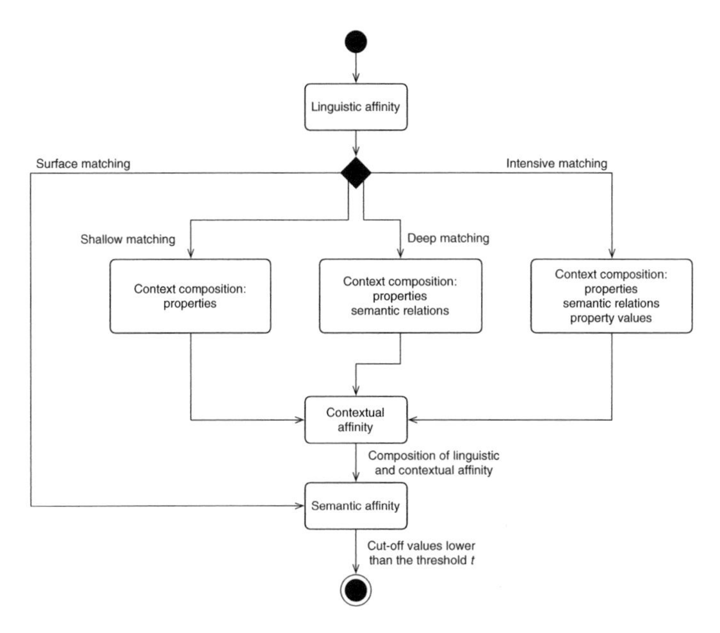

Fig. 2 The matching process of H-Match

5 Matching Techniques for Exploratory Analysis

Available data on the web are not only big and rich, but also dynamic, and the capability to deal with time becomes a crucial requirement for matching and analysis of this continuous flow of information [1]. On the one side, users need to perform a thematic exploratory analysis of the underlying document flow, driven by representative, significant topics extracted from the information flow itself. On the other side, featuring topics must be correctly located in the timeline, to easily get fresh and emergent topics and to enable users to study and understand topic evolution along time.

In this context, we proposed an incremental approach to exploratory analysis based on a combination of keyword similarity and clustering techniques for enabling topic-based exploration of an underlying information flow according to thematic and temporal perspectives [8]. Given a topic of interest T, the thematic analysis perspective exploits the topic keywords to perform an "in-depth analysis" of T and its related document collection, by highlighting the most prominent keywords and, possibly, of other highly-correlated topics. The temporal analysis perspective reconstructs the trend of T by showing how the keyword-set of T evolves in time, to capture the "variants" and "invariants" keyword portions as well as the level of popularity and specificity of T in the different time periods of the trend. Temporal analysis is also useful to discriminate between different kinds of topics, such as *persistent topics*, which are always present in the document flow, with some modifications of the

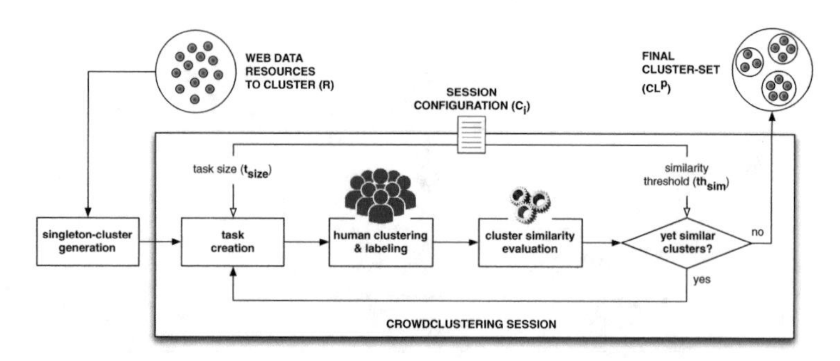

Fig. 3 Crowdclustering process pattern

keyword-set across different periods of time to reflect the variations of the arguments/perspectives in the underlying documents, and *spot topics*, which are bound to a limited and well defined time period.

In the proposed approach to exploratory analysis, a continuous flow of documents coming from one or more documental datasources is considered. Documents are characterized by heterogeneous textual content and format, such as for example the text of a tweet, the title and the abstract of a PDF document, the content of a web page. The need to deal with documents with different format motivates the use of a *structureless matching techniques*, meaning that each document is associated with a keyword-set (i.e., bag of words) extracted from the textual document content through the execution of conventional tokenization and normalization techniques [13]. For document matching and similarity evaluation, we rely on a keyword-based similarity measure called Jaccard coefficient [9], in which the similarity result for two considered documents is proportional to the number of common elements in their keyword-set representations.

For exploratory analysis, a topic-based view is created through the execution of document clustering techniques. In this respect, our contribution to the matching process pattern is the specification of multiple aggregation approaches that can be chosen to enforce different notions of "topic". A first approach is characterized by a fully-automated clustering process based on the HC^{f+} algorithm, a feature-based clustering algorithm which extends the hierarchical algorithm of agglomerative type with capability of tracking features that are common to cluster elements [10]. Topics created through HC^{f+} are similarity-based and they derive from the application of the Jaccard coefficient to the document keywords. An alternative approach to the construction of a topic-view is based on the use of topic modeling techniques, such as for example the Latent Dirichlet Allocation (LDA) algorithm [3]. In LDA, topics are defined as a probabilistic distribution over a fixed vocabulary and they are considered as hidden structures or latent variables to be inferred (i.e., discovered) by observing keyword distributions over documents. A further approach is based on the use of a *crowdclustering* algorithmic pattern in which document classification and topic labeling tasks are addressed through a human-in-the-loop process (see Fig. 3).

Table 1 Overview of the SEBD 1993–2016 dataset

	1993	1994	1995	1996	1997	1998	1999	2000	2001	2002	2003	2004
#papers	19	21	24	26	19	24	27	31	31	40	49	45
#keywords	114	130	152	162	118	158	187	209	198	251	324	298
#authors	56	61	69	63	58	66	73	74	89	105	133	137
	2005	2006	2007	2008	2009	2010	2011	2012	2013	2014	2015	2016
#papers	42	40	59	52	46	51	51	38	52	49	48	47
#keywords	278	248	403	347	313	508	449	251	327	338	369	315
#authors	129	119	153	157	141	171	151	103	184	160	155	157

In crowdclustering, human workers are actively engaged in the topic-view construction by recognizing and aggregating similar documents into clusters and by subsequently labeling the created clusters with featuring keywords (i.e., topics) on the basis of her own skill/perception/judgment [7].

6 The SEBD Case-Study

As a case study of application of matching techniques for exploratory analysis, we consider the dataset of papers published in the history of SEBD from 1993 to 2016. In particular, we collect the titles and authors of SEBD papers[1] and we apply our techniques for data clustering and exploration in order to select the most relevant keywords from paper titles and derive from them the mains topics and topic trends in the SEBD history. An overview of the dataset in terms of number of papers, distinct keywords, and distinct authors per year is shown in Table 1.

6.1 Exploring Keyword Evolution

The first goal of SEBD dataset exploration is to understand how terminology and topics have changed in the SEBD history. To this end, given an year y and a keyword k, we associate k with a measure τ_k^y of its time specificity for y as follows:

$$\tau_k^y = \frac{f_k^y - f_k^*}{\sqrt{f_k^*}},$$

[1]Data have been collected from the DBLP database (http://dblp.org), except for the year 2013 that is missing from DBLP. 1993 data have been collected from the Scopus DB (https://www.scopus.com).

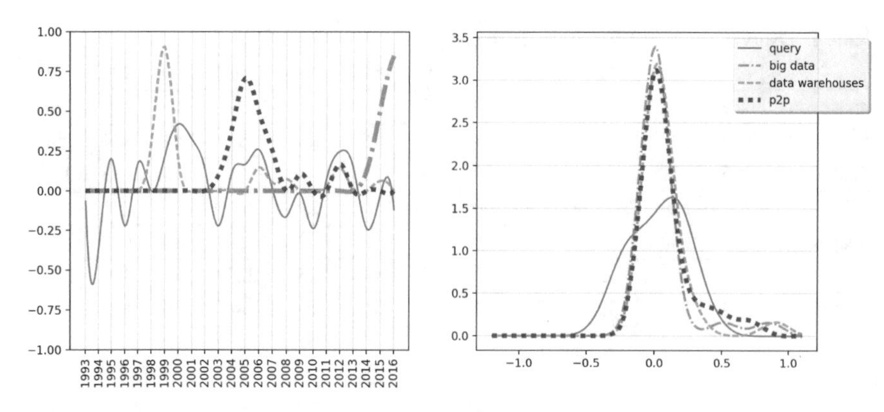

Fig. 4 Example of keyword time series and their kernel density function

where f_k^y denotes the relative frequency of k in the year y and f_k^* denotes the relative frequency of k in the whole dataset. Negative values of τ_k^y represent the fact that k is underused in y with respect to the whole dataset, while positive values of τ_k^y represent the opposite. By exploiting τ_k^y, we can associate each keyword with a time series t_k of length 24 (i.e., the years of SEBD), which describes the use of k in the SEBD editions. An interesting property of these time series is the degree of concentration of their values. Given a time series t_k, if we have a tight concentration of values in a short range of years, we can assume that the keyword k has been specifically used in a relatively short period. On the opposite, widely dispersed values represent the fact that k is constantly and homogeneously used along the whole SEBD timeline. In order to measure the degree of concentration C^{t_k} of a time series t_k, we transformed t_k by means of a Gaussian kernel-density estimation function and we took the peak of the transformed series as C^{t_k}. A graphical example of this approach is shown in Fig. 4, where we compare the time series and kernel-density estimation of the keywords "query", "big data", "data warehouse" and "p2p".

The keyword "query" is an example of a keyword with low concentration depending on the fact that it has been used quite uniformly along all the 24 years of SEBD. Keywords "big data", "data warehouse" and "datalog" have all a high concentration of values, denoting the fact that these keywords are tied to one or more specific periods of time. In particular, "big data" becomes one of the most popular keywords starting from 2014 and in all the last three editions of SEBD; on the opposite, "data warehouse" is highly specific of the time frame 1998–2000; "p2p" was pretty popular between 2003 and 2007 but then we register a sort of renaissance in 2012 and 2013. According to the concentration degree C^{t_k}, we can explore the dataset to find the most specific keywords for different periods of time. In particular, we split the SEBD history in 5 periods of 5 years each (the last covering only the last 4 years) and we collect for each of them the 10 most specific keywords, as reported in Table 2.

The keywords time framing can be exploited in order to compose a keyword map of the SEBD dataset as shown in Fig. 5.

Table 2 Most specific keywords per time frame in the SEBD history

Time frame	Keywords
1993–1997	Database, basi dati, applications, object, schemi, relational, interrogazioni, constraints, language, sistema
1998–2002	Query, web, xml, data warehouse, data integration, association rules, temporal, classification, aggregate, knowledge
2003–2007	System, semantic, model, service, documents, schema, p2p, patterns, clustering, access
2008–2012	Mining, search, ontologies, olap, similarity, hierarchical, discovery, mapping, extraction, trajectory
2013–2016	Big data, social network, linked, analytics, performance, graph, learning, prediction, cloud, visual

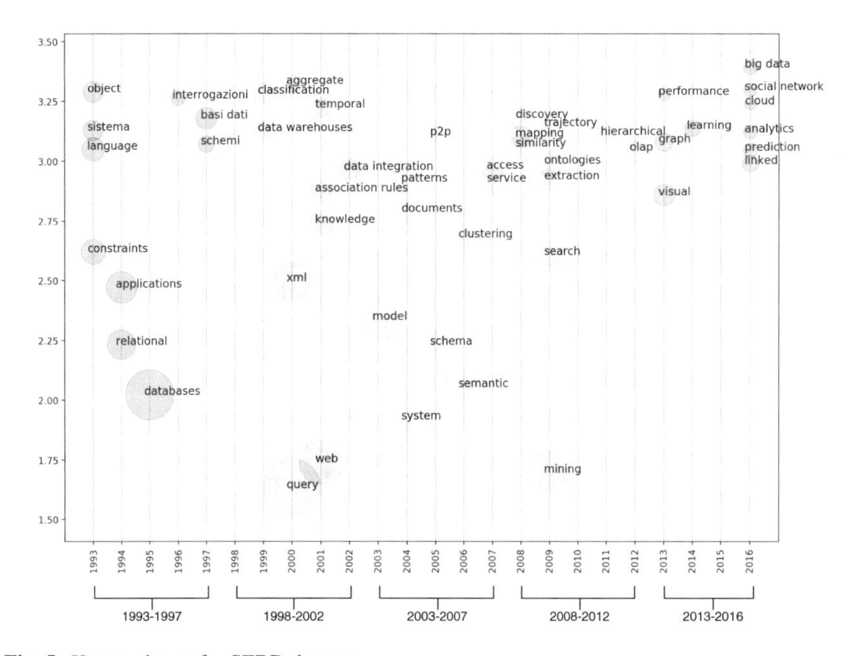

Fig. 5 Keyword map for SEBD dataset

The map is created by determining the position of each keyword in a 2-dimensional space where one dimension represents years and the other represents the concentration coefficients. In particular, given a keyword k, we determine the corresponding year as the year in which the concentration of k is highest. A third dimension is represented by the size of circles associated with keywords, which is proportional to the number of occurrences of the keyword in the whole dataset. Due to the way the map is composed, we visualize in the top the keywords that are most specifically correlated to one period (e.g., "basi dati", "classification", "p2p", "mapping", "big data"). We note also that in the first period of SEBD (1993–1997) was characterized by a good number of papers in Italian and this motivates the presence of Italian words as keywords (e.g., "sistema", "interrogazioni", "basi dati", "schema"). They are a special case in the analysis, in that their specificity is mainly due to the fact that after 1998 the number of papers written in Italian presented at SEBD is constantly decreasing in favor of papers written in English. The most used keywords (e.g., "database", "query", "web", "system", "mining") are less specific, because they are used not exclusively in a period but all along the SEBD timeline. Moreover, they are mainly specifically related to the early periods, from 1993 to 2007. This is due to the fact that the introduction of a keyword in the SEBD terminology is usually corresponding to a research trend in the Italian database community. As a consequence, the keyword is then used also in the following periods. On the opposite, (relatively) new keywords such as "big data" and "social network" do not still have a research tradition, resulting is a high specificity with respect to the last period of the SEBD.

6.2 Topic Trends

The goal of topic trends is to represent the shifts in terminology and topics that we observed in the map of Fig. 5. In particular, we see from the map that the SEBD keywords moved from relational database topics in the early period (e.g., "database", "relational", "interrogazioni") to web, data integration, and data warehouse (e.g., "web", "data integration", "data warehouse", "xml", "aggregate", "association rules"), to semantics and discovery (e.g., "semantic", "ontologies", "discovery", "mapping", "mining"), up to the big and social data era (e.g., "big data", "social network", "analytics", "prediction"). An interesting and specific characteristic of SEBD is that the research community participating in the conference is quite stable. Thus, the shifting of topics is not due to a change in the SEBD population, but rather to a change in the focus of research of the same researchers. According to this assumption, we can link keywords featuring consecutive periods based on the fact that they are used in papers having the same authors. Formally, given a keyword k_i that is specific of period p_i, we can link it to the keyword k_j of period p_{i+1} if the intersection between the set of author using k_i in p_i and the set of authors using k_j in p_{i+1} is not empty. Since the same authors use k_i and k_j in two consecutive periods of time, we can interpret the link $(k_i, p_i) \rightarrow (k_j, p_{i+1})$ between k_i and k_j as a step of an evolution trend, represented by a shift in terminology from k_i to k_j. Through this approach, we can answer to two exploration demands: (i) given a keyword k and its period p_i, which other keywords of the following periods have been used consistently by the same authors? (ii) given a group of researchers, which is the evolution in their terminology and topics? In order to answer, we define two kinds of evolution trends, namely *topic trend* and *author trend*, respectively. A topic trend $\mathcal{T}_k^{p_i}$ for the keyword k of the period p_i is a sequence of inter-related keywords over consecutive periods that have been used by authors in those periods, that is links of the form $\mathcal{T}_k^{p_i} = (k, p_i) \rightarrow (k_{i+1}, p_{i+1}) \rightarrow (k_{i+2}, p_{i+2}) \rightarrow \cdots \rightarrow (k_{n-1}, p_{n-1}) \rightarrow (k_n, p_n)$ such that: $\bigcap_i^n \alpha_{p_i} k_i \neq \emptyset$, where $\alpha_{p_i} k_i$ denotes the set of authors using k_i in period p_i. In exploring the SEBD dataset, we exploited topic trends to study the periods and keywords reported in Table 2. As an example, we take into account the topic trend $\mathcal{T}^{1998\text{–}2002}_{\text{data integration}}$. Starting from the keyword "data integration" in 1998–2002, we explore the dataset in order to look for any sequence in consecutive periods such that there is a common set of authors who use them. The higher the number of common authors, the higher the strength of the topic. An example of such sequences is:

1998–2002	2003–2007	2008–2012	2013–2016
3		ontologies	
	↗		↘
data integration →	schema		cloud
	↘		↗
		mapping	

Figure 6 shows the 9 strongest topic trends (i.e., those with the highest number of authors in the intersection), where topic trends are represented as links connecting the keywords of the SEBD keyword map. The strength of the link is proportional to the number of common authors between the connected keywords.

Topic trends represent evolution threds of research groups in the SEBD community. Some of these trends start from widely used keywords, such as in the case of $\mathcal{T}^{1998\text{–}2002}_{\text{query}}$. In these cases, the research group is usually heterogeneous and originates different threads, such as "query" → "system" → "semantic" → "search", from 2000 up to 2009. In other cases, the starting keyword is less used and thus it originates trends with a more homogeneous research group, resulting in more homogeneous threads, such as in case of $\mathcal{T}^{1998\text{–}2002}_{\text{association rules}}$ and the thread "association rules" → "patterns" → "discovery" → "big data", from 2001 up to 2016.

The second exploration need was to discover the evolution of topics for specific groups of authors. For this purpose, we define the notion of author trend. An author trend $\mathcal{A}^t_G$, where G is a set of authors, is a sequence $\mathcal{A}^t_G = K_{p_i} \rightarrow K_{p_{i+1}} \rightarrow K_{p_{i+2}} \rightarrow \cdots \rightarrow K_{p_{n-1}} \rightarrow K_{p_n}$, where K_{p_i} represents the set of the top-t specific keywords used by authors in G in the period p_i. As an example, we take the authors of this paper (i.e., the ISLab research group at the University of Milan) as a group. Table 3 shows the ISLab group trend, obtained by setting a number $t = 5$ of keywords per period.

The table shows the evolution in time of the research activity of the ISLab group in the SEBD dataset. We observe that moving from left to right table columns, the keywords sequence of this trend describes topic evolution which matches well with the research evolution described in previous sections of this paper.

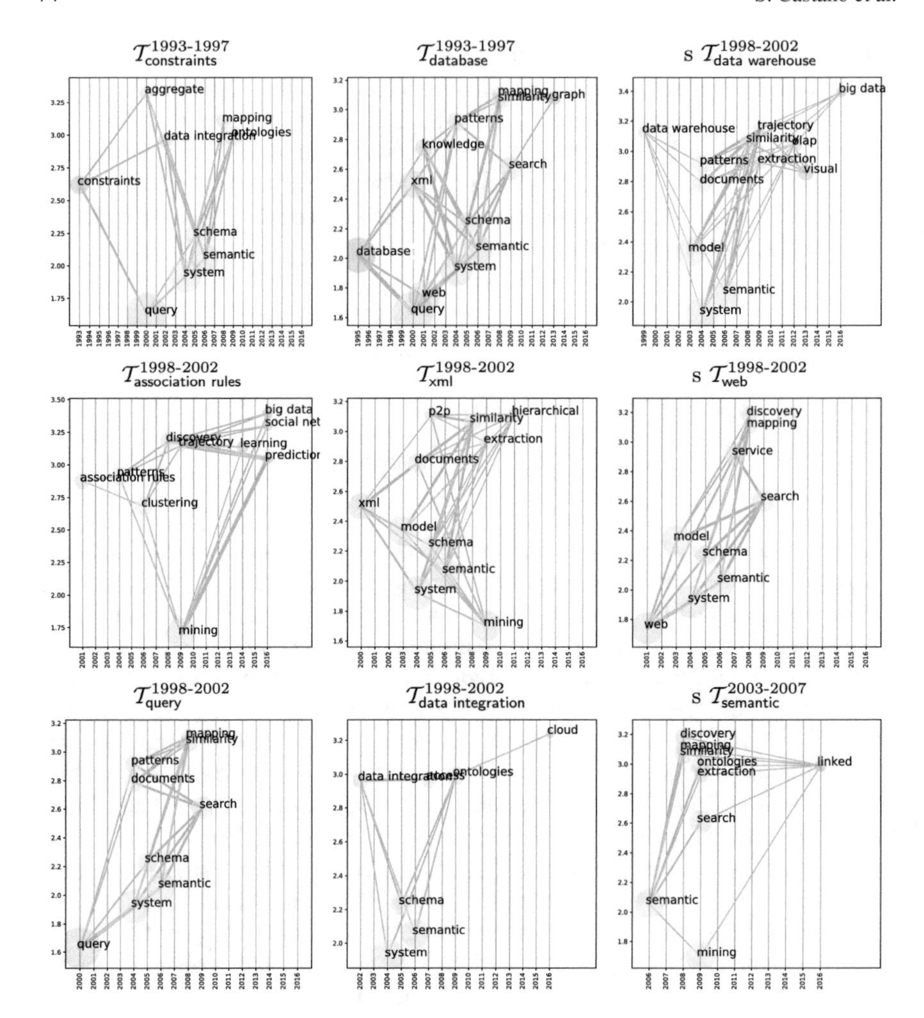

Fig. 6 Exploratory analysis of the SEBD dataset by topic trends

Table 3 Example of author trend for the ISLab research group

1993–1997	1998–2002	2003–2007	2008–2012	2013–2016
Conceptual schema analysis classification neurali	Integration heterogeneous information knowledge xml	Semantic resources propagation helios retrieval	Ontologies matching micro-data instance multimedia	Crowdclustering web information analysis digital resources exploration

7 Concluding Remarks

In this paper, we described the evolution of research on matching techniques for data integration and exploration at the ISLab group of the Università degli Studi di Milano. We analyze the matching techniques according to the structure of target data, the algorithmic pattern of the matching process, and the application focus, and we discussed some main results of applying our techniques for exploratory analysis of a real dataset composed by all the SEBD proceedings publications in the timeframe 1993–2016.

The proposed exploratory analysis framework shows that the combination of structureless matching/clustering techniques of SEBD textual data and of statistical analysis techniques for keyword/topic distribution, enables the exploratory analysis of the underlying dataset along different perspectives over time (e.g., research topic trends). Current and future research is devoted to the extension of the crowdclustering and exploratory analysis in the direction of big data collections. Moreover, we are working to the implementation of the exploratory analysis operators in the inWalk prototype.

References

1. C.C. Aggarwal, S.Y. Philip, On clustering massive text and categorical data streams. Knowl. Inf. Syst. **24**(2), 171–196 (2010)
2. P. Berkhin, Grouping multidimensional data, *A Survey of Clustering Data Mining Techniques* (Springer, Berlin, 2006)
3. D.M. Blei, A.Y. Ng, M.I. Jordan, Latent dirichlet allocation. J. Mach. Learn. Res. **3**(4–5), 993–1022 (2003)
4. S. Castano, V. De Antonellis, Global viewing of heterogeneous data sources. IEEE Trans. Knowl. Data Eng. **13**(2), 277–297 (2001)
5. S. Castano, A. Ferrara, S. Montanelli, Matching ontologies in open networked systems: techniques and applications. J. Data Semant. **V**, 25–63 (2006)
6. S. Castano, A. Ferrara, S. Montanelli, Structured data clouding across multiple webs. Inf. Syst. **37**(4), 352–371 (2012)
7. S. Castano, A. Ferrara, S. Montanelli, Human-in-the-loop web resource classification, in *Proceedings of the On the Move to Meaningful Internet Systems: OTM 2016 Conferences* (Rhodes, Greece, 2016), pp. 229–244
8. S. Castano, A. Ferrara, S. Montanelli, Exploratory analysis of textual data streams. Future Gener. Comput. Syst. **68**, 391–406 (2017)
9. A. Ferrara, A. Nikolov, F. Scharffe, Data Linking for the Semantic Web. Semantic Web: Ontology and Knowledge Base Enabled Tools, Services, and Applications **169** (2013)
10. A. Ferrara, L. Genta, S. Montanelli, S. Castano, Dimensional clustering of linked data: techniques and applications. Trans. Large-Scale Data- Knowl.-Centered Syst. **XIX**, 55–86 (2015)
11. A.Y. Halevy, Answering queries using views: a survey. VLDB J. **10**(4), 270–294 (2001)
12. A. Halevy, A. Rajaraman, J. Ordille, Data integration: the teenage years, in *Proceedings of the 32nd International Conference on Very Large Data Bases*, VLDB Endowment (2006), pp. 9–16
13. C.D. Manning, P. Raghavan, H. Schütze, *Introduction to Information Retrieval*, vol. 1 (Cambridge university press Cambridge, Cambridge, 2008)

14. E. Rahm, P.A. Bernstein, A survey of approaches to automatic schema matching. VLDB J. **10**(4), 334–350 (2001)
15. P. Shvaiko, J. Euzenat, A Survey of Schema-based Matching Approaches. J. Data Semant. IV (2005)

25+ Years of Query Processing - From a Single, Stored Data Set to Big Data (and Beyond)

Barbara Catania and Giovanna Guerrini

Abstract From the late 1970s, the database research community has devoted huge efforts in devising more and more effective and efficient solutions for processing queries against data. In this chapter, we retrace some of challenges that have been faced in the last 25 years to enable data to be effectively and efficiently queried and how the notions of queries and query processing themselves have evolved during these five lusters.

1 Introduction

Query Processing. Query processing denotes one of the fundamental processes of a Database Management System (DBMS), by which a user request, typically specified by means of a declarative query language, such as SQL, is first compiled and then executed against a data set, reflecting the neat separation between the specification of 'what' we are searching for from 'how' these searches are processed by the system. Traditionally, data have a completely known structure and are stored on persistent storage. Thus, data items are entirely available before the query is specified and processing starts. At the same time, queries precisely represent user needs since the user has familiarity with data on querying, and she is thus able to express information needs in terms of constraints, on the structure and the values, that characterize the items of interest.

Traditional query processing consists of a compile-time phase and a runtime phase. At compile-time, the system transforms a declarative query into an *execution plan*, which is statically determined before the result computation starts. This transformation is made in several steps: lexical, syntactical, and semantical analysis of the query are first performed, then query optimization is applied in order to select the best execution plan among those available, and then the final code is generated.

B. Catania (✉) · G. Guerrini
University of Genoa, Genoa, Italy
e-mail: barbara.catania@unige.it

G. Guerrini
e-mail: giovanna.guerrini@unige.it

S. Flesca et al. (eds.), *A Comprehensive Guide Through the Italian Database Research Over the Last 25 Years*, Studies in Big Data 31, DOI 10.1007/978-3-319-61893-7_5

The generated code usually consists of physical operators for a database engine, e.g., in the relational context, they implement data access, join, selection, projection, grouping, aggregation, and ordering. In implementing such operators, besides the naïve semantics, auxiliary data structures (e.g., indexes), if available, can be used for achieving a more efficient implementation. At runtime, the database engine interprets and executes the selected query execution plan and computes the final result, corresponding to the set of items which *exactly* satisfy the specified query, i.e., meet the constraints it states on values and structure.

Since the late 1970s, this approach has been proved to be highly effective and efficient and it has played a major role in the relational DBMSs success over the last decades. *Effectiveness* is guaranteed by the usage of simple declarative languages. *Efficiency* is guaranteed by the usage of several sophisticated optimization techniques. These techniques heavily rely on the existence of metadata about the data which have to be processed, such as the distribution of values and the selectivity of the relational operators. Metadata are easily collected since queries are typically executed in *stable environments*, for which a reasonable set of statistical information on data is usually available.

Unfortunately, the advanced and fundamentally elegant concepts and ideas underlying the query processing language-to-engine stack of classical DBMSs do not obviously extend to much more unorganized and unpredictable data providers, typical of data intensive applications and novel processing environments which started to appear from the middle 1990s and continuously challenged query processing till the current Big Data dare.

Evolution of Data and Processing Contexts. If we observe the evolution of application and data processing environments in the last 25 years, we can identify some factors leading to the rise of new data processing needs and challenges, we will refer to as *themes*,which characteristics are summarized in Table 1 and further discussed in the following.

Multiple Data Sets. The high network connectivity brought by the Internet advent (and the rise of the Web) caused, starting from the late 1990s, a shift from single to multiple data sets. Queries can now be targeted to multiple data sources in that

Table 1 Themes in data management

	Multiple data Sets	Streamed data	Semantic enrichment	Data at Scale
From	late 1990s, early 2000s	early 2000s	early 2000s	late 2000s
Volume	✓	(✓)	✓	✓
Variety	✓		(✓)	✓
Veracity	✓		✓	✓
Velocity		✓		✓
Data model			graph model	key-value models

information resources are shared and often not completely and statically integrated in a single data set. Rather, they are only federated or managed through a mediator approach [34]. In terms of data characteristics, this results in high data heterogeneity, incompleteness, and uncertainty, in limited user knowledge about the data to be processed, and in extremely high variability and unpredictability of data characteristics during processing. Referring to the Big Data V's, main query processing challenges arise from data *variety* and *veracity*; as a consequence of data integration, data *volume* is also increasing.
Streamed Data. Almost in the same years, the progress in hardware made it possible for organizations to record large streams of transactional data while the development of sensor technology enabled the monitoring many events in real time. Such data sets, which continuously and rapidly grow over time, are referred to as *data streams* [4]. In the data stream scenario, input arrives very rapidly and there is limited memory to store it. This new type of data availability resulted in limited resources with respect to the data volumes under processing: the response time should be low and the space is not only not sufficient to store the entire data stream, which is unbounded, but also limited with respect to the size of data under current processing. The characteristics of data under processing become extremely variable and unpredictable. Referring to the Big Data V's, the data *velocity* issue is raised (and the *volume* issue is also relevant).

Semantic Enrichment. The increasing amount of information resources available on the Web motivated in the early 2000s the need to enhance usability and usefulness of interconnected resources by enriching data with semantic annotations [15]. Common metadata vocabularies (ontologies) and maps between vocabularies shift interoperability and data integration from a syntactic to a semantic level. By interconnecting machine-readable data, a Web of Linked Data that can be processed by machines is created [16]. Referring to the Big Data V's, the *volume* issue now raises to a further scale, since we are facing with the entire Web as a gigantic database. Semantic *variety* and data *veracity* are also issues, since we deal with autonomous, loosely controlled data sources, constituting a *dataspace* rather than an integrated database, and with user generated content. Graphs become the reference model for interconnected data.

Data at Scale. In the distributed system world, Big Data started to become a major issue in the late 2000s due to the unprecedented scale and elasticity data management requirements posed by Web applications and user-facing Web sites (e.g., e-commerce applications and social networks). Developers of very large scale Web applications opted for horizontal scalability, i.e., clusters consisting of a large number of connected computers. Since database technology was found to be neither well-suited nor cost-effective for their purposes, solutions were developed based on MapReduce and Hadoop [28] for analytics, key issue in this era to extract useful knowledge from such enormous data sets, and on simple key-value stores [27], that are fast, highly scalable, and reliable for (fairly simple) data storage and retrieval needs. All the well known Big Data V's, namely, *volume*, *variety*, *veracity*, and *velocity*, have raised.

Challenges to Traditional Query Processing. The characteristics discussed above make traditional query processing not feasible for most of the new processing environments [21]. Indeed:

- Due to heterogeneous, incompleteness, and uncertainty of the, often shared, resources and the usually limited knowledge about data when specifying queries, it is difficult for the user to exactly characterize what she is looking for. Additionally, even by relying on sophisticated but traditional query processing techniques, the time required to compute a precise answer is often too high. As a consequence, the effectiveness of declarative languages and of logical data independence decreases.
- The instability of the new environments, due to the high network connectivity, the heterogeneity of the shared resources, and the consequent extremely high variability and unpredictability of their characteristics during the processing, lead to the impossibility of statically devising optimization strategies that will be optimal even beyond the very short term. At the same time, new processing environments have to deal not only with stored data but also with data continuously arriving as a stream. Statically detected execution plans thus do not constitute a suitable approach; rather, the processing has to be adapted to dynamic conditions that may change during the query execution. As a consequence, efficiency of traditional optimization techniques is undermined and the overall query processing should be revised.

We therefore assisted to a deep revision of the query processing approach, which, starting from the middle 1990s and guided by the Big Data V's, has been enriched by two main features, namely *approximation* and *adaptivity*, and a bunch of additional ones, to maintain both its effectiveness and its efficiency.

Goal and Outline of the Chapter. In the following, we discuss in more details query processing features characterizing novel applications and environments, with a special emphasis on approximation and adaptivity. Specifically, we will focus on the processing of *queries* and distinguish between transactional and analytical workload only when the distinction is relevant. We attempt to do an analysis which is as much as possible independent from the peculiarities of the data model. The reminder of the chapter is organized in four sections, each devoted to one of the themes introduced above. At the end of the chapter, we provide some conclusions and we outline the role of the SEBD conference in the last 25 years of query processing research.

2 From Single to Multiple Data Sources

In the middle 1990s, twenty years after the rise of the relational model, large enterprises owned a multitude of data sources and scientific data sets were being produced independently by multiple researchers. From one hand, this situation further confirmed the strengthening of the relational model in the market, from the other hand, the research community had to face with new problems: the need of combining and querying data residing in multiple autonomous and heterogeneous data sources and providing users with a unified view of those data [34], exploiting the high network connectivity provided by Internet.

Various architectures have been proposed for addressing this issue, often relying on federation or mediator approaches instead of single data sets obtained by static integration of the sources [34]. The availability of multiple data sets and the ability to query them in an integrated way gave a great impact to analytical data processing to drive decision-making processes, which found in data exchange architectures and materialized integration approaches the right background for its development [24]. In the same period, the increasing need of data interchange lead to the rise of the semi-structured data model and of XML as a common syntactic format for data representation and sharing among data sources, and query languages were exposed to a contamination with Web-style information retrieval and full-text search [14, 48].

The management of multiple data sources led to several new issues for what concerns query processing, which, starting from the late 1990s, has been revised along two main directions, namely *approximation* and *adaptivity*, to take care of *variety*, *veracity*, and *volume* of new data sets and environments.

Approximation: the impact of variety and veracity. Data coming from multiple sources are highly heterogeneous as well as often uncertain, incomplete, and mutually inconsistent. Due to *variety* and *veracity*, the user might not have enough knowledge of data for precisely specifying her information needs at query time. As a result, query answers computed according to the precise query semantics are not always satisfactory, since they may contain too less or too many answers for being semantically significant. In order to increase user satisfaction, starting from the early 2000s, solutions have been devised which give up the computation of 'precise' answers in favour of an *approximate result computation*: instead of getting a precise but unsatisfactory answer, new data characteristics, and contamination with Web-style information retrieval and full-text search, suggested it might be preferred to relax the query, returning an approximate result set with quality guarantees [5]. Relaxation can be applied either by query rewriting or by replacing equality conditions with similarity-based ones, according to some distance function defined for data values and data structure. The resulting approximation is often referred to as *query-driven*.

Query relaxation by rewriting has been proposed in the relational context [45, 50, 62] and more recently extended to cope with data integration architectures [43]. Query-relaxation based on similarity extends to structured and semi-structured data concepts taken from multimedia information retrieval, by applying the minimum amount of relaxation to a query with the aim of returning a non-empty result close enough to the user request. For the relational model, the proposed techniques focus on suitable functions for ranking tuples with respect to a query [5] or specific operators, with a special emphasis on join [46], facing approximate matches for strings [23] or numeric values [56]. For XML documents, structure is not that prescriptive and both content and structural constraints can be approximated through similarity [6, 7].

In the same period, preference-based queries, such as top-k or skyline operators [18, 38], have been defined with the aim of addressing the many-answer problem. The top-k operator restricts the number of query results to a fixed number (k), based on some ranking function. Since the choice of a suitable ranking function could be cumbersome, the skyline operator returns the best items taking into account only the attributes of interests. Preference-based queries are inherently more complex than traditional query operators and require sophisticated query processing techniques to guarantee a reasonable response time.

Approximation: the impact of volume. Data *volume* contributes to difficulties in querying multiple data sources, especially in data exchange architectures, like data warehousing. Of course, executing complex operations, like join and aggregation, over huge amount of data, and possibly limited or constrained resource availability, might impact performance and further reduce user satisfaction due to a too high response time. In order to optimize resource usage, one option is to optimize the input data set, leading to *data-driven* approximation solutions. Data-driven approximation refers to all techniques in which input data is reduced, simplified or summarized into a synopsis, significantly smaller than the input data set, for which ad hoc-efficient query processing algorithms for expensive operators, like join and aggregation, can be defined. Historically, summaries have been used for selectivity estimation and only in the late 1990s they have been proved to be very useful for approximate query answering.

Various types of summaries have been proposed (see [13, 33] for a survey). When using summaries for query processing, the main challenges are [33]: (*i*) what summary to maintain in a limited space in order to maximize accuracy and confidence of the computed approximate result; (*ii*) how the summary can be efficiently updated as soon as data it refers to change. Different types of summaries may be suitable for different operations. For example, table-level summaries, such as histograms or wavelets, are usually not suitable for capturing join-based correlations since their aim is to summarize the content of a single table. In this case, schema-level synopses can be used [3].

Adaptivity: the impact of variety. The processing context of large federated databases and wide-area data integration is very dynamic and the optimizer has much less information on individual sources. As a result: (*i*) the optimizer may not have enough information to decide on a good plan, and (*ii*) a plan that looks good at optimization time may be arbitrarily bad if the sources do not respond exactly as expected. The first issue is related to unavailability or unreliability of statistics metadata a query optimizer relies on, the second one to the highly variable and uncontrolled processing conditions induced by distribution. Among others, a key factor is how to maximize CPU utilization, given the (highly variable and unpredictable) rate data are received from the distributed data sources. In addition, the processor needs to consider overlap and redundancy among sources, to minimize the access to redundant sources and respond flexibly when some sources are unavailable. To cope with the widely fluctuating characteristics of resources, query processing cannot be neatly divided into a query optimization step followed by a query execution step. The idea of combining optimization and execution is applied to different extents and different modalities, resulting in a number of *adaptive* processing proposals. A query processing technique is said to be adaptive if the way in which a query is executed may change on the basis of the feedbacks obtained from the environment during evaluation [29]. Adaptive query processing techniques do not rely on a priori information, rather they incrementally gather current information, that may be less complete but is up to date, in parallel with the query execution.

According to [10], adaptation may be *plan-change based* [19] if the execution of a current plan is monitored and re-optimization triggered whenever observed plan properties (e.g., intermediate result size) or system conditions (e.g., available memory) differ significantly from the estimates. Adaptation, at the other extreme, may be *routing-based* when the notion of query plan disappears and query processing is seen as routing of tuples through operators, realizing plan changes by changing the order tuples are routed. Eddies [8] control the execution plan at run-time at the level of each single tuple. Most tuples exploit the route that is currently more efficient, while the rest explore other routes.

Adaptivity first faces different aspects of processing in isolation. For instance, to minimize idle time during query processing, [60] deals with startup delay and burstiness of remote data sources. Unifying architectures have been proposed later [40, 41]. Adaptive data partitioning [41] divides source data into regions, each executed by different, complementary plans. This allows the processor to correct badly estimated cardinality and selectivity values, but also to discover and exploit order in source data as well as source data that can be effectively pre-aggregated.

3 From Stored to Stream Data

Several applications naturally generate data streams rather than finite stored data sets: measurements in network monitoring and traffic management, sensor applications, log records or click-streams in Web tracking and personalization, financial tickers, manufacturing processes, call detail records in telecommunications, email messages, and others. At the beginning of 2000s, due to the lack of support by database systems to any kind of special storage management or query processing for data streams, heavily stream-oriented applications tended to use a DBMS largely as an offline storage system, or not at all.

Several aspects of data management need to be reconsidered in the presence of data streams, offering a new research direction for the database community. Processing queries on streaming data requires to reconsider most of the basics of queries on stored data: not only transient, but also persistent *continuous* queries need to be processed; query answers are necessarily *approximate* due to the unboundedness of the stream, query operators and plans are necessarily *adaptive*: reacting to changes in input characteristics and system conditions is a major requirement for long-running query processing over data streams.

Continuous queries. Continuous queries [12] are queries that are issued once and then logically run continuously over the database (in contrast to traditional one-time queries which run once to completion). As opposed to the traditional context, in which data are stored and queries come in, now the roles are inverted: queries are registered and data flow in. Continuous queries also stress query processors in that: (*i*) when new tuples arrive in a data stream, they generally must be 'consumed' immediately; (*ii*) the amount of storage required for answering a continuous query, or to ensure that the answer can be always computed, may be unbounded. To cope with continuous queries in 'traditional' architectures, we are forced either to restrict the expressiveness of traditional query operators (e.g., join) and/or impose constraints on characteristics of the data stream, so that we can guarantee that answer size, continuously computed at each instant of time, is bounded, or that the amount of extra storage needed to continuously compute the answer is bounded. At the query language level, this is usually done by introducing the *window* operator, by which a region of definite cardinality is superimposed over a stream of unknown cardinality at each instant of time.

Processing continuous queries over data streams entails making fundamental tradeoffs among *efficiency*, *accuracy*, and *storage*. A query plan may consist of *stateless* as well as *stateful* operators. A stateless operator does not need to maintain intermediate data nor other auxiliary state information to be able to generate complete and correct results. A stateful operator, by contrast, at intermediate points of its execution stores data (or auxiliary information extracted from them) that have been processed so far. Only pipelined plans can be exploited, and operators realizing joins and aggregates need to be stateful. Care must be taken to reduce the state continuous queries accumulate.

Approximation: the impact of velocity and volume. The tradeoff between potentially unbound streams of data, which very rapidly arrive in the system, and limited resources with respect to the data volumes under processing, made traditional query processing approaches unsatisfactory. In this context, a 'precise' computation might be unsatisfactory since it can be obtained only at the price of a unacceptably high response time, communication overhead, or occupied space. To address this problem, both *query-driven* and *data-driven* approximation approaches have been provided.

Query-driven approximation approaches rewrite the query as soon as it is submitted to the system, with the aim of using less resources at execution time [52]. This can be done by statically rewriting some query operators, like windows, that may have an impact on resource usage. For what concerns data-driven approximation, summaries, here called *sketches*, should be incrementally computed and used either for efficiently generating approximate results in presence of limited resources (especially for window-based operations [42]) or for allowing stateful computations over past data [26, 61]. *Load shedding* can also be interpreted as a type of data reduction, with the aim of dropping excess load from the system and process only a subset of input data for performance issues [9, 57].

Adaptivity: the impact of velocity. Continuous queries and the data streams on which they operate are long-running. Unlike during the processing of a simple one-time query, during the lifetime of a continuous query parameters, such as the amount of available memory, stream data characteristics, and stream flow rates, may vary considerably. For instance, stream arrival may be bursty, unpredictably alternating periods of slow arrival and periods of very fast arrival. The system conditions as well, e.g., the memory available to a single continuous query, may vary significantly over the query running time. In such a situation, all the relevant statistics are estimated during execution. To minimize the overhead, sampling based techniques are used for statistics tracking, which is combined with query execution whenever possible.

In a streaming context, the overall objective is to maximize the output rate for a query, rather than devising the least cost plan. The rate of the input stream is further taken into account during optimization rather than the input cardinality. Another optimization goal is minimizing resource (memory) consumption. Both plan-based [11] and routing-based (i.e., eddies) [22, 47] approaches have been considered for continuous queries. Adaptivity has targeted the adaptive sharing of common sub-expressions [25] and distributed stream processing [2, 55], in which run-time load is monitored and operators are moved across machines to improve performance, relying on coordinated load shedding for detecting and eliminating CPU overload from multiple machines.

Finally, approximation and adaptivity are not 'exclusive': many adaptive query processing techniques for stream (e.g., [9, 57]) introduce approximation; other, employ approximate data summaries, that are incrementally updated during processing, to collect data characteristics needed to drive adaptation.

4 From Data to Semantically Rich Data

In the new millennium, the space of interconnected data resources and of information sharing continues to grow. In 2001, the vision of *Semantic Web* came out [15]: to enhance usability and usefulness of interconnected resources, data are, often automatically, 'marked up' with machine-understandable semantic information about the human-understandable content. Servers expose existing data using the so called Semantic Web standards, such as RDF, and semantically enriched data can be queried through the SPARQL language. This semantic enrichment aims at facilitating the integration of information from mixed sources, dissolving ambiguities in terminology, improving information retrieval thereby reducing information overload, and identifying relevant information with respect to a given domain.The reference data model for semantically enriched data is graph-based: subjects, objects, and the predicates relating them.

The Semantic Web view is further enhanced by *Linked Data* [16] (the term first appeared in 2006), providing a publishing paradigm in which not only documents, but also data, can be a first class citizen of the Web, thereby enabling the extension of the Web with a global dataspace [37] based on open standards - the Web of Data. As in traditional Web, publishers can make their data available using a small set of standard technologies, and consumers can search for and browse published data using generic tools. Consumers frequently consume data in broadly the form in which they were published; this will be satisfactory in some cases, but the diversity of publishers means that the data required to support a task may be stored in many different sources and described in many different ways. Linking data distributed across the Web through a standard mechanism transforms a set of data islands to a dataspace. The notion of *dataspace* was introduced in 2005 [31] for information systems that provide for the coexistence of heterogeneous data and do not require an upfront investment into a unifying schema. In such systems, data integration is achieved in a 'pay-as- you-go' manner: as long as no or only a small number of mappings has been added to the system, applications can only display data in an unintegrated fashion and can only answer simple queries, or even only provide text search. Once more effort is invested over time in generating mappings, applications can further integrate the data and provide better query answers.

The impact of variety and veracity. Although RDF provides a syntactically homogeneous language for describing data, sources typically manifest a wide range of heterogeneities, in terms of how data on a concept are represented. Data stored in an RDF source does not conform to any specific structure analogous to the schema of a relational database. However, query processing requires an understanding of how concepts are represented. For instance, in [54] sources are assumed to provide structural summaries. Techniques that infer the structures found in sources in ways that support downstream tasks, such as query routing, are being developed and exploited, thus blurring the distinction between query processing and data integration. Also at the instance matching level, the identification of structures representing the same real world objects, usually part of the data integration and cleaning tasks, has been incorporated in query processing [39]. Another relevant aspect of these new contexts is, indeed, that many data sources are unreliable, in that their data is typically dirty. The need to keep into account the quality of data to cope with its increasing quantity decisively emerge. Adaptive techniques to cope with variable data quality during processing have been proposed (see, e.g., [51]).

Semantically enhanced data sets shift interoperability and data integration issues from a syntactic to a semantic level. Therefore, all problems faced with the integration of multiple data sets are addressed taking into account semantic annotations. The user may have a limited knowledge of data she is going to query, and specifying her retrieval needs may be difficult. The notion of query itself thus becomes more flexible: we may generally talk of 'entity-relationship' (graph-based) queries expressed at different levels of formalization and query interpretation becomes a first step of the processing [49]. Query-driven approximation can be used with the aim of assisting users in formulating their information needs even when they lack full knowledge of the data structure and of its irregularities, relying on inference rules for query relaxation [36, 53].

The impact of volume. Typical systems operating on Linked Data collect (crawl) and pre-process (index) large amounts of data, and evaluate queries against a centralised repository. Given that crawling and indexing are time-consuming operations, the data in the centralised index may be out of date at query execution time. An ideal query answering system for *live querying* Linked Data should return current answers in a reasonable amount of time, even on corpora as large as the Web. In such a live query system, *source selection*, determining which sources contribute answers to a query, is a crucial step of query processing [35] that requires knowledge of the contents of data sources. Lightweight data summaries for determining relevant sources during query evaluation have been proposed and are exploited for this task [58].

To cope with the volume issues raised by these new contexts, moreover, several practical techniques can be exploited [30], including distributed query processing with partial evaluation, data compression, view-based query answering, and bounded incremental computation. Approximate algorithms, i.e., heuristics to find answers that are not far to the exact query answers have been proposed, both in the form of query-driven approximation (i.e., by approximating costly subgraph isomorphism with graph simulation) and data-driven approximation.

5 From Data to Big Data

In the late 2000s, Big Data arose due to the confluence of three major trends [1]. First, due to inexpensive storage, sensors, smart devices, social software, and the Internet of Things, which connects daily usage 'things', the generation of a wide variety of data has become much cheaper. At the same time, due to advances in multicore CPUs, solid state storage, cluster computing, and inexpensive cloud computing, the cost of processing massive amounts of data has considerably decreased as well. Additionally, nowadays, the process of generating, processing, and consuming data is no longer just for few database professionals, but for a much wider range of users, including everyday consumers. As a consequence, all previously discussed V's, namely *volume*, *variety*, *velocity*, and *veracity*, are raised and amplified: Big Data applications typically deal with multiple data sources, often coming as a data stream, which contain a large volume of data and are frequently updated; they may have different formats, may not come with a schema, and are represented according to an aggregate data model (as it happens in NoSQL systems) or a graph data model. Many data sources are unreliable: their data is typically dirty [30]. New challenges have therefore to be faced, which, never as before, require to rethink at all previously addressed data management issues, including query processing.

The impact of volume. The reference architecture for Big Data management consists in clusters of a large number of (unreliable) commodity machines upon which distributed data processing is achieved through parallel programming models such as MapReduce. Higher-level languages, such as Pig and Hive, have then been layered on top, enabling a broader audience of developers to use scalable big data platforms.

Such languages bring request specification and processing for Big Data close to those of more traditional data. As a consequence, the significant expertise in declarative and efficient data processing, coming from the database community, should now contribute to the design of more appropriate and efficient future Big Data platforms [17]. As an example, there is a growing recognition that more powerful cost-aware query optimizers and set-oriented query execution engines are needed, to fully exploit large clusters of many-core processors, scaling both 'up' and 'out'. This creates challenges for progress monitoring (of queries that are running too slowly or consuming excessive resources), for adapting to the characteristics of previously unseen data, and for reducing the cost of data movement between stages of processing [1]. Some of the previously devised solutions for coping with the impact of volume have, in the meantime, been customized to the new environments; this is the case of data-driven approximation [59].

The impact of variety and veracity. Big Data applications are characterized by many levels of heterogeneity [1]: data comes at different degrees of veracity, are represented according to different models, managed by different software systems, with different application programming interfaces, query processors, and analysis tools. As a consequence, a single, one-size-fits-all, big data system cannot meet the requirements of this degree of diversity. Rather, multiple classes of different systems will certainly emerge, raising new issued in data integration, further emphasizing the need for pay-as-you-go approaches. Processing and optimizing data requests which span diverse big data systems (possibly managing data residing on different data centers where disconnected devices will be more and more common) and flows that move data between them are open issues that in the next future should be addressed [1]. To handle data diversity, lazy computation could be beneficial and may lead to new data processing solutions.

Variety and veracity make data processing very challenging also because guaranteeing user satisfaction with respect to conflicting requirements on result quality and response time in such complex environments is not a trivial task. To this aim, a revisitation of query-driven approximation and source selection approaches may help, possibly exploiting data quality information, request context, and information on similar requests recurring over time to guide the overall process and reduce the need of user involvement for the interpretation of the request (see, e.g., [44] for a MapReduce approach for preference-based algorithms and [20] for a preliminary approach relying on recurring retrieval needs).

The impact of velocity. Big Data applications may need to access and analyze high-speed data streams [32]. Due to the high input rate, processing solutions developed in the past for data streams should be revised and tuned to take care of non-uniform memory access and limited transfer rates across layers of the memory hierarchy [1]. As an example, continuous query processing can be used for data coming from very high-speed but not highly informative data sources, while reduction of such data might be stored for future accesses.

6 Conclusions

Query processing has been one of the hot topics in the last 25 years of data management. In the same period, the Italian Symposium on Advanced Database Systems (SEBD) has followed the advances in the field by proposing, besides regular papers, a high number of keynote talks on this topic, ranging from query processing issues in P2P systems and preference queries, in 2004 (by Amr El Abbadi and Jan Chomicki, respectively), to query processing issues for massive social networks, in 2011 (V.S. Subrahmanian), dynamic big data, in 2013 (M. Garofalakis), and optimization techniques for semantic queries, in 2013 (I. Manolescu). Query processing issues have also been deeply discussed by SEBD tutorials, including data stream query processing, in 2004 (N. Koudas), similarity join algorithms, in 2008 (W. Wang), and querying for large-scale ontologies, in 2009 (R. Möller). New exciting query processing challenges lie ahead of us: we are ready to follow them together with the SEBD community.

References

1. D. Abadi et al., The Beckman report on database research. Commun. ACM **59**(2), 92–99 (2016)
2. D. Abadi et al, The design of the borealis stream processing engine, in *CIDR* (2005), pp. 277–289
3. S. Acharya et al. Join synopses for approximate query answering, in *SIGMOD Conference* (1999), pp. 275–286
4. C.C. Aggarwal, *Data Streams: Models and Algorithms* (Springer, Berlin, 2006)
5. S. Agrawal et al., Automated ranking of database query results, in *CIDR* (2003)
6. S. Amer-Yahia, S. Cho, D. Srivastava, Tree pattern relaxation, in *EDBT* (2002), pp. 496–513
7. S. Amer-Yahia et Al. Structure and content scoring for XML, in *VLDB* (2005), pp. 361–372
8. R. Avnur, J.M. Hellerstein, Eddies: continuously adaptive query processing, in *SIGMOD Conference* (2000), pp. 261–272
9. B. Babcock, M. Datar, R. Motwani, Load shedding for aggregation queries over data streams, in *ICDE* (2004), pp. 350–361
10. S. Babu, P. Bizarro, Adaptive query processing in the looking glass, in *CIDR* (2005), pp. 238–249
11. S. Babu et al. Adaptive ordering of pipelined stream filters, in *SIGMOD Conference* (2004), pp. 407–418
12. S. Babu, J. Widom, Continuous queries over data streams. ACM SIGMOD Rec. **30**(3), 109–120 (2001)
13. D. Barbará et al., The New Jersey data reduction report. IEEE Data Eng. Bull. **20**(4), 3–45 (1997)
14. C.K. Baru et al., Xml-based information mediation with MIX, in *SIGMOD Conference* (1999), pp. 597–599
15. T. Berners-Lee et al., The semantic web. Sci. Am. **284**(5), 28–37 (2001)
16. C. Bizer, T. Heath, T. Berners-Lee, Linked data - the story so far. *Semantic Services, Interoperability and Web Applications: Emerging Concepts* (2009), pp. 205–227
17. V. Borkar, M.J. Carey, C.Li, Inside big data management: ogres, onions, or parfaits? in *EDBT* (ACM, 2012), pp. 3–14
18. S. Börzsönyi, D. Kossmann, K. Stocker, The skyline operator, in *ICDE* (2001), pp. 421–430

19. L. Bouganim et al., A dynamic query processing architecture for data integration systems. IEEE Data Eng. Bull. **23**(2), 42–48 (2000)
20. B. Catania, F. De Fino, G. Guerrini, Recurring retrieval needs in diverse and dynamic dataspaces: issues and reference framework, in *GraphQ Workshop*, associated with *EDBT* (2017)
21. B. Catania, L. Jain, Advanced query processing, Volume 1: issues and trends. Intelligent Systems Reference Library, vol. 36 (Springer, 2013)
22. S. Chandrasekaran et al., TelegraphCQ: continuous dataflow processing for an uncertain world, in *CIDR* (2003)
23. S. Chaudhuri, V. Ganti, R. Kaushik, A primitive operator for similarity joins in data cleaning, in *ICDE*, vol. 5 (2006)
24. S. Chaudhuri, D. Umeshwar, An overview of data warehousing and OLAP technology. ACM, SIGMOD Rec. **26**(1), 65–74 (1997)
25. J. Chen, D et al., NiagaraCQ: a scalable continuous query system for internet databases, in *SIGMOD Conference* (2000), pp. 379–390
26. J. Considine et al., Robust approximate aggregation in sensor data management systems. ACM Trans. Database Syst. 34(1) (2009)
27. A. Corbellini et al., Persisting big-data: the NoSQL landscape. Inf. Syst. **63**, 1–23 (2017)
28. J. Dean, S. Ghemawat, MapReduce: simplified data processing on large clusters. Commun. ACM **51**(1), 107–113 (2008)
29. A. Deshpande, Z.G. Ives, V. Raman, Adaptive query processing. Found. Trends Databases **1**(1), 1–140 (2007)
30. W. Fan, J. Huai, Querying big data: bridging theory and practice. J. Comput. Sci. Technol. **29**(5), 849–869 (2014)
31. M. Franklin, A. Halevy, D. Maier, From databases to dataspaces: a new abstraction for information management. ACM SIGMOD Rec. **34**(4), 27–33 (2005)
32. M. Garofalakis, J. Gehrke, R. Rastogi, *Data Stream Management: Processing High-Speed Data Streams* (Springer, Berlin, 2007)
33. P.B. Gibbons, Y. Matias, Synopsis data structures for massive data sets, in *External Memory Algorithms* (American Mathematical Society, 1999), pp. 39–70
34. A.Y. Halevy, A. Rajaraman, J.J. Ordille, Data integration: the teenage years, in *VLDB* (2006), pp. 9–16
35. A. Harth et al., Data summaries for on-demand queries over linked data, in *WWW* (ACM, 2010), pp. 411–420
36. O. Hartig, M.T. Özsu, Linked data query processing, in *ICDE* (IEEE, 2014), pp. 1286–1289
37. T. Heath, C. Bizer, *Linked Data: Evolving the Web into a Global Data Space. Synthesis Lectures on the Semantic Web* (Morgan & Claypool Publishers, 2011)
38. I.F. Ilyas, G. Beskales, M.A. Soliman, A survey of top-*k* query processing techniques in relational database systems. ACM Comput. Surv. 40(4) 2008
39. E. Ioannou et al., On-the-fly entity-aware query processing in the presence of linkage. PVLDB **3**(1–2), 429–438 (2010)
40. Z.G. Ives et al., An adaptive query execution system for data integration, in *SIGMOD Conference* (1999), pp. 299–310
41. Z.G. Ives, A.Y. Halevy, D.S. Weld, Adapting to source properties in processing data integration queries, in *SIGMOD Conference* (2004), pp. 395–406
42. Y. Jiao, Maintaining stream statistics over multiscale sliding windows. ACM Trans. Database Syst. **31**, 1305–1334 (2006)
43. V. Kantere et al., Query relaxation across heterogeneous data sources, in *SIGMOD Conference* (2015), pp. 473–482
44. J. Koh et al., MapReduce skyline query processing with partitioning and distributed dominance tests. Inf. Sci. **375**, 114–137 (2017)
45. N. Koudas et al., Relaxing join and selection queries, in *VLDB* (2006), pp. 199–210
46. N. Koudas, D. Srivastava, Approximate joins: concepts and techniques, in *VLDB*, vol. 1363 (2005)

47. S. Madden et al., Continuously adaptive continuous queries over streams, in *SIGMOD Conference* (2002), pp. 49–60
48. I. Manolescu, D. Florescu, D. Kossmann, Answering XML queries on heterogeneous data sources, in *VLDB* (2001), pp. 241–250
49. Y. Mass et al., IQ: the case for iterative querying for knowledge, in *CIDR* (2011), pp. 38–44
50. C. Mishra, N. Koudas, Interactive query refinement, in *EDBT* (2009), pp. 862–873
51. P. Missier et al., Data quality support to on-the-fly data integration using adaptive query processing, in *SEBD* (2009), pp. 213–220
52. R. Motwani et al., Query processing, approximation, and resource management in a data stream management system, in *CIDR* (2003)
53. A. Poulovassilis, P. Selmer, P.T. Wood, Approximation and relaxation of semantic web path queries. J. Web Sem. **40**, 1–21 (2016)
54. B. Quilitz, U. Leser, Querying distributed RDF data sources with SPARQL, in *ESWC* (Springer, 2008), pp. 524–538
55. E.A. Rundensteiner et al., CAPE: continuous query engine with heterogeneous-grained adaptivity, in *VLDB* (2004), pp. 1353–1356
56. Y.N. Silva, W.G. Aref, M.H. Ali, The similarity join database operator, in *ICDE* (2010), pp. 892–903
57. N. Tatbul et al., Load shedding in a data stream manager, in *VLDB* (2003), pp. 309–320
58. J. Umbrich et al., Comparing data summaries for processing live queries over linked data. World Wide Web **14**(5–6), 495–544 (2011)
59. M.H. ur Rehman et al., Big data reduction methods: a survey. Data Sci. Eng. 1(4), 265–284 (2016)
60. T. Urhan, M.J. Franklin, L. Amsaleg, Cost based query scrambling for initial delays, in *SIGMOD Conference* (1998), pp. 130–141
61. K. Yi et al., Small synopses for group-by query verification on outsourced data streams. ACM Trans. Database Syst. 34(3) (2009)
62. X. Zhou et al., Query relaxation using malleable schemas, in *SIGMOD Conference* (2007), pp. 545–556

From Star Schemas to Big Data: 20+ Years of Data Warehouse Research

M. Golfarelli and S. Rizzi

Abstract Data Warehouses are the core of the modern systems for decision making. They store integrated information extracted from various and heterogeneous data sources, making it available in multidimensional form for analyses aimed at improving the users' knowledge of their business. Though the first use of the term dates back to the 80s, only during the late 90s data warehousing has emerged as a research area on its own, though in strict correlation with several other research topics as database integration, view materialization, data visualization, etc. This paper surveys more than 20 years of research on data warehouse systems, from their early relational implementations (still widely adopted in corporate environments), to the new architectures solicited by Business Intelligence 2.0 scenarios during the last decade, and up to the exciting challenges now posed by the integration with big data settings. The timeline of research is organized into three interrelated tracks: techniques, architectures, and methodologies.

Keywords Data warehouse · OLAP · Big data

1 Introduction

Data Warehouses (DWs) are the core of the modern systems for decision making. They store integrated information extracted from various and heterogeneous data sources through ETL tools, making it available in multidimensional form for analyses aimed at improving the users knowledge of their business.

Though the term "data warehouse" has appeared during the late 80s, the official year of birth of DWs is considered to be 1992, when Inmon defined a DW as *"a subject-oriented, integrated, time-variant, and non-volatile collection of data in support of managements decision making process"* [58]. A few years later, the reference book by Kimball laid the fundamentals of DW design by introducing the star

M. Golfarelli · S. Rizzi (✉)
DISI, University of Bologna, Bologna, Italy
e-mail: stefano.rizzi@unibo.it

S. Flesca et al. (eds.), *A Comprehensive Guide Through the Italian Database Research Over the Last 25 Years*, Studies in Big Data 31, DOI 10.1007/978-3-319-61893-7_6

schema as the basic solution for modeling multidimensional data on relational DWs [64]. During the late 90s, data warehousing has progressively emerged as a research area on its own, though in strict correlation with several other research topics as database integration, view materialization, data visualization, etc.

This paper surveys more than 20 years of research on DW systems, from their early relational implementations (still widely adopted in corporate environments), to the new architectures solicited by Business Intelligence (BI) 2.0 scenarios during the last decade, and up to the exciting challenges now posed by the integration with big data settings. For space reasons the topics related to database integration and ETL, though strictly related to DWs, are not covered here. The timeline of research is organized into three interrelated tracks: techniques (Sect. 2), architectures (Sect. 3), and methodologies (Sect. 4).

2 Techniques

Multidimensional modeling. The multidimensional model, based on the concepts of *fact*, *measure*, *dimension*, and *hierarchy*, is the core of data representation in DWs. Multidimensional modeling requires specialized techniques, both at the conceptual and at the logical level [99]. In relational implementations of DWs, the (denormalized) *star schema* and the (partially or completely normalized) *snowflake schema*, both introduced in [64], are still the most widely adopted solutions for multidimensional modeling at the logical level. In the 00s, some more advanced modeling solutions have been proposed since then to handle specific situations (e.g., [87, 93] for modeling irregular hierarchies and, more recently, [40] for modeling topic hierarchies in social BI). Concerning multidimensional implementations of DWs, several efficient multidimensional data structures such as *condensed cubes* [128], *dwarfs* [110], and *QC-Trees* [66] have been proposed to manage data cubes. A comprehensive survey of logical models for multidimensional data can be found in [125].

A lot of research in multidimensional modeling has focused on the conceptual level. The existing approaches may be framed into three categories: extensions to the Entity-Relationship model (e.g., [38, 105]), extensions to UML (e.g., [1, 73]), and ad hoc models (e.g., [42, 57]). While all models have the same core expressivity, in that they all allow the basic concepts of the multidimensional model to be represented, they significantly differ as to the possibility of representing more advanced concepts such as irregular hierarchies, many-to-many associations, and additivity. A good survey of the different proposals is provided in [103].

We finally mention that a relevant role in multidimensional modeling is taken by the conditions of *summarizability*, which ensure that aggregates are correctly computed avoiding for instance double counting (see [68, 81]).

Performance optimization and tuning. Interactive querying capability is one of the main goals of DWs, and in presence of huge amount of data it can be achieved only coupling effective modeling solutions with appropriate performance optimization techniques. Research efforts in this direction were mainly focused on *aggregated view materialization* and *indexing*.

Materialized views physically store pre-aggregate multidimensional data to reduce the computational effort for OLAP queries. Since the number of materialized views for each cube is exponential in the number of hierarchy levels [10], choosing the optimal subset in presence of constraints is a difficult task. Besides the minimization of the workload execution time and space constraints, typical constraints include minimizing the update time [54] and respecting an upper bound of the query response time [117].

DWs are read-only databases based on standard schemas (mainly star or snowflake), which enables the adoption of specific indexes on the one hand, the revamping of underused solutions such as *bitmap indexes* and *join indexes* on the other. Bitmap indexes, first proposed in [111], associate one column of bits to each possible attribute value and are specifically tailored for indexing attributes with low cardinality. Join indexes have been proposed in the 80s [123] and consist in pre-computing the join operation between two relations. In the context of relational DWs, join indexes have been extended to multi-table join indexes called *star indexes*. A relevant variant of the join index, called *bitmap join index*, is proposed in [88] as a bitmap index on a fact table based on column(s) of dimension table(s). The index selection problem consists in selecting an appropriate set of indexes for a DW given a workload and some constraints related for instance to storage and maintenance. The solutions proposed are mainly based on heuristic enumeration techniques [44, 55]. The problem of index selection is often coupled with that of materialized views selection, since they compete on the same shared resources [7]. More recently, index selection has been faced with reference to non-classical DW dimensions such as spatial ones [92].

Query processing. Efficient query processing is a classic research theme in the DW area. OLAP navigation relies on operators applied to cubes; starting with the well-known *data cube* operator [53], several proposals have extended the traditional OLAP expressiveness either based on data mining techniques [50, 83] or exploiting the specificity of the underlying architecture [32].

Research aimed at reducing the query processing time typically comes in parallel with research related to new architectures, applications, and data structures. In the context of relational databases, a large effort has been devoted to devise execution plans that keep into account the peculiar features of DW systems and OLAP queries, such as the presence of materialized views [27], the use of standard logical schemas (e.g., the star schema) [129], and the need for efficiently aggregating data [132]. Specific processing solutions have been devised for different, either logical or physical, data models. For example, when OLAP queries are run against an XML repository, specific modeling and query processing solutions are needed to ensure a good level of performance [95]; conversely, column-oriented DBMSs typically ensure high performance and enable further optimizations to be applied to query plans [36]. Specific approaches to query answering have also been developed for DWs including complex data such as graphs [89], streams [34, 84], and spatial data [133].

Security. Security is an important issue in databases, and is even more relevant in DWs due to the strategic importance of the stored information. Among the different aspects of security, the ones that have been mainly investigated in DWs are *confidentiality* and *privacy-preserving queries*. Confidentiality refers to the capability of ensuring that users can only access the information they have privileges for. Unfortunately, the grant models used in transactional databases are unsuitable for DWs. While in the former protection is expressed in terms of tables and attributes, in the latter it relies on dimensions, hierarchical paths, and granularity levels. To ensure a proper definition of grants and privileges, security design is embedded in the overall design process [120] and the security goals are elicited as non-functional goals at the conceptual level [96]. Differently from the previous solutions, in [104] the permission policies are derived from the source databases by treating the source exported tables and the DW itself as part of the same distributed database.

As to privacy-preserving queries, the main issue emerging in OLAP analysis is how to preserve the confidential information in individual data cells while still providing an accurate estimation of the original aggregated values for range queries [127]. Most of the approaches relies on either data distortion techniques [114] or data sampling [30].

Quality. Ensuring quality is a recurrent problem in DW projects. The early foundations in this area have been established in the late 90s by the *Data Warehouse Quality* project [60]. In the following years, the research has been mostly focused on measuring the quality of the design process. In particular, at the conceptual level there have been attempts towards defining metrics that allow the intuitive notions of quality of conceptual schemas to be replaced with quantitative measures [108]. At the logical/physical level, some works were focused on quantitatively evaluating the complexity of dimensional models [25]. Other relevant research directions include normal forms for DW [67] and quality-driven view selection [22].

As to the quality of data, meant as the ability to understand, use, access, and trust the data, some early work listing the main quality dimensions can be found in [94]. More recently, the use of logic rules has been proposed as a way to assess the quality of measures in a DW, accounting for the context in which these measures are considered [79].

Evolution and versioning. During the 00s, as several mature implementations of DW systems were going fully operational, the continuous evolution of the application domains brought to the forefront the dynamic aspects related to describing how the information stored in the DW changes over time. As concerns changes in data values (often referred to as *DW maintenance*), a number of approaches have been devised, starting with Kimball's *slowly-changing dimensions* [64] and up to the object-oriented approach of [97] and, more recently, the *slowly-changing measures* of [51].

The problem of managing changes at the schema level (which may be required by changes either in the business domain or in the user requirements or in the data sources) is more challenging. The approaches to management of schema changes in

DWs can be framed into two categories, namely *evolution* [21, 122] and *versioning* [11, 45]: while both categories support schema changes, only the latter keeps track of previous versions. In some versioning approaches, besides "real" versions determined by changes in the application domain, also "alternative" versions to be used for what-if analysis are considered [11]. Finally, the TOLAP temporal query language is proposed in [82] to express typical OLAP queries in presence of schema evolution and versioning.

Visualization. OLAP is the main paradigm for flexibly and friendly accessing the information stored in a DW. OLAP front-ends consolidated as an indispensable technique to support routine reporting and interactive data analysis in the mid-90s; traditionally, they use visualization merely for expressive presentation of the data. To achieve deeper insight in data and enable a more comprehensive user interaction, *visual OLAP* has emerged as a framework of advanced visualization techniques for representing the retrieved data set along with a powerful navigation and interaction scheme for specifying, refining, and manipulating the subset of interest. While some proposals of querying paradigms for visual access to large data sets were proposed already in the 90s (e.g., VisDB [63]), only in the following years advanced techniques specifically tailored for multidimensional data were devised; prominent examples are *Advizor* [35], the *Cube Presentation Model* [75], *HDDV* [115], and *Enhanced Decomposition Trees* [76]. Specifically, some multi-scale visualization techniques were also proposed [109, 113].

Personalization and recommendation. In line with the goals of BI 2.0, the research in this field goes in the direction of improving the user experience in analyzing data by fitting it to the user's tastes and context. An attempt to relate preferences to the multidimensional model was made in [112], where user preferences are attached to contexts parameterized by a set of hierarchical attributes. Another work situating preferences in the context of multidimensional databases is [131], which enables efficient computation of preference expressions on numerical domains. Finally, in [46] a wider expressiveness is achieved by including preferences on the aggregation level of queries and on categorical domains.

Recently recommender systems have gained interest in the database community, with approaches ranging from content-based to collaborative query recommendation, especially to cope with the problem of interactively analyzing large databases. This problem is particularly important in a DW context, where the peculiarities of the multidimensional model and queries can be leveraged. A recent survey of query recommendation approaches for OLAP is offered in [77]. Among the most prominent approaches we mention [106], where a query answer is analyzed to detect "surprising" values to recommend; [59], which proposes a content-based approach that synthesizes a recommendation by enriching the current query answer with elements extracted from a user's profile; and [6], whose goal is not to recommend single OLAP queries, but whole OLAP sessions.

3 Architectures

Basic architectures. The foundations of classical DW architectures were laid in the early 90s by Inmon and Kimball. In Inmon's *hub-and-spoke* architecture, the DW is a normalized, enterprise-wide repository from which distinct *data marts* are derived to meet the needs of single departments [58]; the suggested approach to design is top-down. Conversely, in Kimball's *data mart bus architecture*, data marts are first-class citizens and are built in a bottom-up fashion following an iterative approach [64]; in the absence of a centralized physical repository of data, enterprise-wide integration is ensured by the presence of *conformed dimensions*. Conformed dimensions have exactly the same meaning and content when being referred from different facts; they are defined once in collaboration with the business data governance representatives and reused across all data marts. Later on, *federated architectures* have been proposed to support the integration of heterogeneous data marts into a global, logical schema [17], followed by *peer-to-peer architectures* where no global schema needs be defined and integration is based on inter-peer mappings [48].

Orthogonally to this classification, architectures have also been evolving to cope with the new requirements posed by emerging settings and sophisticated applications of BI 2.0. For instance, modern enterprise architectures often include a *data lake*, meant as a repository of unstructured or loosely-structured data, typically stored on NoSQL databases, to be incorporated in the decision-making process following a schema-on-read approach [71]. In the case of social BI applications, the architecture should be enhanced to include, besides classical ETL, also a semantic enrichment component relying on a domain ontology to extract semantic information hidden in the user-generated content [37]. In the context of on-demand BI, sophisticated architectures are needed to transparently accommodate situational data extracted from external sources (e.g., see *fusion cubes* [2]). Similarly, the *text OLAP* framework is about providing specific models and operators for analyzing textual data [91].

We finally mention that, while traditional DW architectures are inherently characterized by some non-negligible data latency, *real-time architectures* are adopted to reduce (virtually to 0) the data latency in settings where operative decisions must be taken based on real-time monitoring of performance indicators [65]. It is also worth mentioning that some research has been recently devoted to devising comprehensive models for analytical metadata [124].

Spatial architectures. Today it is estimated that over 80% of the data companies work with geo-localized information. In the context of BI 2.0, the term *location intelligence* identifies all the BI applications that leverage spatial data to deliver strategic information. Location intelligence applications are typically based on a *spatial DW*, that is, a DW that is capable of representing, storing, and manipulating spatial data by coupling GIS and DW capabilities [49].

SOLAP is an extension of the OLAP paradigm aimed at exploring spatial data in spatial DWs [98]. A critical factor when building a SOLAP application is how to ensure summarizability between spatial attributes. To this end spatial hierarchies have been classified in non-geometric, geometric to non-geometric, and fully geometric [12]. The topic of ensuring proper aggregations in different contexts has been further studied in [19]. While SOLAP copes with discrete spatial data only, sophisticated analyses require dealing also with continuous spatial data, called *continuous fields*, which describe physical phenomena that change continuously in time and/or space (e.g., temperature, land elevation, and population density). Some solutions have been proposed for modeling [4, 121] and processing [20, 52] continuous fields, but no commercial solution effectively supports this feature yet.

The availability of large amounts of mobility data coming from GPS-enabled devices such as smart phones and life-logging devices pushed spatial DWs one step forward to track and analyze trajectories and movements. A *trajectory DW* [90] enables OLAP-like analysis of trajectories; the main issues here are related to defining a proper technique to model trajectories [23], an efficient pre-processing for raw data [90], and a set of OLAP-like operators for their manipulation [69, 78].

Parallel and big data architectures. DWs were born in environments dominated by relational databases running on traditional servers. The answer to the more demanding requirements of big data has been a shift towards parallel architectures. Specifically, either shared-memory, shared-disk, shared-nothing, or shared-everything architectures can be adopted, with different pros and cons. In this complex context two main solutions are followed. On the one hand, several vendors proposed closed, proprietary, and pre-configured solutions, called *appliances* (e.g., Teradata, Oracle Exadata, and Netezza), as a robust and scalable solution to store, process, and analyze data. On the other hand several open source solutions, typically cloud-provided and based on the Hadoop framework, have emerged. *Hive* [118] has been the first engine providing a SQL interface to MapReduce, while HadoopDB extends Hive by coupling Hadoop and PostgreSQL. Shark and Spark [8] overcome the batch limitation of MapReduce and support interactive queries on big data. To further reduce the response time, BlinkDB [3] supports interactive aggregation querying with approximated results on very large amounts of data. Another solution that overcomes the MapReduce limitations is Presto, a distributed SQL engine optimized for interactive use with low-latency on thousands of nodes.

Of course, the adoption of a parallel architecture has a relevant impact on design [14], optimization [13], and query processing [5]. In particular, as to design, additional stages are made necessary, namely *data partitioning* [39] and *data allocation and replication* [70].

4 Methodologies

DW design. Designing a DW is quite different from designing an operational database. Though few comprehensive ad-hoc design methods have been devised so far (e.g., [42, 72]), a lot has been written about how a DW should be designed. Most methods distinguish between a phase of *conceptual design* and one of *logical design*:

conceptual design aims at deriving an implementation-independent and expressive conceptual schema for the DW, while logical design creates a corresponding logical schema on the chosen platform by considering some set of constraints. Several methods (e.g., [42]) also support a phase of *physical design*, that addresses all the issues specifically related to the suite of tools chosen for implementation – such as indexing and allocation. Conversely, the phase of *testing* has received little attention [43]. Several techniques for automating single phases of DW design have been proposed in the literature: for instance, [42] for conceptual design from E/R and relational sources, [116] for logical design on a relational platform, [44] for relational physical design, [126] for designing the ETL process. More recent approaches are focused on designing implementations of a DW based on NoSQL databases, e.g. [28, 107].

The approaches to DW design can be classified in five categories. *Data-driven* approaches, the first ones to be devised, design the DW starting from a detailed analysis of the data sources (e.g., for relational sources, [42, 57]). In the early 00s, *requirement-driven* approaches emerged as a way to give more weight to the information requirements of end users [80, 130]. *Mixed approaches* consider both (i.e. availability of data and user requirements) at the same time [26, 41]. *Query-based approaches* design the DW starting from a specification of the workload [86, 101]. More recently, *pattern-based approaches* look for typical multidimensional patterns in the source data [61, 100]. Independently of the specific approach followed, some attempts were made to apply the agile principles to DW design [47, 56].

Design from non-relational data sources. Though most enterprise operational data are stored in relational databases, alternative types of data sources have been emerging over the years. The first attempts to feed a DW with non-relational data sources were made in the early 00s using XML sources to create so-called *XML warehouses* by discovering functional dependencies (a survey is offered in [74]). With the advent of the semantic web on the one hand, of big data on the other, the community has moved its attention towards data sources that are richer in semantics and possibly have huge volumes. In this direction, a semi-automatic method to discover multidimensional concepts from a domain ontology describing heterogeneous data sources is proposed in [102]. More recently, both [85, 100] propose approaches to automate the discovery of dimension hierarchies on linked data, while [62] investigates the problem of executing OLAP queries via SPARQL on linked data. A comprehensive approach for designing DWs based on RDF knowledge bases is outlined in [18]. Finally, an iterative methodology for designing and maintaining social BI applications fed from user-generated content crawled from social networks is presented in [37].

Discovering multidimensional structure in NoSQL data sources is currently an object of research; among the first approaches we mention [31], which proposes a new aggregation operator called MC-CUBE to compute OLAP cubes from a column-oriented database, and [29], which describes an interactive, schema-on-read approach for finding multidimensional structures in document stores aimed at enabling OLAP querying.

DW integration and interoperability. This concept, which refers to the ability of combining the content of distinct DWs aimed at cross-analysis, has been informally introduced by Kimball with his conformed dimensions [64], then later formalized in [24] by defining *dimension compatibility* for drill-across. During the following years, three categories of approaches were devised: *warehousing* approaches, where the integrated data are physically materialized [119], *federative* approaches, where integration is virtual but still based on a global schema [9, 16], and *peer-to-peer* approaches, that do not rely on a global schema to integrate the component DWs [48].

In particular, in the tightly-coupled approach of [119], a technique that combines the contents of the dimensions to be integrated is used to derive a materialized view that includes the component DWs. In [9], schema matching of different DWs in a federation is based on linguistic and structural comparison between multidimensional concepts. The approach in [16] relies on a distributed architecture where users are enabled to directly access the heterogeneous schemas of the component DWs, and a query language is proposed to transform a cube and make it compatible with a global schema and ready for integration. A P2P approach is the one introduced in [48], where a query formulated on a peer DW is reformulated onto the other peers using semantic mappings that mediate between the different multidimensional schemas. In the same direction, [15] proposes a method for the semi-automatic discovery of mappings between dimension hierarchies of heterogeneous DWs using topological properties of dimensions and semantic techniques. Finally, in the context of interoperability, we mention a recent semantics-based approach for dynamic calculation of key performance indicators to support query rewriting over heterogeneous DWs [33].

5 Conclusion

In this paper we have condensed more than 20 years of research in DWs, aimed at giving the reader an idea of the main topics within each track and of how the focus of research has shifted in time. To close the paper we show in Fig. 1 the distribution

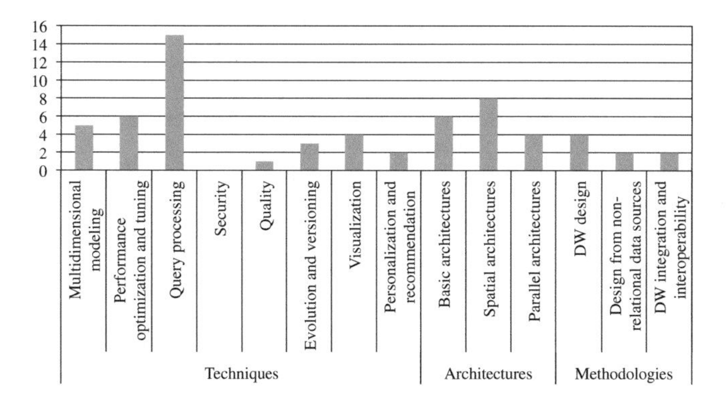

Fig. 1 Distribution of papers for DaWaK and DOLAP since 2013

by topic and track of papers published in the two main specific forums for DW research, i.e., the DaWaK Conference and the DOLAP Workshop, since 2013. It clearly appears that query processing and optimization issues still capture most of the interest in the community, and that a relevant portion of research is devoted to architectural issues.

References

1. A. Abelló, J. Samos, F. Saltor, YAM^2: a multidimensional conceptual model extending UML. Inf. Syst. **31**(6), 541–567 (2006)
2. A. Abelló et al., Fusion cubes: towards self-service business intelligence. IJDWM **9**(2), 66–88 (2013)
3. S. Agarwal, B. Mozafari, A. Panda, H. Milner, S. Madden, I. Stoica, BlinkDB: queries with bounded errors and bounded response times on very large data, in *Proceedings of Eurosys* (2013), pp. 29–42
4. T.O. Ahmed, M. Miquel, Multidimensional structures dedicated to continuous spatiotemporal phenomena, in *Proceedings of the BNCOD* (2005), pp. 29–40
5. F. Akal, K. Böhm, H. Schek, OLAP query evaluation in a database cluster: a performance study on intra-query parallelism, in *Proceedings of the ADBIS* (2002), pp. 218–231
6. J. Aligon, E. Gallinucci, M. Golfarelli, P. Marcel, S. Rizzi, A collaborative filtering approach for recommending OLAP sessions. Decis. Support Syst. **69**, 20–30 (2015)
7. K. Aouiche, J. Darmont, Data mining-based materialized view and index selection in data warehouses. JIIS **33**(1), 65–93 (2009)
8. M. Armbrust et al., Spark SQL: relational data processing in spark, in *Proceedings of the SIGMOD* (2015), pp. 1383–1394
9. M. Banek, B. Vrdoljak, A.M. Tjoa, Z. Skocir, Automating the schema matching process for heterogeneous data warehouses, in *Proceedings of the DaWaK* (2007), pp. 45–54
10. E. Baralis, S. Paraboschi, E. Teniente, Materialized views selection in a multidimensional database, in *Proceedings of the VLDB* (1997), pp. 156–165
11. B. Bębel, J. Eder, C. Koncilia, T. Morzy, R. Wrembel, Creation and management of versions in multiversion data warehouse, in *Proceedings of the SAC* (2004), pp. 717–723
12. Y. Bédard, T. Merrett, J. Han, Fundamentals of spatial data warehousing for geographic knowledge discovery. Geogr. Data Min. knowl. Discov. **2**, 53–73 (2001)
13. L. Bellatreche, A. Cuzzocrea, S. Benkrid, Query optimization over parallel relational data warehouses in distributed environments by simultaneous fragmentation and allocation, in *Proceedings of the ICA3PP* (2010), pp. 124–135
14. S. Benkrid, L. Bellatreche, A. Cuzzocrea, A global paradigm for designing parallel relational data warehouses in distributed environments. TLDKS **15**, 64–101 (2014)
15. S. Bergamaschi, M. Olaru, S. Sorrentino, M. Vincini, Dimension matching in peer-to-peer data warehousing, in *Proceedings of the DSS* (2012), pp. 149–160
16. S. Berger, M. Schrefl, Analysing multi-dimensional data across autonomous data warehouses, in *Proceedings of the DaWaK* (2006), pp. 120–133
17. S. Berger, M. Schrefl, From federated databases to a federated data warehouse system, in *Proceedings of the HICSS-41* (2008), p. 394
18. N. Berkani, L. Bellatreche, B. Benatallah, A value-added approach to design BI applications, in *Proceedings of the DaWaK* (2016), pp. 361–375
19. S. Bimonte, A. Tchounikine, M. Miquel, Geocube, a multidimensional model and navigation operators handling complex measures: application in spatial OLAP, in *Proceedings of the IAIT* (Springer, Berlin, 2006), pp. 100–109

20. S. Bimonte, A. Tchounikine, M. Miquel, F. Pinet, When spatial analysis meets OLAP: multidimensional model and operators, *Exploring Advances in Interdisciplinary Data Mining and Analytics* (2011), pp. 249–277
21. M. Blaschka, C. Sapia, G. Höfling, On schema evolution in multidimensional databases, in *Proceedings of the DaWaK* (1999), pp. 153–164
22. M. Bouzeghoub, Z. Kedad, A quality-based framework for physical data warehouse design, in *Proceedings of the DMDW* (2000)
23. F. Braz, S. Orlando, R. Orsini, A. Raffaetà, A. Roncato, C. Silvestri, Approximate aggregations in trajectory data warehouses, in *Proceedings of the ICDE* (2007), pp. 536–545
24. L. Cabibbo, R. Torlone, On the integration of autonomous data marts, in *Proceedings of the SSDBM* (2004), pp. 223–231
25. C. Calero, M. Piattini, C. Pascual, M.A. Serrano, Towards data warehouse quality metrics, in *Proceedings of the DMDW* (2001)
26. D. Calvanese, L. Dragone, D. Nardi, R. Rosati, S.M. Trisolini, Enterprise modeling and data warehousing in Telecom Italia. Inf. Syst. **31**(1), 1–32 (2006)
27. S. Chaudhuri, R. Krishnamurthy, S. Potamianos, K. Shim, Optimizing queries with materialized views, in *Proceedings of the ICDE* (1995), pp. 190–200
28. M. Chevalier, M.E. Malki, A. Kopliku, O. Teste, R. Tournier, Implementation of multidimensional databases in column-oriented NoSQL systems, in *Proceedings of the ADBIS* (2015), pp. 79–91
29. M.L. Chouder, S. Rizzi, R. Chalal, Enabling self-service BI on document stores, in *Proceedings of the DOLAP* (2017, to appear)
30. A. Cuzzocrea, D. Saccà, Balancing accuracy and privacy of OLAP aggregations on data cubes, in *Proceedings of the DOLAP* (2010), pp. 93–98
31. K. Dehdouh, Building OLAP cubes from columnar NoSQL data warehouses, in *Proceedings of the MEDI* (2016), pp. 166–179
32. K. Dehdouh, F. Bentayeb, O. Boussaid, N. Kabachi, Columnar NoSQL CUBE: aggregation operator for columnar NoSQL data warehouse, in *Proceedings of the SMC* (2014), pp. 3828–3833
33. C. Diamantini, D. Potena, E. Storti, Semantics-based multidimensional query over sparse data marts, in *Proceedings of the DaWaK* (2015), pp. 190–202
34. A. Dobra, M. Garofalakis, J. Gehrke, R. Rastogi, Processing complex aggregate queries over data streams, in *Proceedings of the SIGMOD* (2002), pp. 61–72
35. S. Eick, Visualizing multi-dimensional data. SIGGRAPH Comput. Graph. **34**(1), 61–67 (2000)
36. F. Färber, S.K. Cha, J. Primsch, C. Bornhövd, S. Sigg, W. Lehner, SAP HANA database: data management for modern business applications. SIGMOD Record **40**(4), 45–51 (2012)
37. M. Francia, M. Golfarelli, S. Rizzi, A methodology for social BI, in *Proceedings of the IDEAS* (2014), pp. 207–216
38. E. Franconi, A. Kamble, A data warehouse conceptual data model, in *Proceedings of the SSDBM* (2004), pp. 435–436
39. C. Furtado, A.A.B. Lima, E. Pacitti, P. Valduriez, M. Mattoso, Physical and virtual partitioning in OLAP database clusters, in *Proceedings of the SBAC-PAD* (2005), pp. 143–150
40. E. Gallinucci, M. Golfarelli, S. Rizzi, Advanced topic modeling for social business intelligence. Inf. Syst. **53**, 87–106 (2015)
41. P. Giorgini, S. Rizzi, M. Garzetti, GRAnD: a goal-oriented approach to requirement analysis in data warehouses. Decis. Support Syst. **45**(1), 4–21 (2008)
42. M. Golfarelli, S. Rizzi, *Data Warehouse Design: Modern Principles and Methodologies* (McGraw-Hill, New York, 2009)
43. M. Golfarelli, S. Rizzi, Data warehouse testing: a prototype-based methodology. IST **53**(11), 1183–1198 (2011)
44. M. Golfarelli, S. Rizzi, E. Saltarelli, Index selection for data warehousing, in *Proceedings of the DMDW* (2002), pp. 33–42

45. M. Golfarelli, J. Lechtenbörger, S. Rizzi, G. Vossen, Schema versioning in data warehouses: enabling cross-version querying via schema augmentation. DKE **59**(2), 435–459 (2006)
46. M. Golfarelli, S. Rizzi, P. Biondi, myOLAP: an approach to express and evaluate OLAP preferences. TKDE **23**(7), 1050–1064 (2011)
47. M. Golfarelli, S. Rizzi, E. Turricchia, Modern software engineering methodologies meet data warehouse design: 4WD, in *Proceedings of the DaWaK* (2011), pp. 66–79
48. M. Golfarelli, F. Mandreoli, W. Penzo, S. Rizzi, E. Turricchia, OLAP query reformulation in peer-to-peer data warehousing. Inf. Syst. **37**(5), 393–411 (2012)
49. M. Golfarelli, M. Mantovani, F. Ravaldi, S. Rizzi, Lily: a geo-enhanced library for location intelligence, in *Proceedings of the DaWaK* (2013), pp. 72–83
50. M. Golfarelli, S. Graziani, S. Rizzi, Shrink: an OLAP operation for balancing precision and size of pivot tables. DKE **93**, 19–41 (2014)
51. M. Goller, S. Berger, Slowly changing measures, in *Proceedings of DOLAP* (2013), pp. 47–54
52. L.I. Gómez, S.A. Gómez, A. Vaisman, Modeling and querying continuous fields with OLAP cubes. IJDWM **9**(3), 22–45 (2013)
53. J. Gray et al., Data cube: a relational aggregation operator generalizing group-by, cross-tab, and sub-totals. Data Min. Knowl. Discov. **1**(1), 29–53 (1997)
54. H. Gupta, I.S. Mumick, Selection of views to materialize under a maintenance cost constraint, in *Proceedings of the ICDT* (1999), pp. 453–470
55. H. Gupta, V. Harinarayan, A. Rajaraman, J.D. Ullman, Index selection for OLAP, in *Proceedings of the ICDE* (1997), pp. 208–219
56. R. Hughes, Agile data warehousing: delivering world-class business intelligence systems using Scrum and XP, IUniverse (2008)
57. B. Hüsemann, J. Lechtenbörger, G. Vossen, Conceptual data warehouse design, in *Proceedings of the DMDW* (2000), pp. 3–9
58. B. Inmon, *Building the Data Warehouse* (Wiley, New York, 1992)
59. H. Jerbi, F. Ravat, O. Teste, G. Zurfluh, Preference-based recommendations for OLAP analysis, in *Proceedings of the DaWaK* (2009), pp. 467–478
60. M.A. Jeusfeld, C. Quix, M. Jarke, Design and analysis of quality information for data warehouses, in *Proceedings of the ER* (1998), pp. 349–362
61. M.E. Jones, I. Song, Dimensional modeling: identification, classification, and evaluation of patterns. Decis. Support Syst. **45**(1), 59–76 (2008)
62. B. Kämpgen, S. O'Riain, A. Harth, Interacting with statistical linked data via OLAP operations, in *Proceedings of the Semantic Web Satellite Events* (2015), pp. 87–101
63. D.A. Keim, H. Kriegel, VisDB: a system for visualizing large databases, in *Proceedings of the SIGMOD* (1995), p. 482
64. R. Kimball, *The Data Warehouse Toolkit* (Wiley, New York, 1996)
65. A. Kotopoulis, Best practices for real-time data warehousing. Technical report, Oracle Corporation (2014)
66. L.V.S. Lakshmanan, J. Pei, Y. Zhao, QC-Trees: an efficient summary structure for semantic OLAP, in *Proceedings of the SIGMOD* (2003), pp. 64–75
67. J. Lechtenbörger, G. Vossen, Multidimensional normal forms for data warehouse design. Inf. Syst. **28**(5), 415–434 (2003)
68. H.J. Lenz, A. Shoshani, Summarizability in OLAP and statistical data bases, in *Proceedings of the SSDBM* (1997), pp. 132–143
69. L. Leonardi et al., T-warehouse: visual OLAP analysis on trajectory data, in *Proceedings of the ICDE* (2010), pp. 1141–1144
70. A.A.B. Lima, C. Furtado, P. Valduriez, M. Mattoso, Parallel OLAP query processing in database clusters with data replication. Distrib. Parallel Databases **25**(1–2), 97–123 (2009)
71. Z.H. Liu, D. Gawlick, Management of flexible schema data in RDBMSs - opportunities and limitations for NoSQL, in *Proceedings of the CIDR* (2015)
72. S. Luján-Mora, J. Trujillo, A comprehensive method for data warehouse design, in *Proceedings of the DMDW* (2003)

73. S. Luján-Mora, J. Trujillo, I. Song, A UML profile for multidimensional modeling in data warehouses, in *DKE* (2006, in press)
74. H. Mahboubi, XML warehousing and OLAP, *Encyclopedia of Data Warehousing and Mining*, 2nd edn. (IGI Global, Hershey, 2009), pp. 2109–2116
75. A.S. Maniatis, P. Vassiliadis, S. Skiadopoulos, Y. Vassiliou, G. Mavrogonatos, I. Michalarias, A presentation model & non-traditional visualization for OLAP. IJDWM **1**(1), 1–36 (2005)
76. S. Mansmann, M.H. Scholl, Extending visual OLAP for handling irregular dimensional hierarchies, in *Proceedings of the DaWaK* (2006), pp. 95–105
77. P. Marcel, E. Negre, A survey of query recommendation techniques for data warehouse exploration, in *Proceedings of the EDA* (2011), pp. 119–134
78. G. Marketos, Y. Theodoridis, Ad-hoc OLAP on trajectory data, in *International Conference on Mobile Data Management (MDM)* (2010), pp. 189–198
79. A. Marotta, A.A. Vaisman, Rule-based multidimensional data quality assessment using contexts, in *Proceedings of the DaWaK* (2016), pp. 299–313
80. J. Mazón, J. Trujillo, M. Serrano, M. Piattini, Designing data warehouses: from business requirement analysis to multidimensional modeling, in *Proceedings of the International Workshop on Requirements Engineering for Business Needs and IT Alignment* (2005)
81. J.N. Mazón, J. Lechtenbörger, J. Trujillo, A survey on summarizability issues in multidimensional modeling. DKE **68**(12), 1452–1469 (2009)
82. A.O. Mendelzon, A.A. Vaisman, Temporal queries in OLAP, in *Proceedings of the VLDB* (2000), pp. 242–253
83. R.B. Messaoud, S. Rabaséda, O. Boussaid, F. Bentayeb, OpAC: a new OLAP operator based on a data mining method, in *Proceedings of the DB & IS* (2004), pp. 417–420
84. M.A. Naeem, G. Dobbie, G. Weber, S. Alam, R-MESHJOIN for near-real-time data warehousing, in *Proceedings DOLAP* (2010), pp. 53–60
85. V. Nebot, R.B. Llavori, J.M. Pérez-Martínez, M.J. Aramburu, T.B. Pedersen, Multidimensional integrated ontologies: a framework for designing semantic data warehouses. J. Data Semant. XIII **13**, 1–36 (2009)
86. T. Niemi, J. Nummenmaa, P. Thanisch, Constructing OLAP cubes based on queries, in *Proceedings of the DOLAP* (2001), pp. 9–15
87. T. Niemi, J. Nummenmaa, P. Thanisch, Logical multidimensional database design for ragged and unbalanced aggregation, in *Proceedings of the DMDW* (2001), p. 7
88. P. O'Neil, G. Graefe, Multi-table joins through bitmapped join indices. SIGMOD Record **24**(3), 8–11 (1995)
89. C. Ordonez, A. Gurram, N. Rai, Recursive query evaluation in a column DBMS to analyze large graphs, in *Proceedings of the DOLAP* (2014), pp. 71–80
90. S. Orlando, R. Orsini, A. Raffaetà, A. Roncato, C. Silvestri, Trajectory data warehouses: design and implementation issues. J. Comput. Sci. Eng. **1**(2), 211–232 (2007)
91. L. Oukid, O. Asfari, F. Bentayeb, N. Benblidia, O. Boussaid, CXT-cube: contextual text cube model and aggregation operator for text OLAP, in *Proceedings of the DOLAP* (2013), pp. 27–32
92. D. Papadias, P. Kalnis, J. Zhang, Y. Tao, Efficient OLAP operations in spatial data warehouses, in *Proceedings of the SSTD* (2001), pp. 443–459
93. T.B. Pedersen, C.S. Jensen, C.E. Dyreson, A foundation for capturing and querying complex multidimensional data. Inf. Syst. **26**(5), 383–423 (2001)
94. L. Pipino, Y.W. Lee, R.Y. Wang, Data quality assessment. Comm. ACM **45**(4), 211–218 (2002)
95. J. Pokorný, XML data warehouse: modelling and querying, in *Proceedings of the DB & IS* (2002), pp. 267–280
96. T. Priebe, G. Pernul, A pragmatic approach to conceptual modeling of OLAP security, in *Proceedings of the ER* (2001), pp. 311–324
97. F. Ravat, O. Teste, A temporal object-oriented data warehouse model, in *Proceedings of the DEXA* (2000), pp. 583–592

98. S. Rivest, Y. Bedard, M.J. Proulx, M. Nadeau, SOLAP: a new type of user interface to support spatiotemporal multidimensional data exploration and analysis, in *Proceedings of the ISPRS Joint Workshop on Spatial, Temporal and Multi-Dimensional Data Modeling and Analysis* (2003)
99. S. Rizzi, A. Abelló, J. Lechtenbörger, J. Trujillo, Research in data warehouse modeling and design: dead or alive?, in *Proceedings of the DOLAP* (2006), pp. 3–10
100. S. Rizzi, E. Gallinucci, M. Golfarelli, A. Abelló, O. Romero, Towards exploratory OLAP on linked data, in *Proceedings of the SEBD* (2016), pp. 86–93
101. O. Romero, A. Abelló, Multidimensional design by examples, in *Proceedings of the DaWaK* (2006), pp. 85–94
102. O. Romero, A. Abelló, Automating multidimensional design from ontologies, in *Proceedings of the DOLAP* (2007), pp. 1–8
103. O. Romero, A. Abelló, A survey of multidimensional modeling methodologies. IJDWM **5**(2), 1–23 (2009)
104. A. Rosenthal, E. Sciore, View security as the basis for data warehouse security, in *Proceedings of the DMDW* (2000), p. 8
105. C. Sapia, M. Blaschka, G. Höfling, B. Dinter, Extending the E/R model for the multidimensional paradigm, in *Proceedings of the ER Workshop on Data Warehousing and Data Mining* (1998), pp. 105–116
106. S. Sarawagi, G. Sathe, i^3: Intelligent, interactive investigation of OLAP data cubes, in *Proceedings of the SIGMOD* (2000), p. 589
107. L.C. Scabora, J.J. Brito, R.R. Ciferri, C.D. de Aguiar Ciferri, Physical data warehouse design on NoSQL databases, in *Proceedings of the ICEIS* (2016), pp. 111–118
108. M. Serrano, C. Calero, J. Trujillo, S. Luján-Mora, M. Piattini, Empirical validation of metrics for conceptual models of data warehouses, in *Proceedings of the CAiSE* (2004), pp. 506–520
109. M. Sifer, A visual interface technique for exploring OLAP data with coordinated dimension hierarchies, in *Proceedings of the CIKM* (2003), pp. 532–535
110. Y. Sismanis, A. Deligiannakis, Y. Kotidis, N. Roussopoulos, Hierarchical dwarfs for the rollup cube, in *Proceedings of the DOLAP* (2003), pp. 17–24
111. I. Spiegler, R. Maayan, Storage and retrieval considerations of binary data bases. Inf. Process. Manag. **21**(3), 233–254 (1985)
112. K. Stefanidis, E. Pitoura, P. Vassiliadis, Adding context to preferences, in *Proceedings of the ICDE* (2007), pp. 846–855
113. C. Stolte, D. Tang, P. Hanrahan, Polaris: a system for query, analysis, and visualization of multidimensional relational databases. TVCG **8**(1), 52–65 (2002)
114. S.Y. Sung, Y. Liu, H. Xiong, P.A. Ng, Privacy preservation for data cubes. KAIS **9**(1), 38–61 (2006)
115. K. Techapichetvanich, A. Datta, Interactive visualization for OLAP, in *Proceedings of the ICCSA* (2005), pp. 206–214
116. D. Theodoratos, T. Sellis, Designing data warehouses. DKE **31**(3), 279–301 (1999)
117. D. Theodoratos, M. Bouzeghoub, Data currency quality satisfaction in the design of a data warehouse. IJCIS **10**(03), 299–326 (2001)
118. A. Thusoo et al., Hive: a warehousing solution over a map-reduce framework. Proc. VLDB Endow. **2**(2), 1626–1629 (2009)
119. R. Torlone, Two approaches to the integration of heterogeneous data warehouses. Distrib. Parallel Databases **23**(1), 69–97 (2008)
120. J. Trujillo, E. Soler, E. Fernández-Medina, M. Piattini, An engineering process for developing secure data warehouses. IST **51**(6), 1033–1051 (2009)
121. A. Vaisman, E. Zimányi, A multidimensional model representing continuous fields in spatial data warehouses, in *Proceedings of the SIGSPATIAL* (2009), pp. 168–177
122. A. Vaisman, A. Mendelzon, W. Ruaro, S. Cymerman, Supporting dimension updates in an OLAP server, in *Proceedings of the CAiSE* (2002), pp. 67–82
123. P. Valduriez, Join indices. TODS **12**(2), 218–246 (1987)

124. J. Varga, O. Romero, T.B. Pedersen, C. Thomsen, SM4AM: a semantic metamodel for analytical metadata, in *Proceedings of the DOLAP* (2014), pp. 57–66
125. P. Vassiliadis, T.K. Sellis, A survey of logical models for OLAP databases. SIGMOD Record **28**(4), 64–69 (1999)
126. P. Vassiliadis, A. Simitsis, P. Georgantas, M. Terrovitis, S. Skiadopoulos, A generic and customizable framework for the design of ETL scenarios. Inf. Syst. **30**(7), 492–525 (2005)
127. L. Wang, D. Wijesekera, S. Jajodia, Cardinality-based inference control in data cubes. J. Comput. Secur. **12**(5), 655–692 (2004)
128. W. Wang, H. Lu, J. Feng, J.X. Yu, Condensed cube: an efficient approach to reducing data cube size, in *Proceedings of the ICDE* (2002), pp. 155–165
129. A. Weininger, Efficient execution of joins in a star schema, in *Proceedings of the SIGMOD* (2002), pp. 542–545
130. R. Winter, B. Strauch, A method for demand-driven information requirements analysis in data warehousing projects, in *Proceedings of the HICSS* (2003), pp. 1359–1365
131. D. Xin, J. Han, P-cube: answering preference queries in multi-dimensional space, in *Proceedings of the ICDE* (2008), pp. 1092–1100
132. W.P. Yan, P.B. Larson et al., Eager aggregation and lazy aggregation, in *Proceedings of the VLDB*, vol. 95 (1995), pp. 345–357
133. J. Zhang, S. You, L. Gruenwald, High-performance online spatial and temporal aggregations on multi-core CPUs and many-core GPUs, in *Proceedings of the DOLAP* (2012), pp. 89–96

A Short Account of Techniques for Assisting Users in Mastering Big Data

Davide Martinenghi, Elisa Quintarelli, Fabio A. Schreiber and Letizia Tanca

Abstract One of the most challenging problems faced by the database community is to assist inexperienced or casual users, who need the support of a sophisticated system that guides them in making sense of the data. This problem becomes especially relevant in the case of Big Data, where the amount of data may quickly overwhelm users and discourage them from leveraging the richness of the data patrimony. In the last years, often in collaboration with other members of the Italian database community, we have developed several different techniques whose aim is both to reduce the size of the problem and to focus on the information that is most relevant to the user. To this end, most of these techniques fruitfully extract and exploit data semantics, for example by succinctly characterizing data via intensional properties such as integrity constraints or by tailoring the answer to the user context or preferences. Other techniques support the users in information exploration, for instance by extracting data not readily accessible (such as the Hidden Web) or by presenting them with appropriate summaries and suggesting possible exploration paths.

1 Introduction

One of the most challenging problems faced by the database community is to assist inexperienced or casual users. Such users typically need the support of a sophisticated system that guides them in making sense of data. Indeed, the growth of information, if not properly controlled, leads to a data overload that may cause confusion rather than knowledge, and dramatically reduce the benefits of a rich information ecosystem.

D. Martinenghi (✉) · E. Quintarelli · F.A. Schreiber · L. Tanca
DEIB - Politecnico di Milano, Piazza Leonardo da Vinci 32, 20133 Milano, Italy
e-mail: Davide.Martinenghi@polimi.it

E. Quintarelli
e-mail: Elisa.Quintarelli@polimi.it

F.A. Schreiber
e-mail: Fabio.Schreiber@polimi.it

L. Tanca
e-mail: Tanca.Letizia@polimi.it

S. Flesca et al. (eds.), *A Comprehensive Guide Through the Italian Database Research Over the Last 25 Years*, Studies in Big Data 31, DOI 10.1007/978-3-319-61893-7_7

This problem becomes especially relevant in the case of Big Data, where the amount of data itself quickly overwhelms users and discourages them from leveraging the knowledge potential of the data patrimony.

In the last years, often in collaboration with other members of the Italian database community, we have developed different techniques whose aim is both to reduce the size of the problem and to focus on the information that is most relevant to the user. To this end, most of these techniques fruitfully extract and exploit data semantics, either by succinctly characterizing data via *intensional* aspects such as integrity constraints or statistical properties, or by *taming the Big data beast* by reducing, in various ways, the amount of data provided by the query answer.

The techniques described in Sect. 2 fall in the first category. In mathematics, a description is called *intensional* when it exhibits the properties of the data instead of the data themselves. Intensional descriptions are often much more compact than extensional ones, and are typically employed to express semantic properties of the data such as *integrity constraints*, i.e., conditions that must always be satisfied for the data to be considered consistent. Such conditions may be as simple as primary and foreign keys, or involve nontrivial requirements that capture complex dependencies and "business logic", as often happens in real-world applications, and even more so with Big Data.

Preserving the compliance of data with integrity constraints is a crucial issue: without it, answers to queries cannot be trusted. Usually, integrity constraints are intentionally stated by the data designer or analyst to convey the semantics of the data better; as such, their validity ought to be maintained in the face of data updates. This can be achieved in two different ways: (i) by acting on the data, preventing updates that might impair consistency or repairing inconsistencies that have occurred; (ii) by acting on the constraints themselves, modifying them in the face of a change in the modeled reality that requires a change in the constraints too. This research is described in Sect. 2.1.

On the other hand, being a mirror of the data semantics, intensional information may be used to improve query answering in terms of both performance and informativeness. In some cases, constraints may be extracted from the data to gain useful insight on the information at hand, for instance under the form of succinct characterizations of the data, explanations, or informative answers. Constraints may also be of a structural kind, as is done, e.g., to model access to Web sources, and to support users in information exploration, for instance by extracting data not readily accessible (such as the Hidden Web) or by presenting them with appropriate summaries and suggesting possible exploration paths. All these topics are discussed in Sect. 2.2, while Sect. 2.3 introduces yet another way to leverage intensional information: database exploration and exploratory computing. This approach may be described [7] as *the step-by-step conversation* of a user with the system, where each step can refine the previous ones incrementally, gathering new knowledge that fulfills the user needs, sometimes even unveiling new ones. At each step, the answer to a user query is an *approximate, intensional* description of the data, which highlights relevant aspects at a glance, permitting to see the data at a higher level of abstraction and to decide to take one or more further actions.

Big Data are usually associated with large Databases in Information Systems, with Web pages and other rather static multimedia information. However, the new information sources that are emerging add new perspectives to this scenario: the Internet of Things (IoT), Wireless Sensor Networks (WSN), RFIDs, and social networks are generating large amounts of data that continuously and autonomously flow from the sources in streams and must be processed on the fly [22]. The following sections deal with the attempt to provide users, who issue queries to large masses of such diverse data, with reduced, manageable answers.

Section 3 discusses those cases in which datasets are very large and queries are not sufficiently selective, thereby flooding users with results. Section 3.1 describes methods to efficiently focus on the best ("top-k") answers to a query, according to a user's needs and preferences. Fitting the data to the application and user needs can be compared to the activity of fitting a dress to a person, and thus we refer to it as *data tailoring*. The current user context and her preferences can be used as scissors to tailor data, possibly assembled and integrated from many data sources. In Sect. 3.2 we describe how the information about the current user context and preferences supports the system in providing users exactly with the information they need in the different situations, when they are alone or in groups.

Wireless sensor networks produce high frequency data requiring smart processing methods. A smart sensor query-processing architecture can dramatically improve, energy, space and time efficiency of sensor data processing, by embedding the query processing power in the network itself, thus relieving the external database from the burden of processing (and often of storing) the raw data. In Sect. 3.3 we describe some advances on data management in these *Pervasive Systems*.

2 Intensionally, Intentionally

Integrity constraints are a very good example of intensional information that is intentionally stated at design time and contributes to supporting the "health" of a database; other intensional information, on the other hand, may be extracted a-posteriori, by analyzing the data to the end of providing the users with succinct descriptions of the database content. In this section we provide a short account of the main techniques for leveraging intensional information.

2.1 Managing Integrity Constraints

As data evolve through updates, integrity may be compromised. Traditional systems *check* whether updates preserve data integrity and then *maintain* it by rejecting harmful updates. Fully checking each update for consistency is typically too time consuming; fortunately, a number of techniques exist for optimizing integrity checking. The main idea is that, if integrity holds before the update, then checking integrity after the update can be done *incrementally* [44]. As an example, consider a functional dependency stating that the ISBN of a book univocally determines its title. Checking that no two books have the same ISBN and different titles typically takes quadratic time in the number of books. Consider now the insertion of a new book b in the collection. If the constraint holds right before the insertion, then it suffices to check whether the updated collection contains a book with the same ISBN as b but a different title, which can be done in linear time. Extensions of this idea are surveyed in [34]. Notably, some works [18, 20, 33] check integrity even *before* executing the update itself, thus saving unnecessary effort in case the update violates integrity. For instance, in the previous example, we do not need to execute the insertion to check whether the collection already contains another book with the same ISBN as b but a different title.

In spite of a long-standing presence in standard query languages and a wide acceptance of their importance, non-trivial constraints are frequently disregarded at the data management level and, often, their maintenance is relegated to the application level. One reason for this discrepancy between theory and practice in the field is that, in real scenarios with large amounts of data, it is not always possible to check constraints at all times, and some violations of integrity are unavoidable. This may happen, e.g., if data come from different, partly conflicting sources integrated over a network [19]. Consider, for instance, a large database containing VAT percentages: in Italy, for each kind of goods this percentage is the same all over the country, while in India it is different in different states thus, in the integrated database, the pair *(COUNTRY, PRODUCT KIND)* does not univocally determine the VAT percentage. In this and similar contexts, full integrity before each update is too strong a requirement.

An important step to mitigate this issue was taken by showing how optimized integrity checking can be performed in an *inconsistency-tolerant* way, i.e., without requiring full data integrity before the update [23]. Several traditional methods used for optimizing integrity checking turn out to be inconsistency-tolerant. Any such method can then be used to guarantee that, starting from a possibly inconsistent dataset, the update will not increase the measured amount of inconsistency [28], although some violations of integrity might still be present.

A common problem with very large datasets is that part of the data may be missing. To this end, schema models provide a useful characterizations that may help to infer relevant information. Common schema models, including the ER model and extensions thereof, can be suitably represented by integrity constraints. With *incomplete data*, schema reasoning may provide some of the missing pieces to answer a query. As an example, consider the sets player (player-team pairs) and team (team-city pairs), and a constraint stating that the teams in player ought to be a subset of those in team. The following pairs are known: ⟨*Abate*, *ACMilan*⟩, ⟨*Totti*, *Roma*⟩ (for player), and ⟨*ACMilan*, *Milan*⟩ (for team). From the constraint, we infer that *Roma* is the name of a team associated with some city not mentioned in the dataset: if we ask the names of the teams in team, the *certain answers* are then *ACMilan* and

Roma, although the latter is never mentioned in the team relation. Query answering under dependencies usually resorts to techniques such as query rewriting [12] and the so-called *chase* [32], which is intuitively a representative of all datasets that satisfy the constraints and are a superset (i.e., a possible completion) of the initial data.

Once a database designer or administrator is able to understand that a constraint no longer holds, s/he can decide what to do. While the techniques described above re-establish consistency by changing the data that violate the constraints, another possible approach is based on the fact that frequent constraint violations may suggest that the represented reality is changing, and thus the database does not reflect it any longer. Therefore, once the DB administrator has ascertained that this is the case, s/he will be able to re-establish consistency by appropriately *modifying the violated constraints* instead of modifying the data.

A typical case where a constraint modification might be needed is a change in government policies: for example, in Italy a recent change in the school regulations allowed 5-year-old children to access primary school, which was forbidden before. Accordingly, the constraint saying that only 6-year-old pupils can be enrolled must be relaxed.

Different kinds of constraints may need different repair techniques, which however share the basic principle of applying data mining and machine learning techniques. For instance, a violated functional dependency (FD) like the one about VAT is repaired by adding, to its antecedent, a minimal set of attributes that makes them consistent with the data [17, 41]; in the case of the example, the dependency $COUNTRY,\ PRODUCTKIND \rightarrow VAT\%$ would be replaced with $COUNTRY,\ PRODUCTKIND,\ REGION \rightarrow VAT\%$. On the other hand, the violation of a tuple constraint like the one about primary school pupils can be repaired by finding appropriate disjuncts that, added to the constraint, relax the condition on the relation tuples [39]. In the specific example, this disjunct might specify the condition that the child be born after a certain year.

2.2 Leveraging Integrity Constraints

Access patterns [49] are structural constraints indicating the input/output mode of the fields of a relation. In this respect, access patterns suitably characterize several contexts relevant to Big Data, such as the Hidden Web [13], i.e., that part of the Web (the largest one) that is reachable only after filling in and submitting a Web form. For instance, a search form for hotels commonly includes input fields for the desired city and check-in date, and, after completion, outputs tuples with hotel name, price and other details. The Hidden Web may be simply modeled by decorating standard relations with access patterns; query processing, however, requires specialized techniques. Consider a Web site listing musical events, with a form (events) that receives a city as input and returns the name of the artist(s) performing in that city, and another form (bio) that receives an artist name as input and returns the city and country of birth of the artist. To request the names of Italian artists performing in Milan, one can first use events with input "Milan", then feed bio's input with the output returned by events, and finally check that the country output by bio is indeed Italy. Sometimes, however, query answering is more involved. Take, e.g., a query requesting the names of artists born in Milan. There is no direct way to feed bio's input field with a known value; yet, the city mentioned in the query ("Milan") can be fed to events, thus providing artist names as output; in turn, these names can be fed to bio to retrieve values for nations and cities; the new city names can be used to access events again, and so on. This possibly lengthy discovery process consists in disclosing all the content of the relations that can be extracted with the given initial knowledge; at the end of the process, when no new values can be discovered, the original query can be evaluated over the data retrieved so far. The answers obtained in this way, called *reachable certain answers*, are in general only a subset of the answers that would have been found without access patterns. Several optimization techniques may be exploited to make this lengthy process converge to a result faster [9–11], even when queries are expressed by means of simple keywords [14].

Traditional information search is shifting towards scenarios in which schemas are vague or even absent, and data comes from heterogeneous sources. Query answering needs then to be adapted to match user requests with accessible data. When additional information about the data is available (e.g., in the form of a taxonomy or contextual information), extensions of relational algebra addressing these issues become possible [38]. For instance, if a user looks for concerts in Milan, but locations are stored at the level of concert halls, suitable results may only be returned if we have extra knowledge about, say, La Scala being in Milan.

2.3 Database Exploration

The challenges posed by the Big Data phenomenon have promoted the discipline of database exploration [29, 56], a paradigm that takes different shapes, among which Exploratory Computing [4] and Faceted Search [57].

The aim of exploratory computing – a kind of database exploration – is to assist mainly inexperienced or casual users, the so-called data enthusiasts [43], who need the support of a sophisticated system that guides them in the inspection of a dataset that is complex and very large, thus difficult to grasp. The process may start from simple input queries, and guide the user in a reconnaissance path where, at each step, the system presents the users with relevant and concise facts about the data, in order for them to understand better and progress to the next exploration step. For example, while exploring a historical medical dataset, the distribution of the age values in the result of the query Q1 "find all patients whose thyroid function tests are out of range" might be different from that in the original dataset. This is a relevant fact, maybe allowing the investigator to spot some relationship between age and thyroid disorders. After this step the user might ask another query, like Q2 "find all patients whose thyroid function tests are out of range and who are over 60 years old", to see whether the patients of that age group exhibit any special characteristics. We may thus describe our exploration [7] as *the step-by-step conversation of a user with the system, where each step can refine the previous ones incrementally, gathering new knowledge that fulfills the user needs, sometimes even unveiling new ones.* In this kind of exploration, the notion of relevance is of paramount importance, since the main system task is to call the user attention to relevant (or surprising) differences or similarities with the other datasets encountered along the route.

The use of statistical summaries like distributions may well be employed to highlight relevant aspects at a glance, permitting to see the data at a higher level of abstraction and decide to take one or more further actions. In fact, a distribution can describe a dataset in an *intensional, though approximate way*. Computing data distributions is one of many ways to assess relevance by looking at concise descriptions: we may add other measures, like entropy, which establishes the "level of variety" of a set of values, or describe the data by means of mined patterns like association rules [3], used to discover relationships between attribute values, etc. For instance, in the result of query Q2, the system might discover that (i) "80% of the patients from Lombardia lived between the Second World War and the sixties in the area of Valtellina" and (ii) "70% of patients from Southern Italy are from a seaside location". In the same fashion as in Inductive DBs, these intensional descriptions can be stored to be queried when needed, or computed on the fly to obtain answers that are fast and concise, though potentially partial and approximate. Of course, computing them on the fly entails the need of fast computation techniques, which constitutes one of the big challenges for DB researchers.

Following the Exploratory Computing paradigm, we have contributed to realizing two systems, INDIANA [8][1] and IQ4EC [40], both aiming at supporting the exploration of large databases by non-expert or *data-enthusiastic* users. While sharing a common goal, these two systems differ in the adopted approaches, core mechanisms and outcomes, hence they represent two valuable and up-to-date points of view on Exploratory Computing. The INDIANA project [8] uses statistical summaries – like empirical distributions, or entropy, or histograms – to exhibit the data properties instead of the data themselves, permitting the user to see the data at a higher level of abstraction. In fact, a distribution is an example of how we can describe a dataset in an intensional, though approximate way.

With IQ4EC [40], instead, data mining techniques are applied to mine intensional knowledge presented in the form of association rules that are extracted and stored into a relational database, which acts as an *intensional repository*. Users can engage the IQ4EC system in a highly interactive dialog, where the IQ4EC answers may be intensional or extensional depending on the user needs.

[1]This particular research is carried out in collaboration with Università della Basilicata.

3 Taming the Beast

When datasets are very large and queries are not sufficiently selective, users are flooded with results. In this case, the richness of the data may produce a paradoxical effect, confounding the users, who are not able to choose among the plethora of data that is presented to them. This section gives an account of some techniques that can be used to "tame the Big-Data beast".

3.1 Crowdsourcing and top-k Queries

To avoid these "flooding" situations, users sometimes set a limit on the number of results, focusing on the k most relevant matches for their queries. Relevance is usually expressed as a combination of different numeric attributes into an aggregate score (think, e.g., of ranking hotels by considering both rating and price). It is very expensive to compute *all* the query results first, and then to sort them by relevance, especially if sources are remote. However, if the sources are endowed with special access modes (e.g., return the data in batches sorted by a given attribute), then the top k results may be found by retrieving only a small fraction of the available tuples (for examples and foundational works, see [30]).

Several extensions of the "top-k" model are available. Spatial proximity queries [35, 37] regard objects equipped with both a score and a feature vector, which may be used to represent, e.g., the object location or other aspects. An example of this scenario is a search for a restaurant and a theater that are close to the user, to each other, and recommended, respectively, in terms of price and rating. In the same setting, one may additionally wish to diversify the result set, yet retaining only results with high scores [15, 26]. When multiple sources are joined, and special access modes are available, suitable execution strategies can be devised so as to further speed up the computation of the top k results [36].

Often, users are unable to precisely specify the criteria used to rank the results of a query. For instance, we may want to rank hotels by a weighted sum of rating and price without knowing the precise weights; yet, we might consider the price more important than the rating. Albeit easier to express, such partial specifications may determine uncertainty and give rise to multiple possible rankings of the results, each associated with a probability of being correct [54]. Whenever the results are too vague or uncertain, one can try to leverage the so-called "wisdom of the crowd" to eliminate ambiguities in the ranking of objects. This can be done on a crowdsourcing platform by assigning simple tasks (such as selecting the best of two objects) to a crowd of "workers" [16] and then discarding the rankings that are incompatible with their indications [21].

3.2 Taming Data by Means of Contexts and Preferences

In the last decade we worked at helping end users, by means of a different approach to the problem of information overload,[2] proposing a design model and associated methodology for context-aware databases [5]. By context-awareness [2], each user is served, in each situation of use (context), the information that is most appropriate, thus avoiding disorientation in the face of an answer that is too large to be of use. Natural examples of context-aware tools are those apps that suggest a restaurant that is located around you, that is, in your *physical context*; however, taking a richer perspective on context, such an app might also suggest restaurants that are appropriate for children when you are with your family, or jazz concerts after having learned that you are a jazz fan.

In the early years of this research, we first proposed a context model that allows the reduction of large datasets on the basis of some perspectives (dimensions) describing the situation (i.e., the *context*) in which the user is involved; typical such dimensions are, for example, the user's role (it may be directly asked to the user), and her location (collected through the GPS sensor of the user's device). More sophisticated context parameters can also be introduced, like the current activity of the user, or her main interest topic. For example, in the scenario of apps suggesting restaurants the conjunction $\texttt{user} = \texttt{adult} \wedge \texttt{situation} = \texttt{withFriends} \wedge \texttt{time} = \texttt{daytime}$ describes the context of an adult and her friends looking for a restaurant where to have lunch. The activity of selecting the relevant information (over a global, centralized or not, database) for a target application or user, in a specific context, is called *data tailoring*; the CARVE methodology [6] requires that the designer manually specifies, at design time, both a *context schema* – identifying the possible contexts the user may find herself in at run-time –, and the way the information relevant to each context must be tailored. This is done by associating each context with a database view that is materialized at run time, when the system presents the user only with data useful for her currently active context.

However, with a large variety of data and a considerable number of possible contexts, the manual specification of these *contextual views* may be a trying experience, discouraging the designer; moreover, such an approach cannot take into account the evolution of user tastes or situations, which may generate, during the system life, changes in the user interests and needs.[3]

A way around this problem is to apply data mining algorithms to the results of the user past querying activity, in order for the system to learn which data are appropriate for each possible context [27]. Data mining is applied to extract association rules, which describe frequent co-occurrence of data items in the collected data. We have investigated two methodologies that approach the discovery of contextual views following two opposite philosophies: the first one mines rules of the form *context*

[2]Information overload is the difficulty of people in understanding an issue in the presence of too much information. The term is used by Alvin Toffler in his books *Future Shock* and *The Third Wave*.

[3]*Context evolution* [47] is the research topic that takes this into account; however, if this task is performed by the designer, it makes his or her burden even heavier.

→ *data features*, with the aim of learning which data features have been the most popular in the various contexts. The second approach mines rules in the opposite form, i.e. *data features* → *context*: it starts from the past accessed data trying to understand whether their features are or are not relevant to the different contexts.

In general, data personalization based on context may be only a partial solution, since the tailoring of the available dataset may still be too coarse-grained; indeed, the designer specifies views for classes of users, situations, interest topics. For example, if we consider Bob, an adult interested in having lunch with friends, a contextual system will suggest restaurants close to Bob's location, but will not be able to rank this contextual data according to Bob's personal tastes: for example, ranking gluten-free restaurants as first if Bob is allergic to gluten. To obtain a more effective personalization, the notion of context is enriched with the user personal preferences [55]. In particular, we use data mining to learn the contextual preferences of the users on both tuples and attributes of databases [42], thus the set of data associated with each context, is ordered by their relevance for the user to "recommend" the highest-ranked data. Since a lot of activities are inherently social, generally carried out by groups of users (e.g. people rarely eat out alone), we have also investigated the problem of inferring contextual preferences for ephemeral groups [1], i.e. groups constituted by people together for the first time. In this scenario, when considering the preferences of the members of a group, not all the users should have the same weight: some users might indeed have a greater influence in determining the final decision. For example, in a group of people going out for lunch, a person with allergies can have more influence in the decision process. In [48] we have proposed a novel aggregation function, taking influence into account but considering also contextual influence, because influence weights may vary with respect to the situation, i.e., the context, that the group is currently living in.

3.3 Reducing the Abundance of Sensor Data

In Pervasive Systems, such as those used in health-care applications, smart buildings, access control, and others, large amounts of data are autonomously produced by peripheral devices and must be filtered and processed, either in batch, after it has been collected in a database, or in real-time, through data stream management systems [46].

The first layer of a Pervasive System, such a WSN, which can be seen as the backbone of an IoT, consists of sensors, whose goal is to measure physical quantities from the environment and transform them into signals (either analog or digital) that drive outputs/actuators or feed Data analysis tools [24, 25]. The choice of sensors in a WSN is application dependent and strictly related to the physical phenomena to be monitored, thus resulting in the complexity and the heterogeneity of the devices that have been designed and developed to acquire data.

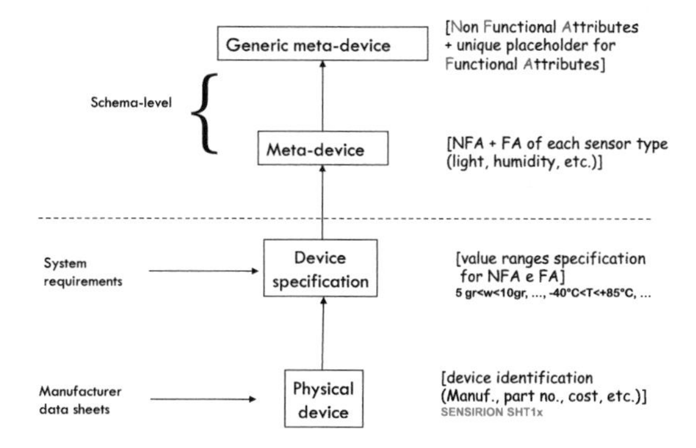

Fig. 1 The abstraction process to define the generic meta-device

To address this problem, we refer to a typical practice in the Database world: *Abstraction*. The data-centric view approach presented in Fig. 1 allows one to hide the heterogeneity of devices to the application designers by means of sensor attribute generalization. To achieve this goal, both functional and non functional attributes are abstracted from the specific physical sensor, whose description is given in the sensor data-sheet, to a generic meta-device. This process results in the formulation of an XML Document Type Definition (DTD) for the Generic meta-device [52].

Note that the word "sensor" does not necessary refer only to physical hardware devices, but it can be extended to every source of data providing information of interest. For instance, the location of people can be determined by means of tracking systems through GPS location (physical devices) as well as by looking at their electronic calendars or by analyzing travel bookings or emails (non-physical sources of information), which are also referred to as *virtual sensors*, without requiring a real measurement of their positions. *Logical sensors* can be defined, to aggregate data coming from both a physical sensor and a virtual sensor as well as information coming from a database [2].

The abstraction process can go further, from single components to the entire monitoring system, producing relational tables that describe the sensors and the values they measure, along with their interactions and the network topology [52].

In principle, data could be directly collected from the sensor sources and stored in a Database for further processing; however, operating in this way would be very inefficient owing to the poor quality of incoming data (redundancy, noise, data losses, etc.) which also results in energy waste, energy being a precious resource when tiny, battery powered devices are involved. Practically, the most frequent query mode in sensor networks is continuous query operation over the data streams; queries are downloaded to the peripheral devices together with some time- or event-based measurement conditions and data are collected accordingly with no need for further polling actions. In [31], Madden et al. clearly state the critical issues to be faced for processing a query in a WSN: (i) find out the sensors that have data relevant to the query; (ii) when sampling for answering the query should occur; (iii) in which order the samples should be taken and how to schedule sampling with respect to other operations; (iv) evaluating the trade-off between the lower energy consuming local computations and the higher energy required to send data to a collecting station.

Given the specific characteristics of the data handled by WSNs [46], a number of efforts have been made to define and implement High Level Languages for managing data in WSN applications, privileging the "what to do" with respect to the "how to do" approach and preferring declarative languages to procedural ones; such a choice also allows an easier in-network distributed data processing managed at the middleware level.

Referring to a simple RFID system [58], let us suppose that, by entering a large archaeological area, each visitor is endowed with a bracelet containing an RFID tag and that readers be located at the most interesting points of the site. The site manager could be interested both in monitoring the visit paths followed by each visitor (i.e.: the readers encountered by a specific tag) and how long do visitors stand by a point of interest (i.e.: the time between two consecutive readings of the same tag by the same reader). Two simple SQL queries can very easily answer the manager needs [52].

In order to query a pervasive system as if it were a database, we designed PerLa (PERvasive LAnguage),[4] a SQL like language. The language is supported by a middleware that masks the idiosyncrasies of the nodes employed in complex sensing networks and performs the run-time management of the system [51].

The goal of the language is to enable the end user to collect data in a fast and easy way, without dealing with low-level programming issues. The idea behind PerLa is to extend the database-like abstraction from a single WSN unit to a whole pervasive system and to provide full support for heterogeneity, both at runtime and at deployment time.

The language allows the user to interact with logical objects, called Functional Proxy Component (FPC), which implement the previously discussed abstraction process of the peripheral devices and data sources. An FPC can abstract both a single node and a set of devices (e.g. a whole WSN); FPCs have common and homogeneous interfaces, and are used by PerLa queries to access the data gathered from the network nodes. No knowledge of the nodes hardware and computational characteristics is needed to perform a PerLa query. Moreover, by means of the FPC abstraction, the language is not tied to any particular type of sensing device, including social networks and messaging systems.

PerLa support for pervasive systems extends to node developers as well. The addition of new sensing devices in an existing network is facilitated by a Plug and Play connection system, i.e. a runtime factory that generates all the software components needed to query new sensor nodes. The information required to automatically assemble a device driver is stored in an XML file provided by the node developer. This file - the Device Descriptor - details all the node characteristics in terms of data structures, protocols of communication, computational capabilities, and behavioral patterns.

Low Level Queries (LLQ) are the most original part of the language. In order to provide the needed expressiveness for managing the interaction with logical objects

[4]The PerLa web site - http://perlawsn.sourceforge.net/index.php.

(e.g. definition of the sampling parameters), LLQs are composed of three main sections [50]: (i) the *Sampling Section* specifies how and when the sampling operation should be performed; (ii) the *Data Management Section* manages sampled data to compute the query results. The syntax of this query section is similar to the standard SQL one, but some important differences exist due to the need of managing the data stream, through the definition of windows on an ideally infinite buffer, where all the sampled records are appended. (iii) The *Execution Conditions Section* defines the rules to establish if a certain logical object should participate to the query. This section is optional and, if not specified, all the logical objects in the system will be involved in the query execution.

Born as a simple WSN query language, the functionality of PerLa has evolved toward a true pervasive systems language. As a first step, PerLa has been extended in order to create and maintain contexts [53]. The relationship between context and data is mastered in two directions: while the sensors contribute by gathering the context related values, they themselves can be influenced by their context and react accordingly.

The following context manager expression creates a context named *HumidityTemperatureMonitoring*, associated with the `user = operator` and `risk = artwork_damage` dimension-value pairs. When this context occurs, since the artworks could be subject to possible injuries, the operator starts receiving monitoring data about humidity and temperature.

```
CREATE CONTEXT HumidityTemperatureMonitoring
ACTIVE IF user=operator AND risk=artwork_damage
ON ENABLE (HumidityTemperatureMonitoring)
    SELECT humidity, temperature, device_location
    SAMPLING EVERY 1m
ON DISABLE
    DROP HumidityTemperatureMonitoring
REFRESH EVERY 10m
```

Intuitively, the `ACTIVE IF` clause is a list of dimensions associated with their values, defining a context, the `ON ENABLE` clause specifies the behavior associated with the context while the `ON DISABLE` clause specifies the actions to be taken when the context is deactivated. The `REFRESH EVERY` clause specifies the rate with which these conditions have to be checked.

Further work is being carried on to enhance the expressive power of PerLa by including also calls to layers of a Context Oriented Programming language and to web services [45].

4 Conclusions

In this chapter we have proposed different methods to assist inexperienced or casual users in making sense of massive quantities of data. We have developed several techniques, either exploiting information about the data semantics, succinctly represented by integrity constraints or statistical summaries, or by tailoring or ranking query answers on the basis of various criteria, or even by processing data on the fly before they are definitely stored. We believe that these researches pave a long and interesting way towards more fruitful and pleasant interactions of the users with database systems.

References

1. S. Amer-Yahia, S. Basu Roy, A. Chawla, G. Das, C. Yu, Group recommendation: Semantics and efficiency. PVLDB **2**(1), 754–765 (2009)
2. M. Baldauf, S. Dustdar, F. Rosenberg, A survey on context-aware systems. Int. J. Ad Hoc Ubiquitous Comput. **2**(4), 263–277 (2007)
3. E. Baralis, P. Garza, E. Quintarelli, L. Tanca, Answering XML queries by means of data summaries. ACM Trans. Inf. Syst. **25**(3), 10 (2007)
4. N.D. Blas, M. Mazuran, P. Paolini, E. Quintarelli, L. Tanca, Exploratory computing: a challenge for visual interaction, in *AVI* (2014), pp. 361–362
5. C. Bolchini, C. Curino, E. Quintarelli, F.A. Schreiber, L. Tanca, A data-oriented survey of context models. SIGMOD Record **36**(4), 19–26 (2007)
6. C. Bolchini, E. Quintarelli, L. Tanca, Carve: Context-aware automatic view definition over relational databases. Inf. Syst. **38**(1), 45–67 (2013)
7. M. Buoncristiano, G. Mecca, E. Quintarelli, M. Roveri, D. Santoro, L. Tanca, Database challenges for exploratory computing. SIGMOD Record **44**(2), 17–22 (2015)
8. M. Buoncristiano, G. Mecca, E. Quintarelli, M. Roveri, D. Santoro, L. Tanca, Exploratory computing: What is there for the database researcher?, in *23rd Italian Symposium on Advanced Database Systems, SEBD 2015, Gaeta, Italy, 14–17 June 2015* (2015), pp. 128–135
9. A. Calì, D. Calvanese, D. Martinenghi, Dynamic query optimization under access limitations and dependencies. J. Univers. Comput. Sci. **15**(21), 33–62 (2009)
10. A. Calì, D. Martinenghi, Conjunctive query containment under access limitations, in *ER 2008* (2008), pp. 326–340
11. A. Calì, D. Martinenghi, Querying Data under Access Limitations, in *ICDE 2008* (2008), pp. 50–59
12. A. Calì, D. Martinenghi, Querying Incomplete Data over Extended ER Schemata. TPLP **10**(3), 291–329 (2010)
13. A. Calì, D. Martinenghi, Querying the deep web (tutorial), in *EDBT 2010* (2010), pp. 724–727
14. A. Calì, D. Martinenghi, R. Torlone, Keyword queries over the deep web, in *ER 2016* (2016), pp. 260–268
15. I. Catallo, E. Ciceri, P. Fraternali, D. Martinenghi, M. Tagliasacchi, Top-k diversity queries over bounded regions. TODS **38**(2), 10 (2013)
16. I. Catallo, S. Coniglio, P. Fraternali, and D. Martinenghi. A workload-dependent task assignment policy for crowdsourcing. *WWW J.*, to appear, 2017
17. F. Chiang, R. Miller, A unified model for data and constraint repair, in *ICDE 2011* (2011), pp. 446–457
18. H. Christiansen, D. Martinenghi, Simplification of database integrity constraints revisited: A transformational approach, in *LOPSTR 2003* (2004), pp. 178–197

19. H. Christiansen, D. Martinenghi, Simplification of integrity constraints for data integration, in *FoIKS* (2004), pp. 31–48
20. H. Christiansen, D. Martinenghi, On simplification of database integrity constraints. Fundam. Inform. **71**(4), 371–417 (2006)
21. E. Ciceri, P. Fraternali, D. Martinenghi, M. Tagliasacchi, Crowdsourcing for top-k query processing over uncertain data. TKDE **28**(1), 41–53 (2016)
22. F. Colace, M.D. Santo, V. Moscato, A. Picariello, F.A. Schreiber, L. Tanca (eds.), *Data Management in Pervasive Systems*, Data-Centric Systems and Applications (Springer, 2015)
23. H. Decker, D. Martinenghi, Inconsistency-tolerant integrity checking. TKDE **23**(2), 218–234 (2011)
24. K.R. Fowler, J. Schmalzel, Why do we care about measurement? IEEE Instrum. Meas. Mag. **7**(1), 38–46 (2004)
25. K.R. Fowler, J.L. Schmalzel, Sensors: The first stage in the measurement chain. IEEE Instrum. Meas. Mag. **7**(3), 60–65 (2004)
26. P. Fraternali, D. Martinenghi, M. Tagliasacchi, Top-k bounded diversification, in *SIGMOD 2012* (2012), pp. 421–432
27. P. Garza, E. Quintarelli, E. Rabosio, L. Tanca, Reducing big data by means of context-aware tailoring, in *New Trends in Databases and Information Systems - ADBIS 2016 Short Papers and Workshops, BigDap, DCSA, DC, Proceedings, Prague, Czech Republic, 28–31 August 2016* (2016), pp. 115–127
28. J. Grant, A. Hunter, Measuring inconsistency in knowledgebases. J. Intell. Inf. Syst. **27**(2), 159–184 (2006)
29. S. Idreos, O. Papaemmanouil, S. Chaudhuri, Overview of data exploration techniques, in *Proceedings of the 2015 ACM SIGMOD* (2015), pp. 277–281
30. I.F. Ilyas, G. Beskales, M.A. Soliman, A survey of top-*k* query processing techniques in relational database systems. ACM Comput. Surv. **40**(4) (2008)
31. S.R. Madden, M.J. Franklin, J.M. Hellerstein, W. Hong, Tinydb: an acquisitional query processing system for sensor networks. ACM Trans. Database Syst. **30**(1), 122–173 (2005)
32. D. Maier, A.O. Mendelzon, Y. Sagiv, Testing implications of data dependencies. TODS **4**, 455–469 (1979)
33. D. Martinenghi, Simplification of integrity constraints with aggregates and arithmetic built-ins, in *FQAS 2004* (2004), pp. 348–361
34. D. Martinenghi, H. Christiansen, H. Decker, Integrity checking and maintenance in relational and deductive databases - and beyond, in *Intelligent Databases: Technologies and Applications*, ed. by Z. Ma (2006), pp. 238–285. Chap. X
35. D. Martinenghi, M. Tagliasacchi, Proximity rank join. PVLDB **3**(1), 352–363 (2010)
36. D. Martinenghi, M. Tagliasacchi, Cost-aware rank join with random and sorted access. TKDE **24**(12), 2143–2155 (2012)
37. D. Martinenghi, M. Tagliasacchi, Proximity measures for rank join. TODS **37**(1) (2012)
38. D. Martinenghi, R. Torlone, Taxonomy-based relaxation of query answering in relational databases. VLDB J. **23**(5), 747–769 (2014)
39. M. Mazuran, E. Quintarelli, R. Rossato, L. Tanca, Mining violations to relax relational database constraints, in *DaWaK* (2009), pp. 339–353
40. M. Mazuran, E. Quintarelli, L. Tanca, IQ4EC: intensional answers as a support to exploratory computing, in *2015 IEEE International Conference on Data Science and Advanced Analytics, DSAA 2015, Campus des Cordeliers, Paris, France, 19-21 October 2015* (2015), pp. 1–10
41. M. Mazuran, E. Quintarelli, L. Tanca, S. Ugolini, Semi-automatic support for evolving functional dependencies, in *Proceedings of the 19th International Conference on Extending Database Technology, EDBT 2016, Bordeaux, France, 15-16 March 2016* (2016), pp. 293–304
42. A. Miele, E. Quintarelli, E. Rabosio, L. Tanca, A data-mining approach to preference-based data ranking founded on contextual information. Inf. Syst. **38**(4), 524–544 (2013)
43. K. Morton, M. Balazinska, D. Grossman, J.D. Mackinlay, Support the data enthusiast: Challenges for next-generation data-analysis systems. PVLDB **7**(6), 453–456 (2014)

44. J.-M. Nicolas, Logic for improving integrity checking in relational data bases. Acta Informatica **18**, 227–253 (1982)
45. E. Panigati, E.A. Schreiber, Context-aware software approaches: a comparison and an integration proposal (discussion paper), in *Proceedings of the 22nd Italian Symposium on Advanced database Systems* (2014), pp. 175–184
46. E. Panigati, E.A. Schreiber, C. Zaniolo, Data streams and data stream management systems and languages, in ed. by Colace, et al. [22], pp. 93–111
47. E. Quintarelli, E. Rabosio, L. Tanca, A principled approach to context schema evolution in a data management perspective. Inf. Syst. **49**, 65–101 (2015)
48. E. Quintarelli, E. Rabosio, L. Tanca, Recommending new items to ephemeral groups using contextual user influence, in *Proceedings of the 10th ACM Conference on Recommender Systems* (2016), pp. 285–292
49. A. Rajaraman, Y. Sagiv, J.D. Ullman, Answering queries using templates with binding patterns, in *PODS* (1995), pp. 105–112
50. F.A. Schreiber, R. Camplani, M. Fortunato, and M. Marelli, Design of a declarative data language for pervasive systems. Art Deco Technical Report R. A. 11.1b (2008), http://perlawsn.sourceforge.net/documentation.php?official=1
51. F.A. Schreiber, R. Camplani, M. Fortunato, M. Marelli, G. Rota, Perla: A language and middleware architecture for data management and integration in pervasive information systems. IEEE Trans. Software Eng. **38**(2), 478–496 (2012)
52. F.A. Schreiber, M. Roveri, Sensors and wireless sensor networks as data sources: Models and languages, in ed. by Colace, et al. [22], pp. 69–92
53. F.A. Schreiber, L. Tanca, R. Camplani, D. Viganó, Pushing context-awareness down to the core: more flexibility for the perla language, in *Proceedings of the 6th PersDB 2012 Workshop (Co-located with VLDB 2012)* (2012), pp. 1–6
54. M.A. Soliman et al., Ranking with uncertain scoring functions: semantics and sensitivity measures, in *SIGMOD 2011* (2011), pp. 805–816
55. K. Stefanidis, E. Pitoura, P. Vassiliadis, Managing contextual preferences. Inf. Syst. **36**(8), 1158–1180 (2011)
56. J.W. Tukey, *Exploratory Data Analysis* (Addison-Wesley, Reading, 1977)
57. D. Tunkelang, *Faceted Search*, Synthesis Lectures on Information Concepts, Retrieval, and Services (Morgan & Claypool Publishers, 2009)
58. R. Want, An introduction to rfid technology. IEEE Pervasive Comput. **5**(1), 25–33 (2006)

Part II
Incomplete/Inconsistent Data and Uncertain Reasoning

Multimedia, Similarity, and Preferences: Adding Flexibility to Your Information Needs

Ilaria Bartolini, Paolo Ciaccia and Marco Patella

Abstract Starting from the 90's, it was easily recognized that commonly adopted search paradigms were not enough to deal with at-the-time emerging novel DB applications, in which the presence of multimedia data and high dimensionality were both key aspects. In this paper we survey the research activity of our group in the last 25 years, therefore going through issues such as indexing, approximate query processing, and support for preference queries, which are now quite well understood. In doing this we also consider the need to provide the users with simple but powerful tools, able to smooth the processes of query creation/customization and of result interpretation. We complete with a look to the novel issues that the "Big Data" era brings to us.

1 Introduction

Efficient retrieval of relevant data has always been a key issue for database systems since their introduction. Dynamic indexing, caching, and, more in general, query evaluation and optimization, have been the enabling technologies that, in the 70's and 80's, allowed relational database systems to find their place in the DBMS market. Starting from efficient solutions for locating records based on their primary key value, the issue of dealing with queries on non-key attributes was then dealt with, and efficiently solved thanks to the "ubiquitous" B^+-tree. The 80's witnessed the proliferation of multi-dimensional indices, whose main purpose was to deal with both the management of 2D and 3D objects, as found in Spatial DB's applications, and the support of multi-attribute queries. However, it was only in the 90's that a major shift occurred, for two major reasons. First, the extension of DB technology to non-conventional data types, notably multimedia (MM) data (images, videos, speech, etc.); second, the widespread adoption of database systems, which made it possible also for non-experienced users to interact with large data repositories. While the first aspect called for the introduction of novel query paradigms, remarkably *similarity*

I. Bartolini (✉) · P. Ciaccia · M. Patella
DISI - Università di Bologna, Bologna, Italy
e-mail: ilaria.bartolini@unibo.it

S. Flesca et al. (eds.), *A Comprehensive Guide Through the Italian Database Research Over the Last 25 Years*, Studies in Big Data 31, DOI 10.1007/978-3-319-61893-7_8

queries, the second fostered the search of more user-friendly interaction modalities, among which *preference queries*.

In this paper we survey the research activity of our group on above themes over the last 25 years, with the aim of providing a concise view of the evolution of such topics in the period. More in detail, in Sect. 2 we sketch the challenges that non-conventional data (such as MM) brought to the DB arena (90's-early 00's). Section 3 discusses two novel issues that arose in the 00's, namely user interaction and approximate searching. In Sect. 4 we consider more recent trends coupling semantic-based to content-based retrieval of MM objects, and in Sect. 5 we take a look at the adoption of preference queries in DB systems. Finally, in Sect. 6 we complete by sketching a set of novel issues that the"Big Data" era brings to the areas covered in the paper.

2 First Steps (90's-Early 00's)

This section introduces the peculiarities of non-conventional data, starting from the realization that information needs, in presence of multimedia and high-dimensional data, are inherently different from those originally required to standard (relational) database management systems. This introduces a number of issues, including the so-called semantic gap and the problem of indexing these novel data types.

2.1 Representing Non-conventional Data

Starting from the 90's, the diffusion of inexpensive tools for capturing images, videos, audio, and other MM data allowed the creation of massive collections of digital MM content. It was immediately clear that the relational model was inappropriate to deal with such data, basically because their digitized representation does not immediately convey their intrinsic reality [74] (this was also the case for other data types, like Spatial DBs, that emerged in the 80's, or text retrieval, starting as early as the 50's).

Two paradigms evolved for the retrieval of MM content:

Concept-Based Retrieval: Using annotation that have been manually (or semi-automatically, see Sect. 4) attached to the datum.

Content-Based Retrieval: Using characteristics automatically extracted from each object (*features*) to characterize its content.

Both approaches have their pros and cons, the former depending on human annotation (which brings issues of cost, ambiguity, and annotation quality), while the latter introduces the *semantic gap*, i.e., the difference between the internal system representation and the user-perceived nature of an object. The success of Content-Based Retrieval is due to its dramatic reduction in cost and effort for satisfying user needs and to the development of a number of techniques able to reduce the semantic gap (see Sects. 3.1 and 4).

The fundamental concept in Content-Based Retrieval is that of similarity, which is used to compare the content of MM objects. Evaluating the similarity between two objects involves comparing their automatically extracted features and assessing a *score* in [0, 1], with the understanding that a value 1 signifies an exact match between the objects, while lower values indicate decreasing degrees of similarity. The most common query technique used by this model is *query by example*, where the user provides the system with an example object, representing her current information need. This could be done by either choosing a preexisting object (e.g., supplied by the user or chosen from a random set) or "creating" a new object with the intended content (e.g., by drawing an image or humming a melody).

Given a MM query object q, two query types are commonly provided:

- Range queries, where the user specifies a minimum similarity threshold θ, so that only objects whose similarity with q is not lower than θ are returned (note that this requires a knowledge of the distribution of similarity values between objects, in order to limit the cardinality of the result).
- Top-k queries, where the user asks for k results, and only the k objects most similar to the query are returned.

To exemplify the evolution of Content-Based Retrieval during the years, it is useful to retrace the history of research in retrieval of images (Content-Based Image Retrieval, CBIR) [67]; research on other media are surveyed in [44] (video), [53] (audio), and [50] (text).

The earliest commercial CBIR system was developed by IBM and was called QBIC (Query By Image Content) [39]. In QBIC, each image was represented as an histogram, identifying the amount of image pixels holding a specific color (colors were clustered together to reduce the size of histograms to a few dozen bins), and a cross-talk distance was used to implement the similarity between image features. Such representation also allowed, thanks to the so-called GEMINI approach [38], efficient indexing of image collections by using multi-dimensional index structures (see Sect. 2.2).

Several systems (e.g., see [62, 68]) followed this *global approach* to characterize image content, where features represent the content of the image as a whole. At the end of the 90's, this emerged as a limitation, since CBIR systems were requested to support more specific queries (e.g., find all those images containing a small grass region under a big blue sky region). Moreover, it was also proved that better results can be obtained when *local*, rather than global, features are extracted to characterize images' content. According to this *Region-Based Image Retrieval* (RBIR) paradigm, each image is decomposed (segmented) into a set of homogeneous regions, each of which is then characterized with a set of low-level features (such as color, texture, and region shape). Relevant examples of RBIR systems are Blobworld [24] and Windsurf [1]. Besides leading to more precise results, RBIR also allows for novel query types that are not possible in the unsegmented case, such as partial-match queries. This increased effectiveness did not came at no price: The similarity model of RBIR systems is clearly more complicated, since the similarity between images now depends on the similarity between component regions (as computed using local features) [19]. This (only apparently) simple modification to the definition of similarity introduces a number of issues when efficient retrieval is considered,[1] and index-based algorithm for correct retrieval under the RBIR similarity model have been devised [8, 15], exploiting multiple index scans (one for each query region) and a middleware algorithm, aggregating results of individual scans.

Finally, in recent times, the use of local descriptors, borrowed from computer vision [49], has gained a large popularity. The use of such features perfectly fits the RBIR similarity model, but efficient indexing is still an open research issue, since the number of "local" features (and thus of index scans) is two orders of magnitude higher than in the RBIR case, making techniques devised for the latter scenario hardly scalable and thus unsuitable for large data sets.

2.2 Indexing Non-conventional Data

Although the problem of indexing relational data can be considered efficiently solved, thanks to access structures like B^+-trees and hash files, such techniques cannot be directly applied to the data representations described in Sect. 2.1. This is due to two factors: (a) The inherently multi-dimensional nature of non-conventional data representations and (b) the type of queries issued on such data. Indeed, conventional index types do not efficiently handle spatial queries such as which points are the closest ones to a given data object, or whether points fall within a spatial area of interest. To overcome this handicap of one-dimensional indices, a number of Spatial (multi-dimensional) Access Methods (SAMs) have been introduced (see [64] for a comprehensive survey).

The idea behind SAMs is *local order preservation*: Points that are close in the multi-dimensional space should be stored together in the index (e.g., in the same node if the index is tree-based). For this, the use of linear orders, as commonly exploited by one-dimensional indices, is prevented, due to the fact that such orders do not preserve spatial proximity. For example, in R-tree [41] (the best-known representative of a SAM), points are grouped together and represented in tree nodes using their Minimum Bounding Box (i.e., the smallest product of orthogonal intervals containing all points stored in the sub-tree rooted at the node) in the next higher level of the tree. Since all objects lie within this bounding rectangle, a query not intersecting it leads to prune the whole sub-tree (because objects contained therein cannot be part of the query result).

For a large amount of applications, however, the kind of indexing provided by SAMs proved to be not efficient enough or not appropriate, e.g., when the data representation is more general than a simple point in a multi-dimensional space. The

[1] The (in-)famous "two tigers" example considers a user asking for an image containing two regions each representing a tiger: If a database image contains a single "tiger" region, it is not correct to match both query regions to the single "tiger" region of the database image, since, in this case, information on the number of query regions is lost.

90's saw the insurgence of the novel paradigm of similarity searching in a metric space [75]. Formally, a *metric space* is a pair, $\mathcal{M} = (\mathcal{D}, d)$, where $\mathcal{D}$ is a domain of feature values and $d : \mathcal{D} \times \mathcal{D} \rightarrow \Re^+$ is a non-negative distance function satisfying the properties of symmetry, identity, and triangle inequality.

Given a database $X \subseteq \mathcal{D}$ and a query object $q \in \mathcal{D}$, a range query returns all DB object whose distance to q does not exceed a user-specified value r, i.e., $\{x \in X, d(x, q) \leq r\}$. On the other hand, in metric spaces, top-k queries correspond to k-nearest neighbor queries, returning the k DB objects having minimum distance to q.

Metric Access Methods (MAMs) [26] rely on the triangle inequality for efficient resolution of such queries and have been successfully exploited in several application scenarios.[2] The M-tree [35], arguably the best-known MAM, was first presented at SEBD [34] and uses the ball decomposition principle for organizing the metric space into a set of (possibly overlapping) regions, to which the same principle is recursively applied. The QIC-M-tree [32] is an evolution of M-tree that allows for queries using a distance different from the one used to build the index (a possibility that was overlooked for many years).

3 New Problems (the 00's)

In this section we discuss two novel issues that arose in the 00's: The introduction of user interaction in the retrieval loop and the use of approximation techniques for speeding up searches.

3.1 User Interaction

As pointed out in Sect. 2, similarity queries are a powerful way to retrieve interesting information from large MM repositories. However, the very nature of MM objects often complicates the user's task of choosing an appropriate query and a suitable distance criterion to retrieve from the database the objects which best match her needs. This can be due both to limitations of the query interface and to the objective difficulty, from the user's point of view, to properly understand how the retrieval process works in high-dimensional spaces, which are typically used to represent the relevant features of the MM objects. For instance, the user of an image retrieval system will hardly be able to predict the effects that the modification of a single parameter of the distance function used to compare the individual objects can have on the result of a query [66].

---

[2] Clearly, also SAMs can be used for k-nearest neighbor queries, provided they can index objects' features.

To obviate this unpleasant situation, several methods have been proposed for implementing interactive similarity queries in MM DBs. Common to all these methods is the idea to exploit user feedback (i.e., user evaluation of the relevance of the result objects) in order to progressively adjust the query parameters and to eventually converge to an "optimal" parameter setting.

By analyzing such relevance judgments, the system can then generate a new, refined query, which will likely improve the quality of the result, as confirmed by experimental evidence [63]. This interactive retrieval process, which can be iterated several times until the user is satisfied with the results, gives rise to a feedback loop during which the default parameters used by the query engine are gradually adjusted to fit the user's needs [57].

Although relevance feedback has been recognized as a highly effective tool, its applicability suffers from two major problems:

1. Depending on the query, several iterations might occur before an acceptable result is found, thus convergence can be slow.
2. Once the feedback loop of a query is terminated, no information about this particular query is retained for re-use in future processing.

Note that both problems concern the efficiency of the feedback process, whereas the effectiveness of retrieval will depend on the specific feedback mechanisms used by the system, on the similarity model, and on the features used to represent the objects.

With the aim of overcoming above limitations, FeedbackBypass, a new approach to interactive similarity query processing, which complements the role of relevance feedback engines, was proposed in [10]. FeedbackBypass is based on the idea that, by storing and maintaining the information on query parameters gathered from past feedback loops, it is possible to either "bypass" the feedback loop completely for already-seen queries or to "predict" near-optimal parameters for new queries. In both cases, as an overall effect, the number of feedback and DB search iterations is greatly reduced, thus resulting in a significant speed-up of the interactive search process, as experimental evidence confirms [10].

An effective interaction with large MM DB's cannot be based only on querying, rather it has also to include advanced browsing facilities [65]. These are needed, say, to determine a "good" starting point for searching (because a suitable query object might not be available at the beginning of a user session), to get an overall view of the DB contents (because the user does not know yet what she is looking for), and so on.

Clearly, since manually organizing large MM collections for browsing purposes is untenable due to scalability issues, several systems based on clustering techniques were developed for automatically structuring large (image) collections [2, 27, 40, 47, 65, 72]. However, the applicability of such systems is limited due to some major problems, in particular:

1. With some notable exceptions, most image browsers are based on a static browsing structure, i.e., the organization of images is based on a fixed (usually hierarchical) layout that cannot be altered by the user. This means that no personalization at all is possible.

2. Browsing systems that allow the user to modify the browsing structure through interaction typically do so by reorganizing the whole DB or, at least, a large part of it. Clearly, this becomes unfeasible with very large data sets.
3. All the browsing systems for which personalization is an issue do not consider the possibility to make persistent the so-modified browsing structure. Thus, each time the user starts a new session she is faced with the original (default) image DB organization.

With the aim to overcome above limitations, in [12] the authors proposed an adaptive image browsing engine called PIBE (Personalizable Image Browsing Engine). PIBE provides the user with a customizable hierarchical browsing structure (called the Browsing Tree), whose changes persist across different sessions, and with a set of graphical personalization actions that can be used to modify the Browsing Tree. The effects of such actions are defined so as to guarantee that only a "local" reorganization of the image DB is required. This is possible since PIBE maintains specific similarity criteria for each portion (sub-tree) of the Browsing Tree.

3.2 Approximate Searching

Indexing of non-conventional data by way of SAMs and MAMS (see Sect. 2.2) was questioned at the end of the second millennium by studies pointing out the fact that using such access structures is sometimes not very efficient, e.g., when the feature space is a high-dimensional vector space [43]: In such cases, the most efficient way to exactly solve similarity queries is to sequentially scan the entire data set, comparing each object against the query object q. Obviously, such solution is not viable for very large DBs.

In order to accelerate the search, the user is commonly offered a quality/time trade-off: For saving search time, she has to accept a degradation in the quality of the result, i.e., an error with respect to the exact case. The goal of approximate similarity search is to reduce search times for similarity queries by possibly, but not necessarily, introducing an error in the result. The main rationale for providing the user with approximate techniques is (at least) threefold:

- First of all, a gap exists between the user-perceived similarity and the one actually implemented via the distance function (see Sect. 2.1). The "exact" result of a query, in many cases, might actually be deemed incorrect by the user, which would rather obtain a (possibly still not correct) result in much less time.
- For the same reason, the process of similarity search is typically iterative, because the user may be searching, using a feedback cycle, for the "correct" query object or the "perfect" distance function for her current information needs (see Sect. 3.1). In early stages of this process, the user may just want to have a quick feel of what the data set contains.

- Finally, even when both the distance function and the query object are adequate, the user may still prefer to quickly obtain a (good enough) approximate result rather than to wait longer for the exact answer.

The success of an approximate technique relies in solving the quality/time trade-off: Cost should be reduced as much as possible, while still keeping a high quality of the result.

A comprehensive survey of approximate searching techniques for both multi-dimensional and metric spaces is included in [59], where 30+ approaches are classified according to a schema organized along the following independent coordinates:

1. The type of space the approach applies to;
2. How approximation is obtained;
3. The guarantees on the result quality;
4. The degree of interaction with the user.

For instance, the PAC approach [31], first presented at SEBD in 1999 [30], is a method that applies to metric spaces and provides probabilistic guarantees on the quality of approximation by exploiting information on the distance distribution of indexed objects.

4 Adding Semantics (Late 00's-Nowadays)

Although content-based techniques, possibly assisted with user relevance feedback (see Sect. 3.1), can be completely automatized and can indeed attain a very good effectiveness, in several cases they still stay below the optimal 100% precision value [67]. This is especially true when the user is looking for objects matching some high-level concept, which is hardly representable by means of low-level features only. In such cases, a possible way to fill the semantic gap is to assign meaningful terms to objects, so as to really allow a high-level, concept-based, retrieval.

Terms associated to objects can, indeed, be considered as a different type of metadata, one which is not included in the MM datum, but which is associated with it by way of a manual or automatic process. Clearly, such terms can be seamlessly used into predicates for boolean search, as traditional metadata are, and, more in general, to enable text-based techniques (search, browsing, clustering, classification, etc.) to be applied also to objects that otherwise could only be dealt with by relying on feature-based similarity assessment.

The enrichment of MM objects through semantic labels can basically be performed in three distinct ways:

Annotation by an expert: This is the solution commonly used in libraries, where an expert provides labels for every datum. Such labels are expected to be of great quality, but the process is lengthy and expensive; moreover, the problem of subjectivity can plague the whole data collection, because the final user can have a different view on the data with respect to the user providing labels.

Social annotation: This approach is the one exploited by "social" DBs, like YouTube (www.youtube.com) and Flickr (www.flickr.com), where users accessing a MM object can also provide labels for it. The so-obtained annotation is usually of low quality, due to problems of ambiguity (because a label could carry different meanings due to polysemy or homonymy), lack of information (because an object could never have been labeled), and problems of synonymy/mistyping.

(Semi-)Automatic annotation: These techniques exploit the similarity among multimedia objects, computed by way of low-level features, to extract labels relevant for a non-annotated object, with the assumption that objects sharing similar features also convey the same semantic content, and can thus be annotated with the same labels.

Approaches to automatic object annotation include a variety of techniques, and they differ in what "annotation" actually means, ranging from enriching images with a set of keywords (or tags) [4, 36, 46, 48] to providing a rich semantic description of image content through the concepts of a full-fledged RDF ontology [61]. Further, solutions may differ in what kind of tags/concepts they ultimately provide, in this case the difference being among general-purpose systems and others that are tailored to discover only specific concepts/classes [60, 71].

Several automatic annotation techniques based on tags have been proposed in the last decade [3, 4, 21, 36, 46, 48] and the first prototypes are now available on the Internet (e.g., ALIPR (www.alipr.com) and Behold (www.behold.cc) for images).

State-of-the-art automatic annotation solutions can be grouped into two main classes: *Semantic propagation* and *statistical inference*. In both cases, the problem to be solved is the same: Given a training set of annotated multimedia objects, discover affinities between low-level features and terms that describe the object content, with the aim of predicting "good" terms to annotate a new object. With propagation models [51], a supervised learning technique is adopted that evaluates content similarity at a low-level and then annotates new objects by propagating terms assigned to their most similar objects. Working with statistical inference models [58], an unsupervised learning approach tries to capture correspondences between low-level features and terms by estimating their joint probability distribution. Both approaches improve the annotation process and the retrieval on large MM collections.

The typical approach for annotating MM objects exploits user-defined textual labels [4, 46, 73]. However, this is commonly performed by drawing tags from an unstructured set, thus possibly missing their actual semantics, which depends on the context where their associated MM data are found; this gives a sort of meaning vagueness to labels [56]. To deal with this, in order to connect each tag with its intended meaning, the coexistence of multiple, independent classification criteria can be exploited [37]. According to this "multidimensional" approach, labels belonging to different dimensions may have separate meanings, while each dimension will represent the meaning of high-level concepts contained therein, providing a disambiguation of their semantics. In the Scenique model [5], each dimension takes the shape of a tree and terms are linked with a parent/child relationship. Each concept is called a *semantic tag* and corresponds to a path in a tree. For instance, the semantic tags `landscape/sky` and `landscape/sky/rainbow` are associated to two tree nodes of the dimension `landscape`.

Semantic tags can be regarded as means to describe objects that is more precise and powerful than “free” tags (with no inherent semantics), yet not so complex to derive as concepts of RDF-like ontologies (whose semantics might not be so easy to grasp by end-users).

Each object can be assigned a variable number of semantic tags. If a dimension is not relevant for a object, then no semantic tag from such dimension is used to characterize content. On the other hand, an object could be characterized by multiple semantic tags from the same dimension, if this is appropriate. For instance, an image depicting a dog and a cat might be assigned the two semantic tags `animal/dog` and `animal/cat`, both from the `animal` dimension. Thus, although each dimension provides a means to classify objects, this classification is not exclusive at the instance level, a fact that provides the necessary flexibility to organize objects.

The Ostia algorithm [6, 7] enriches above model by taking the advantages of both MM object features and keywords (extracted through text analysis procedures, such as stemming, stoplist, and NLP techniques, from associated metadata), in order to predict for a novel object a high-quality set of concepts taken from “light-weight” ontologies (or classification hierarchies). Unlike approaches based on machine learning techniques [36, 48], which require a new classifier to be built from scratch whenever a new class/concept is needed, new concepts can be freely added since Ostia does not require a learning phase.

5 Adding User Preferences (00’s-Nowadays)

At the beginning of this century it became apparent that similarity queries and, more in general, *ranking* queries were not powerful enough to discover all “interesting” data hidden in large databases, and that a more flexible approach was needed. For instance, consider a complex multimedia query, which is usually processed by splitting it into a set of simpler sub-queries, each one dealing with only some of the query features. In order to determine which are the overall best-matching objects, some rule is needed to integrate the results of such sub-queries. Clearly, the same problem arises whenever an overall ranking is sought for a query over a multi-dimensional data set, in which attribute values are to be somehow combined together. The common approach of using as integration rule a scoring function, e.g., a weighted linear function, has the drawback that one is necessarily forced to compromise between the different sub-queries/attributes and can easily lead to miss relevant results. Further, this approach is by its very nature suitable only for numerical attributes and can hardly be extended to categorical ones.

Since distance/scoring functions are just a mean to express one’s preferences, it seemed quite natural to look at more general preference models, able to provide users with increased flexibility and ease of use. In 2002 three works [28, 45, 69] (one of them presented at SEBD), analyzed the use of *qualitative preferences* for

DB querying purposes, paving the way for many subsequent studies. Under the so-called preference (or dominance) relation model, preferences among objects/tuples are pairs of a binary relation $\succ$, where $t \succ s$ is interpreted as "t is (strictly) preferred to s", i.e., *t dominates s*. Preference relations include scoring functions as a special case, since for any scoring function f one can define a corresponding $\succ_f$ as: $t \succ_f s$ iff $f(t) > f(s)$. On the other hand, not all preference relations can be represented by a scoring function, a necessary condition being that $\succ$ is a weak order.

As an example, given a set of used cars with schema Cars(ID,Model,Price,Year), the preferences "*I would like to buy a recent VW Golf that costs less than 8,000€*" can be expressed as:

$$t \succ s \Leftrightarrow (t.\mathsf{Model} = \text{`VWGolf'} \wedge s.\mathsf{Model} \neq \text{`VWGolf'}) \otimes \\ (t.\mathsf{Year} > s.\mathsf{Year}) \otimes (t.\mathsf{Price} < 8,000 \wedge s.\mathsf{Price} \geq 8,000)$$

where $\otimes$ is the *Pareto accumulation* operator, which can be understood as stating that basic preferences over the single attributes are equally important. With above preferences the tuple t_1 : ('C1','VW Golf',7500, 2014) would dominate t_2 : ('C2','VW Golf',6000, 2013), yet it would *not* dominate t_3 : ('C3','VW Golf',9000, 2015). Since it is commonly accepted that the result of a preference query is the set of undominated objects in the database (this is always a non-empty set when $\succ$ is a strict partial order), t_1 and t_3 are the only alternatives to consider here. Notice that, unlike scoring functions, there is no need to trade-off one attribute for another (Price for Year in the example).

Skyline queries are undoubtedly the specific case of preference queries most studied by the DB community, as made evident by the high number of citations that the original paper [25] has ($>$ 2100 according to Google Scholar, February 2017). Given a set of numerical attributes of interest, a tuple t is a skyline tuple iff it is *Pareto-undominated*, i.e., there is no other tuple s that is as good as t on all attributes of interest and strictly better than t on at least one attribute. Skyline queries have gained full relevance as a valid support in multi-criteria decision analysis, due to their ability to extract the "most interesting" results. Indeed, they are a remarkable alternative to top-k ranking queries, since they require no parameters to be specified and are insensitive to attributes' scales. Further, since the skyline consists of all the top-1 tuples that would result from using *all* scoring functions that are monotone in the skyline attributes, it can provide users with an overall view of all potential best-ranked objects.

The research on skyline queries has covered several aspects, ranging from efficient evaluation (to which the authors contributed with the SaLSa algorithm [14]) to generalizations and extensions of the basic model (see [29] for a recent survey). Although most of the work on skylines considers relational data, the idea has also been applied to other data types. For instance, in [13] the skyline model and a variant of it are applied to image retrieval and contrasted to the commonly used top-k ranking model. Results show that the skyline approach has superior performance in terms of both classical precision-recall measures and efficiency. Similar results have been also observed in [15] for the case of Region-Based Image Retrieval discussed in Sect. 2.1.

Among the many directions that the research on preference queries has taken in the DB field, is worth mentioning the ones dealing with uncertain [16, 18, 20, 22] and distributed [11, 70] data, and preference elicitation [52].

6 New Issues that Big Data Bring to Us

Nowadays, thanks to the world-wide diffusion of cheap information-sensing devices (like sensors, cameras, RFID readers, mobile phones) and the growth of storage capacity, data generation has greatly increased, reaching several exabytes per day [42]. Such data deluge has made relational DBMSs unable to handle the storage, analysis, and searching it. In order to cope with what has been called *Big Data*, the use of massively parallel architectures is required, bringing a number of novel challenges to deal with the four V's characterizing Big Data (Volume, Velocity, Variety, and Veracity) in order to extract the fifth V (Value) from the analysis of such data. Concerning the types of data/queries we have described so far, the following is a (non-exhaustive) list of new challenges related to the very nature of Big Data:

- Accurate description of the content of MM objects is still an active research area. However, feature extraction can be a very time consuming process, thus posing scalability issues [23]. To this end, the challenge is to devise new *multi-step* query processing approaches which adopt approximate, cheaper-to-extract MM descriptions that can be refined on-the-fly as needed.
- There is a pressing need for new technologies able to deal with multiple Big Data V's at the same time. For instance, real-time (*velocity*) analysis of massive (*volume*) multimedia (*variety*) data streams is still far to be satisfactorily solved [9].
- Next Generation Sequencing technology is nowadays producing huge volumes of heterogeneous data, thus opening many interesting practical and theoretical computational problems. Besides well-known problems related to *primary analysis* (production of sequences in the form of short DNA segments, or "reads") and *secondary analysis* (alignment of reads to a reference genome and extraction of specific genomic features), *tertiary analysis* deals with discovering how multiple genomic regions, representing heterogeneous genomic features, interact with each other. For this kind of problems, similarity search is still in its infancy [54, 55].
- User-generated data, possibly originating from mobile devices, due to their inherent heterogeneous and often unreliable nature, pose completely new challenges to preference modeling, management, and processing. For instance, aggregating preferences of many users, possibly expressed using different modalities (e.g., numerical vs. qualitative) [17], dealing with time- and/or space-varying preferences [33], and producing relevant results for groups of users, to name a few, are all open important research issues.

References

1. S. Ardizzoni, I. Bartolini, M. Patella, Windsurf: region-based image retrieval using wavelets, in *IWOSS* (Florence, Italy, 1999)
2. I. Bartolini, P. Ciaccia, MuSIQUE: a multi-system image querying user interface, in *SEBD* (Cetraro, Italy, 2003)
3. I. Bartolini, P. Ciaccia, Towards an effective semi-automatic technique for image annotation, in *SEBD* (Torre Canne, Italy, 2007)
4. I. Bartolini, P. Ciaccia, Imagination: exploiting link analysis for accurate image annotation, in *AMR* (Paris, France, 2007)
5. I. Bartolini, P. Ciaccia, Scenique: a multimodal image retrieval interface, in *AVI* (Naples, Italy, 2008)
6. I. Bartolini, P. Ciaccia, Multi-dimensional keyword-based image annotation and search, in *KEYS* (Indianapolis, IN, 2010)
7. I. Bartolini, P. Ciaccia, Automatically joining pictures to multiple taxonomies, in *SEBD* (Rimini, Italy, 2010)
8. I. Bartolini, M. Patella, Correct and efficient evaluation of region-based image search, in *SEBD* (L'Aquila, Italy, 2000)
9. I. Bartolini, M. Patella, A general framework for real-time analysis of massive multimedia streams (Submitted for publication, 2017)
10. I. Bartolini, P. Ciaccia, F. Waas, FeedbackBypass: a new approach to interactive similarity query processing, in *VLDB* (Rome, Italy, 2001)
11. I. Bartolini, P. Ciaccia, M. Patella, *Distributed Aggregation Strategies for Preferences Queries*, in *SEBD* (Portonovo, Italy, 2006)
12. I. Bartolini, P. Ciaccia, M. Patella, Adaptively browsing image databases with PIBE. MTAP **31**(3), 269–286 (2006)
13. I. Bartolini, P. Ciaccia, V. Oria, M.T. Özsu, Flexible integration of multimedia sub-queries with qualitative preferences.MTAP **33**(3), 275–300 (2007)
14. I. Bartolini, P. Ciaccia, M. Patella, Efficient sort-based skyline evaluation. ACM TODS **33**(4), 1–45 (2008)
15. I. Bartolini, P. Ciaccia, M. Patella, Query processing issues in region-based image databases. KAIS **25**(2), 389–420 (2010)
16. I. Bartolini, P. Ciaccia, M. Patella, Getting the best from uncertain data, in *SEBD* (Maratea, Italy, 2011)
17. I. Bartolini, Z. Zhang, D. Papadias, Collaborative filtering with personalized skylines. TKDE **23**(2), 190–203 (2011)
18. I. Bartolini, P. Ciaccia, M. Patella, Getting the best from uncertain data: the correlated case, in *SEBD* (Venice, Italy, 2012)
19. I. Bartolini, M. Patella, G. Stromei, Efficiently managing multimedia hierarchical data with the windsurf library, in *CCIS Series*, vol. 314 (Springer, Berlin, 2012)
20. I. Bartolini, P. Ciaccia, M. Patella, The skyline of a probabilistic relation. IEEE TKDE **25**(7), 1656–1669 (2013)
21. I. Bartolini, M. Patella, C. Romani, SHIATSU: tagging and retrieving videos without worries. MTAP **63**(2), 357–385 (2013)
22. I. Bartolini, P. Ciaccia, M. Patella, Domination in the probabilistic world: computing skylines for arbitrary correlations and ranking semantics. ACM TODS **39**(2), 14:1–14:45 (2014)
23. M. Batko, F. Falchi, et al., Building a web-scale image similarity search system. MTAP **47**(3), 599–629 (2010)
24. S. Belongie, C. Carson, H. Greenspan, J. Malik, Color- and texture-based image segmentation using EM and its application to content-based image retrieval, in *ICCV* (Mumbai, India, 1998)
25. S. Börzsönyi, D. Kossmann, K. Stocker, The skyline operator, in *ICDE* (Heidelberg, Germany, 2001)
26. E. Chávez, G. Navarro, R. Baeza-Yates, J.L. Marroquín, Searching in metric spaces. ACM CSUR **33**(3), 273–321 (2001)

27. J. Chen, C. Bouman, J. Dalton, Active browsing using similarity pyramids, in *SPIE* (San Jose, CA, 1999)
28. J. Chomicki, Querying with Intrinsic Preferences, in *EDBT* (Prague, Czech Republic, 2002)
29. J. Chomicki, P. Ciaccia, N. Meneghetti, Skyline queries, front and back SIGMOD Record **42**(3), 6–18 (2013)
30. P. Ciaccia, M. Patella, PAC nearest neighbor queries: using the distance distribution for searching in high-dimensional metric spaces, in *SEBD* (Como, Italy, 1999)
31. P. Ciaccia, M. Patella, PAC nearest neighbor queries: approximate and controlled search in high-dimensional and metric spaces, in *ICDE* (San Diego, CA, 2000)
32. P. Ciaccia, M. Patella, Searching in metric spaces with user-defined and approximate distances. ACM TODS **27**(4), 398–437 (2002)
33. P. Ciaccia, R. Torlone, Modeling the propagation of user preferences, in *ER* (Brussels, Belgium, 2011)
34. P. Ciaccia, M. Patella, F. Rabitti, P. Zezula, Indexing metric spaces with M-tree, in *SEBD* (Verona, Italy, 1997)
35. P. Ciaccia, M. Patella, P. Zezula, M-tree: an efficient access method for similarity search in metric spaces, in *VLDB* (Athens, Greece, 1997)
36. R. Datta, W. Ge, J. Li, J.Z. Wang, Toward bridging the annotation-retrieval gap in image search. IEEE Multimed. **14**(3), 24–35 (2007)
37. R. Fagin, R. Guha et al., Multi-structural databases, in *PODS* (Baltimore, MD, 2005)
38. C. Faloutsos, R. Barber, et al., Efficient and effective querying by image content. JIIS **3**(3/4), 231–262 (1994)
39. M. Flickner, H. Sawhney, et al., Query by image and video content: the QBIC system. IEEE Comput. **28**(9), 23–32 (1995)
40. A. Graham, H. Garcia-Molina, A. Paepcke, T. Winograd, Time as essence for photo browsing through personal digital libraries, in *JCDL* (Portland, OR, 2002)
41. A. Guttman, R-Trees: a dynamic index structure for spatial searching, in *SIGMOD* (Boston, MA, 1984)
42. M. Hilbert, P. López, The world's technological capacity to store, communicate, and compute information. Science **332**(6025), 60–65 (2011)
43. A. Hinneburg, C.C. Aggarwal, D.A. Keim, What is the nearest neighbor in high dimensional spaces? in *VLDB* (Cairo, Egypt, 2000)
44. W. Hu, N. Xie, et al., A survey on visual content-based video indexing and retrieval. IEEE TSMC-C **41**(6), 797–819 (2011)
45. W. Kießling, Foundations of preferences in database systems, in *VLDB* (Hong Kong, China, 2002)
46. J. Kleban, E. Moxley, J. Xu, B.S. Manjunath, Global annotation of geo-referenced photographs, in *CIVR* (Santorini, Greece, 2009)
47. J. Laaksonen, M. Koskela, S. Laakso, E. Oja, Self-organising maps as a relevance feedback technique in content-based image retrieval. PAA **2**(4), 140–152 (2000)
48. J. Li, J.Z. Wang, Real-time computerized annotation of pictures, in *MM* (Santa Barbara, CA, 2006)
49. D.G. Lowe, Distinctive image features from scale-invariant keypoints. IJCV **60**(2), 91–110 (2004)
50. C.D. Manning, P. Raghavan, H. Schütze, *Introduction to Information Retrieval* (Cambridge University Press, Cambridge, 2008)
51. O. Maron, A.L. Ratan, Multiple-instance learning for natural scene classification, in *ICML* (San Francisco, CA, 1998)
52. N. Meneghetti, D. Mindolin, P. Ciaccia, J. Chomicki, Output-sensitive evaluation of prioritized skyline queries, in *SIGMOD* (Melbourne, Australia, 2015)
53. D. Mitrović, M. Zeppelzauer, C. Breiteneder, Features for content-based audio retrieval. Adv. Comput. **78**, 71–150 (2010)
54. P. Montanari, I. Bartolini et al., Looking for similar patterns in genomic sequences, in *SEBD* (Ugento, Italy, 2016)

55. P. Montanari, I. Bartolini, et al., Pattern similarity search in genomic sequences. TKDE **28**(11), 3053–3067 (2016)
56. R. Navigli, Word sense disambiguation: a survey. ACM CSUR **41**(2), 10 (2009)
57. M. Ortega, Y. Rui et al., Supporting similarity queries in MARS, in *MM* (Seattle, WA, 1997)
58. J.-Y. Pan, H. Yang, C. Faloutsos, P. Duygulu, Automatic multimedia cross-modal correlation discovery, in *KDD* (Seattle, WA, 2004)
59. M. Patella, P. Ciaccia, Approximate similarity search: a multi-faceted problem. JDA **7**(1), 36–48 (2009)
60. A. Payne, S. Singh, A benchmark for indoor/outdoor scene classification, in *ICAPR* (Bath, UK, 2005)
61. A. Penta, A. Picariello, L. Tanca, Multimedia knowledge management using ontologies, in *MS* (Vancouver, BC, 2008)
62. A. Pentland, R.W. Picard, S. Sclaroff, Photobook: content-based manipulation of image databases. IJCV **18**(3), 233–254 (1996)
63. Y. Rui, T.S. Huang, M. Ortega, S. Mehrotra, Relevance feedback: a power tool for interactive content-based image retrieval. IEEE TCSV **8**(5), 644–655 1998
64. H. Samet, *Foundations of Multidimensional and Metric Data Structures* (Morgan Kaufmann, San Francisco, 2006)
65. S. Santini, R. Jain, Integrated browsing and querying for image databases. IEEE Multimed. **7**(3), 26–39 (2000)
66. T. Seidl, H.-P. Kriegel, Efficient user adaptable similarity search in large multimedia databases, in *VLDB* (Athens, Greece, 1997)
67. A.W.M. Smeulders, M. Worring, et al., Content-based image retrieval at the end of the early years. IEEE TPAMI **22**(12), 1349–1380 (2000)
68. M. Stricker, M. Orengo, Similarity of color images, in *SPIE* (San Jose, CA, 1995)
69. R. Torlone, P. Ciaccia, Finding the best when it's a matter of preference, in *SEBD* (Portoferraio, Italy, 2002)
70. G. Trimponias, I. Bartolini, D. Papadias, D. Yang, Skyline processing on distributed vertical decompositions. IEEE TKDE **25**(4), 850–862 (2013)
71. R. Tye, G. Nathaniel, N. Mor, Towards automatic extraction of event and place semantics from flickr tags, in *SIGIR* (Amsterdam, The Netherlands, 2007)
72. M. Wallace, K. Karpouzis, et al., The electronic road: personalized content browsing. IEEE Multimed. **10**(3), 49–59 (2003)
73. L. Wang, L. Khan, Automatic image annotation and retrieval using weighted feature selection. MTAP **29**(1), 55–71 (2006)
74. A. Yoshitaka, T. Ichikawa, A survey on content-based retrieval for multimedia databases. IEEE TKDE **11**(1), 81–93 (1999)
75. P. Zezula, G. Amato, V. Dohnal, M. Batko, *Similarity Search: The Metric Space Approach* (Springer, Berlin, 2006)

Dealing with Inconsistency in Databases: An Overview

Marco Calautti, Luciano Caroprese, Bettina Fazzinga, Sergio Flesca, Filippo Furfaro, Sergio Greco, Cristian Molinaro, Francesco Parisi, Andrea Pugliese, Domenico Saccà, Irina Trubitsyna and Ester Zumpano

Abstract There is a growing number of applications where inconsistent information arises. In the last two decades, the emerging approach for dealing with such scenarios is to "tolerate" inconsistency and provide appropriate reasoning mechanisms. In particular, consistent query answering has been widely accepted as a principled approach for query answering. Several practical and theoretical issues regarding the

M. Calautti · L. Caroprese · S. Flesca · F. Furfaro · S. Greco · C. Molinaro · F. Parisi (✉) ·
A. Pugliese · D. Saccà · I. Trubitsyna · E. Zumpano
DIMES - Università Della Calabria, Via Bucci - Rende (CS), Rende, Italy
e-mail: fparisi@dimes.unical.it

M. Calautti
e-mail: mcalautti@dimes.unical.it

L. Caroprese
e-mail: lcaroprese@dimes.unical.it

S. Flesca
e-mail: flesca@dimes.unical.it

F. Furfaro
e-mail: furfaro@dimes.unical.it

S. Greco
e-mail: greco@dimes.unical.it

C. Molinaro
e-mail: cmolinaro@dimes.unical.it

A. Pugliese
e-mail: pugliese@dimes.unical.it

D. Saccà
e-mail: sacca@dimes.unical.it

I. Trubitsyna
e-mail: itrubitsyna@dimes.unical.it

E. Zumpano
e-mail: ezumpano@dimes.unical.it

B. Fazzinga
ICAR-CNR, via Castellino - Napoli, Napoli, Italy
e-mail: bettina.fazzinga@icar.cnr.it

S. Flesca et al. (eds.), *A Comprehensive Guide Through the Italian Database Research Over the Last 25 Years*, Studies in Big Data 31, DOI 10.1007/978-3-319-61893-7_9

consistent query answering framework have been widely investigated in literature and different techniques for evaluating consistent answers have been proposed. In this work, we provide a brief survey of the research on techniques for repairing and querying inconsistent databases developed by the database research group of the DIMES Department at the University of Calabria.

1 Introduction

Ideally databases and knowledge bases should be completely free of inconsistency. This means that integrity constraints, the mechanism employed in databases to guarantee that available data correctly model the outside world, should be always satisfied. In fact, the traditional approach implemented by commercial database management systems is that of avoiding inconsistency by aborting updates or transactions yielding to an integrity constraint violation.

However, there is an increasing number of applications where integrity constraints satisfaction cannot be guaranteed. In fact, when knowledge from multiple sources is integrated, as in the contexts of data warehousing, data integration and automated reasoning systems, it is not possible to guarantee that the integrated data remain consistent. Moreover, often there is no update that can be rejected in order to guarantee the consistency of the integrated database, as the database instance is not resulting from a single update performed on a single database, but from merging multiple data sources.

In the last two decades, the emerging approach for dealing with inconsistency is to "tolerate" it and provide appropriate mechanisms to handle inconsistent data. As opposed to traditional approaches, the key idea to handle inconsistency is that of living with an inconsistent database and modifying query semantics in order to obtain only consistent information as answers to queries posed on inconsistent databases. Consistent information is then defined as the one that is invariant or persists under all possible "minimal" ways of restoring consistency of the database. Indeed, restoring consistency should be accomplished by trying to preserve as much information as possible. Every database instance corresponding to a "minimal" way of restoring consistency is called a *repair*. There may be several alternative minimality criteria. Therefore, what is consistently true is what is true with respect to every repair. This concept has been formalized with the notion of *consistent query answer*, i.e., query answers that can be obtained from every repair.

The following example describes a typical inconsistency-prone data integration scenario: data from two consistent source databases are integrated, and the resulting database turns out to be in an inconsistent state, even if the sources are individually consistent.

Example 1 Consider the relation *teaches(Course,Professor)*, where attribute *Course* is a key. The following two instances of *teaches* (leftmost tables), provided by different sources, satisfy the integrity constraint. However, from their union we derive an inconsistent relation (the rightmost table) that does not satisfy the constraint since there are two distinct tuples with the same value for the attribute *Course*.

$Course$	$Professor$
CS	$Mary$
$Math$	$John$

Source 1

$Course$	$Professor$
CS	$Frank$
$Math$	$John$

Source 2

$Course$	$Professor$
CS	$Mary$
$Math$	$John$
CS	$Frank$

Integrated database

We can consider as minimal ways for restoring consistency of the integrated relation, the two relations obtained by deleting one of the two tuples that violate the constraint, thus obtaining again the two source relations as repairs. Intuitively, on the basis of these repairs, what is consistently true is that '$John$' is the teacher of the course '$Math$' and that there is a course whose name is 'CS'.

Consider the query $Q(x, y) = teaches(x, y)$ which intends to retrieve the names of courses with their relative teachers. The consistent query answers to Q are those which would be returned posing the query on every repair, that is, $\langle Math, John\rangle$. □

A repair for an inconsistent database (with respect to a set of integrity constraints) is a consistent database which is "as close as possible" to the original instance. Different notions of closeness can be defined, each of them corresponding to a different repair semantics. For instance, in the example above repairs were obtained by performing minimal sets of insertion and deletion of (whole) tuples. Another possible notion of repair is that allowing updates of values in the tuples. Considering the example above, this means that the value of the attribute $Professor$ in one of the two conflicting tuples must be changed in such a way that a consistent status is obtained. For instance, we may update either the value '$Mary$' to '$Frank$' in the tuple $\langle CS, Mary\rangle$ or the value '$Frank$' to '$Mary$' in $\langle CS, Frank\rangle$, obtaining again the two repairs of Example 1.

Several practical and theoretical issues regarding the consistent query answers problem have been widely investigated in literature and different techniques for evaluating consistent answers have been proposed. The problem of computing consistent answers has been studied along several dimensions, such as the repair semantics, the classes of queries and the type of constraints adopted. In this paper, we aim at providing a brief overview of some techniques for repairing and querying inconsistent databases investigated by the database research group of the DIMES (previously DEIS) Department at the University of Calabria.

Plan of the paper. After introducing the basic concepts and the notation in Sect. 2, we discuss in Sect. 3 a logic programming approach that allow the set of repairs for a database with respect to a set of full integrity constraints to be computed by rewriting the constraints into disjunctive rules. This approach is the extended in Sect. 4 to consider *active integrity constraints*, a special form of integrity constraints that allow the user to specify the update actions to be performed on an inconsistent database in order to restore consistency. Then, in Sect. 5, we discuss two approximation approaches to compute consistent answers. The former computes sound (but possibly incomplete) consistent answers in polynomial time. The latter computes approximate "probabilistic" consistent query answers. After this, in Sect. 6, we discuss the problem of repairing and querying numerical databases violating *aggregate constraints*, a kind of integrity constraints allowing us to express algebraic relationships among (aggregate) numerical values extracted from databases. Finally, in Sect. 7 we discuss approaches for computing repairs and consistent answers to unions of conjunctive queries based on the evaluation of the query over a *universal solution*. Conclusions are drawn in Sect. 8, where directions for future work are discussed.

2 Preliminaries

Basics. We assume the existence of the following pairwise disjoint sets of symbols: an infinite set *Consts* of *constants*, an infinite set *Nulls* of *labeled nulls*, and an infinite set *Vars* of *variables*. A *term* is a constant, a labeled null, or a variable. A *schema* is a finite set $\mathcal{D}$ of *predicates*, where each predicate P is associated with an arity $ar(P)$, which is a non-negative integer.

An *atom* over $\mathcal{D}$ is an expression of the form $P(t_1, \ldots, t_n)$, where P is an n-ary predicate in $\mathcal{D}$ and each t_i is a term — we denote an atom also as $P(\mathbf{t})$, where $\mathbf{t}$ is understood to be a sequence of n terms. If $t_i \in Consts \cup Nulls$ for every $1 \leq i \leq n$, then the atom is also called a *fact*. Thus, facts consist of tuples of constants and labeled nulls.

Given a set S of atoms, we use $Consts(S)$ (resp. $Nulls(S)$, $Vars(S)$) to denote the set of all constants (resp. labeled nulls, variables) occurring in S, and use $Dom(S)$ to denote the set $Consts(S) \cup Nulls(S) \cup Vars(S)$.

A (database) *instance* over $\mathcal{D}$ is a set of facts over $\mathcal{D}$. An instance where only constants appear is said to be *ground*. We will use K, I and J, possibly subscripted or primed, to denote instances. Instances are often represented using tables (e.g. Example 1). We also assume that our databases are ground and (labelled) nulls can be introduced in the computation of repairs only. That is, we do not deal with the case that the input database contains nulls.

A *tuple generating dependency* (TGD) over $\mathcal{D}$ is a formula r of the form:

$$\forall \mathbf{x} \forall \mathbf{y}\ \varphi(\mathbf{x}, \mathbf{y}) \rightarrow \exists \mathbf{z}\ \psi(\mathbf{x}, \mathbf{z})$$

where $\mathbf{x}, \mathbf{y}, \mathbf{z}$ are lists of variables, and $\varphi(\mathbf{x}, \mathbf{z})$ (resp. $\psi(\mathbf{x}, \mathbf{y})$) is a conjunction of atoms over $\mathcal{D}$ whose variables are exactly $\mathbf{x}$ and $\mathbf{y}$ (resp. $\mathbf{x}$ and $\mathbf{z}$) and is called the *body* (resp. *head*) of r, denoted as $Body(r)$ (resp. $Head(r)$). With a slight abuse of notation, we sometimes treat $Body(r)$ and $Head(r)$ as sets (of atoms). A TGD is said to be *universally quantified* or *full* if all its variables are universally quantified (i.e., $\mathbf{z}$ is empty), otherwise it is *existentially quantified*. An *extended tuple generating dependency* (ETGD) over $\mathcal{D}$ is a (universally quantified) formula r of the form:

$$\forall \mathbf{x} \forall \mathbf{y}\ \varphi(\mathbf{x}, \mathbf{y}) \rightarrow \psi(\mathbf{x})$$

where $\psi(\mathbf{x})$ is a disjunction of atoms. An *equality generating dependency* (EGD) over $\mathcal{D}$ is a (universally quantified) formula of the form:

$$\forall \mathbf{x}\ \varphi(\mathbf{x}) \rightarrow x_1 = x_2$$

where $x_1, x_2 \in \mathbf{x}$, $\varphi(\mathbf{x})$ is a conjunction of atoms over $\mathcal{D}$ whose variables are exactly $\mathbf{x}$. TGDs, ETGDs and EGDs are also called *(data) dependencies* or *integrity constraints*.

Aggregate constraints were introduced in [17] to express algebraic relationships among (aggregate) numerical values extracted from databases. Before providing the formal definition, we introduce some notation. Here, we use a non-positional notation and thus refer to predicates by associating *attribute* names with the positions of their terms. Therefore, a predicate P is associated with a sorted list $(A_1, \dots, A_n)$ where each A_i is an attribute name and $n = ar(P)$. We say that A_i is a numerical attribute if its domain is either $\mathbb{Z}$ (infinite domain of integers) or $\mathbb{Q}$ (rationals). Given a fact $P(\mathbf{t})$, the value of the attribute A_i of $P(\mathbf{t})$ will be denoted as $P(\mathbf{t})[A_i]$.

An *attribute expression* e on P is either a constant or the name of a numerical attribute of P. Given an attribute expression e on P and a fact $P(\mathbf{t})$, we denote as $e(P(\mathbf{t}))$ the value e, if e is a constant, or the value $P(\mathbf{t})[e]$, if e is an attribute. Given a predicate P and a sequence $\mathbf{y}$ of variables, an *aggregation function* $\chi(\mathbf{y})$ on P is a triplet $\langle P, e, \alpha(\mathbf{y})\rangle$, where e is an *attribute expression* on P and $\alpha(\mathbf{y})$ is a (possibly empty) boolean combination of atomic comparisons of the form $X \diamond Y$, where X and Y are constants, attributes names of P, or variables in $\mathbf{y}$, and $\diamond$ is a comparison operator in $\{=, \neq, \leq, \geq, <, >\}$. Given an aggregation function $\chi(\mathbf{y}) = \langle P, e, \alpha(\mathbf{y})\rangle$ and a sequence $\mathbf{a}$ of constants with $|\mathbf{a}| = |\mathbf{y}|$, $\chi(\mathbf{a})$ maps every instance of P to $\sum_{P(\mathbf{t}) \models \alpha(\mathbf{a})} e(P(\mathbf{t}))$, where $\alpha(\mathbf{a})$ is the ground boolean combination of atomic comparisons obtained from $\alpha(\mathbf{y})$ by replacing each variable in $\mathbf{y}$ with the corresponding value in $\mathbf{a}$. If set of facts selected by the evaluation of an aggregation function χ is empty, then χ evaluates to 0.

An *aggregate constraint* over $\mathcal{D}$ is a (universally quantified) expression of the form:

$$\forall \mathbf{x}\ \varphi(\mathbf{x}) \rightarrow \sum_{i=1}^{n} c_i \cdot \chi_i(\mathbf{y}_i) \leq c_{n+1}$$

where $c_1, \dots, c_n, c_{n+1}$ are constants in $\mathbb{Q}$, and each $\chi_i(\mathbf{y}_i)$ is an aggregation function, where $\mathbf{y}_i$ is a list of variables and constants, and every variable that occurs in $\mathbf{y}_i$ also occurs in $\mathbf{x}$. Examples of aggregate constraints will be provided in Sect. 6.

In the following, we will omit the universal quantification in front of dependencies and assume that all variables appearing in the body are universally quantified. Labeled nulls are not allowed to occur in dependencies. Throughout this paper we assume we are given an arbitrary but fixed schema $\mathcal{D}$. Unless otherwise stated, an atom (database, instance, dependency, etc.) is understood to be over $\mathcal{D}$.

An *update atom* is of the form $+P(\mathbf{w})$ or $-P(\mathbf{w})$. The update atoms $+P(\mathbf{w})$ and $-P(\mathbf{w})$ are *duals* of each other. A ground update atom $+P(\mathbf{t})$ states that $P(\mathbf{t})$ will be inserted into the database, whereas $-P(\mathbf{t})$ states that $P(\mathbf{t})$ will be deleted from the database. The symbol $\pm$ will be used as a placeholder for either $+$ or $-$. Given an update action $\alpha = +P(\mathbf{t})$ (resp. $\alpha = -P(\mathbf{t})$), $comp(\alpha)$ denotes the literal *not* $P(\mathbf{t})$ (resp. $P(\mathbf{t})$). The operator $comp(\cdot)$ is extended to sets of update actions in the standard way. We also define the inverse operator $comp(\cdot)^{-1}$.

We use $\mathcal{U}$ to denote a set of updates atoms, and say that $\mathcal{U}$ is *consistent* if it does not contain two update atoms $+P(\mathbf{t})$ and $-P(\mathbf{t})$. For any database I and consistent set of update atoms $\mathcal{U}$, $I \circ \mathcal{U}$ denotes the database I updated by means of $\mathcal{U}$, that is $I \circ \mathcal{U} = (I \cup \{P(\mathbf{t}) | + P(\mathbf{t}) \in \mathcal{U}\}) \setminus \{P(\mathbf{t}) | - P(\mathbf{t}) \in \mathcal{U}\}$.

Definition 1 (*Consistent database*) Given a database scheme $\mathcal{D}$ and a set of integrity constraints Σ on $\mathcal{D}$, a instance I of $\mathcal{D}$ is said to be *consistent* w.r.t. Σ if $I \models \Sigma$ in the standard model-theoretic sense, *inconsistent* otherwise ($K \not\models \Sigma$). □

Repairs. A database instance I may be inconsistent w.r.t. a given set of integrity constraints Σ. The restoration of the consistency in an inconsistent database can be achieved in a (possible infinite) number of ways, each of them yielding a consistent database. We have interest in minimal restorations of consistency, i.e. in performing actions that give us a new database instance J that shares the scheme with the original database I, but minimally differ from I according to some sort of *distance* between the original instance I and the alternative consistent instance J. In literature the notion of minimal restoration of consistency for a database I has been captured in terms of *repair* for I. Different forms of semantics for repairs have been proposed.

Definition 2 (*Repair and Repairing Update*) Let I be a database instance of a database scheme $\mathcal{D}$ and Σ be a set of integrity constraints on $\mathcal{D}$. Let $\mathcal{S}$ be a repair semantics. Given a partial order $\preceq_{\mathcal{S}}$ over databases instances (over $\mathcal{D}$) which depends on the repair semantics $\mathcal{S}$, a *repair* for I w.r.t. Σ under $\mathcal{S}$ is a new instance J such that:

1. J is over the same scheme and domain as I,
2. J satisfies Σ (i.e., $J \models \Sigma$),
3. there is no database instance K such that $K \prec_{\mathcal{S}} J \preceq_{\mathcal{S}} I$, among the instances satisfying the first two conditions.

Given a repair J w.r.t. Σ under $\mathcal{S}$, we call *repairing update* the set of update atoms $\mathcal{U}$ s.t. $I \circ \mathcal{U} = J$. □

We denote as $\mathcal{R}(I, \Sigma, \mathcal{S})$ the set of all repairs for a database I w.r.t. the set of constraints Σ under the semantics $\mathcal{S}$. The semantics $\mathcal{S}$ determines the partial order $\preceq_{\mathcal{S}}$, over the set of consistent database instances for a given database I.

Universal Solutions. For any database instance I and set of dependencies Σ over the same database schema, a *solution* for (I, Σ) is an instance J such that $I \subseteq J$ and $J \models \Sigma$ (i.e. J satisfies all the dependencies in Σ). The set of solutions for (I, Σ) will be denoted by $Sol(I, \Sigma)$.

Definition 3 (*Homomorphism*) Let I and J be two instances over the same database schema with values in $Consts \cup Nulls$. A *homomorphism* $h : I \rightarrow J$ is a mapping from $Consts(I) \cup Nulls(I)$ to $Consts(J) \cup Nulls(J)$ such that: (1) $h(c) = c$, for every $c \in Consts(I)$, and (2) for every fact $P_i(\mathbf{t})$ of I, we have that $P_i(h(\mathbf{t}))$ is a fact of J (where, if $\mathbf{t} = (a_1, ..., a_s)$, then $h(\mathbf{t}) = (h(a_1), ..., h(a_s))$).

J and K are said to be *homomorphically equivalent* if there is a homomorphism $h : K \rightarrow J$ and a homomorphism h': $J \rightarrow K$. □

Similarly to homomorphisms between instances, a homomorphism h from a formula $\rho(\mathbf{x})$, denoting either a conjunctive formula or a data dependency, to an instance J is a mapping from the variables $\mathbf{x}$ appearing in the formula to $Consts(J) \cup Nulls(J)$, such that for every atom $P(x_1, \ldots, x_n)$ of $\rho(\mathbf{x})$ the fact $P(h(x_1), \ldots, h(x_n))$ is in J. The application of homomorphism h to $\rho(\mathbf{x})$, denoted by $h(\rho(\mathbf{x}))$, consists in the replacement of variables in $\mathbf{x}$ with values in $Consts(J) \cup Nulls(J)$, as defined in h.

Example 2 Consider the three instances $I_1 = \{P(a, b), P(a, \eta_1)\}$, $I_2 = \{P(a, b)\}$ and $I_3 = \{P(a, b), P(a, d)\}$. I_1 and I_2 are homomorphically equivalent as there are two homomorphisms $h_1 = \{\eta_1/b\}$ and $h_2 = \emptyset$ such that $h_1(I_1) \subseteq I_2$ and $h_2(I_2) \subseteq I_1$. Although there is a homomorphism from I_1 to I_3 ($h_3 = \{\eta_1/d\}$ and $h_3(I_1) \subseteq I_3$) I_1 and I_3 are not homomorphically equivalent as there is no homomorphism from I_3 to I_1.

Consider now the conjunctive formula $\phi = P(x, y) \wedge P(x, z)$. There is, for instance, a homomorphism h from ϕ to I_1 (namely $h = \{x/a, y/b, z/\eta_1\}$. □

Definition 4 (*Universal solution*) For any database instance I and set of dependencies Σ over the same database schema, a solution for J for (I, Σ) is called *universal solution* if for every solution J' there exists a homomorphism $h : J \rightarrow J'$. The set of universal solutions for (I, Σ) will be denoted by $USol(I, \Sigma)$. □

Therefore, universal solutions are repairs where tuples cannot be deleted and nulls, introduced to repair the database, can be updated. The next theorem states that universal solutions are homomorphically equivalent.

Proposition 1 ([14]) *Let Σ be a set of data dependencies defined over a given database schema. Then*

- *for any database instance I, if J and K are universal solutions for I and Σ, then J and K are homomorphically equivalent,*
- *for any pair of database instances I_1 and I_2, such that J_1 and J_2 are universal solutions for (I_1, Σ) and (I_2, Σ), respectively, J_1 and J_2 are homomorphically equivalent if and only if $Sol(I_1, \Sigma) = Sol(I_2, \Sigma)$.* □

Therefore, universal solutions are unique up to homomorphic equivalence, and each of them embodies the space of solutions. The problems of checking whether universal solutions exist and how to compute them have been recently investigated. In particular, as we shall see next, it has been shown that the chase algorithm can be used to compute universal solutions and that every finite chase, if it does not fail, computes a (*canonical*) universal solution. If the chase fails, then no solution exists.

All universal solutions have the same core (up to isomorphism) which is the smallest universal solution. The complexity and the efficient computation of the core of a universal solution has been studied [15, 25]. Methods for directly computing the core by SQL queries in a data exchange framework where schema mappings are specified by source-to-target TGDs have been developed [48, 57].

Query answering. We now introduce the definition of *certain* or *consistent query answers*, and relate them to universal solutions.

Definition 5 The set of certain (or consistent) answers to a query Q over a database I with data dependencies Σ and repair semantics $\mathcal{S}$ is

$$certain(Q, I, \Sigma, \mathcal{S}) = \bigcap \{Q(J) \mid J \in \mathcal{R}(I, \Sigma, \mathcal{S})\}$$

Whenever the repair semantics $\mathcal{S}$ is understood we simply write $certain(Q, I, \Sigma) = \bigcap \{Q(J) \mid J \in \mathcal{R}(I, \Sigma)\}$ and for the case of universal solution we use the notation $certain(Q, I, \Sigma) = \bigcap \{Q(J) \mid J \in USol(I, \Sigma)\}$.

In the following, for any set of tuples S with null values, we use $S_{\downarrow}$ to denote the set of the tuples in S containing only constants.

Proposition 2 ([14]) *Let I be a database and Σ a set of data dependencies over a given schema. Then,*

- *for every union of conjunctive queries Q, $certain(Q, I, \Sigma) = Q(J)_{\downarrow}$, where J is a universal solution for I and Σ;*
- *if there is a database J such that, for every conjunctive query Q, $certain(Q, I, \Sigma) = Q(J)$, then J is a universal solution for I and Σ.* □

3 A Logic Programming Approach

As shown in [29] (see also [28, 30]), the set of repairs for a database with respect to a set of full integrity constraints can be computed by rewriting the constraints into disjunctive rules. More specifically, given a database I and a set of integrity constraints Σ, the technique derives a disjunctive program $DP(\Sigma)$ so that the repairs for I can be obtained from the stable models of $DP(\Sigma) \cup I$.

Definition 6 Given a full integrity constraint r of the form $\forall\, \mathbf{x}\ \bigwedge_{j=1}^{n} L_j, \varphi \rightarrow$, where L_j is a literal ($j \in [1..n]$) and φ is a conjunction of built-in atoms, $dj(r)$ denotes the logic rule:

$$\bigvee_{j=1}^{n} comp^{-1}(L_j) \leftarrow \bigwedge_{j=1}^{n} (L_j \vee comp^{-1}(not\ L_j)),\ \varphi$$

Given a set Σ of full integrity constraints, $DP(\Sigma) = \{dj(r) \mid r \in \eta\} \cup \{\leftarrow -b(\mathbf{x}), +b(\mathbf{x}) \mid b \text{ is a predicate symbol}\}$. □

The expression presented in the above definition, containing the disjunctive operator in the body, is used as shorthand for a set of disjunctive rules. To clarify the behavior of logic rules derived from constraints consider the following example:

Example 3 Consider the database consisting of the two relations *emp(Name,Dept)* and *dept(Name)* and the integrity constraint $r = emp(x, y) \rightarrow dept(y)$ stating that every department name appearing in the relation *emp* must also appear in the relation *dept*. This constraint can be rewritten into the denial form $emp(x, y) \wedge not\ dept(y) \rightarrow$. The corresponding logic rule $dj(r)$ is:$-emp(x, y) \vee +dept(x) \leftarrow (emp(x, y) \vee +emp(x, y)) \wedge (not\ \ dept(y) \vee -dept(y))$. It states that if a tuple $emp(x, y)$ was in the source database or it is inserted into the database (to satisfy some other constraint) and a tuple $dept(y)$ was not in the database or has been deleted from it, then the constraint is not satisfied and, in order to make it consistent, the fact $emp(x, y)$ has to be deleted or the fact $dept(y)$ has to be inserted. □

Given an interpretation $\mathcal{M}$, $UpdateAtoms(\mathcal{M})$ denotes the set of update actions in $\mathcal{M}$. In [29] has been proved that for each database I and set of full integrity constraints Σ:

- for every stable model $\mathcal{M}$ of $DP(\Sigma) \cup I$, $UpdateAtoms(\mathcal{M})$ is a repairing update for I w.r.t. Σ (*soundness*);
- for every repairing update $\mathcal{U}$ for I w.r.t. Σ there exists a stable model $\mathcal{M}$ of $DP(\Sigma) \cup I$ such that $\mathcal{U} = UpdateAtoms(\mathcal{M})$ (*completeness*).

This technique can be used to compute consistent answers to queries. Given a database I, a set of dependencies Σ and a query $Q = (g, P)$, where P is a set of datalog rules and g is the output predicate defined in $P \cup I$, the consistent answer of Q w.r.t. $\langle I, \Sigma \rangle$ can be computed by considering the stable models of $MP(g, P) \cup DP(\Sigma) \cup I$, where $MP(g, P)$ is obtained from P by replacing every base predicate symbol p with a new predicate symbol p' and by adding the rule: $p'(\mathbf{x}) \leftarrow (p(\mathbf{x}) \wedge not\ \ - p(\mathbf{x})) \vee +p(\mathbf{x})$. The above rule defining $p'(\mathbf{x})$ states whether atom $p(\mathbf{x})$ is true in the revised database. In particular, an atom $p(\mathbf{t})$ is true in the revised database if it was true in the source database and is not deleted or if it is inserted by our revision.

4 Active Integrity Constraints

Active Integrity Constraints are a special form of integrity constraints that allow the user to specify the update actions (insertions/deletions of facts) to be performed on an inconsistent database in order to restore consistency, and its connection with *Revision Programming*. The following example shows a case in which inconsistencies occur.

Example 4 Consider the relation schema *mgr(Name, Dept, Salary)* with the functional dependency *Dept* → *Name*, which can be defined through the first order formula

$$mgr(n, d, s), mgr(n', d, s') \rightarrow n = n'$$

and the inconsistent instance: I = *{mgr(john, cs, 1000), mgr(frank, cs, 2000)}*. There are two repaired databases obtained by means of minimal sets of update atoms: I_1 = *{mgr(frank, cs, 2000)}* obtained by applying the repairing update $\mathcal{D}_1$ = *{ −mgr(john, cs, 1000)}* (i.e. by deleting the tuple *mgr(john, cs, 1000)*) and I_2 = *{mgr(john, cs, 1000)}* obtained by applying the repairing update I_2 = *{−mgr(frank, cs, 2000)}* (i.e. by deleting the tuple *mgr(frank, cs, 2000)*). □

The problem with such a semantics is that all possible repaired databases are computed and no repairing strategy is defined. For instance, if we know that a given relation is sound (resp. complete) we could use active constraints to compute repairs making only insert (resp. delete) operations on this relation. In [12] a special form of integrity constraint, called *active integrity constraint* (*AIC*), has been presented. An AIC has a body and an head. Its body consists of a conjunction of literals which should be *false* and the head contains the atoms which have to be performed if the body is *true* (i.e. the constraint is violated). The following example illustrates the notion of AIC.

Example 5 Consider the database of Example 4 and the active integrity constraint:

$$mgr(n,d,s),\ mgr(n',d,s'),\ n \neq n', s \geq s' \rightarrow -mgr(n,d,s)$$

It is violated in the same cases of the functional dependency reported in the previous example. In addition, it imposes that in the case of conflicting tuples, the one reporting the higher salary has to be removed from the database. In this case, the constraint will repair the database by deleting the tuple *mgr(frank, cs, 2000)*. □

A (full) *Active Integrity Constraint* (*AIC*) is of the form:

$$\forall \mathbf{x} \bigwedge_{j=1}^{m} B_j(\mathbf{x}_j),\ \bigwedge_{j=m+1}^{n} not\ B_j(\mathbf{x}_j),\ \varphi(\mathbf{x}_0) \rightarrow \bigvee_{i=1}^{p} \pm A_i(\mathbf{y}_i)$$

where, $\mathbf{x} = \bigcup_{j=1}^{m} \mathbf{x}_j$, $\mathbf{x}_i \subseteq \mathbf{x}$, for each $i \in [0\,..\,n]$, and $\mathbf{y}_i \subseteq \mathbf{x}$, for each $i \in [1\,..\,p]$. Given an active integrity constraint r, we denote the conjunction $B_1(\mathbf{x}_1) \wedge \ldots \wedge B_m(\mathbf{x}_m) \wedge not\ B_{m+1}(\mathbf{x}_{m+1}) \wedge \ldots \wedge not\ B_n(\mathbf{x}_n) \wedge \varphi(\mathbf{x}_0)$ as $body(r)$ and the disjunction $\pm A(\mathbf{y}_1) \vee \ldots \vee \pm A(\mathbf{y}_p)$ as $head(r)$.
Now we are ready to provide the concept of *Founded Repairing Update*. Let I be a database, Σ a set of active integrity constraints and $\mathcal{U}$ a repairing update for $\langle I, \Sigma \rangle$.

- An update atom $\alpha \in \mathcal{U}$ is *founded* with respect to $\langle I, \Sigma \rangle$ and $\mathcal{U}$ if there exists $r \in ground(\Sigma)$ such that $\alpha \in head(r)$ and $I \circ (\mathcal{U} \backslash \{\alpha\}) \models body(r)$. We say that r supports α.
- $\mathcal{U}$ is *founded* w.r.t. $\langle I, \Sigma \rangle$ if all its update atoms are *founded* w.r.t. $\langle I, \Sigma \rangle$ and $\mathcal{U}$.

Then, an update atom α is founded if two conditions are verified: i) it belongs to the head of an active integrity constraint r, and ii) if we discard it, the database updated by means of the remaining update atoms ($I \circ (\mathcal{U} \setminus \{\alpha\})$), violates r. This means that α is inferred by r (because it belongs to its head) and it is *necessary* to repair the database in order to satisfy r (because if we discard it, the database violates r).

Example 6 Consider the following set Σ of active integrity constraints:

$$mgr(e, p), proj(p, d), not\ emp(e, d) \rightarrow +emp(e, d)$$
$$emp(e, d), emp(e, d'), d \neq d' \rightarrow -emp(e, d) \vee -emp(e, d')$$

The first constraint states that every manager e of a project p, carried out by a department d, must be an employee of d, whereas the second one states that every employee must be in only one department. Consider now the database $I = \{mgr(e_1, p_1), proj(p_1, d_1), emp(e_1, d_2)\}$. There are three repairs: $I_1 = \{-mgr(e_1, p_1)\}$, $I_2 = \{-proj(p_1, d_1)\}$ and $I_3 = \{+emp(e_1, d_1), -emp(e_1, d_2)\}$. I_3 is the only founded repairing update as only the atoms $+emp(e_1, d_1)$ and $-emp(e_1, d_2)$ are derivable from Σ. □

The paper [12] presents the properties of active integrity constraints and shows that, under the proposed semantics, each update atom occurring in the head of an active integrity constraint that cannot falsify the corresponding body is useless and can be deleted. It also shows that active constraints can be normalized so that every rule contains at most one update atom in the head, and that the computation of founded repairs can be done by rewriting the constraints into a Datalog program and computing its stable models [23]; each stable model will represent a founded repairing update. Since, in the general case, the existence of founded repairs is not guaranteed, the paper investigates a different semantics where update atoms defined by active integrity constraints are interpreted as preference conditions on the set of possible repairing updates (*preferable* semantics). An important result presented in [12] is that the complexity of computing founded repairing updates, preferred repairing updates and answers is not harder than computing (general) repairing updates and answers. Active integrity constraints and preferences have been combined into a general framework in [11], which introduced a generalization of active integrity constraints that allows users to express preferences among updates (see also [10, 35]).

In [42, 43] revision programming, a logic-based framework for describing constraints on databases and providing a computational mechanism to enforce them, is introduced. There are many similarities between the approach proposed in [42, 43] and the one defined in [12], but also significant differences as the aim of both formalisms is to design *policies* to manage inconsistent data [9]. The original semantics of AICs and revision programming seemingly cannot be related in any direct way as they have different computational properties. While the semantics for revision programming do not have the minimality of change property, founded repairs with respect to active integrity constraints do. Reference [9] further investigates the relationship between AIC and RP and proposes new alternative semantics for both formalisms.

5 Approximate Consistent Query Answers

Since computing consistent query answers is in general an intractable problem, different (tractable) approximation algorithms have been proposed over the years. In this section, two such approaches are discussed: a framework based on a three-valued semantics [22, 31], and a technique for computing approximate "probabilistic" consistent query answers [32, 34].

A Three-valued Semantics for Querying Inconsistent Databases. The first approach we discuss is a technique for computing a sound (but possibly incomplete) set of consistent query answers in polynomial time. Universal integrity constraints are considered.

The framework is based on a *three-valued* repair strategy, that is, in order to restore consistency, the truth value of tuples can be modified into *true*, *false*, or *undefined*. This repair strategy yields three-valued databases and thus a new semantics of constraint satisfaction is considered. For standard (two-valued) databases the semantics coincides with the classical one. One interesting property of the framework is that the set of three-valued repairs defines a lower semi-lattice whose top elements are standard (two-valued) repairs and whose bottom element is called *deterministic repair*.

In order to compute approximate consistent query answers, it suffices to evaluate queries over the deterministic repair only, which in a sense summarizes all the possible repairs (with some loss of information, though). The so-obtained answers are approximate in the sense that they are sound (true and false atoms in the answers are, respectively, true and false under the classical two-valued semantics), but possibly incomplete. The technique has a polynomial time complexity.

Classes of queries and constraints for which the technique is also complete have been investigated in [22], where it has also been shown how query evaluation can be carried out by means of logic programs, which can be then evaluated using an answer set programming system (e.g. [36]). The following example illustrates the basic idea.

Example 7 Consider a database consisting of a relation *emp*(*Name*, *Salary*) whose instance is $\{emp(carl, 500), emp(john, 1500), emp(bob, 700), emp(bob, 800)\}$. The database is inconsistent w.r.t. the following universal constraints:

$$emp(e, s), s > 1000 \rightarrow$$
$$emp(e, s_1), emp(e, s_2), s_1 \neq s_2 \rightarrow$$

The database admits two repairs: in both of them the second tuple is deleted, in one of them only the third tuple is deleted, and in the other one only the fourth tuple is deleted. The first tuple is kept in both repairs. The deterministic repairs summarizes this information into a unique database where the first tuple has truth value *true*, the second one has truth value *false*, and the last two tuples have truth value *undefined*, as the repairs do not agree on this piece of information. □

Approximate Probabilistic Consistent Query Answers. The second framework we discuss aims at preserving more information in an inconsistent database and providing more informative query answers. Functional dependencies are considered.

The repair strategy relies on value updates, so that the information of an inconsistent database is preserved better than approaches based on tuple deletions. Answers to queries are tuples associated with probabilities, which depend on the number of repairs from which an answer can be derived. This approach provides more informative query answers than the classical notion of consistent query answer, as shown below.

Example 8 Consider a relation *emp*(*Name*, *Dept*) whose instance is {*emp*($john$, cs), *emp*($john$, $math$), *emp*(*bob*, cs), *emp*(*bob*, $physics$). Suppose the functional dependency $fd : Name \rightarrow Dept$ is defined, stating that each employee cannot be associated with different departments. The employee relation above is clearly inconsistent, and has four repairs, each obtained by choosing either cs or $math$ as $john$'s department, and either cs or $physics$ as *bob*'s department.

Consider now a query asking for the departments in the relation. The intuition suggests that cs should be the most probable department as each employee could work for it, whereas $math$ and $physics$ should be less probable as only $john$ could work for the former and only *bob* could work for the latter. The probabilistic query answers in this case are $\{(cs, 3/4), (math, 2/4), (physics, 2/4)\}$. Observe that cs, $math$ and $physics$ are not consistent answers, and there is no discrimination among them using the classical notion of consistent query answer. □

As the computation of probabilistic query answers has been shown to be $FP^{\#P}$-complete, a polynomial time approximation algorithm based on a compact representation of all repairs has been proposed in [32].

6 Aggregate Constraints and Inconsistent Numerical Databases

Classical forms of constraint (such as functional and inclusion dependencies) often do not suffice to manage data consistency, as they cannot be used to define algebraic relations between stored values. In fact, this issue frequently occurs in several scenarios, such as scientific databases, statistical and biological databases, and data warehouses, where numerical values of tuples are derivable by aggregating values stored in other tuples. In this section, we discuss the problem of repairing and querying numerical databases violating *aggregate constraints*.

Example 9 Table 1 represents a relation *balanceSheets* obtained from the balance sheets of two consecutive years of a company. These data were acquired by means of an OCR (*Optical Character Recognition*) tool from paper documents. Values '*det*', '*aggr*' and '*drv*' in column *Type* stand for *detail*, *aggregate* and *derived*, respectively.

Table 1 Relation *balanceSheets*

	Year	Section	Subsection	Type	Value
t_1	2008	Receipts	Beginning cash	drv	50
t_2	2008	Receipts	Cash sales	det	900
t_3	2008	Receipts	Receivables	det	100
t_4	2008	Receipts	Total cash receipts	aggr	1250
t_5	2008	Disbursements	Payment of accounts	det	1120
t_6	2008	Disbursements	Capital expenditure	det	20
t_7	2008	Disbursements	Long-term financing	det	80
t_8	2008	Disbursements	Total disbursements	aggr	1220
t_9	2008	Balance	Net cash inflow	drv	30
t_{10}	2008	Balance	Ending cash balance	drv	80
t_{11}	2009	Receipts	Beginning cash	drv	80
t_{12}	2009	Receipts	Cash sales	det	1110
t_{13}	2009	Receipts	Receivables	det	90
t_{14}	2009	Receipts	Total cash receipts	aggr	1200
t_{15}	2009	Disbursements	Payment of accounts	det	1130
t_{16}	2009	Disbursements	Capital expenditure	det	40
t_{17}	2009	Disbursements	Long-term financing	det	20
t_{18}	2009	Disbursements	Total disbursements	aggr	1120
t_{19}	2009	Balance	Net cash inflow	drv	10
t_{20}	2009	Balance	Ending cash balance	drv	90

Relation *balanceSheets* must satisfy the following aggregate constraints:
(1) For each section and year, the sum of the values of all *detail* items must be equal to the value of the *aggregate* item of the same section and year. That is:

$$\kappa_1 : \textit{balanceSheets}(y, x, \_, \_, \_) \rightarrow \chi_1(x, y, \text{`}det\text{'}) - \chi_1(x, y, \text{`}aggr\text{'}) = 0$$

where $\chi_1(x, y, z) = \langle \textit{balanceSheets, Value, (Section} = x \wedge \textit{Year} = y \wedge \textit{Type} = z) \rangle$ and $\chi_2(x, y) = \langle \textit{balanceSheets, Value, (Year} = x \wedge \textit{Subsection} = y) \rangle$ are aggregation functions on *balanceSheets (Year, Section, Subsection, Type, Value)*. For instance, $\chi_1(x, y, z)$ returns the sum of *Value* of all the tuples where *Section=x*, *Year=y* and *Type=z*, and evaluating χ_1('Receipts', '2008', 'det') on the instance shown in Table 1 results in 900 + 100=1000, whereas χ_1('Disbursements', '2008', 'aggr') returns 1220.
(2) For each year, the *net cash inflow* must be equal to the difference between *total cash receipts* and *total disbursements*. That is:

$$\kappa_2 : \textit{balanceSheets}(x, \_, \_, \_, \_) \rightarrow \chi_2(x, \text{`net cash inflow'}) - \big(\chi_2(x, \text{`total cash receipts'}) - \chi_2(x, \text{`total disbursements'})\big) = 0.$$

(3) For each year, the *ending cash balance* must be equal to the sum of the *beginning cash* and the *net cash inflow*. That is:

$$\kappa_3: \ \textit{balanceSheets}\,(x,\_,\_,\_,\_) \rightarrow \quad \chi_2(x,\ \text{`ending cash balance'}) - \big(\chi_2(x,\ \text{`beginning cash'}) + \chi_2(x,\ \text{`net cash inflow'})\big) = 0.$$

Although the original balance sheet (in paper format) was consistent, its digital version is not, as some symbol recognition errors occurred during the digitizing phase. In fact, constraints κ_1, κ_2 and κ_3 are not satisfied on the acquired data shown in Table 1: For instance, for year 2008, in section *Receipts*, the aggregate value of *total cash receipts* is not equal to the sum of detail values of the same section: $900 + 100 \neq 1250$;

Repairing semantics. Most of the work discussed in the previous sections considers repairs obtained by performing tuple insertions/deletions on the inconsistent database. However, this repairing strategy is not suitable for contexts analogous to that of Example 9. In fact, using tuple insertions/deletions as basic primitives means hypothesizing that the OCR tool skipped/invented a whole row when acquiring the source paper document, which is rather unrealistic. In this scenario, a repairing semantics based on *attribute-update operations* only is more reasonable, as updating single attribute values is the most natural way for fixing inconsistencies resulting from symbol recognition errors. Moreover, in the scenario of Example 9, where inconsistency is due to acquisition errors, repairing the data by means of sets of updates of minimum cardinality is well-suited since it corresponds to the case that the acquiring system made the minimum number of bad symbol-recognition errors, which can be considered the most probable event. Thus, two repairing semantics $\mathcal{S}$, namely *card*- and *set*-minimal semantics are considered in this setting. Both of them rely on updating values of numerical attributes to restore consistency, while *card*- or set *set*-minimality of the set of updated values is used to define the partial order $\preceq_{\mathcal{S}}$ assessing the reasonableness of a repair.
Computational complexity. In [17, 20] a characterization of several data-complexity issues related to repairing data and computing consistent query answers is provided. In particular, the following fundamental decision problems were investigated: (i) *repair existence*: deciding whether a database inconsistent w.r.t. a set of aggregate constraints can be repaired; (ii) *minimal repair checking*: deciding whether a given repair is minimal, under either the *card*- or the *set*- minimality semantics; (iii) *consistent query answer*: deciding whether a boolean query is consistently true. The sensitivity of computational complexity of these problems was investigated considering the domain of numerical attributes (binary, integers or rationals), the minimality semantics (*card*- or *set*-minimality), and the form of the constraints. Besides the general form of aggregate constraints, a restricted but still expressive form of aggregate constraints, namely *steady aggregate constraints*, was explored. It is worth noting that the loss in expressiveness of using steady aggregate constraints is not dramatic, as they suffice to express relevant integrity constraints in many real-life scenarios.

For instance, all the aggregate constraints introduced in Example 9 can be expressed by means of steady aggregate constraints.

Techniques for computing repairs and consistent answers. Computing repairs of inconsistent numerical data is relevant in several applications, since the availability of a consistent version of the data is mandatory for accomplishing a number of analysis tasks. For instance, in Example 9, in order to perform analysis aiming at determining liquidity problems as well as the financial reliability of a company, a consistent version of the balance-sheets of the company should be available. In [16, 20], it is shown that the problem of finding a *card*-minimal repair for a database w.r.t. a set of steady aggregate constraints can be modeled as a MILP (Mixed Integer Linear Programming) problem, thus allowing us to adopt standard techniques addressing MILP problems to accomplish the computation of reasonable repairs. In general, there may be several *card*-minimal repairs for a database violating a given set of aggregate constraints. In [18], well-established information on the application domain of data to be repaired are exploited to choose the most reasonable repairs (namely, *preferred* repairs) among the *card*-minimal ones. Finally, in [19, 21] the problem of evaluating *aggregate queries* (SUM, MIN, MAX) over data inconsistent w.r.t. aggregate constraints is addressed, and a technique for providing *range-consistent answers* of these kinds of queries is introduced. The range-consistent answer of an aggregate query is the narrowest interval containing all the answers of the query evaluated on every possible repaired database [1].

7 Computing Universal Solutions

So far we have considered universally quantified dependencies. We now consider dependencies with existential variables and repairs consisting of universal solutions.

As recalled in Proposition 2, certain answers to a union of conjunctive queries Q can be computed by evaluating Q over an arbitrary universal solution, that is, $certain(Q, I, \Sigma) = Q(J)_{\downarrow}$, where $J \in USol(I, \Sigma)$. This means that to determine the certain answers to a union of conjunctive queries Q over a database I with dependencies Σ, it is not necessary to compute all solutions for (I, Σ), but it suffices to compute just an arbitrary universal solution. Therefore, the computation of a universal solution is particularly relevant. It is worth mentioning that the aforementioned property has applications in query answering under dependencies, query answering in data exchange, and query answering with incomplete and inconsistent data [8, 14].

The chase. The *chase* takes as input a database I and a set Σ of dependencies, and whenever it terminates without failing, it constructs a universal solution for (I, Σ) [13, 14]. A *chase step* consists of enforcing a TGD or an EGD. As detailed later, the chase step is used by different variants of the chase (*standard*, *oblivious*, *semi-oblivious*), each of which relies on a different condition of "applicability" of the chase step. Thus, the following definition does not incorporate a notion of applicability, but it will be combined with different notions of applicability to define the different variants of the chase. A *substitution* γ is either the empty set or a singleton $\{\eta/t\}$, where η is a labeled null and t is either a labeled null or a constant. The result of applying γ to an expression F (e.g., term, atom, set of atoms, etc.), denoted $F\text{ß}\gamma$, is F if $\gamma = \emptyset$, otherwise it is the expression obtained from F by replacing every occurrence of η with t.

Definition 7 (*Chase step*) Let I be an instance, r a dependency, and h a homomorphism from $Body(r)$ to I. An expression of the form $I \stackrel{r,h,\gamma}{\longrightarrow} J$ is a *chase step* if the following conditions hold.

1. If r is a TGD $\varphi(\mathbf{x}, \mathbf{y}) \to \exists \mathbf{z}\, \psi(\mathbf{x}, \mathbf{z})$ then let h' be the homomorphism obtained by extending h so that each variable in $\mathbf{z}$ is assigned a fresh labeled null not occurring in I. Then, $J = I \cup h'(\psi(\mathbf{x}, \mathbf{z}))$. Furthermore, γ is the empty substitution.
2. If r is an EGD $\varphi(\mathbf{x}, \mathbf{y}) \to x_1 = x_2$ then $h(x_1) \neq h(x_2)$. Furthermore,
   (a) If $h(x_1), h(x_2) \in \mathit{Consts}$, then $J = \perp$ and γ is the empty substitution.
   (b) Otherwise, γ and J are defined as follows. If $h(x_1)$ is a labeled null, then $\gamma = \{h(x_1)/h(x_2)\}$; otherwise, $\gamma = \{h(x_2)/h(x_1)\}$. Moreover, $J = I\text{ß}\gamma$.

In a chase step, γ is used to keep track of the substitution performed when an EGD is enforced.

Example 10 Consider the following set of dependencies:
$\Sigma = \{r_1 : \; N(x) \to \exists y \; E(x, y); \; r_2 : \; E(x, y) \to N(y); \; r_3 : \; E(x, y) \to x = y\}$
and the database $I = \{N(a)\}$. Let $h_1 = \{x/a\}$ be the homomorphism. Then, $I_1 \stackrel{r_1,h_1,\gamma_1}{\longrightarrow} I_2$ is a chase step, where $I_2 = I_1 \cup \{E(a, \eta_1)\} = \{N(a), E(a, \eta_1)\}$ and γ_1 is the empty substitution (as r_1 is a TGD). Consider now the homomorphism $h_2 = \{x/a, y/\eta_1\}$. Then, $I_2 \stackrel{r_2,h_2,\gamma_2}{\longrightarrow} I_3$ is a chase step, where $I_3 = I_2 \cup \{N(\eta_1)\} = \{N(a), \; E(a, \eta_1), N(\eta_1)\}$ and γ_3 is the empty substitution. Another possible chase step starting from I_2 is $I_2 \stackrel{r_3,h_2,\gamma_2'}{\longrightarrow} I_3'$, where $\gamma_2' = \{\eta_1/a\}$ and $I_3' = I_2\text{ß}\gamma_2' = \{N(a), E(a, a)\}$. □

A *chase sequence* of (I, Σ) is a (possibly infinite) sequence of chase steps $S = I_1 \stackrel{r_1,h_1,\gamma_1}{\longrightarrow} I_2 \stackrel{r_2,h_2,\gamma_2}{\longrightarrow} I_3 \cdots$ such that $I_1 = I$ and every $r_i \in \Sigma$. Moreover:

- S is a *standard chase sequence* if it is an exhaustive application of chase steps s.t. for each $I_i \stackrel{r_i,h_i,\gamma_i}{\longrightarrow} I_{i+1}$ in S, if r_i is a TGD, then there is no extension of h_i to a homomorphism h_i' from $Body(r_i) \cup Head(r_i)$ to I_i.
- S is an *oblivious chase sequence* if it is an exhaustive application of chase steps s.t. for each $I_i \stackrel{r_i,h_i,\gamma_i}{\longrightarrow} I_{i+1}$ in S, there is no chase step $I_j \stackrel{r_j,h_j,\gamma_j}{\longrightarrow} I_{j+1}$ in S such that $j < i$, $r_j = r_i = r$, and for each variable x occurring in the body of r, we have that $h_i(x) = h_j(x)\text{ß}\gamma_j\text{ß} \cdots \text{ß}\gamma_{i-1}$.
- S is a *semi-oblivious chase sequence* if it is an exhaustive application of chase steps s.t. for each $I_i \stackrel{r_i,h_i,\gamma_i}{\longrightarrow} I_{i+1}$ in S, there is no chase step $I_j \stackrel{r_j,h_j,\gamma_j}{\longrightarrow} I_{j+1}$ in S such that $j < i$, $r_j = r_i = r$, and for each variable x occurring in both the body and the head of r, we have that $h_i(x) = h_j(x)\text{ß}\gamma_j\text{ß} \cdots \text{ß}\gamma_{i-1}$.

The following example shows the different behaviors of standard, oblivious, and semi-oblivious chase sequences.

Example 11 Consider the database $I = I_1 = \{E(a, b)\}$ and a set Σ consisting only of the TGD r: $E(x, y) \rightarrow \exists z\, E(x, z)$. Since $I \models r$, the only standard chase sequence of I with Σ is the empty sequence.

A non-empty (terminating) semi-oblivious chase sequence is $I_1 \xrightarrow{r,h_1,\gamma_1} I_2$, where $h_1(x) = a$, $h_1(y) = b$, γ_1 is the empty substitution, and $I_2 = I_1 \cup \{E(a, \eta_1)\} = \{E(a, b), E(a, \eta_1)\}$. Notice that adding the chase step $I_2 \xrightarrow{r,h_2,\gamma_2} I_3$, with $h_2(x) = a$, $h_2(y) = \eta_1$, $\gamma_2 = \emptyset$, and $I_3 = I_2 \cup \{E(a, \eta_2)\}$, does not result in a semi-oblivious chase sequence, because of the presence of the chase step $I_1 \xrightarrow{r,h_1,\gamma_1} I_2$ in the same chase sequence, with $h_1(x)\gamma_1 = h_2(x) = a$. As for the oblivious chase, the infinite sequence whose first step is $I_1 \xrightarrow{r,h_1,\gamma_1} I_2$ discussed above, and the i-th chase step ($i > 1$) is $I_i \xrightarrow{r,h_i,\gamma_i} I_{i+1}$, with $h_i(x) = a$, $h_i(y) = \eta_{i-1}$, $\gamma_i = \emptyset$, and $I_{i+1} = I_i \cup \{E(a, \eta_i)\}$ is an (infinite) oblivious chase sequence. □

A standard (resp. oblivious, semi-oblivious) chase sequence S can be finite (when no further chase step can be applied) or infinite (when there is always a further chase step that can be applied)—in the former case we also say that the sequence is *terminating*. If S is finite and consists of m chase steps, we say that I_m is the *result* of S. If $I_m = \perp$ then S is *failing*, otherwise it is *successful*. For instance, the first standard chase sequence discussed in Example 10 is terminating, successful, and its result is I'_3.

In the presence of TGDs only, the oblivious (resp. semi-oblivious) chase procedure is equivalent to the computation of the fixpoint of a particular Skolemized version of Σ with I, where Skolemized terms are used in place of labeled nulls. For instance, the Skolemized version of dependency r in Example 11 for the oblivious (resp., semi-oblivious) chase is $E(x, y) \rightarrow E(x, f_z^r(x, y))$ (resp., $E(x, y) \rightarrow E(x, f_z^r(x))$).

As shown in [14], for every database I and set of dependencies Σ, (1) if J is the result of some successful terminating standard chase sequence of I with Σ, then J is a universal solution for (I, Σ), called *canonical*; (2) if some failing standard chase sequence of I with Σ exists, then there is no solution for (I, Σ). In some cases, we cannot produce a universal solution by the chase as there is no terminating sequence, although a solution does exist.

The *core chase* has been proposed to identify a preferable universal solution [13, 15]. To define the core chase, we first need to introduce the notion of a core of an instance. Roughly speaking, the core of an instance J is the smallest subset of J that is also a homomorphic image of J. More precisely, a subset C of an instance J is a *core of* J if there is a homomorphism from J to C, but there is no homomorphism from J to a proper subset of C. Cores of J are unique up to isomorphism and therefore we can talk about "the" core of J, which is denoted as $core(J)$.

A *core chase sequence* is a sequence of *core chase steps*. Basically, a core chase step first applies all possible standard chase steps "in parallel", and then computes the core of the resulting instance. As all standard chase steps are applied in parallel, the core chase eliminates the nondeterminism of the standard chase. More formally, given an instance I and a set of dependencies Σ, a core chase step consists of the following two sub-steps: *(i)* $J = \bigcup_{I \xrightarrow{r,h,\gamma} I'} I'$, where each $I \xrightarrow{r,h,\gamma} I'$ is understood to be a standard chase step; *(ii)* $J' = core(J)$. Then, J' is the result of the core chase step. [13] showed that whenever there is a universal solution for (D, Σ), the core chase is able to construct one, that is, the core chase is a *complete* procedure for finding universal solutions. Moreover, every core chase sequence of D with Σ constructs the same (up to isomorphism) universal solution.

Example 12 Consider the database D and the set of dependencies $\Sigma = \{r\}$ of Example 11. Recall that there is no standard chase step involving D and r. As the core chase starts by applying all standard chase steps, the only core chase sequence is the empty one, similar to the standard chase case. □

Termination Classes. We denote by $\mathsf{CT}^{\mathsf{c}}_{\forall}$, where $\mathsf{c} \in \{\mathsf{std}, \mathsf{obl}, \mathsf{sobl}, \mathsf{core}\}$ stands for the standard, oblivious, semi-oblivious, and core chase, the class of sets of dependencies Σ such that for every database I *all* c-chase sequences of I with Σ are terminating. Analogously, we denote by $\mathsf{CT}^{\mathsf{c}}_{\exists}$ the class of sets of dependencies Σ such that for every database I *there is* a terminating c-chase sequence of I with Σ.

Even focusing on TGDs only, the problem of verifying whether a set of dependencies belongs to $\mathsf{CT}^{\mathsf{c}}_{\forall}$ or $\mathsf{CT}^{\mathsf{c}}_{\exists}$, for $\mathsf{c} \in \{\mathsf{std}, \mathsf{obl}, \mathsf{sobl}, \mathsf{core}\}$, is undecidable [24, 26]. Thus, the best practical approach is to find relevant decidable classes of dependencies included in these classes. For sets of TGDs only, it was shown in [49, 53] that:

$$\mathsf{CT}^{\mathsf{obl}}_{\forall} = \mathsf{CT}^{\mathsf{obl}}_{\exists} \subsetneq \mathsf{CT}^{\mathsf{sobl}}_{\forall} = \mathsf{CT}^{\mathsf{sobl}}_{\exists} \subsetneq \mathsf{CT}^{\mathsf{std}}_{\forall} \subsetneq \mathsf{CT}^{\mathsf{std}}_{\exists} \subsetneq \mathsf{CT}^{\mathsf{core}}_{\forall} = \mathsf{CT}^{\mathsf{core}}_{\exists}$$

The above hierarchy is relevant because if we determine that a set of TGDs belongs to $\mathsf{CT}^{\mathsf{c}}_{\mathsf{q}}$ with $\mathsf{q} \in \{\forall, \exists\}$ and $\mathsf{c} \in \{\mathsf{obl}, \mathsf{sobl}\}$, then Σ belongs to $\mathsf{CT}^{\mathsf{std}}_{\forall}$ (and, of course, $\mathsf{CT}^{\mathsf{std}}_{\exists}$), and, in some cases, the analysis of the oblivious or semi-oblivious chase is easier. In fact, the importance of these chase variants has been widely recognized and their behavior has been studied in different works [4, 27, 40, 44, 49].

Several sufficient conditions for chase termination have been proposed over the years—we call them *termination criteria*. We will use calligraphic style $\mathcal{C}$ to denote the class of sets of dependencies recognized by a criterion C (written in italics).

Static approaches. The first and basic effort concerning the formalization of a (decidable) sufficient condition guaranteeing that all standard chase sequences are terminating, independently from the database, is *weak acyclicity* (*WA*) [14]. Roughly speaking, it checks whether the TGDs do not allow for nulls to cyclically propagate. The approach works for sets of dependencies containing both TGDs and EGDs, even though the latter are ignored in the analysis (as a strong condition is imposed on TGDs).

An extension of weak acyclicity, called *stratification* (Str), has been proposed in [13]. The idea behind stratification is to decompose the set of dependencies into independent subsets, where each subset consists of dependencies that may fire each other, and to check each component separately for weak acyclicity. However, [49] showed that stratification is not able to check whether all standard chase sequences are terminating (as weak acyclicity does), but ensures only that there is a terminating standard chase sequence. A variant of stratification, called *c-stratification* ($CStr$), guaranteeing that all standard chase sequences are terminating, has been proposed in [49]. C-stratification is defined in the same way as stratification, but the oblivious chase is used instead of the standard one to determine whether a dependency fires another. Both Str and $CStr$ allow TGDs as well as EGDs, but the analysis of EGDs is limited to the firing relation only. A different extension of weak acyclicity, called *safety* (SC), has been proposed in [50]. The improvement is obtained by considering only "affected" positions [8], that is, positions which may actually contain null values. The approach works for sets of dependencies containing both TGDs and EGDs, but the latter are neglected altogether in the analysis.

Another extension of weak acyclicity (which indeed strictly extends SC) has been introduced in [44] under the name of *super-weak acyclicity* (SwA). In addition to considering how dependencies may activate each other, SwA also takes into account the fact that the same variable may appear more than once in the body, and thus a dependency is not fired when different nulls are inserted in positions associated with the same variable. The analysis is carried out by using the semi-oblivious chase. The approach is defined for sets of TGDs only, as EGDs are emulated via "substitution-free simulation".

Safe restriction (*SR*) and *inductive restriction* (*IR*) extend c-stratification, but still perform a limited analysis of EGDs [50]. In terms of expressivity, these approaches are not comparable with SwA. Both SwA and *IR* have been extended by the *Local Stratification* (LS) criterion [37]; however, LS neglects EGDs altogether.

As for the relative expressivity of the termination criteria discussed above, [40] showed that $\mathcal{C}Str \subsetneq \mathcal{SR} \subsetneq \mathcal{IR} \subsetneq \mathcal{LS}$ and $\mathcal{S}w\mathcal{A} \subsetneq \mathcal{LS}$.

Semi-dynamic approaches. In [27], the *model-faithful acyclicity* (MFA) and *model-summarising acyclicity* (MSA) techniques have been proposed. The idea is to run the oblivious (or semi-oblivious) chase and then use sufficient checks to identify cyclic computations. Since no sufficient, necessary, and computable test can be given for the latter, [27] adopted an approach of "raising the alarm" and stop the process if a "cyclic" term $f(t)$ is derived, i.e., where f occurs in t. This is done in a declarative way by extending a given set of dependencies Σ into a new set Σ', and then checking whether Σ' does not entail a special predicate. The two aforementioned techniques are defined for TGDs only, as EGDs are assumed to be emulated through substitution-free simulation.

Rewriting approaches. Rewriting techniques for checking chase termination have been proposed in [33, 37, 40]. They consist in rewriting a set of TGDs Σ into a new set Σ^{α} with the aim of verifying structural properties for chase termination on Σ^{α} rather than Σ. These techniques have been defined for TGDs only and perform an analysis of the semi-oblivious chase. [40] showed that most of the termination criteria improve if we consider adorned TGDs rather than the original ones. The rewriting approach has also been used to define the *acyclicity* (AC) criterion.

Observe that the termination problem of the oblivious and semioblivious chase for TGDs consists in checking whether the set of logic rules obtained by skolemized TGDs terminates. Therefore, the problem is strongly related to the termination check for logic programs and termination criteria defined for logic porograms (e.g. [2, 3, 5, 7, 38, 39, 41]) could also be applied to check chase termination. Moreover, as skolemized TGDs have a special form (i.e. function symbols may appear only in the head of rules and each function symbol may occur in exactly one rule), specific criteria designed for checking chase termination are more effective.

Dealing with EGDs. Most of the chase termination criteria proposed in the literature focus on TGDs considering EGDs in a very limited way. More general approaches (including *SwA*, *LS*, *MFA*, *MSA*) as well as rewriting techniques were meant to guarantee termination of TGDs only. An "indirect" way of dealing with EGDs has been proposed in [25, 44]. Specifically, the analysis of a set of dependencies Σ containing both TGDs and EGDs is performed on a set Σ' derived from Σ and containing only TGDs. The aim is to "simulate" the behavior of the EGDs by means of TGDs only. The first approach of this kind, known as *natural simulation*, has been proposed in [25], and further refined by the *substitution-free simulation* in [44].

The termination criteria discussed so far ensure that all standard chase sequences are terminating, except for stratification (which ensures the existence of at least one terminating standard chase sequence), and perform a limited analysis of EGDs (or no analysis altogether), thereby imposing stronger conditions on TGDs. It is worth noticing that constructing one terminating standard chase sequence suffices for the purpose of getting a universal model. By considering this weaker condition to be ensured and by performing a direct analysis of EGDs, the techniques proposed in [6] identify sets of dependencies that are not captured by any of the aforementioned criteria. The paper also exploits the relationships between the classes $\mathsf{CT}^{\mathsf{c}}_{\mathsf{q}}$, where $\mathsf{q} \in \{\forall, \exists\}$ and $\mathsf{c} \in \{\mathsf{obl}, \mathsf{sobl}, \mathsf{std}, \mathsf{core}\}$, when arbitrary sets of dependencies are considered.

Theorem 1 (Hierarchy) *Reference [6] For general dependencies (including TGDs and EGDs), the relationships showed in Table 2 hold, where $C_1 \nparallel C_2$ is a shorthand for $C_1 \nsubseteq C_2 \wedge C_2 \nsubseteq C_1$.* □

Table 2 Relationships among the $\mathsf{CT}^{\mathsf{c}}_{\mathsf{q}}$'s classes.

TGDs	TGDs and EGDs
$\mathsf{CT}^{\mathsf{obl}}_{\forall} = \mathsf{CT}^{\mathsf{obl}}_{\exists}$ $\mathsf{CT}^{\mathsf{sobl}}_{\forall} = \mathsf{CT}^{\mathsf{sobl}}_{\exists}$	$\mathsf{CT}^{\mathsf{obl}}_{\forall} \subsetneq \mathsf{CT}^{\mathsf{obl}}_{\exists}$ $\mathsf{CT}^{\mathsf{sobl}}_{\forall} \subsetneq \mathsf{CT}^{\mathsf{sobl}}_{\exists}$
$\mathsf{CT}^{\mathsf{obl}}_{\exists} \subsetneq \mathsf{CT}^{\mathsf{sobl}}_{\forall}$ $\mathsf{CT}^{\mathsf{sobl}}_{\exists} \subsetneq \mathsf{CT}^{\mathsf{std}}_{\forall}$	$\mathsf{CT}^{\mathsf{obl}}_{\exists} \nparallel \mathsf{CT}^{\mathsf{sobl}}_{\forall}$ $\mathsf{CT}^{\mathsf{sobl}}_{\exists} \nparallel \mathsf{CT}^{\mathsf{std}}_{\forall}$ $\mathsf{CT}^{\mathsf{obl}}_{\exists} \nparallel \mathsf{CT}^{\mathsf{std}}_{\forall}$
$\mathsf{CT}^{\mathsf{std}}_{\forall} \subsetneq \mathsf{CT}^{\mathsf{std}}_{\exists}$	$\mathsf{CT}^{\mathsf{std}}_{\forall} \subsetneq \mathsf{CT}^{\mathsf{std}}_{\exists}$
$\mathsf{CT}^{\mathsf{core}}_{\forall} = \mathsf{CT}^{\mathsf{core}}_{\exists}$	$\mathsf{CT}^{\mathsf{core}}_{\forall} = \mathsf{CT}^{\mathsf{core}}_{\exists}$

8 Conclusions

We have discussed the main approaches we have investigated in the last few years for dealing with inconsistent databases, including logic programming techniques for computing repairs and consistent answers, approximation approaches, techniques dealing with numerical databases, and the evaluation of the queries over universal solutions. Other approaches that we have not discussed for space limitations are the following.

Inconsistency leads to uncertainty as to the actual values within tuples. Thus, a natural question is how incomplete database formalisms can be used when dealing with inconsistency. The problem of representing repairs by means of disjunctive databases has been investigated in [52] for different notions of repairs, while [51] considered the OR-database framework.

The consistent query answering approach has been recently applied to managing inconsistent (probabilistic) spatio-temporal databases in [54, 56], where specific repairing strategies are proposed for fixing spatio-temporal data collected from heterogeneous sources, such as on-board GPS devices and roadside sensors, that represent conflicting information on the monitored scenario. A general form of spatio-temporal integrity constraints has been proposed in [54, 55], which allow us to restrict the set of trajectories that should be considered as admissible for a moving object. Repairs and consistent answers in the presence of such kind of constraints could be profitably used to better manage inconsistent spatio-temporal data.

Finally, an interesting approach related to consistent query answering has been proposed in [45–47], where the aim is to provide the user with a mechanism (named *Inconsistent Management Policies* - IMPs) to specify the desired policies to manage inconsistencies in the data, possibly allowing inconsistency to persist after applying the policy. Thus, in contrast to the classical consistent query answering approach, IMPs allow users to decide to keep inconsistency in the database they are working on.

References

1. M. Arenas, L.E. Bertossi, J. Chomicki, X. He, V. Raghavan, J. Spinrad, Scalar aggregation in inconsistent databases. Theor. Comput. Sci. (TCS) **3**(296), 405–434 (2003)
2. M. Calautti, S. Greco, I. Trubitsyna, Detecting decidable classes of finitely ground logic programs with function symbols, in *PPDP* (2013), pp. 239–250
3. M. Calautti, S. Greco, C. Molinaro, I. Trubitsyna, Checking termination of logic programs with function symbols through linear constraints, in *RuleML* (2014), pp. 97–111
4. M. Calautti, G. Gottlob, A. Pieris, Chase termi- nation for guarded existential rules, in *PODS* (2015)
5. M. Calautti, S. Greco, C. Molinaro, I. Trubitsyna, Logic program termination analysis using atom sizes, in *IJCAI* (2015), pp. 2833–2839
6. M. Calautti, S. Greco, C. Molinaro, I. Trubitsyna, Exploiting equality generating dependencies in checking chase termination. PVLDB **9**(5), 396–407 (2016)

7. M. Calautti, S. Greco, C. Molinaro, I. Trubitsyna, Using linear constraints for logic program termination analysis. TPLP **16**(3), 353–377 (2016)
8. A. Calì, G. Gottlob, M. Kifer, Taming the infinite chase: query answering under expressive relational constraints. JAIR **48**, 115–174 (2013)
9. L. Caroprese, M. Truszczynski, Active integrity constraints and revision programming. TPLP **11**(6), 905–952 (2011)
10. L. Caroprese, S. Greco, I. Trubitsyna, E. Zumpano, Preferred generalized answers for inconsistent databases, in *ISMIS* (2006), pp. 344–349
11. L. Caroprese, S. Greco, C. Molinaro, Prioritized active integrity constraints for database maintenance, in *Proceedings of the International Conference on Database Systems for Advanced Applications (DASFAA)* (2007), pp. 459–471
12. L. Caroprese, S. Greco, E. Zumpano, Active integrity constraints for database consistency maintenance. IEEE Trans. Knowl. Data Eng. **21**(7), 1042–1058 (2009)
13. A. Deutsch, A. Nash, J.B. Remmel, The chase revisited, in *PODS* (2008), pp. 149–158
14. R. Fagin, P.G. Kolaitis, R.J. Miller, L. Popa, Data exchange: semantics and query answering. Theor. Comput. Sci. **336**(1), 89–124 (2005)
15. R. Fagin, P.G. Kolaitis, L. Popa, Data exchange: getting to the core. ACM TODS **30**(1), 174–210 (2005)
16. B. Fazzinga, S. Flesca, F. Furfaro, F. Parisi, Dart: a data acquisition and repairing tool, in *Proceedings of the International Workshop on Inconsistency and Incompleteness in Databases (IIDB)* (2006), pp. 297–317
17. S. Flesca, F. Furfaro, F. Parisi, Consistent query answers on numerical databases under aggregate constraints, in *Proceedings of the International Symposium on Database Programming Languages (DBPL)* (2005), pp. 279–294
18. S. Flesca, F. Furfaro, F. Parisi, Preferred database repairs under aggregate constraints, in *Proceedings of the International Conference on Scalable Uncertainty Management (SUM)* (2007), pp. 215–229
19. S. Flesca, F. Furfaro, F. Parisi, Consistent answers to Boolean aggregate queries under aggregate constraints, in *Proceedings of the International Conference on Database and Expert Systems Applications (DEXA)* (2010), pp. 285–299
20. S. Flesca, F. Furfaro, F. Parisi, Querying and repairing inconsistent numerical databases, ACM Trans. Database Syst. **35**(2) (2010)
21. S. Flesca, F. Furfaro, F. Parisi, Range-consistent answers of aggregate queries under aggregate constraints, in *Proceedings of the International Conference on Scalable Uncertainty Management (SUM)* (2010), pp. 163–176
22. F. Furfaro, S. Greco, C. Molinaro, A three-valued semantics for querying and repairing inconsistent databases. Ann. Math. Artif. Intell. **51**(2–4), 167–193 (2007)
23. M. Gelfond, V. Lifschitz, Classical negation in logic programs and disjunctive databases. New Gener. Comput. **9**(3/4), 365–386 (1991)
24. T. Gogacz, J. Marcinkowski, All-instances termination of chase is undecidable, in *ICALP* (2014), pp. 293–304
25. G. Gottlob, A. Nash, Efficient core computation in data exchange. J. ACM **55**(2) (2008)
26. G. Grahne, A. Onet, Anatomy of the chase. CoRR arXiv:abs/1303.6682 (2013)
27. B.C. Grau, I. Horrocks, M. Krötzsch, C. Kupke, D. Magka, B. Motik, Z. Wang, Acyclicity notions for existential rules and their application to query answering in ontologies. JAIR **47**, 741–808 (2013)
28. G. Greco, S. Greco, E. Zumpano, A logic programming approach to the integration, repairing and querying of inconsistent databases, in *ICLP* (2001), pp. 348–364
29. G. Greco, S. Greco, E. Zumpano, A logical framework for querying and repairing inconsistent databases. IEEE Trans. Knowl. Data Eng. **15**(6), 1389–1408 (2003)
30. S. Greco, E. Zumpano, Computing repairs for inconsistent databases, in *CODAS* (2001), pp. 33–42
31. S. Greco, C. Molinaro, Querying and repairing inconsistent databases under three-valued semantics, in *Proceedings of the International Conference on Logic Programming (ICLP)* (2007), pp. 149–164

32. S. Greco, C. Molinaro, Approximate probabilistic query answering over inconsistent databases, in *Proceedings of the International Conference on Conceptual Modeling (ER)* (2008), pp. 311–325
33. S. Greco, F. Spezzano, Chase termination: a constraints rewriting approach. PVLDB **3**(1), 93–104 (2010)
34. S. Greco, C. Molinaro, Probabilistic query answering over inconsistent databases. Ann. Math. Artif. Intell. **64**(2–3), 185–207 (2012)
35. S. Greco, C. Sirangelo, I. Trubitsyna, E. Zumpano, Preferred repairs for inconsistent databases, in *Proceedings of the IDEAS* (2003), pp. 202–211
36. S. Greco, C. Molinaro, I. Trubitsyna, E. Zumpano, NP datalog: a logic language for expressing search and optimization problems. TPLP **10**(2), 125–166 (2010)
37. S. Greco, F. Spezzano, I. Trubitsyna, Stratification criteria and rewriting techniques for checking chase termination. PVLDB **4**(11), 1158–1168 (2011)
38. S. Greco, C. Molinaro, I. Trubitsyna, Checking logic program termination under bottom-up evaluation, in *IJCAI* (2013), pp. 323–333
39. S. Greco, C. Molinaro, I. Trubitsyna, Logic programming with function symbols: checking termination of bottom-up evaluation through program adornments. TPLP **13**(4–5), 737–752 (2013)
40. S. Greco, F. Spezzano, I. Trubitsyna, Checking chase termination: cyclicity analysis and rewriting techniques. IEEE Trans. Knowl. Data Eng. **27**(3), 621–635 (2015)
41. Y. Lierler, V. Lifschitz, One more decidable class of finitely ground programs, in *ICLP* (2009), pp. 489–493
42. V.W. Marek, M. Truszczynski, Revision programming. Theor. Comput. Sci. **190**(2), 241–277 (1998)
43. V.W. Marek, I. Pivkina, M. Truszczynski, Annotated revision programs. Artif. Intell. **138**(1–2), 149–180 (2002)
44. B. Marnette, Generalized schema-mappings: from termination to tractability, in *PODS* (2009), pp. 13–22
45. M.V. Martinez, F. Parisi, A. Pugliese, G.I. Simari, V.S. Subrahmanian, Inconsistency management policies, in *Proceedings of the International Conference on Principles of Knowledge Representation and Reasoning (KR)* (2008), pp. 367–377
46. M.V. Martinez, F. Parisi, A. Pugliese, G.I. Simari, V.S. Subrahmanian, Efficient policy-based inconsistency management in relational knowledge bases, in *Proceedings of the International Conference on Scalable Uncertainty Management (SUM)* (2010), pp. 264–277
47. M.V. Martinez, F. Parisi, A. Pugliese, G.I. Simari, V.S. Subrahmanian, Policy-based inconsistency management in relational databases. Int. J. Approx. Reason. **55**(2), 501–528 (2014)
48. G. Mecca, P. Papotti, S. Raunich, Core schema mappings, in *SIGMOD* (2009), pp. 655–668
49. M. Meier, On the Termination of the Chase Algorithm. Albert-Ludwigs-Universitat Freiburg (Germany) (2010)
50. M. Meier, M. Schmidt, G. Lausen, On chase termination beyond stratification. PVLDB **2**(1), 970–981 (2009)
51. C. Molinaro, S. Greco, Polynomial time queries over inconsistent databases with functional dependencies and foreign keys. Data Knowl. Eng. **69**(7), 709–722 (2010)
52. C. Molinaro, J. Chomicki, J. Marcinkowski, Disjunctive databases for representing repairs. Ann. Math. Artif. Intell. **57**(2), 103–124 (2009)
53. A. Onet, The chase procedure and its applications in data exchange, in *Data Exchange, Integration, and Streams* (2013), pp. 1–37
54. F. Parisi, J. Grant, Integrity constraints for probabilistic spatio-temporal knowledgebases, in *Proceedings of the International Conference on Scalable Uncertainty Management (SUM)* (2014), pp. 251–264

55. F. Parisi, J. Grant, Knowledge representation in probabilistic spatio-temporal knowledge bases. J. Artif. Intell. Res. (JAIR) **55**, 743–798 (2016)
56. F. Parisi, J. Grant, On repairing and querying inconsistent probabilistic spatio-temporal databases. Int. J. Approx. Reason. (IJAR) **84**, 41–74 (2017)
57. B. ten Cate, L. Chiticariu, P.G. Kolaitis, W.C. Tan, Laconic schema mappings: computing the core with SQL queries. PVLDB **2**(1), 1006–1017 (2009)

First-Order Ontology Mediated Database Querying via Query Reformulation

Diego Calvanese and Enrico Franconi

Abstract We address the problem of query answering with ontologies over databases. We consider first-order ontology systems playing the role of a conceptual model of a database represented as a classical finite relational store, either with an open world or a closed world reading. Queries over the conceptual signature are reformulated into queries over the database signature, so to get the same answer directly via SQL relational database technology. We consider two distinct approaches to reformulation, perfect and exact reformulation. We discuss advantages and disadvantages of each of the two approaches, and we report on some significant results appeared in the literature.

1 Introduction

We address the problem of query answering with ontologies over databases. An ontology provides a conceptual view of the database and consists of constraints on a vocabulary *extending* the basic vocabulary of the data. Querying a database using the terms in such a richer ontology allows for more flexibility than directly using only the basic vocabulary of the relational database. In the proposed framework, we analyse the case when a query expressed in ontology terms can be reformulated as a query expressed directly in database terms, so that it can be evaluated using standard SQL relational technology, which is very efficient in the size of the data.

We consider first-order ontology systems (that is, expressible as first-order theories) playing the role of a conceptual model of a database represented as a classical finite relational store, called the *locally-closed world* [13, 14], *exact views* [15, 25, 26], or, in our case, a *DBox* [19]. A DBox is a set of ground atoms that semantically behaves like a database, i.e., the interpretation of the database predicates in

D. Calvanese · E. Franconi (✉)
KRDB Research Centre, Free University of Bozen-Bolzano, Bolzano, Italy
e-mail: franconi@inf.unibz.it

D. Calvanese
e-mail: calvanese@inf.unibz.it
URL: http://www.inf.unibz.it/krdb/

S. Flesca et al. (eds.), *A Comprehensive Guide Through the Italian Database Research Over the Last 25 Years*, Studies in Big Data 31, DOI 10.1007/978-3-319-61893-7_10

the DBox is exactly equal to the database relations. We say that the DBox predicates are *closed*, i.e., their extensions are the same across all the interpretations. On the other hand, we may have *incomplete* data that behaves as ground atoms in classical first-order logic, or ABoxes in classical description logics: in this case, predicates holding this data are *open*, i.e., their interpretation may be extended in different ways across different interpretations.

As an example, consider an open ABox $\mathcal{A} = \{\mathsf{Person(john)}\}$ and alternatively a closed DBox with the same data $\mathcal{D} = \{\mathsf{Person(john)}\}$. In every model $\mathcal{I}$ of the ABox $\mathcal{A}$ the extension of the predicate Person includes john: namely, $\mathsf{Person}^{\mathcal{I}} \supseteq \{\mathsf{john}\}$; instead, in every model $\mathcal{I}$ of the DBox $\mathcal{D}$ the extension of Person is exactly john: namely $\mathsf{Person}^{\mathcal{I}} = \{\mathsf{john}\}$.

The difference between data held as a DBox and data held as an ABox becomes evident when querying. As an example, consider a Boolean negative query $\neg\mathsf{Person(mary)}$ over a given standard relational database, expressed by the DBox $\mathcal{D} = \{\mathsf{Person(john)}\}$. The answer of the query to the DBox is true, because the only specified person is john. On the other hand, if we consider the ABox $\mathcal{A} = \{\mathsf{Person(john)}\}$ and evaluate the query over it, the answer is false because the ABox specifies only the necessary facts but not all of them, and, hence, mary still may be a person in some model.

The *certain answer* to an open query consists of the substitutions that make the query true in *all* the models of the ontology with the data (ABox or DBox): so, if a substitution makes the query true only in some models but not in others (namely, it would be only a *possible* answer), then it is not part of the certain answer. Notice that it may indeed be the case that the answer to the query is not necessarily the same among all the models of the ontology with the ABox or DBox. In this case, the query is not fully *determined* by the given source data; indeed, given the database, there is some answer that is possible, but not certain. For expressive ontologies and queries, the possibility of indeterminacy of queries with respect to the data, brings about an increase of the worst-case computational complexity of computing certain answers. Moreover, it has been shown that computing arbitrary certain answers with DBoxes may be strictly harder in data complexity than computing certain answers with ABoxes [10, 17, 28]. Alternatively, to gain in efficiency, we could focus only on queries having the same answer over all the models of the ontology with the data, namely, when the information requested by the query is fully available from the source data without ambiguity. In this way, the indeterminacy disappears, and the complexity of query answering may be lower.

For example, consider an ontology stating that the class Person is partitioned into the classes Male and Female, and that the class Person is a subclass of the class Animal, and assume that there is data only for Person and Male (and that such data is closed, i.e., complete). The query asking for all the animals (for which we do not have directly data) is not fully determined by the data, since there may be animals who are not persons. On the other hand, the query asking for all the female persons is fully determined by the data, since the female persons are exactly all the persons who are not male, which we know exactly given the data.

In this chapter we are interested in answering ontology mediated queries via *query reformulation*: namely the original query is reformulated as a query expressed only in database terms, so that it can be efficiently evaluated directly over the database using standard SQL relational technology.

The mainstream research on query reformulation [21] is based on perfect rewritings with ABoxes and relatively inexpressive ontologies and queries (see, e.g., the *DL-Lite* approach in [2, 8]). Given an ontology and an arbitrary query, its *perfect reformulation* is a formula that, when evaluated over an arbitrary database, is guaranteed to return the certain answers of the original query with respect to the ontology and the same database. In the example above, the perfect reformulation of the query asking for all the animals is the query asking for all the persons, since those are the individuals which for sure are in the answer of the original query. On the other hand, we could construct a stronger reformulated query expressed in terms of database predicates, by requiring it to be also logically equivalent to the original query with respect to the ontology. This condition still obviously guarantees to give the same certain answer to an arbitrary database with respect to the ontology as the original query. In addition, these equivalent reformulations (called *exact reformulations*) exist if and only if the answer of the original query itself is fully determined by the given source data [4, 19]. So, in the case of exact reformulations we can deal with more expressive ontologies and queries, at the cost of being able to answer only queries determined by the data. In our example, there is no exact reformulation for the query asking for all the animals, but there is an exact reformulation for the query asking for all the females. It should be noticed that every exact reformulation is perfect, but not vice-versa.

In this chapter we survey and compare within a common formal framework the two approaches to ontology mediated query answering: via perfect reformulations in Sect. 3, and via exact reformulations in Sect. 4. We discuss advantages and disadvantages of each of the two approaches, and we report on the most relevant results appeared in the literature.

2 First-Order Ontologies and Databases

Let $\mathcal{FOL}(\mathbb{C}, \mathbb{P})$ be a classical function-free first-order language with equality over a signature $(\mathbb{C}, \mathbb{P})$, where $\mathbb{C}$ is a countably infinite set of *constants*, and $\mathbb{P}$ is a countably infinite set of *predicates* with associated arities. We consider $\mathbb{P}$ partitioned into a set $\mathbb{P}_D$ of *database predicates*, which intuitively we consider closed, and a set $\mathbb{P}_T$ of so-called *TBox predicates*, which intuitively we consider open. In the rest of the paper we will use $\mathcal{L}$ to denote some chosen fragments of $\mathcal{FOL}(\mathbb{C}, \mathbb{P})$.

We denote with $\mathbb{P}_{\{\varphi_1,\ldots,\varphi_n\}}$ the set of all predicates and with $\mathbb{C}_{\{\varphi_1,\ldots,\varphi_n\}}$ the set of all constants occurring in the formulas $\varphi_1, \ldots, \varphi_n$. For simplicity, we write $\mathbb{P}_\varphi$ (resp., $\mathbb{C}_\varphi$) instead of $\mathbb{P}_{\{\varphi\}}$ (resp., $\mathbb{C}_{\{\varphi\}}$). When we want to make the free variables X of a (possibly open) formula φ explicit, we write the formula as $\varphi_{[X]}$.

A (possibly empty) *finite* set of closed formulas in $\mathcal{L}$ is called an *ontology* (or knowledge base). We consider the following specific components of an ontology $\mathcal{K}$:

- The *database* (or *DBox*) $\mathcal{D}$ is the set of ground atoms in $\mathcal{K}$ of the form $P(c_1, \ldots, c_n)$, where P is an n-ary predicate in $\mathbb{P}_D$, and $c_i \in \mathbb{C}$, for $i \in \{1, \ldots, n\}$.
- The *ABox* $\mathcal{A}$ of $\mathcal{K}$ is defined analogously to the DBox, except that its predicates are in $\mathbb{P}_T$ (instead of $\mathbb{P}_D$). Clearly, the sets of ABox and DBox predicates are disjoint.
- The *TBox* $\mathcal{T}$ is the set of formulas in $\mathcal{K}$ over predicates in $\mathbb{P}_T$ only, but that are not part of the ABox of $\mathcal{K}$.
- The (sound GAV) *mapping* $\mathcal{M}$ is the set of formulas in $\mathcal{K}$ of the form $\forall X.\varphi_{[X]} \rightarrow P(X)$, where $\mathbb{P}_\varphi \subseteq \mathbb{P}_D$ and $P \in \mathbb{P}_T$.
- The *views* $\mathcal{V}$ is the set of formulas in $\mathcal{K}$ of the form $\forall X.\varphi_{[X]} \leftrightarrow P(X)$, where $\mathbb{P}_\varphi \subseteq \mathbb{P}_D$ and $P \in \mathbb{P}_T$.

Notice that, in general, an ontology might contain additional formulas with respect to those in these specific five components. For a DBox, ABox, TBox, mapping, views, or ontology $\mathcal{B}$, we denote with $\mathbb{P}_\mathcal{B}$ the set of all predicates, and with $\mathbb{C}_\mathcal{B}$ the set of all constants appearing in $\mathcal{B}$.

As usual, an *interpretation* $\mathcal{I} = \langle \Delta^\mathcal{I}, \cdot^\mathcal{I} \rangle$ consists of a non-empty set, the *domain* $\Delta^\mathcal{I}$, and an interpretation function $\cdot^\mathcal{I}$ defined over constants and predicates of the signature, such that for every pair of database constants $c_1, c_2 \in \mathbb{C}$, if $c_1 \neq c_2$ then $c_1^\mathcal{I} \neq c_2^\mathcal{I}$ (i.e., *we assume unique names* (UNA)). An interpretation $\mathcal{I}$ *embeds a database* $\mathcal{D}$, if it holds *(i)* $c^\mathcal{I} = c$ for every database constant $c \in \mathbb{C}_\mathcal{D}$ (i.e., we make the *standard name assumption* (SNA) on the database constants), and that *(ii)* $(c_1, \ldots, c_n) \in P^\mathcal{I}$ *if and only if* $P(c_1, \ldots, c_n) \in \mathcal{D}$. We use $\mathcal{E}(\mathcal{D})$ to denote the set of all interpretations embedding a database $\mathcal{D}$. In other words, in every interpretation embedding a database $\mathcal{D}$, the interpretation of every database predicate is always the same, and it is given exactly by its content in the database; this is, in general, not the case for the interpretation of the TBox (i.e., non-database) predicates. We say that the database predicates are *closed*, while the other predicates, i.e., the TBox predicates, are *open* and may be interpreted differently in different interpretations. In an open world, an interpretation $\mathcal{I}$ *soundly embeds* a database $\mathcal{D}$ if it holds that $(c_1, \ldots, c_n) \in P^\mathcal{I}$ if (but *not* only if) $P(c_1, \ldots, c_n) \in \mathcal{D}$.

As usual, an interpretation in which a closed formula is true according to the classical FOL definitions is called a *model* of the formula; the set of all models of a formula φ (resp., ontology $\mathcal{K}$) is denoted as $Mod(\varphi)$ (resp., $Mod(\mathcal{K})$). A database $\mathcal{D}$ is *legal for an ontology* $\mathcal{K}$ if there exists a model of $\mathcal{K}$ embedding $\mathcal{D}$. In the following, we will consider only consistent non-tautological ontologies and legal databases.

Given an interpretation $\mathcal{I}$ of a language $\mathcal{L}(\mathbb{P}, \mathbb{C})$, and $\mathbb{P}' \subseteq \mathbb{P}$ and $\mathbb{C}' \subseteq \mathbb{C}$, we denote with $\mathcal{I}|_{(\mathbb{P}',\mathbb{C}')}$ the interpretation identical to $\mathcal{I}$, except that the interpretation function $\cdot^{\mathcal{I}|_{(\mathbb{P}',\mathbb{C}')}}$ is defined only for the constants and predicates of the smaller signature $(\mathbb{P}', \mathbb{C}')$.

Let X be a set of variable symbols and $\mathbb{S}$ a set. A *substitution* is a total function $\Theta : X \rightarrow \mathbb{S}$ assigning an element in $\mathbb{S}$ to each variable in X, including the empty substitution ε when $X = \emptyset$. Domain and image (or range) of a substitution Θ are written as $dom(\Theta)$ and $rng(\Theta)$ respectively. Given a formula $\varphi_{[X]}$, an interpretation

$\mathcal{I}$, and a substitution $\Theta : X \to \mathbb{C}$, we use $\mathcal{I} \models \varphi_{[X/\Theta]}$ to denote that $\varphi_{[X]}$ is true in $\mathcal{I}$ with its free variables substituted according to $\Theta : X \to \mathbb{C}$. Given a formula $\varphi_{[X]}$, an interpretation $\mathcal{I} = \langle \Delta^{\mathcal{I}}, \cdot^{\mathcal{I}} \rangle$, and a substitution $\Theta : X \to \Delta^{\mathcal{I}}$ (i.e., a variable assignment), we use $\mathcal{I}, \Theta \models \varphi$ to denote that $\varphi_{[X]}$ is true in $\mathcal{I}$ with its free variables interpreted according to $\Theta : X \to \Delta^{\mathcal{I}}$.

Queries. A *query* is a (possibly closed) formula. The number of free variables of the query is called its *arity*. When the arity is 0 (i.e., the query is a closed formula), we call the query *Boolean*. The *(certain) answer* $\mathsf{cert}(Q_{[X]}, \mathcal{K})$ of a query $Q_{[X]}$ over an ontology $\mathcal{K}$ containing a database $\mathcal{D}_{\mathcal{K}}$ is the set of substitutions with constants that make the query true for all models of $\mathcal{K}$ that embed its database $\mathcal{D}_{\mathcal{K}}$. Formally:

$$\mathsf{cert}(Q_{[X]}, \mathcal{K}) = \{\Theta : X \to \mathbb{C} \mid \text{for all } \mathcal{I} \in Mod(\mathcal{K}) \cap \mathcal{E}(\mathcal{D}_{\mathcal{K}}) : \mathcal{I} \models Q_{[X/\Theta]}\}$$

For a Boolean query Q, $\mathsf{cert}(Q, \mathcal{K})$ is either the empty set (corresponding to false), or the empty assignment (corresponding to true).

In this chapter we are interested in the *query answering problem*: given an ontology $\mathcal{K}$ and a query Q, compute $\mathsf{cert}(Q_{[X]}, \mathcal{K})$. To study its computational complexity, we consider the associated decision problem (sometimes called the *recognition problem for query answering*): given an ontology $\mathcal{K}$, a query Q, and a substitution Θ, decide whether $\Theta \in \mathsf{cert}(Q_{[X]}, \mathcal{K})$. The *combined complexity* of the problem is the complexity measured in the combined size of the ontology and the query, while the *data complexity* [38] is the complexity measured only in the size of the data, i.e., the ABox and the DBox/database.

It has been shown that it is possible to weaken the standard name assumption for the database constants without changing the certain answers, by just assuming *unique names* [19]. So, the proposed framework is entirely classical, namely it can be completely represented in classical first-order logic with equality.

In order to capture exactly the expressivity of Relational Algebra and of core query languages used in databases, notably SQL [1], we define now *domain independent* formulas with respect to a given ontology, by adapting the classical database definition of domain independence to our ontology based framework.

A formula $Q_{[X]}$ is *domain independent with respect to an ontology* $\mathcal{K}$ if and only if for every two models $\mathcal{I} = \langle \Delta^{\mathcal{I}}, \cdot^{\mathcal{I}} \rangle$ and $\mathcal{J} = \langle \Delta^{\mathcal{J}}, \cdot^{\mathcal{J}} \rangle$ of $\mathcal{K}$ that agree on the interpretation of the predicates and constants (i.e., $\cdot^{\mathcal{I}} = \cdot^{\mathcal{J}}$), and for every substitution $\Theta : X \to \Delta^{\mathcal{I}} \cup \Delta^{\mathcal{J}}$ we have:

$$rng(\Theta) \subseteq \Delta^{\mathcal{I}} \text{ and } \mathcal{I}, \Theta \models Q_{[X]} \quad \text{if and only if} \quad rng(\Theta) \subseteq \Delta^{\mathcal{J}} \text{ and } \mathcal{J}, \Theta \models Q_{[X]}.$$

The above definition reduces to the database definition of domain independence whenever the ontology is empty. The problem of checking whether a first-order logic formula is domain independent is undecidable [1, 11]. The well known *safe-range* syntactic fragment of first-order logic introduced by Codd is an *equally expressive* language; indeed any safe-range formula is domain independent, and any domain independent formula can be easily transformed into a logically equivalent safe-range formula. This transformation implements the idea that the range of every variable of domain independent formula should be restricted by some "guard" bounding its extension.

An *SQL query* is a domain independent first order formula [1]. A *conjunctive query* (CQ) is an SQL query constructed using conjunction and existential quantification only. A *union of conjunctive queries* (UCQ) is a disjunction of CQs, all of the same arity. While in UCQs disjunction can only be used as the outermost operator, *positive queries* (PQs) allow for the arbitrary use of conjunction, disjunction, and existential quantification (but they rule out negation). In fact, PQs, which are the positive fragment of domain independent first-order queries, have the same expressive power as UCQs, although they may be exponentially more succinct.

Description logics ontologies. Description logics are relevant fragments of first-order logic with good computational properties. We consider here the expressive description logic $\mathcal{ALCHOIQ}$ as a fragment of $\mathcal{FOL}(\mathbb{C},\mathbb{P})$. In $\mathcal{ALCHOIQ}$, only unary and binary predicates are allowed, called respectively *concepts* and *roles*, together with constants and the equality predicate. $\mathcal{ALCHOIQ}$ formulas φ (also called *DL axioms*), concepts C, and roles R are defined according to the following syntax:

$$
\begin{aligned}
\varphi &\Longrightarrow C \sqsubseteq C' \mid R \sqsubseteq R' \mid C(c) \mid R(c_1, c_2)\\
C, C' &\Longrightarrow A \mid \neg C \mid C \sqcap C' \mid C \sqcup C' \mid \{c_1, \ldots, c_n\} \mid\\
& \exists R.C \mid \forall R.C \mid \geq k\, R.C \mid \leq k\, R.C\\
R, R' &\Longrightarrow S \mid S^-
\end{aligned}
$$

where A denotes a concept name (a unary TBox predicate), S a role name (a binary TBox predicate), $c, c_1, \ldots, c_n$ constants (also called *individuals*), and k a positive integer. Other constructs in the language are defined as shortcuts: $\top \equiv A \sqcup \neg A$ for some A; $\bot \equiv \neg\top$; $\exists R \equiv \exists R.\top$; $\geq n\, R \equiv\, \geq n\, R.\top$; $\leq n\, R \equiv\, \leq n\, R.\top$. A *DL ontology* is a set of DL axioms, where the DL axioms of the form $C \sqsubseteq C'$ and $R \sqsubseteq R'$ constitute the *DL TBox*, and the DL axioms of the form $C(c)$ and $R(c_1, c_2)$ constitute the *DL ABox*.

We introduce also some significant sublanguages of $\mathcal{ALCHOIQ}$. The description logic $\mathcal{ALCHOI}$ is the fragment of $\mathcal{ALCHOIQ}$ without the cardinality operators $\geq k\, R.C$ and $\leq k\, R.C$, while $\mathcal{ALCHOQ}$ is the fragment of $\mathcal{ALCHOIQ}$ without the inverse operator for roles P^-.

The lightweight logics of the *DL-Lite* family [2, 8] are description logics specifically tuned for accessing large amounts of data. Here we adopt the simple variant *DL-Lite*$_{\mathcal{R}}$ presented in [8], but all our considerations can be extended also to some of the more expressive variants of *DL-Lite* that are studied in [2] (specifically, those for which query answering is first-order rewritable, see Sect. 3). We use simply *DL-Lite* when our statements apply in general to such logics. In *DL-Lite*$_{\mathcal{R}}$, DL axioms, concepts, and roles are defined according to the following syntax:

Fig. 1 Mapping $\cdot^\dagger$ from an $\mathcal{ALCHOIQ}$ ontology to a first-order ontology

$$
\begin{aligned}
(C_1 \sqsubseteq C_2)^\dagger &= \forall x.\, {C_1}^\dagger(x) \rightarrow {C_2}^\dagger(x)\\
(R_1 \sqsubseteq R_2)^\dagger &= \forall x, y.\, {R_1}^\dagger(x,y) \rightarrow {R_2}^\dagger(x,y)\\
(C(c))^\dagger &= C^\dagger(c)\\
(R(c_1,c_2))^\dagger &= R^\dagger(c_1,c_2)\\
A^\dagger &= \lambda x.\, A(x)\\
S^\dagger &= \lambda x, y.\, S(x,y)\\
(\neg C)^\dagger &= \lambda x.\, \neg C^\dagger(x)\\
(C_1 \sqcap C_2)^\dagger &= \lambda x.\, {C_1}^\dagger(x) \wedge {C_2}^\dagger(x)\\
(C_1 \sqcup C_2)^\dagger &= \lambda x.\, {C_1}^\dagger(x) \vee {C_2}^\dagger(x)\\
(\exists R.C)^\dagger &= \lambda x.\, \exists y. R^\dagger(x,y) \wedge C^\dagger(y)\\
(\forall R.C)^\dagger &= \lambda x.\, \forall y. R^\dagger(x,y) \rightarrow C^\dagger(y)\\
(\geq k\, R.C)^\dagger &= \lambda x.\, \exists^{\geq k} y.\, R^\dagger(x,y) \wedge C^\dagger(y)\\
(\leq k\, R.C)^\dagger &= \lambda x.\, \exists^{\leq k} y.\, R^\dagger(x,y) \wedge C^\dagger(y)\\
\{c_1, \ldots, c_n\}^\dagger &= \lambda x.\, x = c_1 \vee \cdots \vee x = c_n\\
(S^-)^\dagger &= \lambda x, y.\, S(y,x)
\end{aligned}
$$

$$
\begin{aligned}
\varphi &\Longrightarrow B \sqsubseteq B' \mid B \sqsubseteq \neg B' \mid R \sqsubseteq R' \mid R \sqsubseteq \neg R' \mid A(c) \mid S(c_1, c_2)\\
B, B' &\Longrightarrow A \mid \exists R\\
R, R' &\Longrightarrow S \mid S^-
\end{aligned}
$$

Here, B, B' are called *basic concepts*, and they denote either an atomic concept A, or the projection of a role S on its first component ($\exists S$) or on its second component ($\exists S^-$). We observe that in TBox axioms, negation is used only on the right-hand side of inclusions, i.e., to express disjointness between concepts and between roles. Moreover, *DL-Lite*$_\mathcal{R}$ restricts ABox axioms to use concept and role names only, as opposed to complex concept expressions allowed in $\mathcal{ALCHOIQ}$.

We observe that, despite its simplicity, *DL-Lite*$_\mathcal{R}$ is able to capture the essential features of most conceptual modeling formalisms, such as UML Class Diagrams or Entity-Relationship schemata (see, e.g., [7]).

Description logics versus first-order ontologies. A DL ontology can be translated into first-order logic formulas by means of a mapping function $\cdot^\dagger$. In this translation, concept and role names correspond respectively to unary and binary predicate symbols in $\mathbb{P}_T$, and individuals corresponds to constant symbols in $\mathbb{C}$. Specifically, the mapping $\cdot^\dagger$ from an $\mathcal{ALCHOIQ}$ ontology to a first-order ontology is defined in Fig. 1. In our translation, we have made use of first-order logic with counting quantifiers [31], but such quantifiers are just syntactic sugar that can be easily translated away with just a linear blowup in the size of the formula (assuming numbers are represented in unary).

Notice that, since a DL ontology is constructed over TBox (i.e., open) predicates of $\mathbb{P}_T$ only, the translation of a DL ontology to first-order, results in the TBox and ABox part of a first-order ontology. These are obtained respectively from the translation of the DL TBox and the DL ABox. We may then combine such a first-order ontology with additional axioms that also make use of database predicates in $\mathbb{P}_D$, e.g., a database (or DBox) part consisting of ground atoms over $\mathbb{P}_D$, a sound GAV mapping part, a view part, or additional more general formulas.

3 Perfect Query Reformulations

We illustrate now the approach to query reformulation based on *perfect rewriting*, which was first introduced for answering UCQs over *DL-Lite* ontologies through the *PerfectRef* algorithm [8], and then extended to several other DLs [23, 29, 33], including also more expressive members of the *DL-Lite* family [2]. Such DLs share with *DL-Lite* some crucial properties that are necessary to make a rewriting based approach efficient. We illustrate first the approach for the case of UCQs over standard *DL-Lite*$_\mathcal{R}$ ontologies, constituted by a TBox and an ABox only. We come back later to how to extend the approach to take into account the presence of a database and of sound GAV mappings.

We recall that we are interested only in the case where the ontology is consistent. Indeed, the case where the ontology is inconsistent is not really of interest, since then the certain answers of every query is the set of all possible variable substitutions. However, we address afterwards also the problem of checking consistency.

The key idea at the basis of the rewriting approach is to strictly separate the processing done with respect to the TBox (providing the intensional level of the ontology) from the processing done by taking into account the ABox (i.e., the extensional level). More precisely, let $\mathcal{K}$ be a *DL-Lite*$_\mathcal{R}$ ontology with TBox $\mathcal{T}$ and ABox $\mathcal{A}$, and let Q be a UCQ over $\mathcal{K}$.

1. The query Q is first processed and rewritten into a new query Q', by compiling into Q' the *positive inclusion assertions* of the TBox $\mathcal{T}$, i.e., those inclusion assertions that contain no negation in the right-hand side.
2. The rewritten query Q' is then evaluated over the ABox $\mathcal{A}$, as if $\mathcal{A}$ was a (closed) database (and not considering further the TBox).

In this way, computing the certain answers is essentially reduced to query evaluation over a database instance. Since Q' does not depend on the ABox, the data complexity of the whole query answering algorithm is the same as the data complexity of evaluating Q' over the ABox. A crucial property for *DL-Lite* is that, in the case where Q is a UCQ, the query Q' is also a UCQ. Hence, the data complexity of the whole query answering algorithm is in AC^0, which is the complexity of evaluating a UCQ (or, more in general, a first-order query) over a relational database.

Canonical model. The rewriting based approach relies in an essential way on the *canonical model property*, which holds for *DL-Lite* and for the horn variants of many other DLs [12, 24]. Such property ensures that every satisfiable ontology $\mathcal{K}$ admits a canonical model $\mathcal{I}_c$ that is the least constrained model among all models of $\mathcal{K}$, and that can be homomorphically embedded in all other models. This in turn implies that the canonical model correctly represents *all* the models of $\mathcal{K}$ with respect to the

Fig. 2 Rewriting of query atoms in *DL-Lite*$_{\mathcal{R}}$

$A_1 \sqsubseteq A_2$	$\ldots, A_2(x), \ldots$	$\rightsquigarrow$	$\ldots, A_1(x), \ldots$
$\exists S \sqsubseteq A$	$\ldots, A(x), \ldots$	$\rightsquigarrow$	$\ldots, S(x, \_), \ldots$
$\exists S^- \sqsubseteq A$	$\ldots, A(x), \ldots$	$\rightsquigarrow$	$\ldots, S(\_, x), \ldots$
$A \sqsubseteq \exists S$	$\ldots, S(x, \_), \ldots$	$\rightsquigarrow$	$\ldots, A(x), \ldots$
$A \sqsubseteq \exists S^-$	$\ldots, S(\_, x), \ldots$	$\rightsquigarrow$	$\ldots, A(x), \ldots$
$\exists S_1 \sqsubseteq \exists S_2$	$\ldots, S_2(x, \_), \ldots$	$\rightsquigarrow$	$\ldots, S_1(x, \_), \ldots$
$S_1 \sqsubseteq S_2$	$\ldots, S_2(x, y), \ldots$	$\rightsquigarrow$	$\ldots, S_1(x, y), \ldots$
$S_1 \sqsubseteq S_2^-$	$\ldots, S_2(x, y), \ldots$	$\rightsquigarrow$	$\ldots, S_1(y, x), \ldots$
$\ldots$			

problem of answering positive queries (and in particular, UCQs). In other words, for every UCQ Q, we have that $\mathsf{cert}(Q, \mathcal{K})$ is contained in the result of the evaluation of Q over the canonical model.[1] Intuitively, the canonical model $\mathcal{I}_c$ for a *DL-Lite* ontology $\mathcal{K}$ contains the ABox $\mathcal{A}$ of $\mathcal{K}$, and in addition might contain existentially implied objects, whose existence is enforced by the TBox axioms with $\exists R$ in the right-hand side. For example, if the TBox contains an axiom $\mathsf{Student} \sqsubseteq \exists \mathsf{attends}$, expressing that every student should attend something (presumably a course), and the ABox contains the fact $\mathsf{Student}(\mathsf{john})$, then the canonical model will contain a fact $\mathsf{attends}(\mathsf{john}, o_n)$, where o_n is a newly introduced object.

First-Order rewritability. For simplicity, let us consider the case where we want to answer a Boolean UCQ $Q = \bigvee_i q_i$, where each q_i is a Boolean CQ. In order for some q_i to contribute to the answer to Q, there must exist a homomorphic embedding of all atoms of q_i into facts of the canonical model $\mathcal{I}_c$. We call such embedding a *match* of q_i into $\mathcal{I}_c$. However, $\mathcal{I}_c$ is in general infinite, hence we cannot evaluate q_i over $\mathcal{I}_c$ by effectively computing $\mathcal{I}_c$ and then searching for a match of q_i into $\mathcal{I}_c$. Instead, each CQ q_i is rewritten into a UCQ Q_i in such a way that, whenever q_i has a match in some portion of $\mathcal{I}_c$, then there will be a CQ among those in Q_i that has a corresponding match in the ABox part of $\mathcal{I}_c$. Informally, the rewriting algorithm initializes a set *Rew* of CQs to $\bigcup_i \{q_i\}$ (i.e., to the CQs in the input query Q), and processes each yet unprocessed query q_i in *Rew* by adding to *Rew* also all rewritings of q_i. For each atom α in q_i, it checks whether α can be rewritten by using one of the positive inclusions in the TBox, and if so, adds to *Rew* the CQ obtained from q_i by rewriting α. The rewriting of an atom uses a positive inclusion axiom as rewriting rule, applied from right to left, to compile away the knowledge represented by the positive inclusion itself. For example, using the inclusion $A_1 \sqsubseteq A_2$, an atom of the form $A_2(x)$ is rewritten to $A_1(x)$. Alternatively, we can consider this rewriting step as the application of standard resolution between the query and the inclusion $A_1 \sqsubseteq A_2$, viewed as the (implicitly universally quantified) formula $A_1(x) \rightarrow A_2(x)$. Other significant cases of rewritings of atoms are depicted in Fig. 2, where each inclusion axiom in the left-most column accounts for rewriting the atom to the left

[1]Note that, since the domain of the canonical model contains the individuals of the ontology, the evaluation of a query over such model can indeed return a set of individuals.

of $\rightsquigarrow$ into the atom to the right of it. We have used "_" to denote a variable that occurs only once in the CQ (counting separately occurrences as answer variable of the CQ). Besides rewriting atoms, a further processing step applied to q_i is to consider each pair of atoms α_1, α_2 occurring in the body of q_i that *unify*, and replace them with a single atom, also applying the most general unifier to the whole of q_i. In this way, variables that in q_i occur multiple times, might be replaced by an "_", and hence inclusion assertions might become applicable that were not so before the atom-unification step (cf. the rewriting rules in Fig. 2 *requiring* the presence of "_").

The above presented rewriting technique, realized through *PerfectRef*, allows us to establish that answering UCQs over consistent *DL-Lite*$_\mathcal{R}$ ontologies is *first-order rewritable*, i.e., the problem of computing certain answers over a consistent ontology can be reduced to the problem of evaluating a first-order query over the ABox of the ontology viewed as a database (with complete information). Specifically, let $\mathsf{rew}(Q, \mathcal{T})$ denote the UCQ obtained as the result of applying *PerfectRef* to a UCQ Q and a *DL-Lite*$_\mathcal{R}$ TBox $\mathcal{T}$. Then, for every ABox $\mathcal{A}$ such that the ontology $\mathcal{T} \cup \mathcal{A}$ is consistent, we have that

$$\mathsf{cert}(Q, \mathcal{T} \cup \mathcal{A}) = \mathsf{eval}(\mathsf{rew}(Q, \mathcal{T}), \mathcal{A})$$

where $\mathsf{eval}(\mathsf{rew}(Q, \mathcal{T}), \mathcal{A})$ denotes the evaluation of the UCQ $\mathsf{rew}(Q, \mathcal{T})$ over the ABox $\mathcal{A}$ viewed as a database (i.e., a first-order interpretation).

Ontology satisfiability. The rewriting of a UCQ Q with respect to a TBox $\mathcal{T}$ computed by *PerfectRef* depends only on the set of positive inclusion axioms in $\mathcal{T}$, while disjointness axioms (i.e., inclusion axioms containing a negated basic concept on the right-hand side) do not play any role in such a process. Indeed, the proof of correctness of *PerfectRef* [8], which is based on the canonical model property of *DL-Lite*$_\mathcal{R}$, shows that these kinds of axioms have to be considered only when verifying the consistency of the ontology. Once consistency is established, they can be ignored in the query rewriting phase. In fact, inconsistency of a *DL-Lite*$_\mathcal{R}$ ontology is due to the presence of disjointness axioms and their interaction with positive inclusion axioms. Such interaction can itself be captured by constructing a Boolean UCQ encoding the violation of disjointness axioms, rewriting such a UCQ with respect to the positive inclusion axioms, and checking whether its evaluation over the ABox returns true. This in turn shows that also the problem of checking consistency of a *DL-Lite*$_\mathcal{R}$ ontology is first-order rewritable [8].

Accessing a database via sound mappings. We have so far considered only the TBox and ABox components of a *DL-Lite* ontology. This is coherent with the implicit assumption that we are dealing with systems in which the data is directly represented, in terms of unary and binary ABox facts, in a form that is perfectly compatible with the intensional layer provided by the TBox. In many real-world scenarios, however, the data is not provided by ABox facts, but by legacy data sources, in (possibly large) relational tables of arbitrary arity. We represent such data sources through the *database* part $\mathcal{D}$ of an ontology. To provide access to the data sources via the intensional level of the ontology, we can rely on the *sound (GAV) mapping* component

$\mathcal{M}$ of an ontology. This setting is known in the literature as *ontology-based data access* (OBDA), and has been investigated extensively in recent years [6, 7, 30].

We recall that, in the case of a DL ontology, the mapping $\mathcal{M}$ consists of axioms of the form:

$$\forall x.\varphi^{A}_{[x]} \rightarrow A(x) \qquad \text{or} \qquad \forall x, y.\varphi^{S}_{[x,y]} \rightarrow S(x, y),$$

where the predicate symbols in $\varphi^{A}_{[x]}$ and $\varphi^{S}_{[x,y]}$ in the left part of the implication are among the database predicates $\mathbb{P}_D$, while A and S in the right part of the implication are respectively a concept and a role name of the TBox. Also, we can assume without loss of generality, that for each concept or role name N we have exactly one such mapping axiom, where $\varphi^{N}_{[X]}$ is the formula in the antecedent of the implication.

The query answering technique based on perfect rewriting can be extended to deal also with sound GAV mappings of this form, so as to rewrite the query into one that contains only database predicates [30]. Let us denote with $\mathcal{M}(\mathcal{D})$ the set of facts obtained by populating each concept or role name N of the TBox with precisely the facts retrieved through the mapping axiom associated to N, by evaluating φ^N over $\mathcal{D}$, i.e., with $\mathsf{eval}(\varphi^N, \mathcal{D})$. Considering that the mapping formulas are all implications, if such formulas are satisfied in a model $\mathcal{I}$, then they are also satisfied in every model $\mathcal{I}'$ that extends with respect to $\mathcal{I}$ the interpretation of the concept and role names in $\mathbb{P}_{\mathcal{T}}$, i.e., such that $N^{\mathcal{I}} \subseteq N^{\mathcal{I}'}$ for every concept or role name N. It follows that for a positive query Q formulated over the predicates of the TBox $\mathcal{T}$, the certain answers $\mathsf{cert}(Q, \mathcal{T} \cup \mathcal{M} \cup \mathcal{D})$ coincide with the certain answers computed over the TBox $\mathcal{T}$ together with $\mathcal{M}(\mathcal{D})$ considered as an ABox, i.e.,

$$\mathsf{cert}(Q, \mathcal{T} \cup \mathcal{M} \cup \mathcal{D}) = \mathsf{cert}(Q, \mathcal{T} \cup \mathcal{M}(\mathcal{D}))$$

To compute $\mathsf{cert}(Q, \mathcal{T} \cup \mathcal{M}(\mathcal{D}))$, we can in principle follow the approach illustrated previously: compute $\mathcal{M}(\mathcal{D})$ (which is an ordinary finite ABox), and then evaluate over it the perfect reformulation $\mathsf{rew}(Q, \mathcal{T})$. However, this would require a potentially expensive and large materialization of $\mathcal{M}(\mathcal{D})$ from the database $\mathcal{D}$. Instead, we can obtain the same result, avoiding the costly materialization step, by proceeding as follows:

1. Compute the perfect reformulation $Q' = \mathsf{rew}(Q, \mathcal{T})$ of Q with respect to the (positive) inclusion axioms in the TBox part $\mathcal{T}$ of $\mathcal{K}$.
2. Compute the *unfolding* of Q' with respect to $\mathcal{M}$, denoted $\mathsf{unf}(Q', \mathcal{M})$, by replacing each (concept or role) predicate N appearing in Q' with the corresponding formula $\varphi^{N}_{[X]}$ in the left-hand side of the mapping axiom $\forall X.\varphi^{N}_{[X]} \rightarrow N(X)$.
3. Evaluate $\mathsf{unf}(Q', \mathcal{M})$ over the database $\mathcal{D}$.

The correctness of this approach follows from the following result:

Theorem 1 ([8, 30]) *Let $\mathcal{K}$ be a DL-Lite$_{\mathcal{R}}$ ontology consisting of a TBox $\mathcal{T}$, a mapping $\mathcal{M}$ and a database $\mathcal{D}$, and let Q be a UCQ formulated over $\mathcal{T}$. Then:*

$$\mathsf{cert}(Q, \mathcal{T} \cup \mathcal{M} \cup \mathcal{D}) = \mathsf{eval}(\mathsf{unf}(\mathsf{rew}(Q, \mathcal{T}), \mathcal{M}), \mathcal{D})$$

Complexity of query evaluation. Summarizing the above results, and considering that evaluating a first-order query (and hence a UCQ) over a database is in AC^0 in data complexity, one obtains that answering UCQs over *DL-Lite*$_{\mathcal{R}}$ ontologies has the same data complexity as evaluating UCQs in plain databases. By analyzing the overall rewriting-based query answering technique, and by exploiting a correspondence between the *DL-Lite* family and first-order logic with unary predicates [2], we are able obtain also tight complexity bounds in the size of the TBox (*schema complexity*) and of the overall input (*combined complexity*).

Theorem 2 ([2, 8, 30]) *Answering UCQs over DL-Lite*$_{\mathcal{R}}$ *ontologies is in* AC^0 *in data complexity,* NLOGSPACE-*complete in schema complexity, and NP-complete in combined complexity.*

While the above results sound very encouraging from the theoretical point of view, there still remain significant challenges to be addressed to make rewriting based techniques effective also in real world scenarios, where the TBox and/or the data underlying the ABox are very large, and/or queries have a large number of atoms. Indeed, also in the case where one admits rewritings expressed in languages different from UCQs (e.g., arbitrary FOL queries, or non-recursive Datalog), it has recently been shown that the smallest rewritings can grow exponentially with the size of the query [20]. This has led to an intensive and sustained effort aimed at developing techniques to improve query answering over ontologies, such as alternative rewriting techniques [29, 35], techniques combining rewriting with partial materialization of the extensional level [22], and various optimization techniques that take into account also extensional constraints on the underlying data, or in the mapping layer to the relational data sources [32, 34].

4 Exact Query Reformulations

We illustrate in this section the approach to query reformulation with first-order ontologies based on *exact rewritings*, which was first thoroughly analysed in [26] by Nash, Segoufin and Vianu. They addressed the question whether a query can be answered using a set of (exact) views $\mathcal{V}$ by means of an exact rewriting over a database represented as a DBox. The authors defined and investigated the notions of *determinacy* of a query by a set of views and its connection to exact rewriting. Nash, Segoufin and Vianu also studied several combinations of query and view languages trying to understand the expressivity of the language required to express the exact rewriting, and, thus, they obtained results on the *completeness of rewriting languages*. They investigated languages ranging from full first-order logic to conjunctive queries.

The setting considered by Nash, Segoufin and Vianu is weaker than the one studied later by Franconi, Kerhet, and Ngo [18, 19], since in the former setting the ontology consists only of a set $\mathcal{V}$ of view definitions, while in the latter setting any domain independent first-order ontology with DBox predicates is allowed. The authors also considered the special case of exact reformulations with description logic ontologies and DBoxes [9, 28, 36].

After Franconi, Kerhet, and Ngo, the recent work by Benedikt, ten Cate and Tsamoura [3, 4] analyses the generation of plans to answer queries over DBoxes in the presence of domain independent first-order ontologies, where in addition the DBox relations may have limitations on access patterns to the data (called *access restrictions* or binding patterns). This is the case when one must provide values for one of the attributes of a relation in order to obtain tuples. In this setting, DBox predicates in $\mathbb{P}_D$ as defined in this chapter are fully accessible (as a regular database), while all the other predicates in $\mathbb{P}_T$ are not accessible at all (in fact, they do not provide data).

Toman and Weddell have long advocated the use of exact reformulations for automatic generation of plans that implement user queries under system constraints–a process they called *query compilation* [37]. Their published work focuses on additional extra-logical considerations, such as satisfaction of binding patterns, considerations of inherent ordering of data and the influence of plans on such orderings, and the estimated cost of alternative query plans.

This framework has also been applied to devise the formal foundations of the problems of *view update* and of characterising *unique solutions* in data exchange. In the former problem, a target view of some source database is updatable if the source predicates have an exact reformulation given the view over the target predicates [16]. In the latter problem, unique solutions exist if the target predicates have an exact reformulation given the data exchange mappings over the source predicates [27].

DBoxes versus ABoxes. While the perfect reformulation approach works for data represented in an ABox (or for data in a DBox "connected" to the TBox via a sound mapping), the exact reformulation approach assumes that the data is represented in a DBox. As we have already noticed in the introductory section, ABoxes and DBoxes behave differently in query answering. We observe now that this difference may be relevant for queries from legacy relational database applications.

Indeed, a query mentioning only ABox predicates may have a different answer depending on the presence or absence of an ontology. Consider again the example of a Boolean negative query $\neg\mathsf{Person(mary)}$ over a database expressed by the DBox $\mathcal{D} = \{\mathsf{Person(john)}\}$. As we have said already, the answer of the query to the DBox is true, because the only specified person is john. On the other hand, if we consider the ABox $\mathcal{A} = \{\mathsf{Person(john)}\}$ and evaluate the query over it, the answer is false because the ABox specifies only the necessary facts but not all of them, and, hence, Mary still may be a person in some model. However, if we add the ontology $\{\forall x, y.\, \mathsf{Person}(x) \wedge \mathsf{Person}(y) \rightarrow x = y\}$ that says that there is at most one person, the answer to the query will be now true.

This does not happen by using DBoxes: a query mentioning just DBox predicates can only return an answer that depends only on the DBox predicates, which are complete. In other words, a DBox preserves the behaviour of legacy application queries over relational databases, also when an ontology is added on top of it.

Domain Independence. As we have seen, domain independence is an important property of a formula that guarantees that the truth value of the formula in an interpretation remains the same regardless of the underlying domain of the interpretation.

An example of a domain independent Boolean query is $\exists x.\,\mathsf{Person}(x)$: if the answer to the query is true for some DBox with a specific domain, it is also true for the same DBox and any other compatible domain. On the other hand, the Boolean query $\forall x.\,\mathsf{Person}(x)$ is not domain independent: if it is true for some DBox with some domain, it is definitely not true for the same DBox but with a larger domain. Consider now the query $Q(x) = \neg\mathsf{Person}(x)$ over the DBox $\{\mathsf{Person(john)}\}$: with domain $\Delta_1 = \{\mathsf{john, mary}\}$ the query has the answer $\{x \mapsto \mathsf{mary}\}$, while with the extended domain $\Delta_2 = \{\mathsf{john, mary, sue}\}$ it has the different answer $\{x \mapsto \mathsf{mary}, x \mapsto \mathsf{sue}\}$. If an infinite domain was considered, the answer would be infinite even with a finite DBox. Indeed, the above query is not domain independent. The query can be "fixed" with a *guard* restricting the range of the free variable, as in $Q'(x) = \mathsf{Animal}(x) \wedge \neg\mathsf{Person}(x)$. For any given database this query returns the same answer with any compatible domain (even an infinite one). $Q'(x)$ is domain independent.

Ontologies are also required to be domain independent. Consider the ontology $\{\exists x.\neg\mathsf{Student}(x),\ \exists x.\,\mathsf{Person}(x)\}$, and the DBox $\{\mathsf{Student(john)}, \mathsf{Person(john)}\}$. This DBox with domain $\Delta_1 = \{\mathsf{john}\}$ is inconsistent with respect to the ontology, while it is consistent with respect to the same ontology with domain $\Delta_2 = \{\mathsf{john, mary}\}$. This happens because the first sentence of the ontology is not domain independent.

Determinacy and exact reformulations. The exact reformulation (also called *explicit definition* by Beth [5]) of a query over some DBox predicates under the ontology is a logically equivalent query under the ontology which uses only DBox predicates. More formally, the *exact reformulation* over the database predicates $\mathbb{P}_D$ under the ontology $\mathcal{K}$ of a query $Q_{[X]}$ is some formula $\widehat{Q}_{[X]}$ in $\mathcal{FOL}(\mathbb{C}, \mathbb{P})$, such that $\mathcal{K} \models \forall X. Q_{[X]} \leftrightarrow \widehat{Q}_{[X]}$ and $\mathbb{P}_{\widehat{Q}} \subseteq \mathbb{P}_D$.

But how can we characterise the existence of an exact reformulation of a query? We need to check that the query is fully determined by the data in the DBox. A query is determined by the data in the DBox if its truth value in every model of the ontology depends *only* on the finite interpretation of the DBox predicates. This notion is formalised as follows. A query $Q_{[X]}$ is *finitely determined* by the database predicates $\mathbb{P}_D$ under the ontology $\mathcal{K}$ *if and only if* for any two models $\mathcal{I}$ and $\mathcal{J}$ of the ontology $\mathcal{K}$, both with a *finite* interpretation to the database predicates $\mathbb{P}_D$, whenever $\mathcal{I}|_{\mathbb{P}_D,\mathbb{C}} = \mathcal{J}|_{\mathbb{P}_D,\mathbb{C}}$ then for every substitution $\Theta : X \to \Delta^{\mathcal{I}}$ we have:

$$\mathcal{I}, \Theta \models Q_{[X]} \quad \textit{if and only if} \quad \mathcal{J}, \Theta \models Q_{[X]}\ .$$

The determinacy of a query with respect to a source database represented in a DBox corresponds to the notion of *implicit definability* of a formula from a set of predicates (the database predicates) as introduced by Beth [5]. The correspondence between exact reformulations and finite determinacy has been indeed first studied as *projective definability* by Beth himself in 1953 [5].

Intuitively, the answer of a finitely determined query does not depend on the interpretation of non-database predicates. Once the database and a domain are fixed, it is never the case that a substitution would make the query true in some model of the ontology and false in others, since the truth value of a finitely determined query depends only on the interpretation of the database predicates and constants and on the domain (which are fixed). In practice, by focussing on finite determinacy of queries we guarantee that the answers are interpreted as being not only certain, but also *exact*—namely that whatever is not in the answer can never be part of the answer in any possible world.

But how can we characterise the existence of a *domain independent* exact reformulation of a query? This is needed because we want to evaluate the exact reformulation using standard SQL relational technology. The following characterisation theorem answers exactly this question.

Theorem 3 (Semantic characterisation) *Given a set $\mathbb{P}_D$ of database predicates, a domain independent ontology $\mathcal{K}$, and a query $Q_{[X]}$, a domain independent exact reformulation $\widehat{Q}_{[X]}$ of $Q_{[X]}$ over $\mathbb{P}_D$ under $\mathcal{K}$ exists* if and only if *$Q_{[X]}$ is (i) finitely determined by $\mathbb{P}_D$ under $\mathcal{K}$, and (ii) domain independent with respect to $\mathcal{K}$.*

The above theorem shows us the semantic conditions to have an exact domain independent reformulation of a query, but it does not give us a method to compute such reformulation and its equivalent safe-range form. The following theorem gives us sufficient conditions for the existence of an exact safe-range reformulation in any decidable fragment of $\mathcal{FOL}(\mathbb{C}, \mathbb{P})$ and gives us a constructive way to compute it, if it exists.

Below, we use $\widetilde{Q}$ to denote the formula obtained from a formula Q by uniformly replacing every occurrence of each non-DBox predicate P with a fresh new predicate symbol $\widetilde{P}$. We extend this renaming operator $\widetilde{\cdot}$ to any set of formulas in a natural way. Also, we focus on ontologies and queries in those fragments of $\mathcal{FOL}(\mathbb{C}, \mathbb{P})$ for which determinacy under models with a finite interpretation of DBox predicates (finite determinacy) and determinacy under models with an unrestricted interpretation of DBox predicates (unrestricted determinacy) coincide. We say that these fragments have *finitely controllable determinacy*.

Theorem 4 (Constructive) *If all the following conditions hold:*

1. *$\mathcal{K} \cup \widetilde{\mathcal{K}} \models \forall X.\, Q_{[X]} \leftrightarrow \widetilde{Q}_{[X]}$ (that is, $Q_{[X]}$ is determined),*
2. *$Q_{[X]}$ is safe-range (that is, $Q_{[X]}$ is domain independent),*
3. *$\mathcal{K}$ is safe-range (that is, $\mathcal{K}$ is domain independent),*

then there exists an exact reformulation $\widehat{Q}_{[X]}$ of $Q_{[X]}$ as a safe-range query in $\mathcal{FOL}(\mathbb{C}, \mathbb{P})$ over $\mathbb{P}_D$ under $\mathcal{K}$, and $\widehat{Q}_{[X]}$ can be obtained constructively.

We conclude by mentioning the two decidable description logics $\mathcal{ALCHOI}$ and $\mathcal{ALCHOQ}$, which have a well defined and intuitive syntactic fragment (based on a notion of "safe-range" similar to the one in first-order logic) characterising exactly their domain independent fragment, and which have finitely controllable determinacy. So, these two logic are excellent candidates to play the role of ontology languages on top of databases represented as DBoxes.

Theorem 5 (Description logics version of Codd's Theorem) *The domain independent fragments of $\mathcal{ALCHOI}$ and $\mathcal{ALCHOQ}$ are equally expressive to the safe-range fragments of respectively $\mathcal{ALCHOI}$ and $\mathcal{ALCHOQ}$ and they have finitely controllable determinacy.*

Acknowledgements This research has been carried out within the Euregio IPN12 KAOS, funded by the "European Region Tyrol-South Tyrol-Trentino" (EGTC) under the first call for basic research projects, and by unibz. It has also been supported by the unibz CRC projects KENDO and OnProm. We wish to thank Volha Kerhet and Nhung Ngo for their crucial contributions to the results presented in this chapter.

References

1. S. Abiteboul, R. Hull, V. Vianu, *Foundations of Databases* (Addison Wesley Publ Co, Reading, 1995)
2. A. Artale, D. Calvanese, R. Kontchakov, M. Zakharyaschev, The DL-Lite family and relations. J. Artif. Intell. Res. **36**, 1–69 (2009)
3. M. Benedikt, B. ten Cate, E. Tsamoura, Generating low-cost plans from proofs. Proc. PODS **2014**, 200–211 (2014)
4. M. Benedikt, J. Leblay, B. ten Cate, E. Tsamoura, *Generating Plans from Proofs: The Interpolation-based Approach to Query Reformulation*. Synthesis Lectures on Data Management (Morgan & Claypool Publishers, 2016)
5. E. Beth, On Padoa's method in the theory of definition. Indag. Math. **15**, 330–339 (1953)
6. D. Calvanese, B. Cogrel, S. Komla-Ebri, R. Kontchakov, D. Lanti, M. Rezk, M. Rodriguez-Muro, G. Xiao, Ontop: answering SPARQL queries over relational databases. Semantic Web J. **8**(3), 471–487 (2017)
7. D. Calvanese, G. De Giacomo, D. Lembo, M. Lenzerini, A. Poggi, M. Rodriguez-Muro, R. Rosati, Ontologies and databases: the *DL-Lite* approach, in *RW 2009 Tutorial Lectures*, ed. by S. Tessaris, E. Franconi. LNCS, vol. 5689 (Springer, Berlin, 2009)
8. D. Calvanese, G. De Giacomo, D. Lembo, M. Lenzerini, R. Rosati, Tractable reasoning and efficient query answering in description logics: The DL-Lite family. J. of Automated Reasoning **39**(3), 385–429 (2007)
9. B. ten Cate, E. Franconi, İn. Seylan, Beth definability in expressive description logics, in *Proceedings of IJCAI 2011* (2011), pp. 1099–1106
10. B. ten Cate, E. Franconi, I. Seylan, Beth definability in expressive description logics. J. Artif. Intell. Res. **48**, 347–414 (2013)
11. R.A. Di Paola, The recursive unsolvability of the decision problem for the class of definite formulas. J. ACM **16**(2), 324–327 (1969)
12. T. Eiter, M. Ortiz, M. Simkus, T.K. Tran, G. Xiao, Query rewriting for Horn-SHIQ plus rules, in *Proceedings of AAAI 2012* (AAAI Press, 2012), pp. 726–733
13. O. Etzioni, K. Golden, D. Weld, Sound and efficient closed-world reasoning for planning. Artif. Intell. **89**, 113–148 (1996)
14. O. Etzioni, K. Golden, D.S. Weld, Sound and efficient closed-world reasoning for planning. Artif. Intell. **89**, 113–148 (1997)

15. W. Fan, F. Geerts, L. Zheng, View determinacy for preserving selected information in data transformations. Inf. Syst. **37**, 1–12 (2012)
16. I. Feinerer, E. Franconi, P. Guagliardo, Lossless selection views under conditional domain constraints. IEEE Trans. Knowl. Data Eng. **27**(2), 504–517 (2015)
17. E. Franconi, Y.A. Ibanez-Garcia, İ. Seylan, Query answering with DBoxes is hard. ENTCS **278**, 71–84 (2011)
18. E. Franconi, V. Kerhet, N. Ngo, Exact query reformulation with first-order ontologies and databases. Proc. JELIA **2012**, 202–214 (2012)
19. E. Franconi, V. Kerhet, N. Ngo, Exact query reformulation over databases with first-order and description logics ontologies. J. Artif. Intell. Res. **48**, 885–922 (2013)
20. G. Gottlob, S. Kikot, R. Kontchakov, V.V. Podolskii, T. Schwentick, M. Zakharyaschev, The price of query rewriting in ontology-based data access. Artif. Intell. **213**, 42–59 (2014)
21. A.Y. Halevy, Answering queries using views: a survey. VLDB J. **10**, 270–294 (2001)
22. R. Kontchakov, C. Lutz, D. Toman, F. Wolter, M. Zakharyaschev, The combined approach to query answering in DL-Lite. Proc. KR **2010**, 247–257 (2010)
23. A. Krisnadhi, C. Lutz, Data complexity in the $\mathcal{EL}$ family of description logics. Proc. LPAR **2007**, 333–347 (2007)
24. M. Krötzsch, S. Rudolph, P. Hitzler, Complexity boundaries for horn description logics. Proc. AAAI **2007**, 452–457 (2007)
25. M. Marx, Queries determined by views: pack your views. Proc. PODS **2007**, 23–30 (2007)
26. A. Nash, L. Segoufin, V. Vianu, Views and queries: determinacy and rewriting. ACM Trans. Database Syst. **35**, 21:1–21:41 (2010)
27. N. Ngo, E. Franconi, Unique solutions in data exchange under STS mappings, in *Proceedings of AMW 2016* (2016)
28. N. Ngo, M. Ortiz, M. Simkus, Closed predicates in description logics: results on combined complexity. Proc. KR **2016**, 237–246 (2016)
29. H. Pérez-Urbina, B. Motik, I. Horrocks, Tractable query answering and rewriting under description logic constraints. J. Appl. Logic **8**(2), 186–209 (2010)
30. A. Poggi, D. Lembo, D. Calvanese, G. De Giacomo, M. Lenzerini, R. Rosati, Linking data to ontologies. J. Data Semant. X, 133–173 (2008)
31. I. Pratt-Hartmann, Complexity of the two-variable fragment with counting quantifiers. J. Logic Lang. Inf. **14**(3), 369–395 (2005)
32. M. Rodriguez-Muro, D. Calvanese, High performance query answering over DL-Lite ontologies. Proc. KR **2012**, 308–318 (2012)
33. R. Rosati, On conjunctive query answering in $\mathcal{EL}$, in *Proceedings of DL 2007*. CEUR, vol. 250 (2007), pp. 451–458. www.ceur-ws.org
34. R. Rosati, Prexto: query rewriting under extensional constraints in *DL-Lite*, in *Proceedings of ESWC 2012*. LNCS, vol. 7295 (Springer, Berlin, 2012), pp. 360–374
35. R. Rosati, A. Almatelli, Improving query answering over DL-Lite ontologies. Proc. KR **2010**, 290–300 (2010)
36. İ. Seylan, E. Franconi, J. de Bruijn, Effective query rewriting with ontologies over DBoxes, in *Proceedings of IJCAI 2009* (2009), pp. 923–925
37. Toman, D., Weddell, G.: Fundamentals of Physical Design and Query Compilation. Morgan & Claypool Publishers (2011)
38. M.Y. Vardi, The complexity of relational query languages. Proc. STOC **1982**, 137–146 (1982)

Using Ontologies for Semantic Data Integration

Giuseppe De Giacomo, Domenico Lembo, Maurizio Lenzerini, Antonella Poggi and Riccardo Rosati

Abstract While big data analytics is considered as one of the most important paths to competitive advantage of today's enterprises, data scientists spend a comparatively large amount of time in the data preparation and data integration phase of a big data project. This shows that data integration is still a major challenge in IT applications. Over the past two decades, the idea of using semantics for data integration has become increasingly crucial, and has received much attention in the AI, database, web, and data mining communities. Here, we focus on a specific paradigm for semantic data integration, called Ontology-Based Data Access (OBDA). The goal of this paper is to provide an overview of OBDA, pointing out both the techniques that are at the basis of the paradigm, and the main challenges that remain to be addressed.

1 Introduction

Big data analytics is considered as one of the most important paths to competitive advantage of today's enterprises. However, after years of focus on technologies for big data storing and processing, many observers are pointing out that making sense of big data cannot be done without suitable tools for conceptualizing, preparing, and

G. De Giacomo · D. Lembo · M. Lenzerini · A. Poggi (✉) · R. Rosati
Dipartimento di Ingegneria Informatica,
Automatica e Gestionale Sapienza Università di Roma, Rome, Italy
e-mail: poggi@diag.uniroma1.it

G. De Giacomo
e-mail: degiacomo@diag.uniroma1.it

D. Lembo
e-mail: lembo@diag.uniroma1.it

M. Lenzerini
e-mail: lenzerini@diag.uniroma1.it

R. Rosati
e-mail: rosati@diag.uniroma1.it

S. Flesca et al. (eds.), *A Comprehensive Guide Through the Italian Database Research Over the Last 25 Years*, Studies in Big Data 31, DOI 10.1007/978-3-319-61893-7_11

integrating data.[1] Indeed, a common misconception about big data is that it is a black box: you load data and magically gain insight. This is not the case: loading a big data platform with quality data with enough structure to deliver value is a lot of work.

Thus, it is not surprising that data scientists spend a comparatively large amount of time in the data preparation phase of a project. Whether you call it data wrangling, data munging, or data integration, it is estimated that 50–80% of a data scientists' time is spent on preparing data for analysis. If we consider that in any IT (information Technology) organization, data governance is also essential for tasks other than data analytics, we can conclude that the challenge of identifying, collecting, retaining, and providing access to all relevant data for the business at an acceptable cost, is huge.

Data integration is considered as one of the old problems in data management, and the above observations show that it is a major challenge today. Formal approaches to data integration started in the 90's [17, 33, 36, 47]. Since then, research both in academia and in industry has addressed a huge variety of aspects of the general problem. Among them, we want to focus on the idea of using semantics for making data integration more powerful. Using semantics here means conceiving data integration systems where the semantics of data is explicitly specified, and is taken into account for devising all the functionalities of the system. Over the past two decades, this idea has become increasingly crucial to a wide variety of information-processing applications, and has received much attention in the AI, database, web, and data mining communities [40]. In this paper we focus on a specific paradigm for semantic data integration. Indeed, about a decade ago, a new paradigm for modeling and interacting with a data integration systems, called "Ontology-Based Data Access" (OBDA), was proposed [14, 42]. According to such paradigm, the client of the information system is freed from being aware of how data and processes are structured in concrete resources (databases, software programs, services, etc.), and interacts with the system by expressing her queries and goals in terms of a conceptual representation of the domain of interest, called ontology.

OBDA aims at a radical solution to some of the major challenges that the complex information systems software is throwing at us at its current stage of maturity: (i) the increasing complexity of the solutions required, and the impossibility of formally verifying them; (ii) the fact that the lifespan of software technologies (not to talk about hardware ones) is much shorter that the lifespan of the solutions and the applications that use them; (iii) the obvious observation that legacy systems, or simply legacy components, are everywhere, and realistically they will remain everywhere for a long time. For all the above reasons, while the amount of data stored in current information systems and the processes making use of such data continuously grow, turning these data into information, and governing both data and processes are still tremendously challenging tasks for Information Technology. The problem is complicated by the proliferation of data sources both within a single organization, and in cooperating environments. The following factors explain why such a proliferation constitutes a major problem with respect to the governance goal:

[1] http://www.dbta.com/.

- Despite the fact that the initial design of a collection of data sources is adequate, corrective maintenance actions tend to re-shape them into a form that often diverges from the original conceptual structure.
- It is common practice to change a data source (e.g., a database) so as to adapt it to specific application-dependent needs, so that it often becomes a data structure coupled to a specific application, rather than an application-independent database.
- The data stored in different sources and the processes operating over them tend to be redundant, and mutually inconsistent, mainly because of the lack of central, coherent and unified data management tasks. This poses great difficulties with respect to the goal of accessing data in a unified and coherent way.

A system realizing the vision of OBDA is constituted by three components:

- The *ontology*, whose goal is to provide a formal, clean and high level representation of the domain of interest, and constitutes the component with which the clients of the information system (both humans and software programs) interact.
- The *data source* layer, representing the existing data sources in the information system, which are managed by the processes and services operating on their data.
- The *mapping* between the two layers, which is an explicit representation of the relationship between the data sources and the ontology, and is used to translate the operations on the ontology (e.g., query answering) in terms of concrete actions on the data sources.

Thus, OBDA is an advanced approach to semantic data integration, in which the global schema is given in terms of an ontology, i.e., a formal and conceptual view of the application domain, rather than simply a unified view of the data at the sources.

The goal of this paper is to provide an overview of the OBDA paradigm, pointing out both the techniques that are at the basis of the paradigm, and the main challenges that remain to be addressed. The paper is organized as follows. In Sect. 2 we illustrate the general, formal framework underlying the paradigm. In Sects. 3 and 4 we deal with the main computational problem that has been studied so far in OBDA, namely query answering. Section 5 concludes the paper with a discussion on various aspects that are already the subject of current study, and will be increasingly important in the near future.

2 Framework for Ontology-Based Data Access

In this section we first provide a general framework for OBDA, and then we focus on a notable framework instantiation that allows for practical application of the OBDA paradigm.

Before proceeding further, we point out that here we abstract from the problem of dealing with multiple and heterogeneous sources, by assuming that we have access to a single relational database through an SQL interface. In practice, such a database might be obtained through the use of off-the-shelf data federation tools which allow seeing a set of data sources as if they were a single relational database. Note that this relational database does not represent the integrated view of the various sources, but simply a replication of the source schemas expressed in terms of a unique format.

A general framework for OBDA. An *OBDA specification* $\mathcal{J}$ is as a triple $\langle \mathcal{O}, \mathcal{S}, \mathcal{M} \rangle$, where $\mathcal{O}$ is an ontology, $\mathcal{S}$ is a relational schema, called source schema, and $\mathcal{M}$ is a mapping from $\mathcal{S}$ to $\mathcal{O}$. More precisely, $\mathcal{O}$ represents intensional knowledge about the domain, expressed in some logical language. Typically, $\mathcal{O}$ is a lightweight Description Logic (DL) TBox [7], i.e., it is expressed in a language ensuring both semantic richness and efficiency of reasoning, and in particular of query answering. The mapping $\mathcal{M}$ is a set of mapping assertions, each one relating a query over the source schema to a query over the ontology.

An *OBDA system* is a pair $(\mathcal{J}, D)$ where $\mathcal{J}$ is an OBDA specification and D is a database for the source schema $\mathcal{S}$, called source database for $\mathcal{J}$. The semantics of $(\mathcal{J}, D)$ is given in terms of the logical interpretations that are models of $\mathcal{O}$ (i.e., satisfy all axioms of $\mathcal{O}$, and satisfy $\mathcal{M}$ with respect to D). The notion of mapping satisfaction depends on the semantic interpretation adopted on mapping assertions. Commonly, such assertions are assumed to be *sound*, which intuitively means that the results returned by the source queries occurring in the mapping are a subset of the data that instantiate the ontology. The set of models of $\mathcal{J}$ with respect to D is denoted with $Mod_D(\mathcal{J})$.

In OBDA systems, the main service of interest is *query answering*, i.e., computing the answers to user queries, which are queries posed over the ontology. It amounts to return the so-called *certain answers*, i.e., the tuples that satisfy the user query in all the interpretations in $Mod_D(\mathcal{J})$. Query answering in OBDA is thus a form of reasoning under incomplete information, and is much more challenging than classical query evaluation over a database instance.

From the computational perspective, query answering depends on *(1)* the language used for the ontology; *(2)* the language used for user queries; and *(3)* the language used to specify the queries in the mapping. In the following, we consider a particular instantiation of the OBDA framework, in which we choose each such language in such a way that query answering is guaranteed to be tractable w.r.t. the size of the data. We remark that the configuration we get is to some extent "maximal", i.e., as soon as we go beyond the expressiveness of the chosen languages, we lose this nice computational behaviour (cf. Section 3).

A tractable OBDA framework. From the general framework we obtain a tractable one by choosing appropriate languages as follows:

- the ontology language is *DL-Lite*$_A$ or its subset *DL-Lite*$_R$;
- the mapping language follows the *global-as-view* (GAV) approach [33];
- the user queries are unions of conjunctive queries.

Ontology language

DL-Lite$_A$ [42] is essentially the maximally expressive member of the *DL-Lite* family of lightweight DLs [14]. In particular, its subset *DL-Lite*$_R$ has been adopted as the basis of the OWL 2 QL profile of the W3C standard OWL (Ontology Web Language) [38]. As usual in DLs, *DL-Lite*$_A$ allows for representing the domain of interest in terms of *concepts*, denoting sets of objects, and *roles*, denoting binary relations between objects. In fact, *DL-Lite*$_A$ considers also *attributes*, which denote binary relations between objects and values (such as strings or integers), but for simplicity we do not consider them in this paper. From the expressiveness point of view, *DL-Lite*$_A$ is able to capture essentially all the features of Entity-Relationship diagrams and UML Class Diagrams, except for completeness of hierarchies. In particular, it allows for specifying ISA and disjointness between either concepts or roles, mandatory participations of concepts into roles, the typing of roles. Formally, a *DL-Lite*$_A$ TBox is a set of assertions obeying the following syntax:

$$\begin{array}{lll} B_1 \sqsubseteq B_2 & B_1 \sqsubseteq \neg B_2 & \text{(concept inclusions)} \\ R_1 \sqsubseteq R_2 & R_1 \sqsubseteq \neg R_2 & \text{(role inclusions)} \\ & (\mathsf{funct}\ R) & \text{(role functionalities)} \end{array}$$

where B_1 and B_2 are basic concepts, i.e., expressions of the form A, $\exists P$, or $\exists P^-$, and R, R_1, and R_2 are a basic roles, i.e., expressions of the form P, or P^-. A and P denote an *atomic concept* and an *atomic role*, respectively, i.e., a unary and binary predicate from the ontology alphabet, respectively. P^- is the *inverse* of an atomic role P, i.e., the role obtained by switching the first and second components of P, and $\exists P$ (resp. $\exists P^-$), called existential unqualified restriction, denotes the projection of the role P on its first (resp. second) component. Finally $\neg B_2$ (resp. $\neg R_2$) denotes the negation of a basic concept (resp. role). Assertions in the left-hand side (resp. right-hand side) of the first two rows are called positive (resp. negative) inclusions. Assertions of the form $(\mathsf{funct}\ R)$ are called role functionalities and specify that an atomic role, or its inverse, is functional. *DL-Lite*$_A$ poses some limitations on the way in which positive role inclusions and role functionalities interact. More precisely, in a *DL-Lite*$_A$ TBox an atomic role that is either functional or inverse functional cannot be specialized, i.e., if $(\mathsf{funct}\ P)$ or $(\mathsf{funct}\ P^-)$ are in the TBox, no inclusion of the form $R \sqsubseteq P$ or $R \sqsubseteq P^-$ can occur in the TBox. *DL-Lite*$_R$ is the subset of *DL-Lite*$_A$ obtained by removing role functionalities altogether.

A *DL-Lite*$_A$ *interpretation* $\mathcal{I} = (\Delta^I, \cdot^I)$ consists of a non-empty *interpretation domain* Δ^I and an *interpretation function* $\cdot^I$ that assigns to each atomic concept A a subset A^I of Δ^I, and to each atomic role P a binary relation P^I over Δ^I. In particular, for the constructs of *DL-Lite*$_A$ we have:

$$\begin{array}{ll} A^I \subseteq \Delta^I & (\exists R)^I = \{o \mid \exists o'. (o, o') \in R^I\} \\ P^I \subseteq \Delta^I \times \Delta^I & (\neg B)^I = \Delta^I \setminus B^I \\ (P^-)^I = \{(o_2, o_1) \mid (o_1, o_2) \in P^I\} & (\neg R)^I = (\Delta^I \times \Delta^I) \setminus R^I \end{array}$$

Let C be either a basic concept B or its negation $\neg B$. An interpretation $\mathcal{I}$ satisfies a concept inclusion $B \sqsubseteq C$ if $B^I \subseteq C^I$. Similarly for role inclusions. Also, $\mathcal{I}$ satisfies a role functionality $(\mathsf{funct}\ R)$ if the binary relation R^I is a function, i.e., $(o, o_1) \in R^I$ and $(o, o_2) \in R^I$ implies $o_1 = o_2$.

Mapping language

The mapping language in the tractable framework allows mapping assertions of the following the forms,

$$\phi(\boldsymbol{x}) \rightsquigarrow A(f(\boldsymbol{x})) \qquad \phi(\boldsymbol{x}) \rightsquigarrow P(f_1(\boldsymbol{x}_1),\ f_2(\boldsymbol{x}_2)) \tag{1}$$

where $\phi(\boldsymbol{x})$ is a domain independent first-order query (i.e., an SQL query) over $\mathcal{S}$, with free variables $\boldsymbol{x}$, A and P are as before, variables in $\boldsymbol{x}_1$ and $\boldsymbol{x}_2$ also occur in $\boldsymbol{x}$, and f, possibly with subscripts, is a function. Intuitively, the mapping assertion in the left-hand side, called concept mapping assertion, specifies that individuals that are instances of the atomic concept A are constructed through the use of the function f from the tuples retrieved by the query $\phi(\boldsymbol{x})$. Similarly for the mapping assertion in the right-hand side of (1), called role mapping assertion. Each assertion is of type GAV, i.e., it associates a view over the source (represented by $\phi(\boldsymbol{x})$) to an element of the global schema (in this case the ontology). However, differently from traditional GAV mappings, the use of functions is crucial here, since we are considering the typical scenario in which data sources do not store the identifiers of the individuals that instantiate the ontology, but only maintain values. Thus, functions are used to address the semantic mismatch existing between the extensional level of $\mathcal{S}$ and $\mathcal{O}$ [42]. We notice that a mapping using assertions of the form (1) is indeed expressible in R2RML, the W3C recommendation for specifying mappings from relational database to RDF datasets [20]. Formally, we say that an interpretation $\mathcal{I}$ satisfies a mapping assertion $\phi(\boldsymbol{x}) \rightsquigarrow A(f(\boldsymbol{x}))$ with respect to a source database D, if for each tuple of constants $\boldsymbol{t}$ in the evaluation of $\phi(\boldsymbol{x})$ on D, $(f(\boldsymbol{t}))^I \in A^I$, where $(f(\boldsymbol{t}))^I \in \Delta^I$ is the interpretation of $f(\boldsymbol{t})$ in $\mathcal{I}$, that is, $f(\boldsymbol{t})$ acts simply as a constant denoting an object.[2] Satisfaction of assertions of the form $\phi(\boldsymbol{x}) \rightsquigarrow P(f_1(\boldsymbol{x}_1),\ f_2(\boldsymbol{x}_2))$ is defined analogously. We also point out that *DL-Lite*$_A$ adopts the Unique Name Assumption (UNA), that is, different constants denote different objects, and thus different ground terms of the form $f(\boldsymbol{t})$ are interpreted with different elements in Δ^I.[3]

User queries

In our tractable framework for OBDA, user queries are conjunctive queries (CQs) [2], or unions thereof. With $q(\boldsymbol{x})$ we denote a CQ with free variables $\boldsymbol{x}$. A Boolean CQ is a CQ without free variables. Given an OBDA system $(\mathcal{J},\ D)$ and a Boolean CQ q over $\mathcal{J}$, i.e., over the TBox of $\mathcal{J}$, we say that q is *entailed by* $(\mathcal{J},\ D)$, denoted with $(\mathcal{J},\ D) \models q$, if q evaluates to true in every $\mathcal{I} \in Mod_D(\mathcal{J})$. When the user query $q(\boldsymbol{x})$ is non-Boolean, we denote with $cert_D(q(\boldsymbol{x}),\ \mathcal{J})$ the *certain answers* to q with respect to $(\mathcal{J},\ D)$, i.e., the set of tuples $\boldsymbol{t}$ such that $(\mathcal{J},\ D) \models q(\boldsymbol{t})$, where $q(\boldsymbol{t})$ is the Boolean CQ obtained from $q(\boldsymbol{x})$ by substituting $\boldsymbol{x}$ with $\boldsymbol{t}$.

---

[2]As usual, we assume to deal with pre-interpreted data types, and thus we can harmlessly use $\boldsymbol{t}$ to denote both the values and the constants representing them.

[3]In fact, if we restrict to *DL-Lite*$_R$, then UNA becomes immaterial and can be dropped, as done in OWL 2 QL.

Query answering

Although query answering in the general framework may become soon intractable or even undecidable, depending on the expressive power of the various languages involved, the tractable framework has been designed to ensure tractability of query answering. We end this section by illustrating the basic idea for achieving tractability.

In the tractable OBDA framework previously described, one can think of a simple chase procedure [2] for query answering, which first retrieves an initial set of concept and role instances from the data source through the mapping, and then, using the ontology axioms, "expands" such a set of instances deriving and materializing all the logically entailed concept and role assertions; finally, queries can be evaluated on such an expanded set of instances. Unfortunately, in *DL-Lite*$_A$ (and in *DL-Lite*$_R$ already) the instance materialization step of the above technique is not feasible in general, because the set of entailed instance assertions starting from even very simple OBDA specifications and small data sources may be infinite.

As an alternative to the above materialization strategy, most of the approaches to query answering in OBDA are based on query rewriting, where the aim is to first compute the perfect rewriting q' of a query q w.r.t. an OBDA specification $\mathcal{J}$, and then evaluate q' over the source database. Actually, the above described OBDA framework allows for modularizing query rewriting. Indeed, the current techniques for OBDA consist of two phases: a phase of *query rewriting w.r.t. the ontology* followed by a phase of *query rewriting w.r.t. the mapping*. In the first phase, the initial query q is rewritten with respect to the ontology, producing a new query q_1, still over the ontology signature: intuitively, q_1 "encodes" the knowledge expressed by the ontology that is relevant for answering the query q. In the second phase, the query q_1 is rewritten with respect to the mapping $\mathcal{M}$, thus obtaining a query q_2 to be evaluated over the source data. Thus, the mapping assertions are used for reformulating the query into a new one expressed over the source schema signature.

In the following two sections we delve into the details of the two phases constituting the rewriting-based query answering algorithm described above.

3 Query Rewriting with Respect to the Ontology

In order to isolate the properties of ontology rewriting, in this section we assume that all relevant data in the sources have been stored, using the mapping $\mathcal{M}$, in a database whose schema coincides with the ontology signature (in DL jargon, such database is called ABox database). Note that the ABox database can be constructed by computing, for each mapping assertion, the tuples returned by the query on the left-hand side of the assertion, and then inserting into the ABox databases all the facts as sanctioned by the right-hand side of the assertion. According to this scenario, the phase of rewriting a query q with respect to the ontology aims at deriving a query q_1 still expressed over the signature of the ontology such that evaluating q_1 over the ABox database returns the set of certain answers to q with respect to the whole specification. The goal of this section is to discuss the techniques for the ontology rewriting phase, including their computational complexity.

Most of the proposed techniques [14, 18, 41] start from a CQ or a UCQ (i.e., a set of CQs), and end up producing a UCQ that is an expansion of the initial query. They are based on variants of clausal resolution [30]: every rewriting step essentially corresponds to the application of clausal resolution between a CQ among the ones already generated and a concept or role inclusion axiom of the ontology. Each such step produces a new conjunctive query that is added to the resulting UCQ. The rewriting process terminates when a fix-point is reached, i.e., no new CQ can be generated.

A potential bottleneck of the rewriting approach is caused by the size of the rewritten query, and several research works aim at optimization techniques addressing this issue. For example, the first algorithm for query rewriting w.r.t. a *DL-Lite* ontology [14] has been improved in [18, 41] by refining and optimizing the way in which term unification is handled by the above resolution step. Notice that the sentences corresponding to the ontology axioms may be Skolemized (e.g., due to the presence of existentially quantified variables in the right-hand side of a concept inclusion): to compute perfect rewritings, the unification of Skolem terms during resolution can actually be constrained in various ways with respect to standard resolution.

Some recent proposals for optimizing query rewriting w.r.t. the ontology (e.g., [18, 25, 46]) are based on the use of Datalog queries besides CQs and UCQs, to express either intermediate results or the final rewritten query. The same idea has also been used to extend query rewriting to more expressive, not necessarily first-order rewritable (see below) ontology languages [11, 18, 24, 41]. Other approaches take a more radical view, and propose strategies based on partial materialization of instance assertions [29].

The results in [14, 42] show that, following the technique illustrated above, query answering is *first-order rewritable*, i.e., for each union of CQ q over $\mathcal{J}$, it is possible to compute a first-order query q_r such that, for each ABox database D, $\boldsymbol{t} \in cert_D(q, \mathcal{J})$ iff $\boldsymbol{t}$ is in the evaluation of q_r over D. Since q_r can be effectively expressed as an SQL query, this property is actually saying that CQ answering can be reduced to query evaluation over a relational database (thus, it is in AC^0, a subclass of LOGSPACE), for which we can rely on standard relational DBMSs. The above property also implies that CQ answering is in AC^0 in ABox complexity, which is the complexity of evaluating a first-order query over a relational database. Indeed, this is an immediate consequence of the fact that the complexity of the above phase of query rewriting is independent of the data source, and that the final rewritten query is an SQL expression. It can also be shown that conjunctive query answering in the OBDA setting is NP-complete w.r.t. combined complexity, i.e., the complexity of the problem with respect to the size of the whole input (data source, OBDA specification, and query). This is the same as the combined complexity of SQL query answering over the data source.

Finally, an important question is whether we can further extend the ontology specification language of OBDA without losing the above nice computational property of the query rewriting phase. In [15] it is shown that adding any of the main concept constructors considered in Description Logics and missing in *DL-Lite*$_A$ (e.g., negation, disjunction, qualified existential restriction, range restriction) causes a jump of the data complexity of conjunctive query answering in OBDA, which goes beyond the class AC^0. This issue has been further investigated in [6].

As for the query language, we note that going beyond unions of CQs is problematic from the point of view of tractability, or even decidability. For instance, adding negation to CQs causes query answering to become undecidable [27].

4 Query Rewriting with Respect to the Mapping

Next we discuss the second phase of query rewriting in OBDA; namely the phase of rewriting with respect to the mapping. It is well-known by the studies on data integration [33] that rewriting a query w.r.t. a GAV mapping boils down to a simple unfolding strategy, which essentially means substituting every predicate of the input query with the queries that the mapping associates to that predicate [33].

In OBDA, however, query rewriting w.r.t. mappings is complicated by the following two issues: *(i)* OBDA mappings allow for constructing objects that are instances of the ontology predicates from the values stored in the data source, in order to deal with the mentioned impedance mismatch problem; *(ii)* the source queries in the mapping are expressed using the full expressive power of SQL, which is needed to bridge the large cognitive distance that may exist between the ontology and the source schema.

Solutions to issue (*i*) depend on the strategy adopted to construct objects from values. When functors applied to values are used, as in the OBDA framework instantiation we presented above, logic terms constructed through such functors can be treated in the standard way in the unifications at the basis of the unfolding procedure: see, e.g., the algorithm proposed in [42], which relies on techniques from partial evaluation of logic programs. In the R2RML standard [20], functors are realized through templates that construct W3C compliant URIs for objects from the values returned by the SQL query in the mapping assertion.

Instead, issue (*ii*) above heavily affects the performance of query answering. Indeed, current SQL engines have hard times in optimizing the execution of queries expressed over virtual views, like those introduced by the unfolding, that use complex SQL features such as union, nesting, or aggregation. Performance problems are of course amplified when there are several SQL queries mapping the same ontology predicate. Due to the above mentioned limitations, it is not realistic to group all such queries within a single mapping assertion for each predicate. However, without such grouping, the mapping associates several queries to the same predicate, and therefore the size of the query obtained by rewriting w.r.t. the mapping may be exponential in the size of the input query. Indeed, in real-world applications, it may very well happen that the size of the produced rewriting is too large to be handled by current SQL engines. Techniques to avoid or mitigate these issues are currently under investigation (see Sect. 5).

We observe that, beside sound mapping assertions, the literature on data integration has also considered complete or exact mapping assertions, in order to deal with the cases in which the data that satisfy the global schema or ontology are respectively either a superset of or the same set as the data returned by the source queries. However, both such assumptions soon lead to intractable query answering, as shown in [1, 16].

Also, as for the query language to be adopted in the mapping, we notice that while our framework already allows for very expressive queries over the source schema, queries over the ontology are less expressive. Hence, in the rest of this section, we consider the impact of enabling GLAV mapping assertions [33] in our framework, where a GLAV mapping allows CQs (with existentially quantified variables) on the right-hand side of the assertions. At a first glance, a mapping specified through assertions of the form (1) might be considered a pure GAV mapping [33], since no existentially quantified variables occur in the right-hand side of mapping assertions. In the following we show that, in fact, the presence of object terms of the form $f(\boldsymbol{x})$ allows the above mapping to be more expressive, in the sense that, under certain conditions, OBDA specifications with GLAV mappings can be transformed into specifications having mappings of the form (1) shown in Sect. 2, which have an analogous behaviour with respect to query answering.

First of all, we formally define a GLAV mapping from a source schema $\mathcal{S}$ to a TBox $\mathcal{O}$ as a set of assertions of the form

$$\phi(\boldsymbol{x}) \rightsquigarrow \psi(\boldsymbol{x}, \boldsymbol{y}) \tag{2}$$

such that $\phi(\boldsymbol{x})$ is as before, i.e., a domain independent first-order query over $\mathcal{S}$, and $\exists \boldsymbol{y}.\psi(\boldsymbol{x}, \boldsymbol{y})$ is a CQ over $\mathcal{O}$ that may contain terms of the form $f(\boldsymbol{x}')$, where variables $\boldsymbol{x}'$ also occur in $\boldsymbol{x}$. Given a database instance D for $\mathcal{S}$, an interpretation $\mathcal{I}$ satisfies a GLAV mapping assertion if for each tuple $\boldsymbol{t}$ in the evaluation of $\phi(\boldsymbol{x})$ over D, the Boolean CQ $\exists \boldsymbol{y}.\psi(\boldsymbol{t}, \boldsymbol{y})$ evaluates to true in $\mathcal{I}$.

Given a mapping assertion m of form (2), with $\boldsymbol{y} = \{y_1, \ldots, y_n\}$, we compute from m a set $\mathcal{M}_m$ of mapping assertions of the form (1) as follows:

1. substitute each y_i with the term $f_i(\boldsymbol{x})$, such that $f_i \neq f_j$ for each $i, j \in \{1, \ldots, n\}$;
2. for each atom α occurring in m after the above substitution, add to $\mathcal{M}_m$ the mapping assertion $\phi(\boldsymbol{x}) \rightsquigarrow \alpha$.

For example, if m is the GLAV mapping assertion

$$T(x) \rightsquigarrow A_1(f(x)),\ P(f(x), y),\ A_2(y)$$

the above procedure returns the three mapping assertions

$$T(x) \rightsquigarrow A_1(f(x)) \qquad T(x) \rightsquigarrow P(f(x), g(x)) \qquad T(x) \rightsquigarrow A_2(g(x))$$

We denote with $\tau(m)$ the set of mapping assertions of the form (1) obtained from m through the above procedure. Given a GLAV mapping $\mathcal{M}_L$, we define $\tau(\mathcal{M}_L) = \{\tau(m) \mid m \in \mathcal{M}_L\}$.[4]

Theorem 1 *Let $\mathcal{J} = \langle \mathcal{O}, \mathcal{S}, \mathcal{M}_L \rangle$ be an OBDA specification, where $\mathcal{O}$ is a DL-Lite$_R$ TBox and $\mathcal{M}_L$ is a GLAV mapping, let $\mathcal{J}_\tau = \langle \mathcal{O}, \mathcal{S}, \tau(\mathcal{M}_L) \rangle$, and q a Boolean CQ over $\mathcal{O}$. For each source database D for $\mathcal{J}$ we have that $(\mathcal{J}, D) \models q$ iff $(\mathcal{J}_\tau, D) \models q$.*

Proof Given a database instance D, it is easy to see that $Mod_D(\mathcal{J}_\tau) \subseteq Mod_D(\mathcal{J})$. Indeed, to satisfy mapping assertions in $\mathcal{M}_L$, different existential variables can be assigned with the same object of the interpretation domain, whereas this is not possible in $\tau(\mathcal{M}_L)$, due to the fact that different existential variables are denoted in $\tau(\mathcal{M}_L)$ by terms using different functions, and due to the UNA. Thus, if $\mathcal{J} \models q$, then obviously $\mathcal{J}_\tau \models q$. For the other way round, note that for each model $\mathcal{I} \in Mod_D(\mathcal{J})$ there exists a model $\mathcal{I}_\tau \in Mod_D(\mathcal{J}_\tau)$ such that $\mathcal{I}$ satisfies all joins satisfied by $\mathcal{I}_\tau$. Therefore, since each $\mathcal{I}_\tau$ satisfies q then each $\mathcal{I}$ satisfies q, thus showing the thesis. □

For non-Boolean queries, further work is needed. For example, consider the same GLAV mapping assertion as before, an empty TBox $\mathcal{O}$, the non-Boolean CQ $q(x, y) \leftarrow A(x) \wedge P(x, y)$, and the source database $D = \{T(d)\}$. Then, $cert_D(\langle \mathcal{O}, \mathcal{S}, \{m\} \rangle) = \emptyset$, whereas $cert_D(\langle \mathcal{O}, \mathcal{S}, \tau(m) \rangle) = \{\langle f(d), g(d) \rangle\}$. For non-Boolean queries, however, it is easy to see that $cert_D(\langle \mathcal{O}, \mathcal{S}, \mathcal{M}_L \rangle) \subseteq cert_D(\langle \mathcal{O}, \mathcal{S}, \tau(\mathcal{M}_L) \rangle)$, for any GLAV mapping $\mathcal{M}_L$. More precisely, for this setting it is possible to show that in each $\boldsymbol{t} \in cert_D(\langle \mathcal{O}, \mathcal{S}, \tau(\mathcal{M}_L) \rangle)$ such that $\boldsymbol{t} \notin cert_D(\langle \mathcal{O}, \mathcal{S}, \mathcal{M}_L \rangle)$ some function symbol introduced by the transformation τ occurs. Thus, one might think to filter out all such tuples from the set $cert_D(\langle \mathcal{O}, \mathcal{S}, \tau(\mathcal{M}_L) \rangle)$ to obtain the certain answers to the query over the original GLAV system. In our example, this means dropping the tuple $\langle f(d), g(d) \rangle$. These results indeed say that query answering over an OBDA system having GLAV mapping can be reduced to query answering over an OBDA system using mapping of the form (1), where no existential variables occur. This means that systems like Mastro [23] or Ontop [12], which manage mapping assertions of the form (1), can be used (with minimal adaptations) to answer queries in the presence of GLAV mappings.

The above result has been shown for OBDA where the ontology is expressed in *DL-Lite$_R$*. Unfortunately it cannot be extended to full *DL-Lite$_A$* which includes functional roles. Indeed, it is not hard to see that the result stated in Theorem 1 does no longer hold for full *DL-Lite$_A$* ontologies. Consider for example the GLAV mapping $\mathcal{M}_L$ from a source schema $\mathcal{S}$ that contains the assertions

$$T_1(x) \rightsquigarrow P(f(x), y) \quad \text{and} \quad T_2(x) \rightsquigarrow P(f(x), y).$$

In this case, $\tau(\mathcal{M}_L)$ contains the assertions

[4] In $\mathcal{M}_L$, the sets of fresh function symbols introduced in each $\tau(m)$ are pairwise disjoint.

$$T_1(x) \rightsquigarrow P(f(x), g_1(x)) \quad \text{and} \quad T_2(x) \rightsquigarrow P(f(x), g_2(x)).$$

Assume now to have an ontology $\mathcal{O}$ that contains the axiom (funct P), and a source database $D = \{T_1(d), T_2(d)\}$. Then $Mod_D(\langle \mathcal{O}, \mathcal{S}, \tau(\mathcal{M}_L)\rangle) = \emptyset$, since the individuals $g_1(d)$ and $g_2(d)$ have to be interpreted with different objects, due to the UNA, and this leads to the violation of (funct P). On the other hand, it is not difficult to see that instead $Mod_D(\langle \mathcal{O}, \mathcal{S}, \mathcal{M}_L\rangle) \neq \emptyset$. Thus, in this case the transformation τ causes a consistent system to become inconsistent, and, as a consequence, Theorem 1 is invalidated. Of course, one might think to renounce to the UNA on the fresh skolem terms introduced by τ. This however would cause query answering to be no longer first-order rewritable [13], which, as said, is a crucial requirement for practical applicability of the OBDA approach. Indeed, [13] has shown that query answering for OBDA systems with a *DL-Lite*$_A$ ontology and GLAV mapping is NLOGSPACE-hard in data complexity.

5 Current Challenges

In this section we discuss the main challenges related to OBDA that currently deserve investigation.

Query rewriting optimization. As already mentioned in Sect. 3, despite the theoretical low complexity of query answering within the tractable OBDA framework described Sect. 2, experiments carried in real-world scenarios show that the behaviour of current relational DBMS is extremely disappointing when the queries to be executed at the source database are too complex, which turns out to be the case, especially, when the cognitive distance between the ontology and the data is large. Thus, a few works focused on optimizing query rewriting with respect to the mappings in the tractable framework of Sect. 2. Specifically, in [45], the authors propose an optimization based on the idea of compiling the ontology into the mappings. This has the advantage of allowing ignoring redundant rewritings. In [23], the authors propose to introduce *view predicates* over the data sources, to split the mappings into *low-level mappings*, relating view predicates to SQL queries over the source database, and *high-level mappings*, relating the ontology elements to conjunctive queries over the view predicates. Hence, the rewriting process is split into two rewriting phases, one producing a union of conjunctive queries over the views, and another producing an SQL query over the data source. This allows reducing the size of the final rewriting, by optimizing the size of each conjunctive query over the views without reasoning about SQL expressions and, by adding inclusions between views, to eliminate redundant conjunctive queries within the union. As a further optimization, the authors introduce so-called *perfect mappings*, which are assertions logically entailed by the OBDA specification, allowing for handling whole subqueries as single atoms both in the ontology rewriting and in the mapping rewriting process.

Finally, in [10], the authors propose a query rewriting optimization strategy based on searching within a set of alternative equivalent rewritings, one with minimal evaluation cost when evaluated through a relational DBMS, where the cost depends both on properties of the data itself (e.g., value distribution), on the storage model (e.g., the presence of indexes), and on the DBMS optimizer's algorithm. Despite all the above mentioned efforts, the evaluation of the final rewriting by the DBMS still seems the most critical bottleneck of query answering within practical real-world OBDA scenarios.

Metamodeling and metaquerying. Recent papers point out the need of enriching conceptual ontology languages with *metamodeling* and *metaquerying* features, i.e., features for specifying and reasoning about metaclasses (also called metaconcepts) and metarelations (also called metaproperties) [4, 9]. Roughly speaking, a metaclass is a class whose instances can be themselves classes, a metarelation is a relationship (or, property) between metaclasses, and a metaquery is a query possibly using metaclasses and metarelations, and whose variables may be bound to predicates. In OBDA scenarios, metamodeling and metaquerying are essential both for correctly capturing complex domains, and for performing interesting analyses, such as, for example, "find all data sources that contribute, through mappings, to the instances of C, or any of its superclasses". Recently, a new semantics, called *HOS*, has been proposed for allowing metamodeling and metaquerying over `OWL 2 QL` ontologies [34, 35], which is both semantically adequate and exhibits nice computational properties, being answering unions of conjunctive metaqueries still AC^0 in data complexity, under certain realistic restrictions for the queried ontologies. Hence, a new challenge is now ready to be tackled, namely the problem of investigating how HOS can be combined with previous work on OBDA with dynamic ontologies [22], where also the TBox is determined by suitable mappings to a set of (relational) data sources.

Non-relational data sources. Most of the research on OBDA has focused on mappings between ontologies and data sources that either are natively relational, or can be wrapped by means of any data federation or virtualization tool, able to provide access to their contents through an SQL engine. Nevertheless, one may wonder whether the use of such a federation intermediate layer affects query answering performances and it is worth investigating query answering techniques within OBDA settings, where queries in the left-hand side of the mapping assertions are expressed in the native language of a non-relational data source. Among the wide variety of databases used within modern applications, particularly popular are the so-called `NoSQL` (not only SQL) databases, which are non-relational databases usually adopting one of four main data models, namely the column-family, key-value, document, and graph data models. Thus, in [8], after defining a uniform generalized framework for the access to arbitrary databases, the authors propose a query rewriting algorithm by adapting the technique used for relational databases and using relational algebra as an intermediate representation of the queries. Also, they experiment their results by implementing an OBDA system to access MongoDB[5] document databases.

[5]https://docs.mongodb.com/manual/.

OBDA methodology and tools. Devising an OBDA specification is likely to be a hard and time-consuming task. Several different competencies are required to collaborate, and in order for the collaboration to be successful, it is crucial that a well-defined methodology and appropriate tools are devised, for developing both the ontology and the mappings. As for the ontology, several methodologies have been devised (see e.g. [26]). Also, when it comes to OWL 2 ontologies, a visual graphical language, called *Graphol* [31], has been proposed and experimented in practice [5], that drastically supports the ontology construction. Thus, besides the *Protégé* ontology editor [39], ontology developers can currently use *Eddy* [32], an editor for Graphol ontologies. On the contrary, the problem of devising methodologies and tools for developing mappings for OBDA is largely unexplored. Indeed, while we are not aware of any study aiming at defining a methodology for developing mappings in data integration or OBDA scenarios, schema mappings tools have been proposed (e.g., [37, 43]) for supporting the specification of mappings within data integration and data exchange systems. However, none of such work is ready to be used in the OBDA scenario.

OBDA evolution. An OBDA specification is usually considered as a static piece of information. However, it is certainly crucial to investigate how to face changes over the TBox and/or the source schema. A natural assumption is to repair the mapping in such a way that the semantics of the overall system changes "as little as possible". While many approaches exist for both *ontology evolution* [48] and *database schema evolution* [44], to the best of our knowledge, no previous study has analyzed evolution in the presence of a mapping connecting an ontology to a relational data source.

Beyond data access. OBDA is the problem of accessing data through an ontology. However, the theoretical framework presented in Sect. 2 may offer many other challenging capabilities, among which we mention data quality assessment, instance-level update, and open data publishing. As for data quality assessment, in [19] the authors define a general framework for data consistency in OBDA, and present algorithms and complexity analysis for several relevant tasks related to the problem of checking data quality under the consistency dimension. The (instance-level) update over an OBDA framework is the capability of the framework to react to the addition, removal or change of logically implied assertions about ontology instances (aka *individuals*). Instance-level update was tackled for DL knowledge bases (see e.g., [21, 28]) and, recently, a rewriting technique was proposed for SPARQL updates over (extended) RDFS knowledge bases [3]. However, to the best of our knowledge, no work has focused yet on the problem of updating the extensional level of the ontology, within the theoretical framework proposed for OBDA. Finally, a natural use of OBDA is for publishing open data. However, this requires to define a well-founded semantics of the "right open data set" to be exported, which is far from being clear at the moment.

References

1. S. Abiteboul, O. Duschka, Complexity of answering queries using materialized views, in *Proceedings of the PODS* (1998), pp. 254–265
2. S. Abiteboul, R. Hull, V. Vianu, *Foundations of Databases* (Addison Wesley Publ. Co., Reading, 1995)
3. A. Ahmeti, D. Calvanese, A. Polleres, V. Savenkov, Handling inconsistencies due to class disjointness in sparql updates, in *Proceedings of the ESWC*. Lecture Notes in Computer Science, vol. 9678 (Springer, Berlin, 2016), pp. 387–404
4. D. Allemang, J. Hendler, *Semantic Web for the Working Ontologist: Effective Modeling in RDFS and OWL* (Elsevier, Amsterdam, 2011)
5. N. Antonioli, F. Castanò, S. Coletta, S. Grossi, D. Lembo, M. Lenzerini, A. Poggi, E. Virardi, and P. Castracane. Ontology-based data management for the italian public debt, in *Proceedings of the FOIS* (2014), pp. 372–385
6. A. Artale, D. Calvanese, R. Kontchakov, M. Zakharyaschev, The DL-Lite family and relations. JAIR **36**, 1–69 (2009)
7. F. Baader, D. Calvanese, D. McGuinness, D. Nardi, P.F. Patel-Schneider (eds.), *The Description Logic Handbook: Theory, Implementation and Applications*, 2nd edn. (Cambridge University Press, Cambridge, 2007)
8. E. Botoeva, D. Calvanese, B. Cogrel, M. Rezk, G. Xiao, Obda beyond relational dbs: a study for mongodb, in *Proceedings of the DL, CEUR*, vol. 1577 (2016), http://ceur-ws.org
9. F. Brasileiro, J.P.A. Almeida, V.A. Carvalho, G. Guizzardi, Expressive multi-level modeling for the semantic web, in *Proceedings of the Part I, The Semantic Web - ISWC 2016 - 15th International Semantic Web Conference, Kobe, Japan, 17–21 October, 2016* (2016), pp. 53–69
10. D. Bursztyn, F. Goasdoué, I. Manolescu, Teaching an RDBMS about ontological constraints. PVLDB **9**(12), 1161–1172 (2016)
11. A. Calì, G. Gottlob, T. Lukasiewicz, A general Datalog-based framework for tractable query answering over ontologies. J. Web Semant. **14**, 57–83 (2012)
12. D. Calvanese, B. Cogrel, S. Komla-Ebri, R. Kontchakov, D. Lanti, M. Rezk, M. Rodriguez-Muro, G. Xiao, Ontop: answering SPARQL queries over relational databases. Sem. Web J. **8**(3), 471–487 (2017)
13. D. Calvanese, G. De Giacomo, D. Lembo, M. Lenzerini, A. Poggi, R. Rosati, M. Ruzzi, Data integration through $\mathit{DL\text{-}Lite}_{\mathcal{A}}$ ontologies, in *Revised Selected Papers of the 3rd International Workshop on Semantics in Data and Knowledge Bases (SDKB 2008)*, vol. 4925, LNCS, ed. by K.-D. Schewe, B. Thalheim (Springer, Berlin, 2008), pp. 26–47
14. D. Calvanese, G. De Giacomo, D. Lembo, M. Lenzerini, R. Rosati, Tractable reasoning and efficient query answering in description logics: the DL-Lite family. JAR **39**(3), 385–429 (2007)
15. D. Calvanese, G. De Giacomo, D. Lembo, M. Lenzerini, R. Rosati, Data complexity of query answering in description logics. AIJ **195**, 335–360 (2013)
16. D. Calvanese, G. De Giacomo, D. Lembo, M. Lenzerini, R. Rosati, M. Ruzzi, Using OWL in data integration, in *Semantic Web Information Management - A Model-Based Perspective* (Springer, Berlin, 2009), pp. 397–424
17. D. Calvanese, G. De Giacomo, M. Lenzerini, D. Nardi, R. Rosati, Description logic framework for information integration, in *Proceedings of the KR* (1998), pp. 2–13
18. A. Chortaras, D. Trivela, G.B. Stamou, Optimized query rewriting for OWL 2 QL, in *Proceedings of the CADE* (2011), pp. 192–206
19. M. Console, M. Lenzerini, Data quality in ontology-based data access: the case of consistency, in *Proceedings of the AAAI* (2014), pp. 1020–1026
20. S. Das, S. Sundara, R. Cyganiak, R2RML: RDB to RDF mapping language. W3C Recommendation, W3C (2012), http://www.w3.org/TR/r2rml/
21. G. De Giacomo, M. Lenzerini, A. Poggi, R. Rosati, On instance-level update and erasure in description logic ontologies. JLC Spec. Issue Ontol. Dyn. **19**(5), 745–770 (2009)
22. F. Di Pinto, G. De Giacomo, M. Lenzerini, R. Rosati, Ontology-based data access with dynamic TBoxes in *DL-Lite*, in *Proceedings of the AAAI* (2012)

23. F. Di Pinto, D. Lembo, M. Lenzerini, R. Mancini, A. Poggi, R. Rosati, M. Ruzzi, D.F. Savo, Optimizing query rewriting in ontology-based data access, in *Proceedings of the EDBT* (ACM Press, 2013), pp. 561–572
24. T. Eiter, M. Ortiz, M. Simkus, T.-K. Tran, G. Xiao, *Query rewriting for Horn-SHIQ plus rules* (AAAI Press, In Proc. of AAAI, 2012)
25. G. Gottlob, S. Kikot, R. Kontchakov, V.V. Podolskii, T. Schwentick, M. Zakharyaschev, The price of query rewriting in ontology-based data access. AIJ **213**, 42–59 (2014)
26. M. Grüninger, Guide to the ontology of the process specification language, in *Handbook of Ontologies*, ed. by S. Staab, R. Studer (Springer, Berlin, 2003), pp. 575–592
27. V. Gutiérrez-Basulto, Y.A. Ibáñez-García, R. Kontchakov, E.V. Kostylev, Queries with negation and inequalities over lightweight ontologies. J. Web Semant. **35**, 184–202 (2015)
28. E. Kharlamov, D. Zheleznyakov, D. Calvanese, Capturing model-based ontology evolution at the instance level: the case of DL-Lite. JCSS **79**(6), 835–872 (2013)
29. R. Kontchakov, C. Lutz, D. Toman, F. Wolter, M. Zakharyaschev, The combined approach to ontology-based data access, in *Proceedings of the IJCAI* (2011), pp. 2656–2661
30. A. Leitsch, *The Resolution Calculus* (Springer, Berlin, 1997)
31. D. Lembo, D. Pantaleone, V. Santarelli, D.F. Savo, Easy OWL drawing with the graphol visual ontology language, in *Proceedings of the KR* (2016), pp. 573–576
32. D. Lembo, D. Pantaleone, V. Santarelli, D.F. Savo, Eddy: a graphical editor for OWL 2 ontologies, in *Proceedings of the IJCAI* (2016), pp. 4252–4253
33. M. Lenzerini, Data integration: a theoretical perspective, in *Proceedings of the PODS* (2002), pp. 233–246
34. M. Lenzerini, L. Lepore, A. Poggi, Answering metaqueries over hi (OWL 2 QL) ontologies, in *Proceedings of the IJCAI* (2016), pp. 1174–1180
35. M. Lenzerini, L. Lepore, A. Poggi, A higher-order semantics for metaquerying in OWL 2 QL, in *Proceedings of the KR* (2016), pp. 577–580
36. A.Y. Levy, A.O. Mendelzon, Y. Sagiv, D. Srivastava, Answering queries using views, in *Proceedings of the PODS* (1995), pp. 95–104
37. B. Marnette, G. Mecca, P. Papotti, S. Raunich, D. Santoro, ++spicy: an opensource tool for second-generation schema mapping and data exchange. PVLDB **4**(12), 1438–1441 (2011)
38. B. Motik, A. Fokoue, I. Horrocks, Z. Wu, C. Lutz, B. Cuenca Grau, OWL Web Ontology Language profiles. W3C Recommendation, W3C (2009), http://www.w3.org/TR/owl-profiles/
39. M.A. Musen, The Protégé project: a look back and a look forward. AI Matters **1**(4), 4–12 (2015)
40. N.F. Noy, A. Doan, A.Y. Halevy, Semantic integration (editorial). AI Mag. **26**(1), 7 (2005)
41. H. Pérez-Urbina, I. Horrocks, B. Motik, Efficient query answering for OWL 2, in *Proceedings of the ISWC*. LNCS, vol. 5823 (Springer, Berlin, 2009), pp. 489–504
42. A. Poggi, D. Lembo, D. Calvanese, G. De Giacomo, M. Lenzerini, R. Rosati, Linking data to ontologies. J. Data Sem. X, 133–173 (2008)
43. L. Popa, Y. Velegrakis, R.J. Miller, M.A. Hernández, R. Fagin, Translating web data, in *Proceedings of the VLDB* (2002), pp. 598–609
44. E. Rahm, P.A. Bernstein, An online bibliography on schema evolution. SIGMOD Rec. **35**(4), 30–31 (2006)
45. M. Rodríguez-Muro, D. Calvanese, Dependencies: making ontology based data access work in practice, in *Proceedings of the AMW*. CEUR, vol. 749 (2011), http://ceur-ws.org
46. R. Rosati, A. Almatelli, Improving query answering over *DL-Lite* ontologies, in *Proceedings of the KR* (2010), pp. 290–300
47. J.D. Ullman, Information integration using logical views, in *Proceedings of the ICDT*. LNCS, vol. 1186 (Springer, Berlin, 1997), pp. 19–40
48. F. Zablith, G. Antoniou, M. D'Aquin, G. Flouris, H. Kondylakis, E. Motta, D. Plexousakis, M. Sabou, Ontology evolution: a process-centric survey. Knowl. Eng. Rev. **30**(1), 45–75 (2015)

Schema Mappings: From Data Translation to Data Cleaning

Giansalvatore Mecca, Paolo Papotti and Donatello Santoro

Abstract Schema mapping management is an important research area in data transformation, integration, and cleaning systems. The reasons for its success can be found in the declarative nature of its building block (thus enabling clean semantics and easy to use design tools) paired with the efficiency and modularity in the deployment step. In this chapter we cover the evolution of schema-mappings through what we identify as three main ages. We start presenting the foundations of schema mapping tools and the first tools aimed at translating data from a source to a target schema in the first, heroic age. We then discuss the silver age, when schema mapping tools have grown their way into complex systems and have been translated into both commercial and open-source tools. Finally, we show how recent results in schema-mapping are stimulating a third, golden age, with novel research opportunities and a new generation of systems capable of dealing with a significantly larger class of real-life applications.

1 Introduction

There are many applications that need to exchange, correlate, and integrate heterogenous data sources. These information integration tasks have long been identified as important problems and unifying theoretical frameworks have been advocated by database researchers [9].

To solve these problems, a fundamental requirement is that of manipulating *mappings* among data sources. The application developer is typically given two schemas – one called the source schema, the other called the target schema – that can be based on radically different models, technologies, and rules. Mappings, also called *schema mappings*, are expressions that specify how an instance of the source repository

G. Mecca · D. Santoro
Università della Basilicata, Potenza, Italy
e-mail: giansalvatore.mecca@unibas.it

P. Papotti (✉)
Arizona State University, Tempe, AZ, USA
e-mail: ppapotti@asu.edu

S. Flesca et al. (eds.), *A Comprehensive Guide Through the Italian Database Research Over the Last 25 Years*, Studies in Big Data 31, DOI 10.1007/978-3-319-61893-7_12

should be translated into an instance of the target repository. In order to be useful in practical applications, they should have an executable implementation – for example, by means of SQL queries for relational data, or XQuery scripts for XML.[1] This latter feature is a key requirement in order to embed the execution of the mappings in more complex application scenarios, that is, to make mappings a plug and play component of integration systems.

Traditionally, data transformation has been approached as a manual task requiring experts to understand the design of the schemas and write scripts to translate data [40]. As this work is time-consuming and prone to human errors, mapping generation tools have been created to make the process more abstract and user-friendly, thus easier to handle for a larger class of people.

In this paper, we outline a history of the different phases that have characterized the research about automatic tools and techniques for schema mappings and data exchange.

We identify three different ages, as follows.

The Heroic Age The heroic age of schema-mappings research started with the seminal papers about the Clio system [35, 37]. A first generation of tools was proposed to support the process of generating complex logical dependencies – typically *tuple-generating dependencies* [7] – based on a user-friendly abstraction of the mapping provided by the users. Once the dependencies are computed, these tools transform them into executable scripts to generate a target solution in a scalable and portable way.

Early schema-mapping tools proved to be very effective in easing the burden of manually specifying complex transformations, and were successfully transferred, to some extent, into commercial products (e.g., [27]). However, several years after the development of the initial Clio algorithm, researchers realized that a more solid theoretical foundation was needed in order to consolidate practical results obtained on schema mappings systems. This consideration has motivated a rich body of research about *data exchange* that characterizes the next age.

The Silver Age Data exchange [9, 15, 37] formally studies the semantics of generating an instance of a target database given a source database, a set of mappings, and constraints on the target schema. It has formalized the notion of a *data exchange problem* [15], and has established a number of results about its properties.

After the first data exchange studies, it was clear that a key problem in schema-mappings tools was that of the *quality of the solutions*. In fact, there are many possible solutions to a data-exchange problem, and these may largely differ in terms of size and contents. The notion of the *core of the universal solutions* [17] was identified as the "optimal" solution, since it is the smallest among the solutions that preserve the semantics of the mapping.

An intermediate generation of tools [32, 41] emerged to address the problem of generating solutions of optimal quality, while guaranteeing at the same time the portability and scalability of the executable scripts. Nevertheless, despite the solid

[1] Given the importance of XQuery engines in practice, we will treat them as their relational counterpart, even if the two platforms cannot be compared in terms of performance.

results both in system and theory fields, the adoption of mapping systems in real-life integration applications, such as ETL workflows or Enterprise Information Integration (EII), has been quite slow. This is due to three main factors:

(*a*) these systems were not able, at first, to handle functional dependencies over the target, which, as it can be easily understood, is a key requirement in order to obtain solutions of quality;

(*b*) the results were obtained primarily for relational databases, and did not extend to nested models and XML;

(*c*) finally, there was no open-source schema-mapping tool publicly available to the community.

The Golden Age New results [13, 30], along with the public availability of the first open-source mapping tools – like ++SPICY [31] and OpenII [39][2] – created a promising starting point towards the solution of these problems and the beginning of a new age for mapping tools.

These works, along with others [1, 19, 34, 42], have given new vitality to schema-mappings research and suggested new applications, beyond traditional data exchange tasks. An important breakthrough has been the adoption of the mappings for the *data cleaning* problem [20]. While the supported cleaning language is a simple extension of the original constraints used in data exchange, the cleaning problem is much more complex, as updates over the data are needed to find suitable repairs [23, 24]. For this new challenging application, the system focus has moved from the design of the mappings to their execution, as the previous executable scripts cannot handle the more complex algorithm that computes repairs [25].

In this paper we first expose the basics about schema mappings by presenting the early works of the heroic age in Sect. 2. Section 3 introduces the main advancements about data exchange brought forth in the silver age. Then, in Sect. 4 we discuss how recent advances can positively impact several data management problems and become the starting point for a forthcoming golden age. Finally, a conclusion is drawn in Sect. 5.

2 The Heroic Age: Early Mapping Tools

The first mapping generation tools were created to make the process of defining transformations among schemas easier and more effective with respect to manually developed scripts. This first generation of tools includes primarily Clio [21, 27, 35, 37].[3] We may summarize the features of these early mapping tools as follows.

Value Correspondences The goal of simplifying the mapping specification was pursued by introducing a GUI that allows users to draw arrows, or *correspondences*,

[2] Available at http://www.db.unibas.it/projects/spicy/ and http://sourceforge.net/projects/openii/, respectively.

[3] Also SPICY [12] and OPENII [39] incorporate a Clio-like first-generation mapping module.

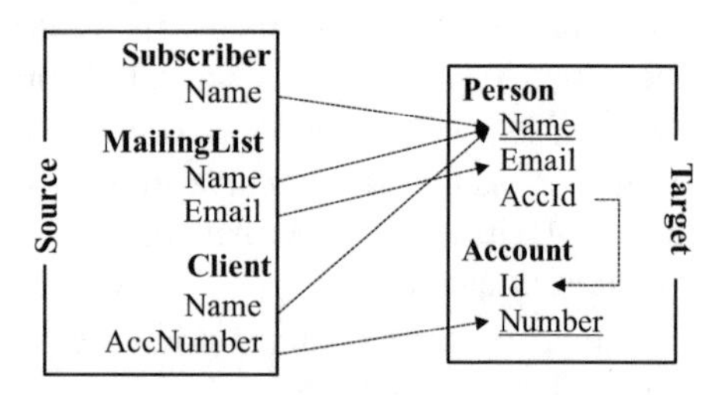

Fig. 1 Schema mapping scenario

between schemas in order to define the desired transformation. Consider the example shown in Fig. 1, where data from multiple sources should be transformed into data for a target schema with a foreign key constraints between two relations. A correspondence maps atomic elements of the source schema to elements of the target schema, independently of the underlying data model or of logical design choices, and can be derived automatically with schema matching components. Notice that, while correspondences are easy to create and understand, they are a "poor" language to express the full semantics of data transformations. For this reason, a schema mapping tool should be able to interpret the semantics the user wants to express with a set of correspondences.

Mapping Generation Based on value correspondences, mapping systems generate logical dependencies to specify the mapping. These dependencies are logical formulas of two forms: *tuple-generating dependencies* (tgds) or *equality-generating dependencies* (egds). There are two classes of constraints. *Source-to-target tgds* (s-t tgds), i.e., tgds that use source relations in the premise and target relations in the conclusion, are used to specify which tuples should be present in the target based on the tuples that appear in the source. In an operational interpretation, they state how to "translate" data from the source to the target. Target schemas are also modeled with constraints: *target tgds*, i.e., tgds that only use target symbols, are used to specify foreign-key constraints on the target; while *target egds* are used to encode functional dependencies, such as keys, on the target database.

Consider, for example, the mapping scenario in Fig. 1. It has three different source tables: (i) a table about subscribers of a service; (ii) a table with the email addresses of the people receiving the company mailing list; (iii) a table about clients and their check accounts. The target schema contains two tables, one about persons, the second about accounts. On these tables, we have two keys: *name* is a key for the persons, while *number* is a key for the accounts. Based on the correspondences drawn in Fig. 1, a Clio-like system would generate the following set of dependencies:

SOURCE – TO – TARGET TGDS
$m_1. \forall n: Subscriber(n) \rightarrow \exists Y_1, Y_2: Person(n, Y_1, Y_2)$
$m_2. \forall n, e: MailingList(n, e) \rightarrow \exists Y_1: Person(n, e, Y_1)$
$m_3. \forall n, acc: Client(n, acc) \rightarrow \exists Y_1, Z: (Person(n, Y_1, Z) \wedge Account(Z, acc))$

TARGET EGDS
e_1. $\forall n, e, a, e', a' : Person(n, e, a) \wedge Person(n, e', a') \rightarrow (e = e') \wedge (a = a')$
e_2. $\forall n, i, i' : Account(i, n) \wedge Account(i', n) \rightarrow (i = i')$

Mapping Execution via Scripts To execute the mappings, schema-mapping systems rely on the traditional *chase procedure* [15]. The chase is a fixpoint algorithm which tests and enforces implication of data dependencies, such as tgds, in a database. To be more specific, a first-generation system, after the mappings had been generated, would discard the target dependencies, and translate the source-to-target ones under the form of an SQL or XQuery script that implements the chase and can be applied to a source instance to return a solution.

Notice, in fact, that the chase of a set of s-t tgds on I can be naturally implemented using SQL. Given a tgd $\phi(\overline{x}) \rightarrow \exists\overline{y}(\psi(\overline{x}, \overline{y}))$, in order to chase it over I we may see $\phi(\overline{x})$ as a first-order query Q_ϕ with free variables $\overline{x}$ over the source database. We execute $Q_\phi(I)$ using SQL in order to find all vectors of constants that satisfy the premise and we then insert the appropriate tuple into the target instance to satisfy $\psi(\overline{x}, \overline{y})$. Skolem functions [37] are typically used to automatically "generate" some fresh nulls for $\overline{y}$.

However, these systems suffer from a major drawback: they did not have a clear theoretical foundation, and therefore it was not possible to reason about the quality of the solutions.

3 The Silver Age: Theory to the Rescue

Data exchange was conceived as an attempt to formalize the semantics of schema mappings. It formalized many aspects of the mapping execution process, as follows.

3.1 Data Exchange Fundamentals

In a data-exchange setting, the source and target databases are modeled by having two disjoint and infinite sets of values that populate instances: a set of *constants*, CONST, and a set of *labeled nulls*, NULLS [15]. Labeled nulls are used to "invent" values according to existential variables in tgd conclusions. The reference data model is the relational one.

A *mapping scenario* (also called a *data exchange scenario* or a *schema mapping*) is a quadruple $\mathcal{M} = (\mathbf{S}, \mathbf{T}, \Sigma_{st}, \Sigma_t)$, where $\mathbf{S}$ is a source schema, $\mathbf{T}$ is a target schema, Σ_{st} is a set of source-to-target tgds, and Σ_t is a set of target dependencies that may contain tgds and egds [15].

Given two disjoint schemas, $\mathbf{S}$ and $\mathbf{T}$, we denote by the pair $\langle \mathbf{S}, \mathbf{T} \rangle$ the schema $\{S_1 \ldots S_n, T_1 \ldots T_m\}$. If I is an instance of $\mathbf{S}$ and J is an instance of $\mathbf{T}$, then the pair $\langle I, J \rangle$ is an instance of $\langle \mathbf{S}, \mathbf{T} \rangle$. A target instance J is a *solution* [15] of $\mathcal{M}$

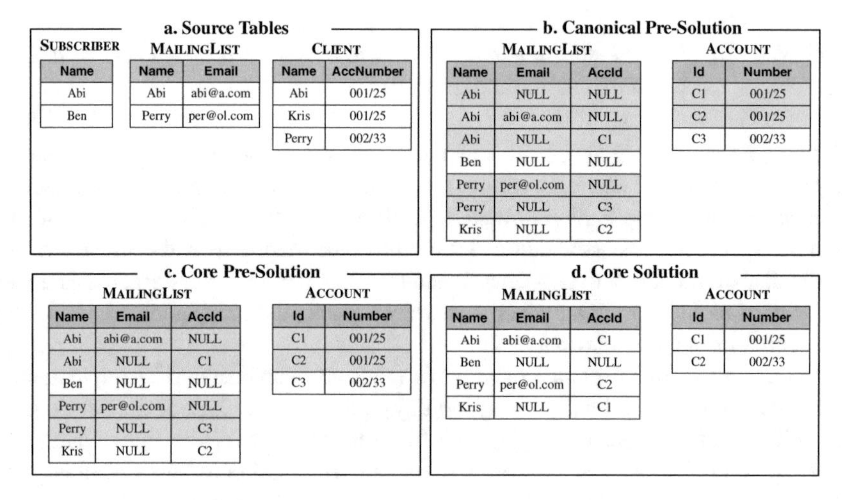

a. Source Tables

SUBSCRIBER

Name
Abi
Ben

MAILINGLIST

Name	Email
Abi	abi@a.com
Perry	per@ol.com

CLIENT

Name	AccNumber
Abi	001/25
Kris	001/25
Perry	002/33

b. Canonical Pre-Solution

MAILINGLIST

Name	Email	AccId
Abi	NULL	NULL
Abi	abi@a.com	NULL
Abi	NULL	C1
Ben	NULL	NULL
Perry	per@ol.com	NULL
Perry	NULL	C3
Kris	NULL	C2

ACCOUNT

Id	Number
C1	001/25
C2	001/25
C3	002/33

c. Core Pre-Solution

MAILINGLIST

Name	Email	AccId
Abi	abi@a.com	NULL
Abi	NULL	C1
Ben	NULL	NULL
Perry	per@ol.com	NULL
Perry	NULL	C3
Kris	NULL	C2

ACCOUNT

Id	Number
C1	001/25
C2	001/25
C3	002/33

d. Core Solution

MAILINGLIST

Name	Email	AccId
Abi	abi@a.com	C1
Ben	NULL	NULL
Perry	per@ol.com	C2
Kris	NULL	C1

ACCOUNT

Id	Number
C1	001/25
C2	002/33

Fig. 2 Source instance (**a**) and three possible solutions (**b–d**)

and a source instance I iff $\langle I, J\rangle \models \Sigma_{st} \cup \Sigma_t$, i.e., I and J together satisfy the dependencies. Given a mapping scenario $\mathcal{M} = (\mathbf{S}, \mathbf{T}, \Sigma_{st}, \Sigma_t)$, a *pre-solution* for $\mathcal{M}$ and a source instance I is a solution over I for scenario $\mathcal{M}_{st} = (\mathbf{S}, \mathbf{T}, \Sigma_{st})$, obtained from $\mathcal{M}$ by removing target constraints. In essence, a pre-solution is a solution for the s-t tgds only, and it does not necessarily enforce the target constraints. Given the source data in Fig. 2a, the canonical pre-solution is reported in Fig. 2b. A mapping scenario may have multiple solutions on a given source instance: each tgd only states an inclusion constraint and does not fully determine the content of the target. Among the possible solutions we restrict our attention to *universal* solutions, which only contain information from I and $\Sigma_{st} \cup \Sigma_t$. Universal solutions have a crucial property: they have a *homomorphism* (i.e., a constant-preserving mapping of values) into all the solutions for a data exchange problem. Intuitively, this guarantees that the solution does not contain any arbitrary information that does not follow from the source instance and the mappings.

Under a condition of weak acyclicity of the target tgds, an universal solution for a mapping scenario and a source instance can be computed in polynomial time by resorting to the classical *chase procedure* [15]. A solution generated by the chase is called a *canonical solution*. In light of this, we may say that early mapping systems were restricted to generate *canonical pre-solutions*, since they chased s-t tgds only.

3.2 Tools of the Intermediate Generation

Once the theory of data-exchange had become more mature, it was clear that producing solutions of quality was a critical requirement. The notion of a *core solution*

[17] was formalized as the "optimal" solution, since it is universal, and among the universal solutions is the one of the smallest size. In our example, the core solution is the one reported in Fig. 2d. Sophisticated algorithms were developed [17, 26] to post-process a canonical solution generated by a schema-mapping tool, and minimize it to find its core [36]. These tools have the merit of being very general, but fail to be scalable: even though the algorithms are polynomial, their implementation requires to couple complex recursive computations with SQL to access the database, and therefore hardly scale to large databases. In fact, empirical results show that they are hardly usable in practice due to unacceptable execution times for medium size databases [32].

It was therefore clear that, in order to preserve the effectiveness and generality of mapping tools, reasoning about the mapping was necessary. First, a number of approaches were proposed to optimize schema mappings in order to improve the efficiency of their execution and manipulation in real-world applications. In fact, schema mappings may present redundancy in their expressions, due for example to the presence of unnecessary atoms or unrelated variables, thus negatively affecting the data management process [16, 36]. Following efforts have aimed at optimizing such dependencies by identifying two kinds of equivalence aside from standard logical equivalence: the relaxed notions of data-exchange (DE) equivalence and conjunctive-query (CQ) equivalence [16]. DE and CQ equivalences coincide with logical equivalence when the mapping scenario is made only of s-t tgds (i.e., $\Sigma = \Sigma_{st}$), but differ on richer classes of equivalences, such as second-order tgds, and scenarios with both Σ_{st} and Σ_t.

A different approach to the generation of core solutions was undertaken in [32, 41]. In these proposals, scalability is a primary concern. Given a mapping scenario composed of source-to-target tgds (s-t tgds), the goal is to rewrite the tgds in order to generate a runtime script, for example in SQL, that, on input instances, materializes core solutions. This is a key requirement in order to embed the execution of the mappings in more complex application scenarios, that is, in order to make data-exchange techniques a real "plug and play" feature of integration applications. +SPICY [32] is an example of mapping tool of this generation. These works exploit the use of *negation* in the premise of the s-t tgds to rewrite them intercepting possible redundancy. Consider again our running example; algorithms for SQL core-generation would rewrite m_1 to make sure that no redundant data are copied to the target from the relation *Subscriber*:

$$m'_1.\ Subscriber(n) \wedge \neg(MailingList(n, E)) \wedge \\ \neg(Client(n, A)) \rightarrow Person(n, Y_1, Y_2)$$

Experiments [32] show that, in the computation of the core solution, with executable scripts there is a gain in efficiency of orders of magnitude with respect to the post-processing algorithms. This is not surprising, as these mapping rewriting approaches preserve the possibility to execute transformations in standard SQL, with the guarantee of scalability to large databases and of portability to existing applications.

However, these tools still have some serious limitations, that prevent their adoption in real-life scenarios. We may summarize these limitations as follows.

(*a*) *They have limited support for target constraints*. Handling target constraints – i.e., keys and foreign keys, represented by *egds* and *target tgds* [15], respectively – is a crucial requirement in many mapping applications. Notice that foreign-key constraints were at the core of the original schema-mapping algorithms, and, under appropriate hypothesis, can always be rewritten as part of the source-to-target tgds [16]. Unfortunately this is not the case for target edgs.

Consider again the running example; the best a tool from this generation can obtain with executable scripts is the core pre-solution reported in Fig. 2c, where the redundancy coming from the source-to-target tgds has been removed, but the solution lacks the enforcement of the target key constraints.

(*b*) *They are limited to relational scenarios, and cannot handle XML or nested datasets*. This is a consequence of the fact that data-exchange research has primarily concentrated on the relational setting, and for a long time no notion of data exchange for more complex models was available. In a way, this is a setback with respect to the early systems, which had supported nested relations since the beginning with a pragmatical approach. In fact, they were able to produce results for XML setting, but without the precise definition of quality that core solutions provide. It is interesting to note that benchmarks for mapping systems have been proposed [2, 6]. However, tools of the intermediate generation cannot be fully evaluated using such benchmarks – for example in order to compare the quality of their solutions – since several scenarios in the benchmarks refer to nested structures, and these systems are not capable to generate core solutions for a nested data model.

4 Time for a Golden Age: Data Cleaning

Recent results have faced these problems and paved the way towards the emergence of a fully-fledged new-generation of schema-mapping and data-exchange tools.

An important aspect is the extension to nested relations and XML. The theoretical properties of data exchange in a general XML setting have been recently studied [3, 4, 13], and, due to the generality, have been shown to exhibit several negative properties. However, important results were established for the fragment of XML data exchange in which the data model is restricted to correspond to nested relations [38]. A very important result was reported in [13]: the authors show that the generation of universal solutions for a nested scenario can be reduced to the generation of solutions for a traditional, relational scenario, even in the presence of target constraints. The authors also provide an algorithm to perform the reduction.

A second important advancement is related to the management of functional dependencies over the target. Although it is not always possible, in general, to enforce a set of egds using a first-order language as SQL, it has been proposed a best-effort algorithm that rewrites the above mapping into a new set of s-t tgds that directly generate the target tuples that are produced by chasing the original tgds first and then the egds [30, 31]. As egds merge and remove tuples from the pre-solution, to correctly simulate their effect the algorithm puts together different s-t tgds and uses negation to avoid the generation of unneeded tuples in the result. It has been recently shown how rewriting the subclass of egds defined by Functional Dependencies (FDs) can also lead to significant reductions of the execution times in the solution computation [11].

The ability to handle functional dependencies paves the way to the application of schema mappings for *Data-fusion*. Consider again the example in Fig. 1 and its tgds. It can be considered as a typical data-fusion scenario, in which it is required to merge together data from three different source databases into a common target. However, by using the s-t tgds only, the best we can achieve is to generate a core pre-solution, as shown in Fig. 2c. This solution can be generated efficiently, but it violates the required key constraints and suffers from an unwanted *entity fragmentation* effect: information about the same entities (e.g., *Abi*, *Perry*, or the account number *001/25*) is spread across several tuples, each of which gives a partial representation of the entity. If we take into account the usual dimensions of data quality [10], it is clear that such an instance must be considered of low quality in terms of compactness (or minimality). Based on these requirements, it is natural to desire the generation of a solution as the one shown in Fig. 2d, where the null values are complemented by constants [10]. Such optimal solution can be materialized by chasing the dependencies above with a post-processing step to minimize the pre-solution, but in practice there are orders of magnitude between the execution time needed to compute the pre-solution and the one needed to achieve the optimal one (e.g., seconds vs hours for the same database) [32].

As discussed above, there is a large class of cases where the best-effort algorithm is able to rewrite the given mapping into a new set of s-t tgds that enforce the target egds [30]. However, these algorithms handle the traditional data-exchange case: when two constant values conflict in the conclusion of an egd, the chase fails as there is not valid solution to the problem. Unfortunately, in practice it is very common to have conflicting values, and a lot of effort is spent for their removal. The challenge of *cleaning data* has motivated a lot of research, and it is very interesting to see how schema mappings and data exchange principles have proven to be fundamental tools to attack also this problem, as we discuss next.

4.1 Mapping and Cleaning

Commercial data cleaning systems are based on approaches in which cleaning actions have to be explicitly specified by users by using transformation operations. They usually focus on data profiling, to identify data quality issues,[4] and record matching, to remove duplicate entities,[5] by using ad-hoc techniques and rules with special

[4]http://www.trifacta.com, http://www.informatica.com/PowerCenter.

[5]http://www.tamr.com, http://www.ibm.com/software/products/en/ibminfoqual.

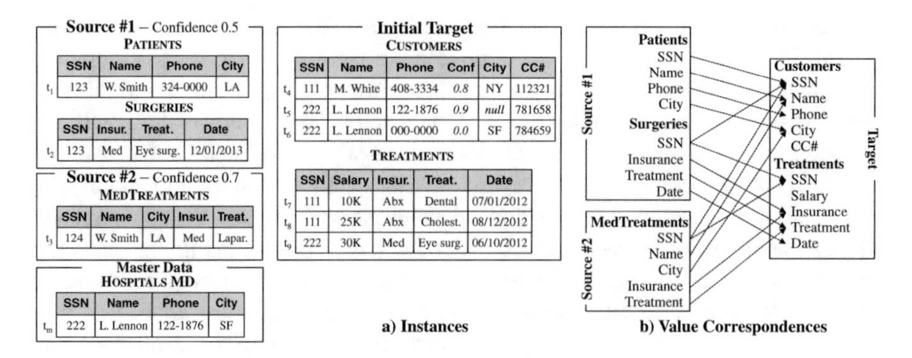

Source #1 – Confidence 0.5

PATIENTS

	SSN	Name	Phone	City
t_1	123	W. Smith	324-0000	LA

SURGERIES

	SSN	Insur.	Treat.	Date
t_2	123	Med	Eye surg.	12/01/2013

Source #2 – Confidence 0.7

MEDTREATMENTS

	SSN	Name	City	Insur.	Treat.
t_3	124	W. Smith	LA	Med	Lapar.

Master Data

HOSPITALS MD

	SSN	Name	Phone	City
t_m	222	L. Lennon	122-1876	SF

Initial Target

CUSTOMERS

	SSN	Name	Phone	Conf	City	CC#
t_4	111	M. White	408-3334	*0.8*	NY	112321
t_5	222	L. Lennon	122-1876	*0.9*	*null*	781658
t_6	222	L. Lennon	000-0000	*0.0*	SF	784659

TREATMENTS

	SSN	Salary	Insur.	Treat.	Date
t_7	111	10K	Abx	Dental	07/01/2012
t_8	111	25K	Abx	Cholest.	08/12/2012
t_9	222	30K	Med	Eye surg.	06/10/2012

a) Instances

Fig. 3 A hospital mapping and cleaning scenario

attention for specific types of data, such as addresses or phone numbers. A principled approach to data cleaning is based on constraints [20, 22]. Data repairing uses declarative constraints, like functional and inclusion dependencies, to detect and repair inconsistencies in the data.

It is natural to think of data exchange and data repairing as two strongly interrelated activities. In fact, the source databases are often structured according to different conventions and rules, and may be dirty. As a consequence, inconsistencies and errors often arise when the sources are brought together into a target schema that comes with its own integrity constraints.

For long time, the database literature has studied these two problems in isolation, with the consequence that there was neither a clear semantics, nor adequate techniques to handle data translation and data repairing in an integrated fashion. One might expect that pipelining data exchange algorithms [15] and data repairing algorithms like those in [20] is sufficient. Unfortunately, this is not the case. In fact, it has been shown that schema mappings and data quality constraints interact in such a way that simply pipelining the two semantics often does not return solutions [24].

Consider the data scenario shown in Fig. 3. Several different hospital-related data sources must be correlated to one another. The first repository has information about Patients and Surgeries. The second one about MedTreatments. Our goal is to move data from the source database into a target database that organizes data in terms of Customers with their addresses and credit-card numbers, and medical Treatments paid by insurance plans. The mappings are informally specified under the form of value correspondences in Fig. 3b, that are then translated into a set of s-t tgds.

Besides deciding how to populate the target to satisfy the mappings above, users must also deal with the problem of generating instances that comply with target constraints, as follows.

(i) *Functional and Inclusion Dependencies*: Traditionally, database architects have specified constraints of two forms: inclusion constraints and functional dependencies. In our example, we have a foreign-key constraint stating that the *SSN* attribute in the Treatments table references the *SSN* of a customer in Customers. The target database also comes with a number of functional dependencies: $d_1 = (SSN, Name \rightarrow Phone), d_2 = (SSN, Name \rightarrow CC\#)$ and $d_3 = (Name, City \rightarrow SSN)$ on table Customers. Here, d_1 requires that a customer's social-security number (*SSN*) and name uniquely determine his or her phone number (*Phone*). Similarly for d_2 and d_3. Differently for traditional data exchange, there is no assumption that the target database is empty. In fact, in Fig. 3a we have reported an instance of the target. There, the pair of tuples $\{t_5, t_6\}$ violates both d_1 and d_2; the database is thus not consistent wrt the constraints, i.e., "dirty".

(ii) *Conditional Dependencies*: Besides standard functional and inclusion dependencies, more advanced forms of constraints are often necessary [20]. Therefore, an expressive data-cleaning tool needs to support a larger class of data-quality rules. Here we mention *conditional functional dependencies* and *conditional inclusion dependencies*. We assume two conditional functional dependencies (CFDs): (i) a CFD $d_4 = (Insur[\text{Abx}] \rightarrow Treat[\text{Dental}])$ on table Treatments, expressing that insurance company *'Abx'* only offers dental treatments (*'Dental'*). Tuple t_8 violates d_4, adding more dirtiness to the target database. (ii) In addition, we also have an *inter-table* CFD d_5 between Treatments and Customers, stating that the insurance company *'Abx'* only accepts customers who reside in San Francisco (*'SF'*). Tuple pairs $\{t_4, t_7\}$ and tuples $\{t_4, t_8\}$ violate this constraint.

(iii) *Master Data and Editing Rules*: Finally, as it is common in corporate information systems, an additional *master-data table* is available; this table contains highly-curated records whose values have high accuracy and are assumed to be clean. We also consider an additional *editing rule*, d_6, stating that whenever a tuple t in Customers agrees on the *SSN* and *Phone* attributes with some master-data tuple t_m in Hospitals MD, then the tuple t must take its *Name* and *City* attribute values from t_m, i.e., $t[Name, City] = t_m[Name, City]$. Tuple t_5 does not adhere to this rule as it has a missing city value (NULL) instead of *'SF'* as provided by the master-data tuple t_m.

As in traditional data exchange, the example requires to map different source databases into a given target, but assumes that the target database may be non-empty, and that the sources and the target instances may generate inconsistencies when the mappings are executed. Given the source and the target instances, the goal is to generate a target instance that satisfies the mappings and the target constraints.

Interestingly, the data cleaning constraints listed above can be all expressed with a specific form of egds. More specifically, besides relation atoms, this form considers *equation atoms* of the form $t_1 = t_2$, where t_1, t_2 are either constants in CONST or variables, and allows for both source and target atoms in the premise. Examples of translation to egds for the constraints in our running example are expressed as follows:

d'_1. Customers($\boldsymbol{s}, \boldsymbol{n}, p, c, k$), Customers($\boldsymbol{s}, \boldsymbol{n}, p', c', cc'$) $\rightarrow p = p'$
d'_4. Treatments($s, a, i =$ *'Abx'*$, t, d$) $\rightarrow t =$ *'Dental'*
d'_6. Customers($\boldsymbol{ssn}, n, \boldsymbol{p}, c, k$), HospMD($\boldsymbol{s}, n', \boldsymbol{p}, c'$) $\rightarrow n = n'$
d''_6. Customers($\boldsymbol{ssn}, n, \boldsymbol{p}, c, k$), HospMD($\boldsymbol{s}, n', \boldsymbol{p}, c'$) $\rightarrow c = c'$

A new unified semantics has been proposed to handle this kind of constraints, together with source-to-target and target tgds [24]. The semantics is a conservative extension of the one of data exchange and incorporates many of the features found in data-repairing algorithms. The new semantics not only preserves the fundamental concepts of data exchange, such as the preference relationship among alternative solutions [29], but also naturally extends the chase procedure to define an algorithms that produces solutions for the new setting. As expected from the negative result on the enforcing of target "classic" egds, executable scripts, such as SQL, cannot express the full recursive power that is needed to solve data cleaning scenarios. This challenge has motivated new recent work on chase engines, with experimental results that show scalability in the computation of solution for mapping scenarios with millions of tuples [8].

4.2 More Applications: ETL

We now briefly discuss an example of a novel applications for the recent advancements in mapping systems.

The most widely used systems in data warehousing environments to express data transformations as a composition of operators in a procedural fashion are known as ETL tools. Operators vary from simple data mappings between tables to more complex manipulations, such as joins, splits of data, and merging of data from different sources. Usually, these tools are used by developers that want to achieve an efficient implementation of a data exchange task. The superior diffusion of ETL systems compared to mapping systems is due to their richer semantics, which allow them to express more operations [14], and to the declarative nature of schema mapping tools that can become a limit with complex transformations where many intermediate steps to manipulate data are needed [33]. For this reason, it became important to study scenarios where flows of mappings, defined using simple intermediate results, are preferable to a single, monolithic mapping with a large number of complex s-t tgds. Preliminary results fill part of this gap by introducing several novel features that allow the possibility to manipulate flows of mappings. This is done by enabling their composition in sequences over intermediate results [18], the ability to invert mappings [5] and to mix data and metadata in them [28], the automatic combination of "parallel" mappings [1], and the reuse of transformations defined in similar settings as components [42]. Moreover, the new results discussed above enable not only to enforce functional dependencies in the target with schema mappings only [30], but also to pair mapping and cleaning specification to ultimately obtain better quality in the solutions [24].

We can give the intuition of how the expression of data exchange scenarios by mapping tools is preferable to ETL systems in terms of easiness of use by comparing a very simple scenario implemented with the two paradigms. Consider the source schema with relations *Students*(name, birthdate, course, program) and *Emps*(name, dept, role), and the target schema *Master*(id, name, birthdate). The desired trans-

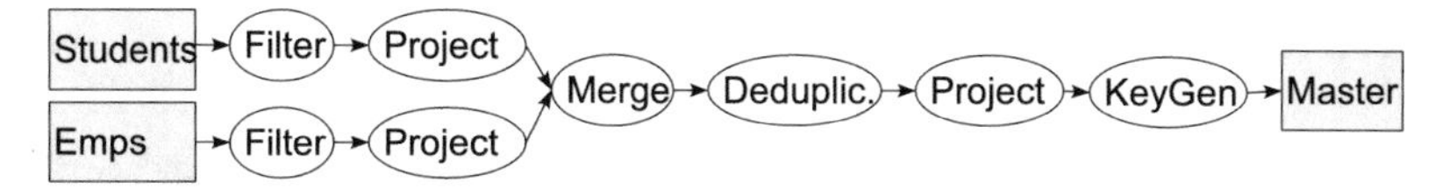

Fig. 4 A simple ETL graph

formation joins employees and students in order to create a new database where supervisors with a master of science are assigned a new id. The scenario is informally depicted as an ETL transformation in Fig. 4 to show how many small procedural steps are needed in order to generate the desired target instance with an ETL approach, while in a mapping system, given the two schemas, only two lines in a GUI and two manually entered strings are required for it, as exemplified by the following s-t tgd:

$$m_a.\ Students(n_1, b_1, c_1, p_1) \wedge Emps(n_1, d_1, r_1)$$
$$\wedge(p_1 = \text{'Msc'}) \wedge (r_1 = \text{'S'}) \rightarrow Master(I_i, n_1, b_1)$$

While, in general, ETL tools are more sophisticated than schema-mapping ones and can handle a larger class of problems, in some cases schema mappings may provide a more abstract, less labor-intensive way of specifying portions of the transformation. Something very similar also happens for other comparable formalisms (like, for example, graphical editors for complex languages as XQuery or XSLT). This does not mean that schema mappings may replace these alternative formalisms, but certainly opens up a number of opportunities to merge these different approaches.

5 Conclusions

Schema mapping management has become an important research area in data transformation, exchange and integration systems. From the early data translation prototypes, important results have been consolidated, but, despite the good results, the adoption of mapping systems in real-life integration applications has been quite slow. We have shown how emerging trends are overcoming the limits of the initial proposal and are going to encourage the developing of more systems based on schema mappings. On one side, novel theoretical results are paving the way to the creation of innovative applications for real world problems. On the other side, a new generation of tools for the creation and optimization of schema mappings are widening the opportunities offered by such technology.

As we discussed in the previous section, we believe that there are quite a lot of interesting research opportunities in this area. Notable examples of the new generation of tools are ++SPICY [31] and LLUNATIC [25]. These tools can deal with different data management tasks, including data fusion, data cleaning and ETL scenarios, which represent very promising areas of application of the latest schema-mappings and data-exchange techniques.

References

1. B. Alexe, M.A. Hernández, L. Popa, W.C. Tan, MapMerge: correlating independent schema mappings. PVLDB **3**(1), 81–92 (2010)
2. B. Alexe, W. Tan, Y. Velegrakis, Comparing and evaluating mapping systems with STBenchmark. PVLDB **1**(2), 1468–1471 (2008)
3. S. Amano, C. David, L. Libkin, F. Murlak, XML schema mappings: data exchange and metadata management. J. ACM **61**(2), 12:1–12:48 (2014)
4. M. Arenas, L. Libkin, XML data exchange: consistency and query answering. J. ACM **55**(2), 1–72 (2008)
5. M. Arenas, J. Pérez, J. Reutter, C. Riveros, Query language-based inverses of schema mappings: semantics, computation, and closure properties. VLDB J. **21**(6), 823–842 (2012)
6. P.C. Arocena, B. Glavic, R. Ciucanu, R.J. Miller, The ibench integration metadata generator. PVLDB **9**(3), 108–119 (2015)
7. C. Beeri, M. Vardi, A proof procedure for data dependencies. J. ACM **31**(4), 718–741 (1984)
8. M. Benedikt, G. Konstantinidis, G. Mecca, B. Motik, P. Papotti, D. Santoro, E. Tsamoura, Benchmarking the chase, in *PODS* (2017)
9. P.A. Bernstein, S. Melnik, Model management 2.0: manipulating richer mappings, in *SIGMOD* (2007), pp. 1–12
10. J. Bleiholder, F. Naumann, Data fusion. ACM Comp. Surv. **41**(1), 1–41 (2008)
11. A. Bonifati, I. Ileana, M. Linardi, Functional dependencies unleashed for scalable data exchange, in *SSDBM* (2016)
12. A. Bonifati, G. Mecca, A. Pappalardo, S. Raunich, G. Summa, Schema mapping verification: the spicy way, in *EDBT* (2008), pp. 85–96
13. R. Chirkova, L. Libkin, J. Reutter, Tractable XML data exchange via relations, in *CIKM* (2011)
14. S. Dessloch, M.A. Hernandez, R. Wisnesky, A. Radwan, J. Zhou, Orchid: integrating schema mapping and ETL, in *ICDE* (2008), pp. 1307–1316
15. R. Fagin, P. Kolaitis, R. Miller, L. Popa, Data exchange: semantics and query answering. TCS **336**(1), 89–124 (2005)
16. R. Fagin, P. Kolaitis, A. Nash, L. Popa, Towards a theory of schema-mapping optimization, in *ACM PODS* (2008), pp. 33–42
17. R. Fagin, P. Kolaitis, L. Popa, Data exchange: getting to the core. ACM TODS **30**(1), 174–210 (2005)
18. R. Fagin, P. Kolaitis, L. Popa, W. Tan, Composing schema mappings: second-order dependencies to the rescue. ACM TODS **30**(4), 994–1055 (2005)
19. R. Fagin, P.G. Kolaitis, L. Popa, W.C. Tan, *Schema matching and mapping*, chapter Schema Mapping Evolution Through Composition and Inversion (Springer, Berlin, 2011), pp. 191–222
20. W. Fan, F. Geerts, *Foundations of Data Quality Management* (Morgan & Claypool Publishers, San Rafael, 2012)
21. A. Fuxman, M.A. Hernández, C.T. Howard, R.J. Miller, P. Papotti, L. Popa, Nested mappings: schema mapping reloaded, in *VLDB* (2006), pp. 67–78
22. H. Galhardas, D. Florescu, D. Shasha, E. Simon, C.-A. Saita, Declarative data cleaning: language, model, and algorithms, in *VLDB* (2001), pp. 371–380
23. F. Geerts, G. Mecca, P. Papotti, D. Santoro, The LLUNATIC data-cleaning framework. PVLDB **6**(9), 625–636 (2013)
24. F. Geerts, G. Mecca, P. Papotti, D. Santoro, Mapping and cleaning, in *ICDE* (2014), pp. 232–243
25. F. Geerts, G. Mecca, P. Papotti, D. Santoro, That's all folks! LLUNATIC goes open source. PVLDB **7**(13), 1565–1568 (2014)
26. G. Gottlob, A. Nash, Efficient core computation in data exchange. J. ACM **55**(2), 1–49 (2008)
27. L.M. Haas, M.A. Hernández, H. Ho, L. Popa, M. Roth, Clio grows up: from research prototype to industrial tool, in *SIGMOD* (2005), pp. 805–810
28. M.A. Hernández, P. Papotti, W.C. Tan, Data exchange with data-metadata translations. PVLDB **1**(1), 260–273 (2008)

29. B. Kimelfeld, E. Livshits, L. Peterfreund, Detecting ambiguity in prioritized database repairing, in *ICDT* (2017)
30. B. Marnette, G. Mecca, P. Papotti, Scalable data exchange with functional dependencies. PVLDB **3**(1), 105–116 (2010)
31. B. Marnette, G. Mecca, P. Papotti, S. Raunich, D. Santoro, ++Spicy: an opensource tool for second-generation schema mapping and data exchange. PVLDB **4**(11), 1438–1441 (2011)
32. G. Mecca, P. Papotti, S. Raunich, Core schema mappings, in *SIGMOD* (2009), pp. 655–668
33. G. Mecca, P. Papotti, S. Raunich, D. Santoro, What is the IQ of your data transformation system? in *CIKM* (2012), pp. 872–881
34. G. Mecca, G. Rull, D. Santoro, E. Teniente, Semantic-based mappings, in *Proceedings of the Conceptual Modeling - 32th International Conference, ER 2013, Hong-Kong, China, 11–13 November, 2013* (2013), pp. 255–269
35. R.J. Miller, L.M. Haas, M.A. Hernandez, Schema mapping as query discovery, in *VLDB* (2000), pp. 77–99
36. R. Pichler, V. Savenkov, DEMo: data exchange modeling tool. PVLDB **2**(2), 1606–1609 (2009)
37. L. Popa, Y. Velegrakis, R.J. Miller, M.A. Hernandez, R. Fagin, Translating web data, in *VLDB* (2002), pp. 598–609
38. A. Roth, M.F. Korth, A. Silberschatz, Extended Algebra and calculus for nested relational databases. ACM TODS **13**, 389–417 (1988)
39. L. Seligman, P. Mork, A. Halevy, K. Smith, M.J. Carey, K. Chen, C. Wolf, J. Madhavan, A. Kannan, D. Burdick, OpenII: an open source information integration toolkit, in *SIGMOD* (2010), pp. 1057–1060
40. N.C. Shu, B.C. Housel, R.W. Taylor, S.P. Ghosh, V.Y. Lum, EXPRESS: a data EXtraction, processing and REstructuring system. ACM TODS **2**(2), 134–174 (1977)
41. B. ten Cate, L. Chiticariu, P. Kolaitis, W.C. Tan, Laconic schema mappings: computing core universal solutions by means of SQL queries. PVLDB **2**(1), 1006–1017 (2009)
42. R. Wisnesky, M.A. Hernández, L. Popa, Mapping polymorphism, in *ICDT* (2010), pp. 196–208

Part III
Data Modeling and Querying

Data Modeling Across the Evolution of Database Technology

Paolo Atzeni, Luca Cabibbo and Riccardo Torlone

Abstract Data modeling has always been a fundamental issue across the whole evolution of database technology, as it is witnessed by the vast literature addressing this issue across the last thirty years. In this paper, we present the studies done by members of the database and big data research group of Roma Tre University on the issue of data modeling in the context of several milestones in field of databases: logic-based and object-oriented databases, data warehousing, and NoSQL systems. We also discuss our proposal for a general framework that allows the description of different models and the translation of schemas from one to another.

1 Introduction

Database technology has evolved dramatically in the past decades but, indisputably, the ability of modeling and managing data at different levels of abstractions has remained a fundamental requirement of any database application, for various reasons. First of all, beside being crucial in the conceptual and logical design phases, it offers support throughout the whole lifecycle: from requirement analysis, where it helps in giving a structure to the process; to coding and maintenance, where it gives valuable documentation. Moreover, it provides support to communication and to individual comprehension. Finally, it provides support to performance management, as physical database design is also based on data structures, and query processing efficiency is often based on reference to the regularity of data.

In this paper, we present an overview of the contributions provided by the database and big data research group of Roma Tre University to data modeling across the evolution of database systems. The goal is to show that models and modeling have always had a fundamental role in this area, and we believe they will keep on having it.

A discussion on data modeling can be tackled along the various level of abstractions that are present in database systems. According to a widely accepted vision in

P. Atzeni · L. Cabibbo (✉) · R. Torlone
Dipartimento di Ingegneria, Università Roma Tre, Rome, Italy
e-mail: cabibbo@dia.uniroma3.it

S. Flesca et al. (eds.), *A Comprehensive Guide Through the Italian Database Research Over the Last 25 Years*, Studies in Big Data 31, DOI 10.1007/978-3-319-61893-7_13

the field of databases, at the lowest level we have raw data, which cannot provide any information without suitable interpretation. At a higher level we usually have *schemas*, which describe the structure of the instances and provide a basic tool for the interpretation of data. Recently, with the advent of NoSQL databases, many developers are led to believe that the importance of schemas gets reduced or even disappears, but we try to argue in this paper that, indeed, data modeling plays a relevant role also in this context. At this level of abstraction we also have various forms of *metadata*, that is, information that describes or supplements raw data. Notable examples are the *constraints*, which allow the specification of properties that must be satisfied by actual data. Finally, we have a third level of abstraction which involves precise formalisms for the description of schemas called *data models*. A data model is a fundamental ingredient of database applications, as it provides the designer with the basic means for the representation of reality in terms of schemas.

Following this three-level view of data representation in databases, in the rest of this paper we address the issue of data modeling in the context of some important milestones in the evolution of database systems. Specifically, in Sect. 2 we illustrate a data model for logic-based and object-oriented databases; in Sect. 3 we discuss the relevance of data modeling in the context of OLAP systems and multidimensional databases; and in Sect. 4 we illustrate, with the same purpose, an abstract model for NoSQL databases. We then present, in Sect. 5, our proposal for a general framework that allows the description of different models and the translation of schemas from one to another. Finally, we draw some conclusions in Sect. 6.

2 Complex-Object Data Models

LOGIDATA+ [1] has been, in the Nineties, a research action aimed at the definition of advanced database systems, with the goal of extending in a significant way the (then-current) relational systems. LOGIDATA+ was intended to support the definition of data with complex structures, and to provide powerful query and update languages, based on a combination of techniques originating from relational databases, logic programming, as well as object-oriented programming.

The *LOGIDATA+ data model* [4] involves three main constructs: classes, relations, and functions. A *class* is a set of *objects*, each with a unique *object identifier* (or *oid*) and a complex value. A *relation* is a relation as in the (nested) relational model, with the further possibility of including references to classes and objects by means of oid's. A *function* is a (partial) function in the ordinary sense; its domain and range may have nested structures. Complex types and their values are based on base types (values), classes (oid's, to reference objects), as well as records, sets, and sequences.

Thus, the LOGIDATA+ model is at the same time value-based, object-based, and functional. This choice supported the spirit of the research project, that is, to allow investigations in various directions. From a modeling viewpoint, these redundant structures provide flexibility and ease of use. In fact:

- classes can be used to model sets of identifiable objects (to represent entities in the Entity-Relationship or other semantic models, or classes of objects in object-oriented models); generalizations or *is-a* relationships can be specified on classes;
- relations can handle value-based data (such as tables in the relational model) and relationships between classes (relationships in the Entity-Relationship model);
- functions can be used in various ways; for example: (i) to specify relationships between classes or between classes and their attributes in a different way; (ii) to embed built-in functions in rules, regardless of the actual implementations of functions; (iii) to attach functions and procedures to classes; and (iv) to manipulate sets and sequences in complex objects.

As an example, consider the LOGIDATA+ schema shown in Fig. 1, to represent students and courses (using classes), classes (using a function), and grades (using a relation). The schema is described using the LOGIDATA+ DDL, which is quite self-explanatory. Furthermore, Fig. 2 shows a sample instance of this schema.

We also participated in the specification of the *LOGIDATA+ query language* [3], which is an extension of Datalog, intended to handle to rich modeling features offered by the LOGIDATA+ data model. In particular, it is a rule-based language, allowing for:

```
TYPE Date = TUPLE d, m, y: integer END;

CLASS Person: TUPLE
                    name: string;
                    birthdate: Date
              END;

CLASS Student ISA Person: TUPLE
                    stud-id: integer;
              END;

CLASS Course: TUPLE
                    name: string;
                    credits: integer
              END;

FUNCTION Class: FROM TUPLE course: Course END
              TO TUPLE students: SET OF Student END;

RELATIONS Grades: TUPLE
                    course: Course;
                    exam-session: Date;
                    student: Student;
                    grade: integer
              END.
```

Fig. 1 A sample LOGIDATA+ schema

class Person

oid	name	birthdate		
		dd	mm	yy
#p1	Tom	[1,	6,	1963]
#p2	John	[12,	3,	1994]
#p3	Mary	[4,	10,	1992]
#p4	Tom	[1,	6,	1963]
#p5	Paul	[25,	3,	1993]

class Student (is-a Person)

oid	stud-id
#p2	12345
#p3	98765
#p5	12321

class Course

oid	name	credits
#c1	Databases	12
#c2	Op. Systems	9
#c3	Compilers	6

function Class

course	students
#c1	{ #p2, #p3, #p5 }
#c2	{ #p2, #p3 }
#c3	{ #p3, #p5 }

relation Grades

course	exam-session			student	grade
	dd	mm	yy		
#c1	[20,	2,	2012]	#p2	A
#c1	[20,	2,	2012]	#p3	B
#c1	[20,	2,	2012]	#p5	F
#c1	[12,	7,	2012]	#p5	C
#c1	[24,	2,	2012]	#p2	A

Fig. 2 A sample LOGIDATA+ instance of the schema of Fig. 1

- the management of complex objects and structures, by means of set, record, and sequence constructors, with built-in functions and predicates;
- creation and manipulation of objects identifiers (oid's), taking isa-hierarchies into account;
- user defined data functions.

The language has a fixpoint semantics, which is deterministic and stratified with respect to negation and (grouping for) multivalued data functions.

A main semantic matter connected with the declarative manipulation of objects is the need for *oid invention*; this aspect is managed by means of Skolem functors, which make the semantics of oid invention truly declarative.

2.1 Discussion

The LOGIDATA+ project was funded by CNR (Consiglio Nazionale delle Ricerche), as a research line within "Sottoprogetto 5: Sistemi Evoluti per Basi di Dati" of "Progetto Finalizzato Sistemi Informatici e Calcolo Parallelo." "Sottoprogetto 5: Sistemi Evoluti per Basi di Dati" played a fundamental role in the history of the SEBD conference series, as the first edition of the Italian Symposium on Advanced Database Systems (SEBD), in 1993, was organized as one of its events.

The research on LOGIDATA+ has been especially important also for our database and big data group at Università Roma Tre, since some of its members started their activity within this project. We also studied many other aspects related to deductive or object-oriented databases (e.g., [5, 10]). Furthermore, other research topics that we investigated (such as model management and data exchange) partially benefited from the knowledge developed in this area.

3 Data Models for Multidimensional Databases

Starting from the end of the Nineties, the separation of the enterprise information architecture into two separate environments has quickly become popular. Alongside the traditional On Line Transaction Processing (OLTP) environment, a new On Line Analytical Processing (OLAP) component was introduced, dedicated to decision-oriented analysis of historical data. The central element of the new environment is the *data warehouse*, a read-only archive of historical snapshots of operational data, rearranged into a multidimensional format more suitable for decision support.

In an OLAP architecture, a system-independent data model serves for two main reasons: (i) to provide an intermediate representation in the design of a data warehouse and (ii) to allow users and applications to manipulate multidimensional data ignoring implementation details.

The *MultiDimensional* data model [13, 14] ($\mathcal{MD}$ for short) is an example of logical model in this context: it provides a few constructs modeling, at a high-level of abstraction, the basic concepts that can be found in any OLAP system. An $\mathcal{MD}$ *dimension* is a collection of *levels* and corresponds to a business perspective under which the analysis can be performed. The levels of a dimension are data domains at different granularity and are organized into a hierarchy. Within a dimension, values of different levels are related through a family of *roll-up functions*, according to the hierarchy defined on them. If a roll-up function associates a value v_1 of a certain level to a value v_2 of an upper level in the hierarchy, we say that v_1 *rolls up to* v_2. A level can have *descriptions* associated with it, which provide further information on it. The main construct of the $\mathcal{MD}$ model is the *f-table*: this is a (partial) function that associates a collection of level values, called *symbolic coordinate*, with one or more *measures*. Components of coordinates are called *attributes*. An *entry* of an f-table f is a coordinate over which f is defined. Thus, an f-table is used to represent factual data on which the analysis is focused (the measures) and the perspective of analysis (the coordinate).

As an example, consider the $\mathcal{MD}$ scheme `Retail`, shown in Fig. 3. This scheme can be used by a marketing analyst of a chain of toy stores and is organized along dimensions **time**, **product**, and **location** (shown on top of the figure). The **time** dimension is organized in a hierarchy of levels involving day, month, quarter, and year. Similarly, the **location** dimension is based on a hierarchy of levels involving store, city, and area. A description of the level store, in the **location** dimension, can be its *address*. Finally, the **product** dimension contains levels item, category, and brand. Finally, there are two further *atomic* dimensions (that is, having just one level) that are used to represent **numeric** values and **string**s.

In this framework, we can define the f-tables SALES and COSTOFITEM. The former describes summary data for the sales of the chain, organized along dimensions **time**

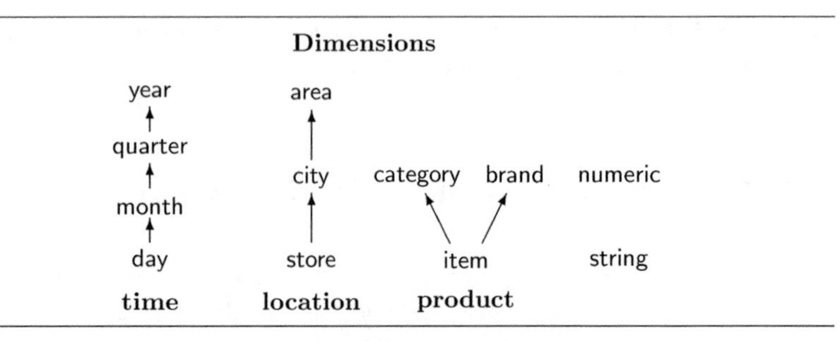

F-tables

$\textsc{Sales}[\mathit{Period} : \text{day}, \mathit{Product} : \text{item}, \mathit{Location} : \text{store}] \rightarrow [\mathit{NSales} : \text{numeric}, \mathit{Income} : \text{numeric}]$

$\textsc{CostOfItem}[\mathit{Product} : \text{item}, \mathit{Month} : \text{month}] \rightarrow [\mathit{Cost} : \text{numeric}]$

Level descriptions

Address (store) : string, *Name* (item) : string

Fig. 3 The sample $\mathcal{MD}$ scheme `Retail`

SALES

Period	*Product*	*Location*	*NSales*	*Income*
Jan 5, 98	*Scrabble*	*Navona*	32	$543.^{68}$
Jan 5, 98	*Risiko*	*Navona*	27	$512.^{73}$
Jan 5, 98	*Lego*	*Sun City*	42	$713.^{58}$
Jan 5, 98	*Risiko*	*Sun City*	22	$439.^{78}$
Feb 19, 98	*Scrabble*	*Navona*	32	$479.^{68}$
Feb 19, 98	*Scrabble*	*Atomium*	26	$422.^{90}$
Feb 19, 98	*Lego*	*Navona*	25	$299.^{75}$
Feb 19, 98	*Lego*	*Colosseum*	11	$142.^{89}$
Mar 10, 98	*Risiko*	*Navona*	5	$69.^{95}$
Mar 10, 98	*Lego*	*Sun City*	6	$71.^{94}$

COSTOFITEM

Cost	*Jan-98*	*Feb-98*	*Mar-98*
Lego	$12.^{99}$	$9.^{99}$	$9.^{99}$
Risiko	$14.^{99}$	$12.^{99}$	$12.^{99}$
Scrabble	$12.^{99}$	$12.^{99}$	$12.^{49}$
Trivia		$18.^{99}$	$17.^{99}$

Fig. 4 An $\mathcal{MD}$ instance for the `Retail` database

(at day level), **product** (at item level), and **location** (at store level). The measures for this f-table are *NSales* (the number of items sold) and *Income* (the gross income), both having type numeric. The f-table COSTOFITEM is used to represents the costs of the various items, assuming that costs may vary from month-to-month.

A possible instance for the `Retail` scheme is shown in Fig. 4.

4 Data Models for NoSQL Databases

NoSQL database systems are today an effective technology to manage large data sets distributed over many servers. A main motivation for the popularity od NoSQL systems is the flexibility they provide in organizing data, as they relax the rigidity provided by the relational model and by the other structured models. In particular, NoSQL systems are intended to support next-generation Web applications, for which relational DBMSs are not well suited. These are simple OLTP applications for which (i) data have a structure that does not fit well in the rigid structure of relational tables, (ii) access to data is based on simple read-write operations, (iii) relevant quality requirements include scalability and performance, as well as a certain level of consistency.

NoSQL technology is characterized by a high heterogeneity; indeed, there are several dozens of NoSQL systems, each with different features. Even though they can be classified into a few main categories (including key-value stores, document stores, and extensible record stores), this heterogeneity is highly problematic to application developers, even within each category.

This flexibility and the heterogeneity that has emerged in this area have led to a little use of traditional modeling techniques, as opposed to what has happened with databases for decades.

However, we believe that traditional notions related to data modeling can be useful in the context of NoSQL databases as well. We argued that model-based approaches can be useful to tackle the difficulties related to heterogeneity, and provided support in the form of abstraction, from at least two perspectives. First, we have proposed an abstract data model for NoSQL databases, and shown how it can be used as an intermediate data representation in the context of a general design methodology for NoSQL applications having initial steps that are independent of the individual target system (Sect. 4.1). Second, we have proposed a framework providing a common programming interface towards NoSQL systems, based on a common data model for NoSQL systems (Sect. 4.2). These approaches are both based on recent contributions of ours [8, 9].

4.1 NoAM: An Abstract Data Model for NoSQL Databases

NoSQL database systems organize their data according to quite different data models. They usually provide simple read-write data-access operations, which also differ from system to system. Despite this heterogeneity, a few main categories of systems can be identified according to their modeling features: key-value stores, extensible record stores, document stores, plus others that are beyond the scope of this paper.

In a *key-value store*, such as Oracle NoSQL or Redis, a database is a schemaless collection of key-value pairs, with data access operations on either individual key-value pairs or groups of related pairs (e.g., sharing part of the key). The key (or part of it, thereof) controls data distribution.

In an *extensible record store*, such as Amazon DynamoDB or Cassandra, a database is a set of tables, each table is a set of rows, and each row contains a set of attributes (columns), each with a name and a value. Rows in a table are not required to have the same attributes. Data access operations are usually over individual rows, which are units of data distribution and atomic data manipulation.

In a *document store*, such as MongoDB, a database is a set of documents, each having a complex structure and value. Documents are organized in collections. Operations usually access individual documents, which are units of data distribution and atomic data manipulation.

To summarize, it is possible to say that each NoSQL system provides a number of "modeling elements" to organize data, which can be considered the "data model" of the system. Moreover, the various systems can be effectively classified in a few main categories, where each category is based on "data models" that, even though not identical, do share some similarities. We now show that it is possible to pursue these similarities, thus defining an "abstract data model" for NoSQL databases.

NoAM (*NoSQL Abstract Data Model*) [9] is a novel data model for NoSQL databases that exploits the commonalities of the data modeling elements available in the various NoSQL systems and introduces abstractions to balance their differences and variations.

The NoAM data model is defined as follows.

- A NoAM *database* is a set of *collections*. Each collection has a distinct name.
- A collection is a set of *blocks*. Each block in a collection is identified by a *block key*, which is unique within that collection.
- A block is a non-empty set of *entries*. Each entry is a pair $\langle ek, ev \rangle$, where ek is the *entry key* (which is unique within its block) and ev is its value (either complex or scalar), called the *entry value*.

In NoAM, a *block* is a construct that models a data access and distribution unit, which is a data modeling element available in all NoSQL systems. By "data access unit" we mean that the NoSQL system offers operations to access and manipulate an individual unit at a time, in an atomic, efficient, and scalable way. By "distribution unit" we mean that each unit is entirely stored in a server of the cluster, whereas different units are distributed among the various servers. With reference to major NoSQL categories, a block corresponds to: (i) a record/row, in extensible record stores; (ii) a document, in document stores; or (iii) a group of related key-value pairs, in key-value stores.

In NoAM, an *entry* models the ability to access and manipulate just a component of a data access unit (i.e., of a block). An entry is a smaller data unit that corresponds to: (i) an attribute, in extensible record stores; (ii) a field, in document stores; or (iii) an individual key-value pair, in key-value stores. Note that entry values can be complex.

Finally, a NoAM *collection* models a collection of data access units. For example, a table in extensible record stores or a document collection in document stores.

Player

mary	*username*	*"mary"*
	firstName	*"Mary"*
	lastName	*"Wilson"*
	games[0]	⟨ *game* : Game:2345, *opponent* : Player:rick ⟩
	games[1]	⟨ *game* : Game:2611, *opponent* : Player:ann ⟩

rick	*username*	*"rick"*
	firstName	*"Ricky"*
	lastName	*"Doe"*
	score	*42*
	games[0]	⟨ *game* : Game:2345, *opponent* : Player:mary ⟩
	games[1]	⟨ *game* : Game:7425, *opponent* : Player:ann ⟩
	games[2]	⟨ *game* : Game:1241, *opponent* : Player:johnny ⟩

Game

2345	*id*	*2345*
	firstPlayer	Player:mary
	secondPlayer	Player:rick
	rounds[0]	⟨ *moves* : ..., *comments* : ... ⟩
	rounds[1]	⟨ *moves* : ..., *actions* : ..., *spell* : ... ⟩

Fig. 5 A sample database in NoAM

Figure 5 shows a sample NoAM database for an on-line social game, representing players, games, and rounds. In the figure, inner boxes show entries, while outer boxes denote blocks. Collections are shown as groups of blocks.

We have also shown how the NoAM data model can be used as an intermediate data representation in the context of a general design methodology for NoSQL applications, which has initial steps that are independent of the individual target system. Specifically, we have proposed a design process that includes a conceptual phase, as common in traditional application, followed (and this is unconventional and original) by a system-independent logical design phase, where the intermediate representation is used, as the basis for both modeling and performance aspects, with only a final phase that takes into account the specific features of individual systems.

4.2 SOS: Uniform Access to NoSQL Databases

SOS [8] is an approach to handle the heterogeneity of the interfaces offered by the various NoSQL systems. Specifically, SOS is a programming environment where different non-relational databases can be uniformly defined, queried, and accessed by an application program.

The SOS programming model is based on a high-level common interface, which is inspired by those of non-relational systems. Indeed, NoSQL systems are based on simple operations for inserting and deleting individual items (objects in an application), mainly one at the time, and retrieving them, one at the time or a set at the time. In correspondence, SOS provides very basic and general operations on objects: put, get and, delete. These operations are defined upon an internal Common Data Model (which is simpler than NoAM) that abstracts the main characteristics of the underlying NoSQL systems.

SOS has been experimented with various systems, such as Redis, MongoDB, and HBase. Indeed, the implementations are transparent to the application, so that they can be replaced at any point in time (and so one NoSQL system can be replaced with another one). Also, the SOS platform allows for a single application to partition the data of interest over multiple NoSQL systems (polyglot persistence), and this can be important if the application has contrasting requirements, satisfied in different ways by distinct systems.

4.3 Discussion

The research on NoSQL databases at Università Roma Tre is one of the first steps in the wider and challenging area of Big Data.

5 A Unified Framework for Managing Multiple Models

As it is apparent from the discussions in the previous sections, many different models have been proposed for organizing data in databases (and in other contexts as well). A natural consequence of this fact is the existence, even within a single organization, of different data sources, often based on different models; therefore, it becomes important to support the integration of heterogeneous schemas and their translation from a model to another.

We have tried to contribute to the solution of this problem by proposing a framework that allows the description of different models and the translation of schemas (and data, but this is beyond the discussion here) from one to the other [2, 6]. The description of models is based on the idea of a *metamodel*, a set of constructs (which is predefined, but extensible). This relies upon an observation made by Hull and King [16] that the constructs used in major models can be expressed by a small set of (model-independent) *metaconstructs*: lexical, abstract, aggregation, generalization, function. In fact, we define a metamodel by means of a set of generic metaconstructs. Each model is defined by its constructs and the metaconstructs they refer to. Without going too much into the details, we can say that basic versions of known database models can be defined as follows:

- an *Entity-Relationship model* involves abstracts (the entities), aggregations of abstracts (relationships), lexical (attributes of entities and, if they are allowed, of relationships);
- an *object-oriented model* has abstracts (classes), reference attributes for abstracts, which are essentially functions from abstracts to abstracts, lexical (fields or properties of classes);
- the *relational model* involves aggregations of lexical (tables), components of aggregations (columns), which can participate in keys, foreign keys defined over aggregations and lexical.

An important concept in our approach, closely related to that of metamodel, is the *supermodel*: it is a model that has constructs corresponding to all the metaconstructs known to the system. Thus, each model is a specialization of the supermodel and a schema in any model is also a schema in the supermodel, apart from the specific names used for constructs.

Let us now discuss the problem of translating a schema from a source model to a target model. Here, the supermodel acts as a "pivot" model, so that it is sufficient to have translations from each model to and from the supermodel, rather than translations for every pair of models. Thus, a linear and not a quadratic number of translations is sufficient. Indeed, since every schema in any model is also a schema of the supermodel (apart from construct renaming), the only needed translations are those within the supermodel with the target model in mind: a translation is composed of (a) a "copy" (with construct renaming) from the source model into the supermodel; (b) an actual transformation within the supermodel, whose output includes only constructs allowed in the target model; (c) another copy (again with renaming into the target model).

Moreover, the supermodel emphasizes the common features of models. So, if two source models share a construct, then their translations towards similar target models could share a portion of the translation as well. In our approach, we follow this observation by defining *basic* translations that refer to single constructs (or specific patterns or variants thereof). Then, actual translations are specified as compositions of basic ones, with significant reuse of them.

For example, assume we have as the source an ER model with binary relationships (with attributes) and no generalizations and as the target a simple OO model. To perform the task, we would first translate the source schema by renaming constructs into their corresponding homologous elements (abstracts, binary aggregations, lexical, generalizations) in the supermodel and then apply the following steps (sketched in Fig. 6, taken from [6, Fig. 6]):

(1) eliminate attributes of aggregations, by introducing new abstracts and one-to-many aggregations
(2) eliminate many-to-many aggregations, by introducing new abstracts and one-to-many aggregations

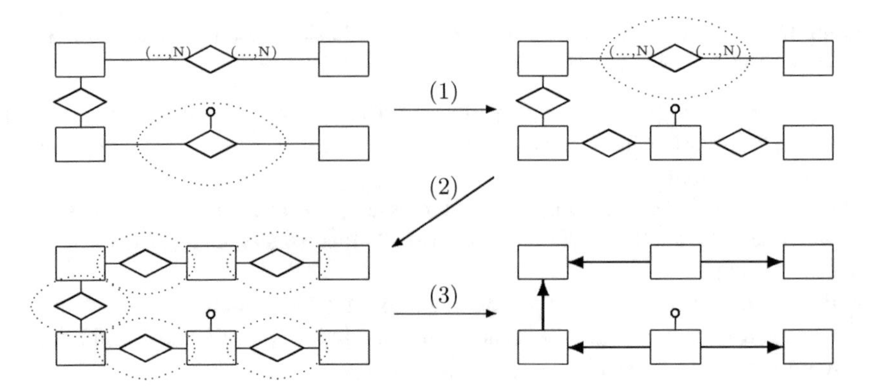

Fig. 6 A translation composed of three steps

(3) replace one-to-many aggregations with references between abstracts

If instead we had a source ER model with generalizations but no attributes on relationships (still binary), then, after the copy in the supermodel, we could apply steps (2) and (3) above, followed by another step that takes care of generalizations:

(4) eliminate generalizations (replacing them with references)

It is important to note that the basic steps are highly reusable. For example, step (2) above can be used in every translation that requires the elimination of many-to-many relationships.

We have developed a library of transformations, so that the approach can become effective. Indeed, it would be impossible to claim (or even to aim at) completeness, but the approach can be extended and refined as needed. Given that complex transformations are built by sequences of elementary ones, we have developed techniques for the automatic selection of translations in the library. Our approach efficiently finds a sequence of transformations, on the basis of (i) a graph ("the space of models") where each model is a node and edges represent basic transformations, and (ii) concise descriptions of models and translation steps.

5.1 Discussion

Research on multiple models was carried out by our group in a somehow intermittent way, with initial activity in the early Nineties, significant contributions a few years later [2], and a further wave in the new millennium, with a new generation of collaborators [6, 7]. Beside the main contribution we have summarized, activity included a view-based approach to support runtime translations and the development of a uniform interface to NoSQL systems (the SOS proposal briefly illustrated in the previous section [8]).

6 Conclusion

In this paper we have summarized some studies on data modeling done by members of the database and big data research group of Roma Tre University. These research activities concern various areas in the context of databases: logic-based and object-oriented databases, data warehousing, NoSQL systems, as well as model management and schema translation.

To conclude, we would like to mention other research activities of the authors involving data modeling issues that have not been discussed here for the sake of space limitation. These include: design of graph databases [20], taxonomies of terms supporting query relaxation [17], Semantic-Web data management [19], modeling of user preferences [15], data and schema exchange [12, 18], object-relational mapping [11].

References

1. P. Atzeni (ed.), *LOGIDATA+: Deductive Databases with Complex Objects*, vol. 701, Lecture Notes in Computer Science (Springer, Berlin, 1993)
2. P. Atzeni, R. Torlone, Management of multiple models in an extensible database design tool, in *EDBT Conference*, LNCS, vol. 1057 (Springer, Berlin, 1996), pp. 79–95
3. P. Atzeni, L. Cabibbo, G. Mecca, L. Tanca, The LOGIDATA+ language and semantics, in ed. by P. Atzeni (Springer, Berlin, 1993), pp. 30–41
4. P. Atzeni, F. Cacace, S. Ceri, L. Tanca, The LOGIDATA+ model, in ed. by P. Atzeni (Springer, Berlin, 1993), pp. 20–29
5. P. Atzeni, L. Cabibbo, G. Mecca, Isalog$^{(\neg)}$: a deductive language with negation for complex-object databases with hierarchies. Data Knowl. Eng. **24**(1), 1–38 (1997)
6. P. Atzeni, P. Cappellari, R. Torlone, P.A. Bernstein, G. Gianforme, Model-independent schema translation. VLDB J. **17**(6), 1347–1370 (2008)
7. P. Atzeni, L. Bellomarini, F. Bugiotti, F. Celli, G. Gianforme, A runtime approach to model-generic translation of schema and data. Inf. Syst. **37**(3), 269–287 (2012)
8. P. Atzeni, F. Bugiotti, L. Rossi, Uniform access to NoSQL systems. Inf. Syst. **43**, 117–133 (2014)
9. F. Bugiotti, L. Cabibbo, P. Atzeni, R. Torlone, Database design for NoSQL systems, in *Proceedings of the Conceptual Modeling - 33rd International Conference, ER-2014, 27–29 October 2014, Atlanta, GA, USA* (2014) pp. 223–231
10. L. Cabibbo, The expressive power of stratified logic programs with value invention. Inf. Comput. **147**(1), 22–56 (1998)
11. L. Cabibbo, Objects meet relations: on the transparent management of persistent objects, in *Advanced Information Systems Engineering, Proceedings of the 16th International Conference, CAiSE 2004, Riga, Latvia, June 7–11, 2004* (2004), pp. 429–445
12. L. Cabibbo, On keys, foreign keys and nullable attributes in relational mapping systems, in *EDBT 2009, Proceedings of the 12th International Conference on Extending Database Technology, Saint Petersburg, Russia, March 24–26, 2009* (2009), pp. 263–274
13. L. Cabibbo, R. Torlone, Querying multidimensional databases, in *Database Programming Languages, Proceedings of the 6th International Workshop, DBPL-6, Estes Park, Colorado, USA, August 18–20, 1997*. Lecture Notes in Computer Science, vol. 1369 (Springer, Berlin, 1997), pp. 319–335

14. L. Cabibbo, R. Torlone, A logical approach to multidimensional databases, in *Advances in Database Technology - EDBT'98, Proceedings of the 6th International Conference on Extending Database Technology, Valencia, Spain, March 23–27, 1998*. Lecture Notes in Computer Science, vol. 1377 (Springer, Berlin, 1998), pp. 183–197
15. P. Ciaccia, R. Torlone, Modeling the propagation of user preferences, in *Conceptual Modeling - ER 2011, Proceedings of the 30th International Conference, ER 2011, Brussels, Belgium, October 31 – November 3, 2011* (2011), pp. 304–317
16. R. Hull, R. King, Semantic database modelling: survey, applications and research issues. ACM Comput. Surv. **19**(3), 201–260 (1987)
17. D. Martinenghi, R. Torlone, Taxonomy-based relaxation of query answering in relational databases. VLDB J. **23**(5), 747–769 (2014)
18. P. Papotti, R. Torlone, Schema exchange: generic mappings for transforming data and metadata. Data Knowl. Eng. **68**(7), 665–682 (2009)
19. R.D. Virgilio, G. Orsi, L. Tanca, R. Torlone, Semantic data markets: a flexible environment for knowledge management, in *Proceedings of the 20th ACM Conference on Information and Knowledge Management, CIKM 2011, Glasgow, United Kingdom, October 24–28, 2011* (2011), pp. 1559–1564
20. R.D. Virgilio, A. Maccioni, R. Torlone, Model-driven design of graph databases, in *Conceptual Modeling - Proceedings of the 33rd International Conference, ER 2014, Atlanta, GA, USA, October 27–29, 2014* (2014), pp. 172–185

Modeling, Modeling, Modeling: From Web to Enterprise to Crowd to Social

Marco Brambilla and Stefano Ceri

Abstract Data management is continuously evolving for serving the needs of an increasingly connected society. New challenges apply not only to systems and technology, but also to the models and abstractions for capturing new application requirements. In this paper, we describe several models and abstractions which have been progressively designed to capture new forms of data-centered interactions in the last twenty five years – a period of huge changes due to the spreading of web-based applications and the increasingly relevant role of social interactions. We initially focus on Web-based applications for individuals, then discuss applications among enterprises, then we discuss how these applications may include rankings which are computed using services or using crowds; we conclude with hints to a recent research discussing how social sources can be used for capturing emerging knowledge.

1 Introduction

Long time ago, in the past century, the International DB Research Community used to meet for assessing new research directions, starting the meetings with 2-min *gong shows* to tell each one's opinion and influencing follow-up discussion. Bruce Lindsay from IBM had just been quoted for his message – very brief: "there are 3 important things in data management: performance, performance, performance". The oldest author of this paper had a chance to speak out immediately after and to give a syntactically similar but semantically orthogonal message: "there are 3 important things in data management: modeling, modeling, modeling".

Of course, if one compares the popularity of 3P and 3M in the data management scientific production, the balance is much in favor of 3P. Query optimization, indexing, parallel and distributed databases, cloud engines are much more popular than

M. Brambilla (✉) · S. Ceri
Politecnico di Milano. Dipartimento di Elettronica, Informazione e Bioingegneria,
Via Ponzio 34/5, 20133 Milano, Italy
e-mail: marco.brambilla@polimi.it

S. Ceri
e-mail: stefano.ceri@polimi.it

S. Flesca et al. (eds.), *A Comprehensive Guide Through the Italian Database Research Over the Last 25 Years*, Studies in Big Data 31, DOI 10.1007/978-3-319-61893-7_14

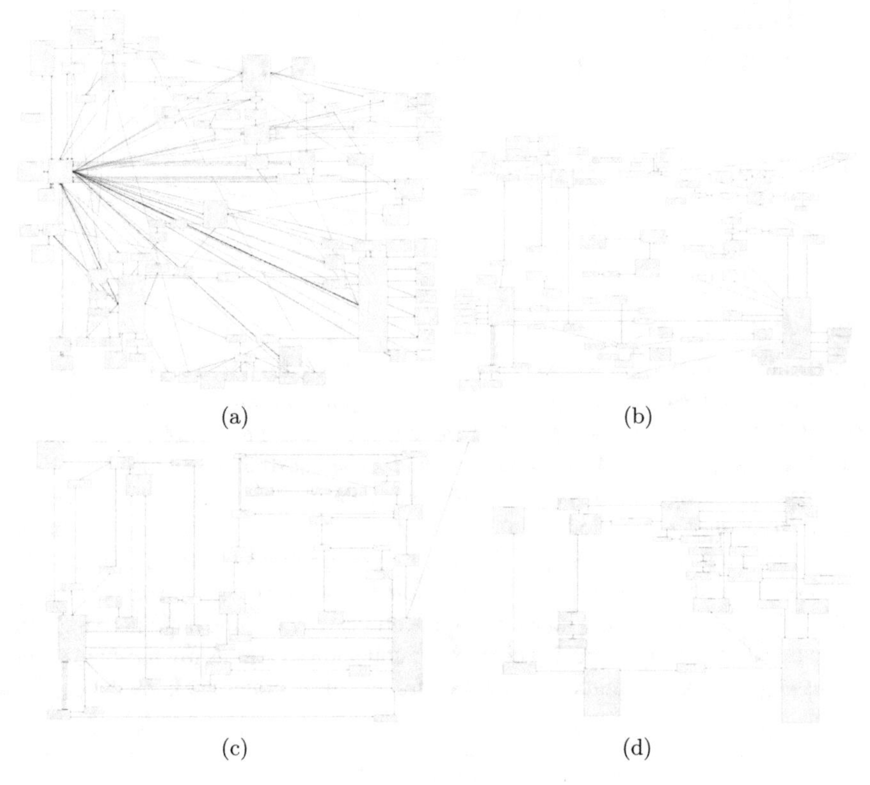

Fig. 1 Schema redesign in three steps. Step 1 eliminates from the diagram the tables which are only used as *active domains*, i.e. carry legal values for given domains. Step 2 isolates small *stars*, i.e. subentities carrying multivalued attributes. Step 3 eliminates *redundant information* from the schema and was the result of long discussions about data semantics

semantic models. Yet, we believe that performance is often attacked without a solid understanding of application needs, resulting in a brute force waste of energies – whereas more modeling could also lead to an overall better performing data system.

As a convincing example, we recall a consultant job for an anonymous Cefriel client,[1] concerned with the overall quality of a very large relational database, originally designed for a commercial DBMS. By adopting systematic good modeling practices for improving the diagram readability and after deep semantic analysis with the designers, the original schema was progressively reduced from Fig. 1a to d – and only then implemented for performance. Our claim is that giving performance to the first schema would not solve its many problems of data redundancy and lack of orthogonality.

[1]Hereby acknowledged for allowing us to publish the anonymized database schemas in Fig. 1; Cefriel is an IT center of excellence linked to Politecnico di Milano.

In this specific real-life case, an initial sub-optimal design had to be rectified though a number of modeling choices before even *thinking* to its performance; we conjecture that this occurs in many other real-life data-centered applications. Hence our "3M" motto. We argue that mastering "3P" and disregarding "3M" could be very dangerous, and therefore equal relevance should be given to data abstractions for semantics – e.g. in the form of high-level or abstract models – and to data structures for implementation – e.g. in the form of specialized persistent data structures or use of parallelism for performance.

We dedicated most of our research to 3M, by inventing new models and by applying modes to real-life scenarios so as to validate and use them. This paper is about model evolution in the last 25 years. Although we also worked on plain data modeling, we keep this aspect outside of our outline of this paper, and refer to Batini's article in this same book. We instead focus on the so called "emerging technologies" – although a 25-years-long period in ICT makes the emerging technologies at the beginning of the period almost obsolete at its end.

Thus, we concentrate on the following technologies: (a) the web, with its evolution throughout all the considered period – from exclusively desktop to mostly mobile; (b) the services, both in their interaction with web models and in their interplay for building search applications; (c) the crowds, and specifically the evolution from strict use of marketplaces such as Amazon Mechanical Turk to the adoption of social networks as sources of work; and (d) social sources themselves, seen as potential repositories of up-to-date knowledge.

This journey of course capitalizes on our results and is very much biased to emphasizing the 3Ms. We are aware that in certain cases some models apply to very few instances, possibly just to our own work. In other cases, however, 3M has been a successful vehicle for commercialization and standardization. Our work is traced by papers that appeared at WWW Conferences of the period, dedicated to WebML [17], to liquid queries over search services [4], to crowd-based search [3] and to knowledge extraction for social sources [9].

2 Web Modeling with WebML

Several researches have applied software engineering and Web engineering techniques to the specification of Web and multi-platform application interfaces and user interaction in broad sense. Among them, we can cite OO-HDM [26], WAE [19], WebDSL [21], OOH-Method [20], WebML [16], RUX-Model [23], HERA [28], and rapid UI development [25] and modeling languages like USIXML [22]. Commercial vendors are nowadays proposing tools for Web development, like Mendix (http://www.mendix.com), Outsystems (http://www.outsystems.com) and Webratio (http://www.webratio.com). However, none of them has managed to become widely adopted in the software industry yet. For this reason, front-end development continues to be a costly and inefficient process, where manual coding is the predominant development approach, reuse is low, and cross-platform portability remains difficult.

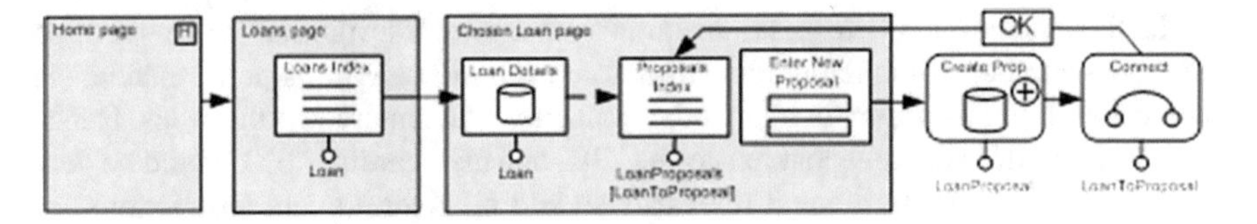

Fig. 2 Example of WebML hypertext

2.1 WebML

WebML was an offspring of several EU-Funded project. It was first presented at the WWW Conference in 2000 [17] and then consolidated in a monography [18]. The specification of a Web application in WebML consists of a set of orthogonal models: the application data model (a standard Entity-Relationship model), one or more hypertext models expressing the navigation paths and the page composition; and the presentation model, describing the visual aspect of the pages. The presentation model is quite interesting, as it enables producing Web pages with the desired layout and look and feel for any rendition technology, but is outside the scope of this paper.

The hypertext model consists of one or more site views, each of them targeted to a specific user role or client device. A site view is a collection of pages (possibly grouped into areas for modularization purposes); the content of pages is expressed by components for data publishing (called content units); the business logic triggered by the users interaction is instead represented by sequences of operation units, which denote components for modifying data or for performing arbitrary business actions (e.g., sending email). Content and operations units are connected by links, which specify the data flow between them and the process flow for computing page content and for enacting the business logic, in reaction to users generated navigation events.

Consider for instance a simple scenario, whose hypertext is shown in Fig. 2; users browse a Home Page, from where they can navigate to a page showing an index of loan products. When the user selects a loan from the index, he is taken to the Chosen Loan page, showing the loan details. In this page, a data unit, labeled Loan Details, displays the attributes of the loan (e.g. the company, the total amount and the rate), and is linked to another index unit, labeled Proposals Index, which displays the plan options. Then, the Enter New Proposal entry unit is used for data entry; the outgoing link of the Enter New Proposal entry unit activates a sequence of create and connect units, which respectively create an instance of the LoanProposal entity and connect it with a relationship instance to the Loan entity.

Syntactically, each type of unit has a distinguished icon and the entity name is specified at the bottom of the unit; below the entity name, predicates (called selectors) express conditions filtering the entity instances to be shown. WebML distinguishes between normal, transport, and automatic links. Normal links (denoted by solid arrows) enable navigation and are rendered as hypertext anchors or form buttons, while transport links (denoted by dashed arrows) enable only parameter passing and are not rendered as navigable widgets. Automatic links are automatically *navigated* by the system on page load.

WebML is associated with a page computation algorithm deriving from the formal definition of the models semantics, which describes how the content of the page is determined after a navigation event produced by the user. Page computation amounts to the progressive evaluation of the various units of a page, starting from input parameters associated with the navigation of a link. This process implies the orderly propagation of the value of link parameters, from an initial set of units, whose content is computable when the page is accessed, to other units, which expect input from automatic or transport links exiting from the already computed units of the page. In WebML, pages are the fundamental unit of computation. A WebML page may contain multiple units linked to each other to form a complex graph, and may be accessed by means of several different links, originating from other pages, from a unit inside the page itself, or from an operation activated from the same page or from another page.

3 Enterprise Modeling with Services and Processes

Five years after we developed it, WebML evolved from model-driven web page generation to model-driven integration of applications within the enterprise. We discuss embedding of web services within WebML, and then use of WebML from within a generic enterprise workflow engine.

3.1 Service Integration

The first WebML extension is towards Service Oriented Architectures, with a focus on provisioning of well-designed services, usable across different Web applications [24]. Extensions to the hypertext model cover both service publication and consumption; service publication is expressed as a *Service View*, which is analogous to a site view, but contains specifications of services instead of pages. A service specification is denoted by a *Port*, which models the operations triggered upon invocation of the service.

Service invocation and reaction to messages are supported by specialized components, called Web Service units. These primitives correspond to the classical WSDL classes of Web service operations and comprise:

- Web service publication primitives: Solicit unit (representing the end-point of a Web service), and Response unit (providing the response at the end of a Web service implementation).
- Web Service invocation primitives: Request-response and Request units; they denote the invocation of remote Web Services from the front-end of a web application.

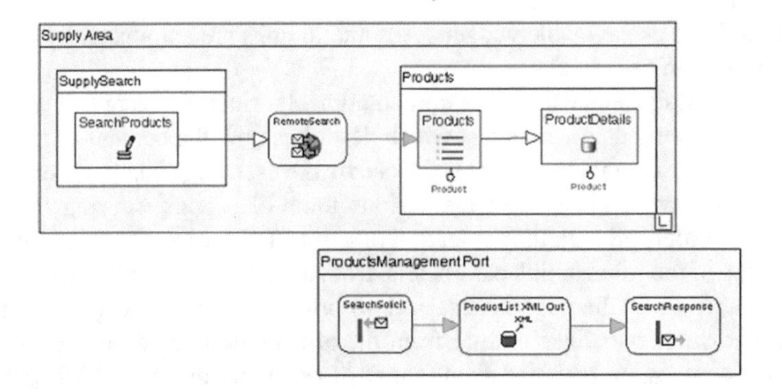

Fig. 3 Example of WebML hypertext model with invocation of remote service

For instance, Fig. 3 shows a hypertext that specifies a front-end for invoking a web Service and the specification of the web Service within a port container. In the former, the user can access the SupplySearch page, in which the Search Products entry unit enables the input of search keywords. The submission of the form, denoted by the navigation of the outgoing link of the entry unit, triggers a request-response operation (Remote Search), which builds the XML input requested by the service and collects the XML response returned by it. From the service response, a set of instances of the Product entity are displayed to the user by means of the Products index unit in the Products page; the user may continue browsing, e.g., by choosing one of the displayed products and looking at its details.

The lower part of Fig. 3 represents the service view that publishes the Remote Search service. The sequence starts with the Search Solicit unit, which denotes the reception of the message. Upon the arrival of the message, an XML-out operation extracts from the service provider's database the list of desired products and formats it as an XML document. The service terminates with the SearchResponse unit, which returns the response message to the invoker.

3.2 *Business Process Integration*

With web services, the Web became a popular implementation platform for B2B applications, whose goal is not only the navigation of content, but also the enactment of intra- and inter-organization business processes. Web-based B2B applications exhibit much more sophisticated interaction patterns than traditional Web applications: they back a structured process, consisting of activities governed by execution constraints, serving different user roles, whose joint work must be coordinated. They may be distributed across different processor nodes, due to organizational constraints, design opportunity, or existence of legacy systems to be reused.

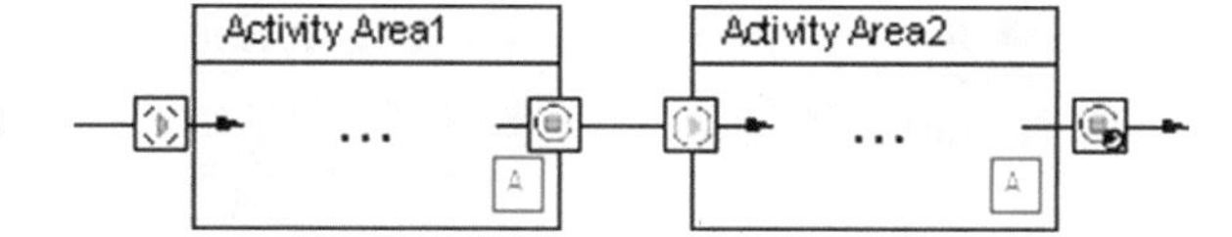

Fig. 4 Two activity areas and the start and end links that denote the initiation and termination of an activity

We extended our approach to cover business process based modeling [10], with a technique that exploits the BPMN notation for the description of the business requirements, and then maps them to hypertext model chunks that describe the user interaction of every task of the business process. The intuition is that the process progresses as the actors navigate the front-end, provided that the hypertext model and the process metadata are kept in synch. To this end, new primitives were added to the hypertext model, for specifying activity boundaries (namely activity areas) and process-dependent navigation (namely workflow links).

Figure 4 shows some of these primitives: Activity Areas denote groups of pages that implement the front-end for executing an activity; specialized links represent the workflow-related side effects of navigation: starting, ending, suspending, and resuming activities. Model transformations were used for translating a business process model into a skeleton of WebML hypertext model; a one-click code generator from the BPMN models generates running prototypes starting from the business processes, without the need of WebML modeling.

4 From WebML to the IFML Standardization and Commercialization

When we designed WebML, we thought that it was strategic to protect it through a US patent; WebML was implemented by WebRatio, a spinoff of Politecnico di Milano, which had unlimited rights of use of the patent. In the following ten years, the world of software tools drastically changed, and within and together with WebRatio we ended up promoting an open standard for enhanced hypertexts called Interaction Flow Modeling Language (IFML) [12], that was largely inspired by WebML; IFML was adopted in 2014 by the Object Management Group as an international standard after a 3-years adoption process. In the course of this operation, we had to give up on the protection of our ideas and completely change our approach [11].

4.1 IFML

From the technical perspective, IFML supports a much wider set of usage scenarios. Indeed, it aims at the platform independent description of graphical user interfaces for applications accessed or deployed on such systems as desktop computers, laptop

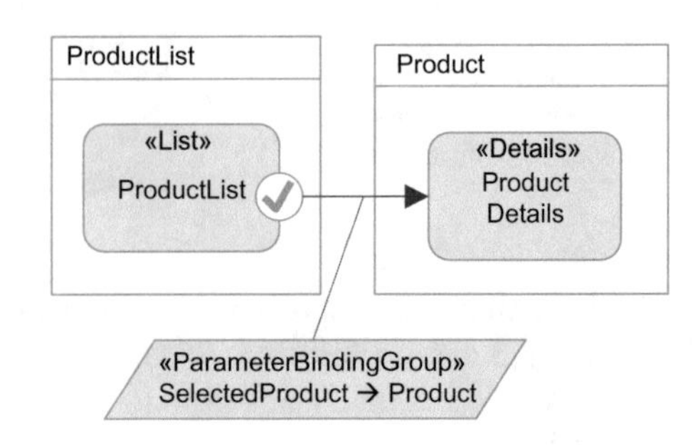

Fig. 5 Example of IFML model showing a list of products and the details view

computers, PDAs, mobile phones, and tablets. IFML adds to WebML several innovations: it increases separation of concerns, completely forbidding the integration of business logic into the user interaction specification; it defines a set of very generic concepts (the core of the language) which can be applied to any kind of user interface; it brings in the concept of event and asynchronous interactions; and it integrates seamlessly with UML and BPMN notations. The focus of the description is on the structure and behavior of the application front-end as perceived by the end user. Hence, with respect to the popular Model-View-Controller (MVC) model of an interactive application, the focus of IFML is mainly on the view part.

IFML models support the following design perspectives: (1) The *view structure specification*, which consists of the definition of view containers; (2) The *view content specification*, which consists of the definition of view components, i.e., content publishing and data entry elements contained within view containers; (3) The *events specification*, which consists of the definition of events (coming from user's interaction, application logic, or external agents) that may affect the state of the UI; (4) The *event transition specification*, which consists of the definition of the effect of an event on the user interface; and (5) The *parameter binding specification*, which consists of the definition of the input-output dependencies between model elements. Furthermore, IFML can be complemented with external models for connecting to any kind of content model (representing databases, ontologies, file systems or other resources) and any kind of dynamic model (describing the business logic behind the application front end).

Figure 5 shows a simple example of IFML diagram, where a starting page displays a list of products and, upon selection by the user, a target page shows the details of the selected product.

4.2 WebRatio

WebRatio is a commercial tool and company, born as spin-off of Politecnico di Milano, that has backed the development of WebML by progessively extending its supported features. Today, the WebRatio Platform[2] is a model-driven development tool based on IFML, which features two editions, respectively focusing on Web and mobile applications [1]. WebRatio provides an integrated environment for supporting the specification of IFML diagrams, including the view description, UML class diagrams for the information design, and optionally the integration with BPMN diagrams for the specification of business process aspects. It also includes a development environment for supporting the implementation of custom components and the layout template and style design environment, which allows the highest possible level of UI sophistication, thanks to full support of HTML 5, CSS and JavaScript based styling.

Based on the input provided through these environments, WebRatio applies a full-fledge model-driven development approach (as described a book of one of the authors of this paper [6], which has become one of the reference readings in the MDD field), which provides model checking, full code generation, group–work support and lifecycle management. The generated code consists of: automatic cloud-deployed Java EE code covering both front-end of back-end of web applications for the Web version of WebRatio; and ready-to-deploy cross-platform mobile applications for the Mobile version of WebRatio. In the deploy, integration and coherency between mobile and web application is granted by a common modeling approach.

5 Search Modeling and Computing

With the start of Search Computing, an ERC-funded project (2008–2012)[3] our interest moved into the integration of search services, i.e. of services capable of extracting ranked responses [13–15]. In this context, modeling interest turns to understanding effective ways of combining services so as to extract only a few answers from them - top answers combine in creating query responses. Although search services can be part of arbitrarily complex applications, their typical usage is to answer queries, such as: *Who is the best doctor to cure insomnia in a nearby hospital? Where can I attend an interesting conference in my field close to a sunny beach?*; such queries are normally geo-referenced and dealing with distances is part of the ranking problem. The Search Computing project devised exploratory user interfaces, service registration tools, query configuration tools, and execution plan optimization techniques.

In particular, at the conceptual level, location-based resources are annotated with labels denoting their geometrical class (point, path, area) and grouped within a specific geo-concept region of the domain diagram. Figure 6 shows a sample domain diagram including general entities and relationships and a geo-referenced region comprising entities such as: Museum, Restaurant, etc. Users can explore concepts by requesting for details of a specific entity or by moving to other, geographically or semantically related entity; we called such search paradigm a *liquid query* [4].

[2] www.webratio.com.

[3] http://search-computing.deib.polimi.it.

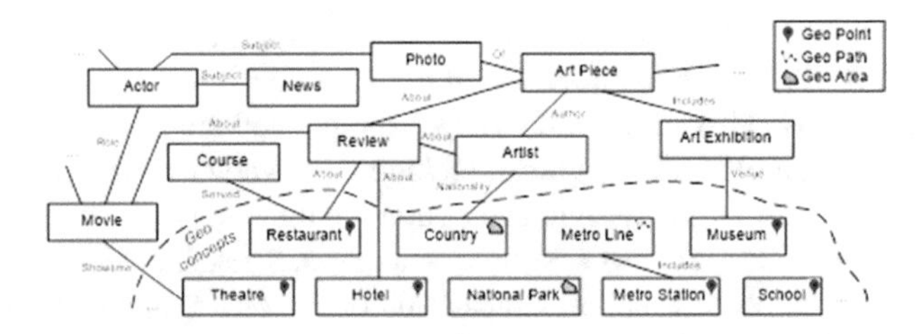

Fig. 6 Conceptual model with geo-referenced concepts

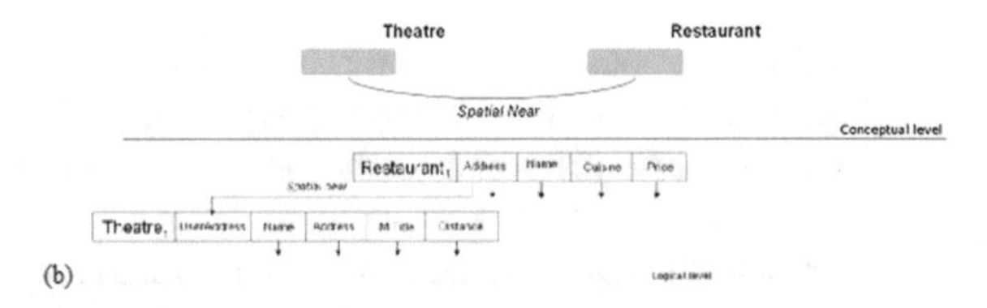

Fig. 7 Example of conceptual and logical service invocations

The exploratory search strategy evolves as follows. Users start by selecting one of the available entities, and submit a query to extract a subset of object instances. Among these, they select the instances they are interested in and then proceed by selecting the next entity to explore; the system retrieves connected object instances and forms several *combinations* with the previously retrieved ones; top-ranked combinations are displayed in ranking order. For instance, they can select a concert nearby their current location, then relate it to close-by transportation and parking facilities (spatial nearness), to other shows taking place the same night in town (spatial nearness and temporal proximity), performing artists (semantic relationships), etc.

The registration of geographical services and their relations used YAGO as reference knowledge base. At the logical level, spatial accesses are supported by services, e.g., GoogleMovies outputs movie shows ranked by distance from an input Location. Alternatively, spatial filters may be supported by ad-hoc services made available in the framework. Figure 7 shows an example of how spatial concepts can be supported. At the conceptual level (Fig. 7(a)) the query searches for theatres close to restaurants. At the logical level, Fig. 7(b) it exploits the a service implementation of Theatre1 supporting distance, by matching the Address output of the service Restaurant1 to the UserAddress attribute of access pattern Theatre1. The nearness function is supported only by the Theatre1 service, therefore the relationship at the logical level is directed, indicating that accesses are possible by first selecting restaurants and then theatres, but not vice-versa.

This process can be particularly useful and efficient for geo-referenced objects that are searched on a mobile device. Figure 8 presents a query combining concerts, restaurants, and hotels in San Francisco. By selecting one or more object combinations (Fig. 8(a)), users can prune the set of available options and look at the details of any object in the map (Fig. 8(b)), including related, non geo-referenced objects. Thanks to the conceptual model representation, the system clusters non geo-referenced items with the semantically closest geo-referenced item. Furthermore,

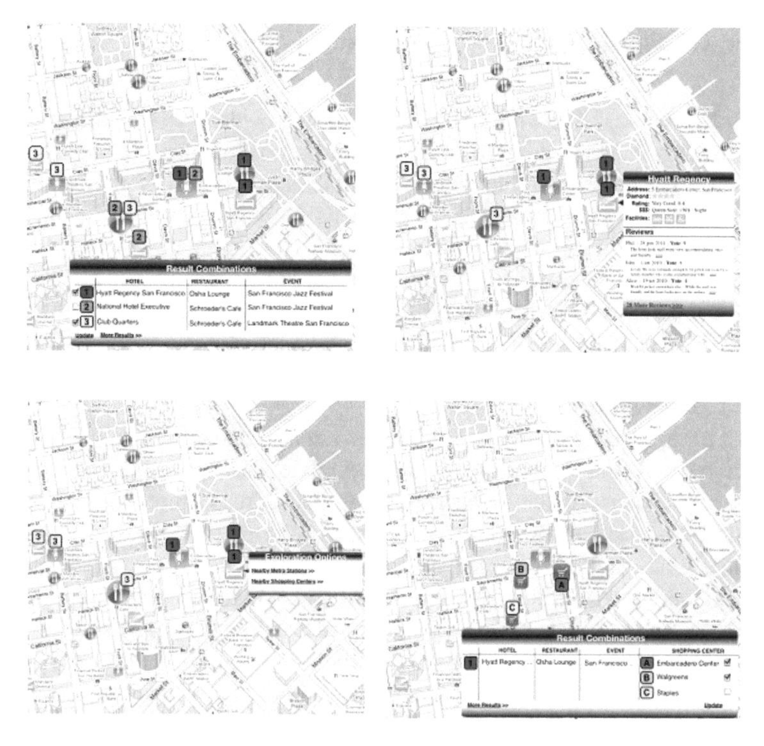

Fig. 8 **a** Visualization of best night-plan as combinations of hotel-restaurant-show; user selects combinations 1 and 3; **b** for the specific hotel Hyatt Regency of combination 1, a popup opens containing guest reviews; **c** a menu suggests exploration options relevant to the selected hotel metros and shopping centers; **d** user explores additional shopping centers near the selected hotel

starting from a given object, users can decide the exploration direction to follow towards other types of items (Fig. 8(c)) based on geographical relationships. In this case, additional objects appear in the map and contribute to the newly calculated combinations (and rankings), as shown in Fig. 8(d). The exploration step can be iterated.

At any stage, users can move forward in the exploration by adding a new object to the query, starting from the available connections and from the objects that have been previously extracted. Users can also move backwards by excluding one of the entities from the query (e.g., removing hotels), or by deselecting previous manually selected objects. Backtracking at the level of individual conditions may help, e.g., in changing the restaurant choice from vegetarian to Japanese.

6 Crowd Modeling Using CrowdSearcher

Amazon Mechanical Turk, the most widely used crowdsourcing marketplace, was created in 2005; but crowd-based computations became much more popular about 5 years later, when social communities (engaged though social platforms) were recognized as much wider and knowledgeable crowds. Humans were found more competent than machines in solving many tasks, ranging from simple ones (such as tagging images) to complex ones (such as finding optimal protein bindings in the 3D space); for what concerns search, the new trend of *asking the crowd* became popular: small *local crowds* could be selected based on geolocalization, expertise, memberships within special interest groups, or simply friendship – thereby complementing the results of search systems.

A that time, we developed CrowdSearcher, a model and tool for engaging crowds in the context of search queries [3]. Our complete paradigm, illustrated in Fig. 9, was alternating steps of search queries, using the search computing platform, and crowd-based queries, using a variety of social systems that could be invoked for providing human rankings. For instance, while planning a move to a new condition which was taking into account job availability, housing, and availability of public transport, experts could be asked to judge the alternative jobs or to comment about public transports.

An example of simple interaction for involving the crowd in suggesting jobs is illustrated in Fig. 10, where friends are asked to provide suggestions on the Facebook personal page of the person looking for jobs; in this case, the page displaying results displays for each result category the command `Ask the friends` which brings to an entry form where the user selects the social network and engagement process, including its duration; answers are automatically extracted and reproduced on a modified UI at the end of the period.

More in general, the input of a CrowdSearcher query QI is a triple $< C, N, S >$ where C is a data collection, N is a textual query expressed in natural language, and

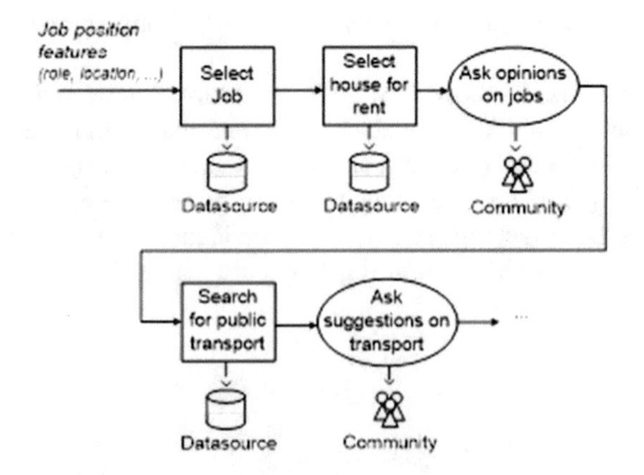

Fig. 9 Search session including two crowd-based search steps

Fig. 10 Example of intertwining crowd-based search within a search session

S is a collection of structured queries. Every component is optional. We next detail each component.

- C is an initial data collection which is proposed to the crowd for crowdsearching. For ease of description, we use the relational model, and therefore C is a collection of tuples. C can be sorted, in which case an attribute POS indicates the position of each tuple in the input sorting.
- N is a natural language query which is presented to the crowd. It can be mechanically generated, e.g. in relationship with specific structured queries, or instead be written by the user who starts the crowd search.
- S is a collection of structured queries that are asked to the crowd, relative to the collection C. Queries allow to express preferences about the elements of C, to rank them, cluster them, and change their content.

Preference queries correspond to typical social interactions (like, dislike, comment, tag); the other structured queries abstract simple and classical primitives of relational query languages which are common in human computation and social computation activities. The preference queries include:

- *Like* query, counting the individuals who like specific tuples of C.
- *Dislike* query, counting the individuals who dislike specific tuples of C.
- *Recommend* query, asking users to provide recommendations about specific tuples of C.
- *Tag* query, asking users to provide either global tags or tags about specific attributes of C.

The rank queries include:

- *Score* query, asking users to assign a (1 … N) score to tuples of C.
- *Order* query, asking users to order the (top N) tuples in C.

The cluster queries include:

- *Group* query, asking users to cluster the tuples in C into (at most N) distinct groups.
- *OrderGroup* query, asking users to cluster the tuples in C into (at most N) distinct groups and then order the (top M) tuples in each group.
- *MergeGroup* query, asking users to merge N sorted groups producing a single ordering.

- *TopGroup* query, asking users to cluster the tuples in C into (at most N) distinct groups and then select the top element of each group.

The modification queries include:

- *Insert* query, asking users to add tuples to C.
- *Delete* query, asking users to delete tuples from C.
- *Correct* queries, asking users to identify and possibly correct errors in the tuples of C.
- *Connect* query, asking users to match pairs of similar tuples.

Our approach was subsequently integrated with a reactive paradigm so as to allow dynamic and continuous evolution of crowdsourcing strategies based on the response of the workers [5], by applying techniques largely inspired by the works on reactive databases and datawarehousing.

7 Using Social Content for Discovering Emerging Knowledge

More recently, we focused on using big data produced by social interactions on platforms such as Facebook, Twitter, LinkedIn, Instagram. Some of our work was dedicated to using social sources for answering questions about Milano, such as understanding the languages being used in the various parts of the city [2], or the geographic spreading of Instagram posts after events such as the Fashion Week [7]. But we also tackled a more general research, consisting of using social content for capturing knowledge.

The most well-developed ontologies, such as DBpedia, Yago, the Knowledge Graphs in Google and Facebook, derive from structured or semi-structured, curated data. This process has involved huge efforts but had a huge payoff: DBpedia is now the crystallization point of linked data, while Google and Facebook saw the business value of this idea and have hugely invested in continuous and manual integration of databases for the development of knowledge graphs. So far, the effort of deriving knowledge has disregarder the contribution of social media, although social content has fueled the new discipline of Social Media Analytics [27], concerned with analyzing real world phenomena using social media.

We started a new research [8] targeted to discovering less popular items, those belonging to the long tail (e.g., the portion of the entity distribution having a large number of occurrences far from the "head" or central part of the distribution itself). Even the largest knowledge bases are largely incomplete for what concerns low-frequency data. It turns out, however, that knowing the long tail has a strong relevance, e.g., in e-commerce or search.[4] While high-frequency entities include well established brands, low-frequency data typically include *emerging* brands, those that

[4]The commercial success of Amazon and Google is due to their ability to discover goods or pages in the long tail.

have a small impact today but may have a high one tomorrow. The early discovery of low-frequency data and their ontological properties is thus a very interesting problem, with economic and practical implications in the innovation process.

Given these premises, our research focuses on the problem of *discovering emerging knowledge* belonging to the long tail, by extracting the low-frequency entities and relationships, with their attributes, from social content, thereby enriching existing domain knowledge [9]. We did so by using the methods for crawling social content and for entity recognition which are well established within social media analytics; our notion of ontology is broad, and includes classic cases, such as DBpedia or PubMed, but also any authoritative source of knowledge, such as the NY Stock Exchange Listings, or software projects in Github, or locations available in Open Street Map. These sources are used to define the ontological content of high-frequency entities.

We approach this problem with general, domain independent methods, but also with a well defined focus. We do not attempt at building full knowledge graphs, but rather we build small graphs, called *enriched domain graphs*, where the emphasis is on a given domain, and the enrichment is concerned with emerging concepts extracted from the long tail. Examples are: discovering emerging fashion designers (their identity/trends/brands)[5]; or discovering bloggers or narrative writers; or scouting emerging startups or products while they are becoming popular.

Domain knowledge is of course very useful in order to extract the relevant facts about the domain, e.g., high-frequency entities or relationships (thus, we know about Gucci or Prada) or structures from existing knowledge graphs (thus, we know that data about fashion designers can be linked to hubs such as fairs or magazines). We use such domain knowledge as the driver to select and organize relevant social content.

The method takes advantage of initial knowledge, that we call *seeds* and is typically provided by domain experts, to scout relevant *candidates* for the various kinds of emerging knowledge, extracted from social content, and ranked according to a variety of mechanisms, from syntactic to semantic ones, from information retrieval to machine learning, possibly helped by crowdsourcing; the first elements in the ranking are new concepts (e.g., entities or relationships), that can be validated by domain experts or, when confidence is sufficient, entered in the enriched domain graph.

In our future work, we plan to use social content to approach the dual problem of *detecting obsolete knowledge*, i.e., of knowledge that may have appeared at a given time but has not been confirmed as it has lost social confirmation. Examples in the medical domain include therapeutic options or theories about diseases which are very popular for a limited amount of time but then they either are ignored or confuted. In this case, we start from domain graphs, i.e., restrictions of knowledge graphs to specific domains, and we solve the dual problems of finding obsolete entities, relationships or attributes, and of discovering that certain types of the domain graph have lost relevance.

---

[5]This problem is particularly relevant in Milano with its well-known fashion industry; it has been presented to us by the Fashion Design research group within Politecnico.

As an intellectual exercise, we are also interested in *detecting and confuting factoids*, i.e., studying the correctness of the domain graph. Specifically, one can search for factoids, i.e., assumptions or speculations that have been reported and repeated so often that they have become commonly accepted "facts," even though they lack any validity or truth. For instance, the belief that the Great Wall of China is visible from the moon is a factoid, as doing so would require a 17,000 times better eye resolution than we actually have.

8 Conclusions

Starting from the 3M motto, we presented several contributions to data-centered modeling of applications in the last twenty-five years. The lessons we learnt throughout our research is that **modeling must adapt to new concepts** and that focusing on the static aspects of conceptual schemas is not enough, as data has its own dynamics within constantly evolving applications. We started with hypertexts, added them services, workflows, and crowd-based computations which embed social contributions. Our recent work turned towards model construction, using social information.

Acknowledgements We acknowledge many contributors to this work; among them, Piero Fraternali has a predominant role, being the main motor in all our works related to WebML and IFML. We also acknowledge the contributions of Alessandro Bozzon, Emanuele Della Valle and Florian Daniel, as well as a continuous interaction with Stefano Butti, Roberto Acerbis and Aldo Bongio from WebRatio.

References

1. R. Acerbis, A. Bongio, M. Brambilla, S. Butti, Model-driven development based on omg's IFML with webratio web and mobile platform, in *Engineering the Web in the Big Data Era - 15th International Conference, ICWE 2015, Rotterdam, The Netherlands, 23–26 June 2015, Proceedings* (2015), pp. 605–608
2. M. Arnaboldi, M. Brambilla, B. Cassottana, B. Ciuccarelli, D. Ripamonti, S. Vantini, R. Volonterio, Studying multicultural diversity of cities and neighborhoods through social media language detection, in *CitiLab, Papers from the 2016 ICWSM Workshop, Cologne, Germany, 17 May 2016* (2016)
3. A. Bozzon, M. Brambilla, S. Ceri, Answering search queries with crowdsearcher, in *21st International Conference on World Wide Web 2012, WWW 2012* (ACM, 2012), pp. 1009–1018
4. A. Bozzon, M. Brambilla, S. Ceri, P. Fraternali, Liquid query: multi-domain exploratory search on the web, in *Proceedings of the 19th International Conference on World Wide Web, WWW 2010, New York, NY, USA* (ACM, 2010)
5. A. Bozzon, M. Brambilla, S. Ceri, A. Mauri, Reactive crowdsourcing, in *22nd International World Wide Web Conference, WWW 2013, Rio de Janeiro, Brazil, 13–17 May 2013* (2013), pp. 153–164
6. M. Brambilla, J. Cabot, M. Wimmer, *Model-Driven Software Engineering in Practice*, vol. 1, 2nd edn., Synthesis Lectures on Software Engineering (Morgan & Claypool Publishers 2017)

7. M. Brambilla, S. Ceri, F. Daniel, G. Donetti, Spatial analysis of social media response to live events, in *LocWeb, Papers from the 2017 WWW Conference Workshop, Perth, Australia. WWW Companion Volume, in print* (2017)
8. M. Brambilla, S. Ceri, F. Daniel, E.D. Valle, On the quest for changing knowledge, in *Proceedings of the Workshop on Data-Driven Innovation on the Web, DDI@WebSci 2016, Hannover, Germany, 22–25 May 2016* (2016), pp. 3:1–3:5
9. M. Brambilla, S. Ceri, E.D. Valle, R. Volonterio, F.A. Salazar, Extracting emerging knowledge from social media, in *International Conference on World Wide Web 2017, WWW 2017, page in print* (ACM, 2017)
10. M. Brambilla, S. Ceri, P. Fraternali, I. Manolescu, Process modeling in Web applications. ACM Trans. Softw. Eng. Methodol. (2006)
11. M. Brambilla, P. Fraternali, Interaction flow modeling language: Model-driven UI engineering of web and mobile apps with IFML. Morgan Kaufmann - The OMG Press (2014)
12. M. Brambilla, P. Fraternali, et al., The interaction flow modeling language (ifml), version 1.0. Technical report, Object Management Group (OMG) (2014), http://www.ifml.org
13. S. Ceri, M. Brambilla (eds.), in *Search Computing - Challenges and Directions*, vol. 5950, Lecture Notes in Computer Science (Springer, 2010)
14. S. Ceri, M. Brambilla (eds.), in *Search Computing - Trends and Developments (outcome of the second SeCO Workshop on Search Computing, Como/Milan, Italy, 25–31 May 2010)*, vol. 6585, Lecture Notes in Computer Science (Springer, 2011)
15. S. Ceri, M. Brambilla (eds.), in *Search Computing - Broadening Web Search*, vol. 7538, Lecture Notes in Computer Science (Springer, 2012)
16. S. Ceri, M. Brambilla, P. Fraternali, The history of webml lessons learned from 10 years of model-driven development of web applications, in *Conceptual Modeling: Foundations and Applications*, LNCS (Springer, 2009), pp. 273–292
17. S. Ceri, P. Fraternali, A. Bongio, Web modeling language (webml): a modeling language for designing web sites. Comput. Netw. **33**(1–6), 137–157 (2000)
18. S. Ceri, P. Fraternali, A. Bongio, M. Brambilla, S. Comai, M. Matera, Morgan Kaufmann Series in Data Management Systems: DesigningSata-Intensive Web Applications (Morgan Kaufmann, 2003)
19. J. Conallen, *Building Web Applications with UML*. Addison Wesley (2002)
20. J. Gómez, C. Cachero, O. Pastor, *Conceptual modeling of device-independent web applications* (2001), pp. 26–39
21. D.M. Groenewegen, Z. Hemel, L.C.L. Kats, E. Visser, WebDSL: A domain-specific language for dynamic web applications, in *OOPSLA Companion*, ed. by G.E. Harris (ACM, 2008), pp. 779–780
22. Q. Limbourg, J. Vanderdonckt, B. Michotte, L. Bouillon, V. Lopez-Jaquero, USIXML: A language supporting multi-path development of user interfaces, in *Engineering Human Computer Interaction and Interactive Systems*, vol. 3425, LNCS (Springer, 2005), pp. 200–220
23. M. Linaje, J.C. Preciado, F. Sánchez-Figueroa, A method for model based design of rich internet application interactive user interfaces, in *Proceedings of International Conference on Web Engineering, 16–20 July 2007, Como, Italy* (2007), pp. 226–241
24. I. Manolescu, M. Brambilla, S. Ceri, S. Comai, P. Fraternali, Model-driven design and deployment of service-enabled web applications. ACM Trans. Inter. Tech. **5**(3), 439–479 (2005)
25. A. Schramm, A. Preussner, M. Heinrich, L. Vogel, Rapid UI development for enterprise applications: Combining manual and model-driven techniques, in *Model Driven Engineering Languages and Systems*, vol. 6394, LNCS (Springer, 2010), pp. 271–285
26. D. Schwabe, G. Rossi, S.D.J. Barbosa, Systematic hypermedia application design with OOHDM, in *Proceedings of Hypertext 1996* (1996), pp. 116–128
27. S. Stieglitz, L. Dang-Xuan, A. Bruns, C. Neuberger, Social media analytics. Bus. Inf. Syst. Eng. **6**(2), 89–96 (2014)
28. R. Vdovják, F. Frăsincar, G.-J. Houben, P. Barna, Engineering semantic web information systems in hera. J. Web Eng. **1**(1–2), 3–26 (2003)

Structural Decomposition Methods: Key Notions and Database Applications

G. Greco, N. Leone, F. Scarcello and G. Terracina

Abstract Many difficult problems that are tractable when restricted to acyclic instances are good candidates to be solved efficiently whenever their structure is not precisely acyclic, but not far from that. This is the case for fundamental database problems such as answering conjunctive queries or counting the number of answers (without actually computing them). The chapter describes structural decomposition methods that guarantee tractability for all such problem instances whose associated hypergraphs have a small degree of cyclicity, called width. In particular, it focuses on the notion of hypertree width, by describing its properties and its applications to the database field, and covering queries with aggregate operators and some recent parallel and distributed implementations.

1 Introduction

Answering conjunctive queries to relational databases is a basic problem in database theory, and it is equivalent to many other fundamental problems, such as conjunctive query containment and constraint satisfaction. Recall that conjunctive queries are defined through conjunctions of atoms (without negation), and are known to be equivalent to Select-Project-Join queries. The problem of evaluating such queries is NP-hard in general, but it is feasible in polynomial time on the class of acyclic queries, which was the subject of many seminal research works since the early ages of database theory. This class contains all queries Q whose associated query hypergraph $\mathcal{H}_Q$ is *acyclic*, where the hypergraph $\mathcal{H}_Q$ associated to Q has as vertices

G. Greco (✉) · N. Leone · F. Scarcello · G. Terracina
University of Calabria, 87036 Rende, Italy
e-mail: ggreco@mat.unical.it

N. Leone
e-mail: leone@mat.unical.it

F. Scarcello
e-mail: scarcello@dimes.unical.it

G. Terracina
e-mail: terracina@mat.unical.it

S. Flesca et al. (eds.), *A Comprehensive Guide Through the Italian Database Research Over the Last 25 Years*, Studies in Big Data 31, DOI 10.1007/978-3-319-61893-7_15

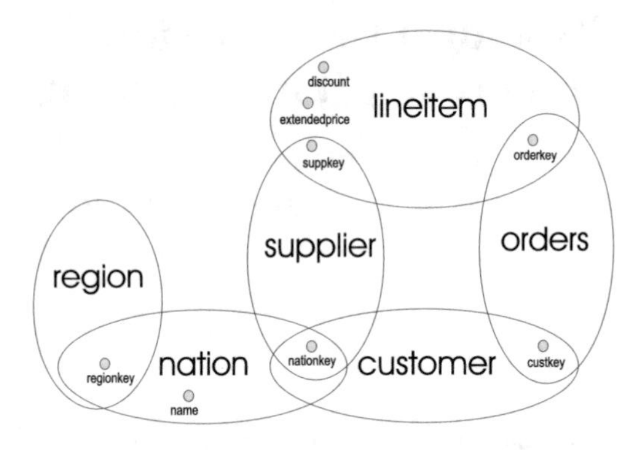

Fig. 1 Hypegraph associated with the query in Example 1

its variables and has, for each query atom, a hyperedge consisting of the set of all variables appearing in that atom.

Example 1 Consider the following conjunctive query Q, adapted from a benchmark SQL query taken from the TPC-H specifications (see [21]):

customer(CustKey, NationKey) $\wedge$ orders(OrdKey, NationKey)
$\wedge$ lineitem(SuppKey, OrdKey, ExtendedPrice, Discount)
$\wedge$supplier(SuppKey, NationKey) $\wedge$ region(RegionKey)
$\wedge$nation(Name, NationKey, RegionKey),

The query consists of the conjunction of 6 atoms and its associated hypergraph $\mathcal{H}(Q)$ is the one shown in Fig. 1. $\triangleleft$

To be precise, there are several notions of hypergraph acyclicity, among which this chapter focuses on the most liberal one, known as α-acyclicity (cf. [18]). According to this notion [7], a hypergraph $\mathcal{H}$ is acyclic if it has a *join tree* $JT(\mathcal{H})$, that is a tree whose vertices are the hyperedges of $\mathcal{H}$ and such that, whenever the same node X occurs in two hyperedges h_1 and h_2 of $\mathcal{H}$, then X occurs in each vertex on the unique path linking h_1 and h_2 in $JT(\mathcal{H})$. Note that deciding whether a hypergraph is acyclic is feasible in linear time [61], and also in deterministic logspace. This latter property follows from the fact that hypergraph acyclicity belongs to symmetric logspace [30], and that this class is equal to deterministic logspace [58].

It is well-known that *Boolean* acyclic conjunctive queries (ACQs) with m atoms, where r is the size of the largest database relation relevant to the query, can be answered in time $O(m \cdot r \cdot \log r)$, while non-Boolean ACQs in time $O(m \cdot N \cdot \log N))$, where N is the size of the output plus r. This is achieved by processing the query efficiently using a smart algorithm, such as Yannakakis' algorithm [65], which reduces the relevant database relations via semi-joins along the edges of the query join tree in such a way that no remaining tuple is superfluous. Combining the fact that acyclic queries can be efficiently answered with the fact that acyclicity is efficiently recognizable, such queries identify a so-called (accessible) "island of tractability" for the query answering problem [51]—and for equivalent problems, such as conjunctive query containment, constraint satisfaction problems, and so on [30].

As a matter of fact, however, conjunctive queries arising in practical applications are not properly acyclic, in many cases. For instance, the query introduced in Example 1 is not acyclic, as it can be checked easily by looking at its hypergraph in Fig. 1. Therefore, significant efforts have been made since the nineties to define appropriate notions of "quasi-acyclicity", leading to identify classes of conjunctive queries over which efficient evaluation algorithms can still be singled out. In this context, the degree of cyclicity of a hypergraph (or of a corresponding query) is usually referred to as its *width* and, for each fixed width $k \geq 1$, one seeks notions enjoying the following fundamental three conditions:

(i) **Generalization of Acyclicity**: Queries of width k include the acyclic ones.
(ii) **Tractable Recognizability**: Queries of width k can be recognized in polynomial time.
(iii) **Tractable Query-Answering**: Queries of width k can be answered in input–output polynomial time (that is, with respect to the size of the input and the output). Moreover, tractability holds even by considering the so-called combined complexity, where both the query and the database are taken into account, and nothing is assumed to be fixed (or small).

The rest of the chapter is devoted to illustrate *structural decomposition methods* proposed in the literature to generalize acyclic queries, in particular, by focusing on *tree decompositions* and *hypertree decompositions*. Moreover, applications in the database area are discussed, and some recent advances and directions for future work are illustrated.

2 Tree Decompositions

The notion of tree decomposition [59] represents a significant success story in Computer Science (see, e.g., [25]). The associated notion of *treewidth* was meant to provide a measure of the degree of cyclicity in graphs and hypergraphs.[1]

Formally, a *tree decomposition* [59] of a hypergraph $\mathcal{H}$, where $nodes(\mathcal{H})$ and $edges(\mathcal{H})$ denotes its set of nodes and of edges, respectively, is a pair $\langle T, \chi \rangle$, where $T = (V, F)$ is a tree, and χ is a labeling function assigning to each vertex $p \in V$ a set of vertices $\chi(p) \subseteq nodes(\mathcal{H})$, such that the following three conditions are satisfied: (1) for each node b of $\mathcal{H}$, there exists $p \in V$ such that $b \in \chi(p)$; (2) for each hyperedge $h \in edges(\mathcal{H})$, there exists $p \in V$ such that $h \subseteq \chi(p)$; and (3) for each node b in $nodes(\mathcal{H})$, the set $\{p \in V \mid b \in \chi(p)\}$ induces a connected subtree of T. The *width* of $\langle T, \chi \rangle$ is the number $\max_{p \in V}(|\chi(p)| - 1)$. The *treewidth* of the hypergraph $\mathcal{H}$, denoted by $tw(\mathcal{H})$, is the minimum width over all its tree decompositions.

[1]Some notions strongly related to the treewidth appeared even before the 80's in the literature. For a detailed story, we refer to [16].

The notion of treewidth enjoys some desirable properties. First, it is efficiently recognizable (cf. condition *(ii)* in the Introduction). Indeed, for any constant k, determining whether a hypergraph has treewidth k is feasible in linear time [8]. Moreover, condition *(iii)* is satisfied, too. In particular, Boolean conjunctive queries (whose associated hypergraphs are) of treewidth k can be answered in time $O(m' \cdot D^{k+1} \cdot \log D)$, where m' is the number of vertices of the decomposition tree T, and D is the number of distinct values occurring in the given database.

However, treewidth is not a proper generalization of hypergraph acyclicity, that is, the notion does not satisfy condition *(i)*. Indeed, the notion of treewidth is essentially aimed at characterizing nearly-acyclic *graphs*, rather than hypergraphs. In fact, it is easy to check that a graph is acyclic if, and only if, it has treewidth 1. More in detail, note that the treewidth of a hypergraph $\mathcal{H}$ coincides with the treewidth of its *primal graph*, which is defined over the same set $nodes(\mathcal{H})$ of nodes of $\mathcal{H}$ and contains an edge for each pair of nodes included in some hyperedge of $edges(\mathcal{H})$. So, the tree decomposition method obscures, in many cases, the actual degree of cyclicity of the query hypergraph. For instance, by looking at the primal graph of $\mathcal{H}$, there is no way to understand whether a given clique over three variables comes from one atom having arity 3 in the original query or, instead, it comes from the interaction of three binary atoms.

3 Hypertree Decompositions

The notion of hypertree decomposition has been proposed in the literature as a generalization of the tree decomposition method, specifically designed to deal with query hypergraphs [32, 33]. The idea is to use the power of hyperedges, which may involve many variables at once, contrasted with tree decompositions, which are based just on single variables.

3.1 Basic Notions

In order to define hypertree decompositions, a more general related notion is first introduced.

A *generalized hypertree decomposition* [32] of a hypergraph $\mathcal{H}$ is a triple $HD = \langle T, \chi, \lambda \rangle$, called a *hypertree* for $\mathcal{H}$, where $\langle T, \chi \rangle$ is a tree decomposition of $\mathcal{H}$, and λ is a function labeling the vertices of T by sets of hyperedges of $\mathcal{H}$ such that, for each vertex p of T, $\chi(p) \subseteq \bigcup_{h \in \lambda(v)} h$. That is, all nodes in the χ labeling are covered by hyperedges in the λ labeling.

The *width* of $\langle T, \chi, \lambda \rangle$ is the number $\max_{p \in V}(|\lambda(p)|)$. The *generalized hypertree width* of $\mathcal{H}$, denoted by $ghw(\mathcal{H})$, is the minimum width over all its generalized hypertree decompositions. A class of hypergraphs has bounded generalized hypertree

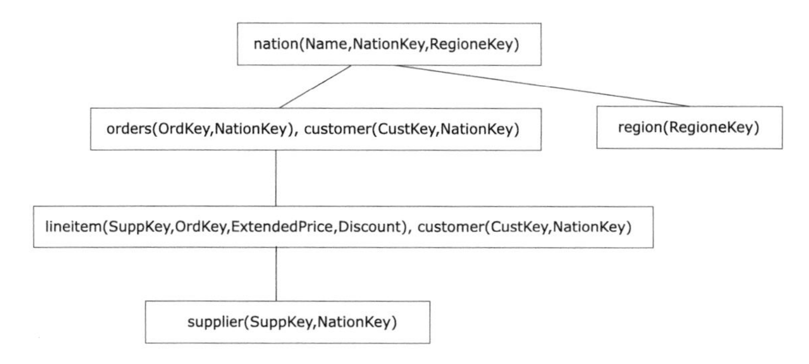

Fig. 2 A hypertree decomposition of $\mathcal{H}(Q)$

width if every hypergraph in the class has generalized hypertree width at most k, for some (finite) natural number k.

Example 2 Consider again the query Q in Example 1 and the generalized hypertree decomposition that is depicted in Fig. 2. We use an intuitive graphical notation: for each vertex p of the decomposition, the figure reports the atoms corresponding to the hyperedges in $\lambda(p)$. It can be checked that this is indeed a generalized hypertree decomposition and its width is 2.

In the example, $\chi(p) = \bigcup_{h \in \lambda(v)} h$, that is, each variable of any atom occurring in p is included in $\chi(p)$, too. If this is not true, an anonymous variable "__" is used in place of any variable occurring in some atom in $\lambda(p)$, but not occurring in the set of *relevant* variables $\chi(p)$. For instance, an alternative decomposition may be obtained by replacing customer(CustKey,NationKey) with customer(__,NationKey) in the vertex covering the atoms lineitem and customer. ◁

The notion of generalized hypertree width is a true generalization of acyclicity, as the acyclic hypergraphs are precisely those hypergraphs having generalized hypertree width 1 [33]. Hence, condition *(i)* is satisfied. Moreover, given a query Q and a width-k generalized hypertree decomposition of its hypergraph, Q can be answered in polynomial time [32]. In particular, if Q is Boolean than it can be answered in time $O(v \cdot r^k \cdot \log r)$, where r is the size of the largest database relation mentioned by the query. In the general case of queries with output variables (whose results may consist of exponentially many tuples), the input–output polynomial time bound is $O(v \cdot (r^k + s) \cdot \log(r + s))$, where s is the number of output tuples. Note that the exponent of the polynomial in the upper bound involves k as a factor, so that in practice the method can be used only for small degrees of cyclicity. With this respect, it is worthwhile noting that the factor k in the exponent cannot be avoided, unless some unlikely collapse occurs in parameterized complexity theory.

However, condition *(ii)* is not satisfied by this method, since it is NP-hard to decide whether a given query has generalized hypertree width bounded by a fixed constant k, even for $k = 2$ [19]. In fact, an additional restriction has to be added in the definition of generalized hypertree decomposition, in order to get a tractable notion.

A *hypertree decomposition* [32] of a hypergraph $\mathcal{H}$ is a generalized hypertree decomposition of $\mathcal{H}$ that satisfies the following additional condition: for each vertex p of T and for each hyperedge $h \in \lambda(p)$, it holds that $h \cap \chi(T_p) \subseteq \chi(p)$, where T_p denotes the subtree of T rooted at p, and $\chi(T_p)$ is the set of all variables occurring in the χ labeling of this subtree. Then, the *hypertree width* of $\mathcal{H}$, denoted by $hw(\mathcal{H})$, is naturally defined as the minimum width over all its hypertree decompositions.

Intuitively, the above technical condition forces variables to be included in a χ label the first time (looking top-down) some atom where they occur is used in the decomposition. A very important property of this notion is that it is not very far apart from the notion of generalized hypertree width. Indeed, for each hypergraph $\mathcal{H}$, $ghw(\mathcal{H}) \leq hw(\mathcal{H}) \leq 3 \cdot ghw(\mathcal{H}) + 1$ [3]. In particular, this entails that a class of queries has bounded generalized hypertree width if, and only if, it has bounded hypertree width.

Example 3 Since $\chi(p) = \bigcup_{h \in \lambda(v)} h$ holds for each vertex p in the generalized hypertree decomposition of Fig. 2, we trivially derive that this also a hypertree decomposition of the (hypergraph associated with the) query in Example 1. ◁

3.2 Desirable Properties of Hypertree Decompositions

Since their introduction, hypertree decompositions have received considerable attention in the literature, due to the fact that they satisfy the fundamental conditions stated in the Introduction (see also, [23, 39]):

(i) Hypertree width properly generalizes hypergraph acyclicity. In particular, a hypergraph $\mathcal{H}$ is acyclic if, and only if, $hw(\mathcal{H}) = 1$ [32].

(ii) For a fixed constant k, it can be checked in polynomial time whether (the hypergraph associated with) a conjunctive query Q has hypertree width $k' \leq k$. Moreover, if so, a hypertree decomposition of width k' can be computed in polynomial time, more precisely in $O(m^{2k}v^2)$, where m and v are the number of atoms and the number of variables in Q, respectively [29]. The decision and computation problems are, moreover, at a very low level of computational complexity and are highly parallelizable, as they belong to LOGCFL (see Sect. 4.2 for some properties of this class). Again, it is deemed very unlikely that we can get rid of the factor k in the exponent, because deciding whether a query has hypertree width k is fixed-parameter intractable [26].

(iii) Queries having bounded hypertree width can be answered in input–output polynomial time. This follows immediately from the same property of generalized hypertree width, however in this case it is not necessary that the decomposition is provided in input with the query, because it can be computed easily, by property (ii).

4 Applications of Hypertree Decompositions

In this section, we overview a number of database applications where the notion of hypertree decomposition has been used profitably.

4.1 Hypertree-Based Plans for Multiway Joins

Recent works have shown that traditional query optimizers are provably suboptimal on large classes of queries, and worst-case optimal algorithms have been developed [56, 64]. Such algorithms, based on a multiway join approach that may look at all atoms at once, have been implemented, e.g., in the LogicBlox system [5] and in the EmptyHeaded relational engine [1, 2]. Unfortunately, these algorithms are not able to recover the polynomial-time worst-case bounds for queries having bounded hypertree width. For instance, they may require exponential time for acyclic queries with an empty output (as long as, in principle, such queries could have an exponential number of answers). To deal with this issue, EmptyHeaded additionally features a query compiler based on hypertrees: it searches for a generalized hypertree decomposition having the minimum possible estimated size for the intermediate results, and then uses this information to determine the order of attributes to be used in the multiway joins. Such an order is also exploited for the multi-level data structures, called *tries*, used to store input and output relations, and to perform the joins efficiently.

We also mention a different approach designed for the Leapfrog Trie Join algorithm [64], where (hyper)tree decompositions are used to guide a flexible caching of intermediate results [44]. The algorithm described in [47] considers also possible functional dependencies, by using the coloring number bound of [27].

Further algorithms based on multiway joins have been defined in order to guarantee the worst-case upper bound that can be obtained by using generalized hypertree decompositions of the given query, without using a dynamic programming approach á la Yannakakis. This is the approach described in [48], where a notion of geometric resolution is defined to support different kinds of indices and even multiple indices per table. By performing such resolutions, the proposed algorithm covers the whole multidimensional (tuple) space by distinguishing the output tuples (if any) and the other infeasible (non-matching) tuples. As opposed to the standard bottom-up computation, this method can be viewed as a backtracking algorithm with memoization.

4.2 Parallel and Distributed Evaluation

From a computational complexity viewpoint, evaluating Boolean queries having bounded hypertree-width is LOGCFL-complete (even for binary acyclic queries) [30]. Combining the result with the techniques discussed by [31], it can be seen easily that the corresponding computation problem (output an answer) is in (functional) LOGCFL. It is known that this class contains highly parallelizable problems: Any problem in LOGCFL is solvable in logarithmic time by a concurrent-read concurrent-write *parallel random access machine* (CRCW PRAM) with a polynomial number of processors, or in $\log^2$-time by an exclusive-read exclusive-write (EREW) PRAM with a polynomial number of processors.

Several efforts have been spent in the literature to translate the above theoretical results into practical implementation of algorithms for parallel (and distributed) query evaluation. In particular, a parallel algorithm, called DB-SHUNT, has been defined in [30] for answering Boolean acyclic queries—an extension, called ACQ, which is able to deal with acyclic queries with output variables is discussed in [28]. This algorithm is well-suited for bounded hypertree-width queries after a suitable preprocessing step where they are transformed in acyclic queries. In this step, for each vertex p in the decomposition tree T, a fresh atom is computed such that its set of variables is $\chi(p)$, and its relation is obtained by projecting on $\chi(p)$ the join of all relations associated with the query atoms whose sets of variables are the hyperedges in $\lambda(p)$. After that, because the χ labeling encodes a tree decomposition, it can be seen that the conjunction of these fresh atoms forms an acyclic conjunctive query. Then, having the (equivalent) acyclic query at hand, DB-SHUNT (and ACQ) uses a special *shunt* operation based on relational algebra for contracting a join tree, akin the well-known shunt operation used for the parallel evaluation of arithmetic expressions [45]. This way, any tree can be contracted to a single node by a logarithmic number of parallel steps, independently of the shape of the tree (e.g., even if the tree is a highly unbalanced chain).

More in detail, the decomposition tree T is preliminary made strictly binary, and its leaves are numbered from left to right. At each iteration, the shunt operation is applied in parallel to all odd numbered leaves of T. To avoid concurrent changes on the same relation, left and right leaves are processed in two distinct steps. Thus, after each iteration, the number of leaves is halved, and the tree-contraction ends within $2 \log m$ parallel steps by using $O(m)$ processors and $O(m)$ intermediate relations having size at most $O(r^2)$, where r is the size of the largest relation of the database instance and m is the number of vertices of T. It can be shown that having a fixed number of processors $c < m/2$, the number of parallel shunt operations becomes $2(\lceil \log c \rceil + 2\lceil m/(4c) \rceil)$. Here, transformations of query atoms and input other than relations are considered costless as long as they are polynomial. Moreover, the above cost refers to the first bottom-up processing of the decomposition tree, which is enough for the evaluation of a Boolean query. For computing the answers of a non-Boolean query, two further processing of the tree are needed, with the same cost (but for the space consumption of the final ascending phase, where we have to consider an additional cost that is linear in the output size).

An implementation of the above strategy has been recently described in [4] for the Valiant's bulk synchronous parallel (BSP) computational model [63], which can simulate the PRAM model. In this model, there are a set of connected machines that do not share any memory. The computation consists in general in a series of rounds, where machines perform some local computation in parallel and communicate messages over the network. Moreover, the same authors use generalized hypertree decompositions for parallel query answering in their GYM algorithm [4], which is a distributed and generalized version of Yannakakis' algorithm specifically designed for MapReduce [15] (as well as for other BSP-based frameworks). In fact, especially after the introduction of Google MapReduce, the problem of evaluating joins efficiently in distributed environments has been attracting much attention in the literature. In a BSP distributed environment, algorithms should mainly deal with the communication costs between the machines, and the number of global synchronizations that are needed to take place between the machines, in particular, the number of rounds of MapReduce jobs to be executed.

4.3 Counting Query Answers

So far, the chapter focused on the use of hypertree decompositions for evaluating conjunctive queries, that is, Select-Project-Join queries. The method, however, found applications also in the evaluation of further kinds of queries involving more complex constructs. A noticeable example is the use of hypertree decompositions to deal with "COUNT" aggregates (see, e.g., [9, 10]).

In this context, the query evaluation problem can be abstractly reformulated as *counting* the number of answers of a given conjunctive query. The challenge is then to compute the right number in polynomial time without actually computing the (possibly exponentially-many) query answers. In particular, it is crucial, both in theory and in practice, to deal with queries where output variables can be specified, so that we are only interested in counting the answers projected on them. Technically, such distinguished variables are free, while all the other variables are existentially quantified. As a matter of fact, in almost all practical applications, there are many of such "auxiliary" (existentially quantified) variables whose instantiations must not be counted.

Whenever all variables are free, having bounded generalized hypertree width is a sufficient condition for the tractability of the counting problem [57]. Moreover, on fixed-arity queries, this condition is also necessary (under widely believed assumptions in parameterized complexity) [14]. However, in presence of projections, classical decomposition methods are not helpful. Indeed, even for acyclic queries [57], counting answers is #P-hard in this case. An algorithm counting the answers of an acyclic query Q in $O(m \cdot r^2 \cdot 4^r)$, with r being the size of the largest database relation and m the number of atoms, has been exhibited in [57]. Therefore, the evaluation is tractable over acyclic instances w.r.t. query complexity (where the database is fixed and not part of the input). The technique can be extended easily to queries having generalized hypertree width at most k, for some fixed number k (cf. [37]). Thus, to get more powerful structural results, we should distinguish the hypergraph nodes associated with free and existentially quantified variables. A sufficient condition for tractability is the existence of a homomorphically equivalent subquery Q' of Q that includes all free variables, and such that there is a width-k generalized hypertree decomposition for Q' which covers all frontiers of the free variables with the existential variables (see [37] for more information). Interestingly, it turns out that for fixed-arity queries this structural condition precisely characterizes the queries where the counting problem is tractable [12].

4.4 Further Aggregations

When evaluating queries, database management systems list answers according to some arbitrary order, which reflects the algorithms internally used for optimization purposes. However, this is often unsatisfactory, so that users often write ORDER BY clauses to force answers to be listed according to their preferences. For instance, interactive queries are typically ORDER BY queries, where users are interested in a few best answers (to be returned very fast). In this context, a crucial observation is that answering ORDER BY queries is intimately related to the problem of computing the best solution according to the given ordering, which is incidentally another important aggregate operator (MAX) in SQL queries. This was first pointed out in the context of ranking solutions to discrete optimization problems [52]. In particular, whenever the MAX problem of computing an optimal solution is feasible in polynomial time, then we can solve the problem of returning all solutions in the ranked order with polynomial delay, as well as the more general problem of returning the best K-ranked solutions over all solutions, i.e., the top-K query evaluation problem [41].

A number of structural tractability results for MAX (so, for top-K and ORDER BY) queries are known in the literature (see, e.g., [24]). For *monotone* functions built on top of one binary aggregation operator (such as the standard $+$ and $\times$), MAX is feasible in polynomial time over queries that have bounded treewidth [17], and over queries that have bounded generalized hypertree width [22, 50]. Structural tractability results for extensions to certain *non-monotone* functions which manipulate "small" (in fact, polynomially-bounded) values have been studied, too [24, 36].

An implementation of some aggregation operators based on hypertree decompositions is described in [20, 21]. It can be used with any system at a logical level. For the sake of efficiency, a semi joinoperator that supports the execution of these operators over decompositions is implemented in a prototypal extension of the open-source DBMS PostgreSQL. Further operators are considered in [42] for the so-called AJAR queries, that is, queries with annotated relations over which (possibly multiple) aggregations can be used. Technically, aggregations are modeled by means of semiring quantifiers that "sum over" or "marginalize out" values. It is argued that such queries can be used to extend conjunctive queries with most SQL-like aggregation operators, as well as to capture data processing problems, such as probabilistic inference via message passing on graphical models [46]. A crucial step with these queries is finding a variable ordering that allows us to manage the aggregations by using a suitable generalized hypertree decompositions ("compatible" with the ordering), together with standard join algorithms. The proposed algorithm can efficiently be implemented in the parallel framework, by using GYM [4] and the degree-based

MapReduce algorithm in [43]. Answering AJAR queries is equivalent to answering so-called Functional Aggregate Queries [49].

4.5 Parallel and Distributed Querying of Deductive Databases

This overview of applications of hypertree decompositions closes by focusing on problems related to querying distributed deductive databases, in particular, by considering a scenario where data *natively* resides on different *autonomous* sources, and it is necessary to deal with reasoning tasks via logic programming. This scenario has been considered in [6] and differs from the classic parallel evaluation of queries (e.g., [28, 30]) in several aspects: *(i)* the focus is the optimized evaluation of (normal, stratified) logic programs [55]; *(ii)* the input data are physically distributed over several databases, and this distribution must be considered in the optimization process; *(iii)* the integration of the reasoning engine responsible of orchestrating the querying process with the distributed database management systems must be tight enough to allow efficient interactions, but general enough to avoid limitations in the kind and location of databases.

In this scenario, the notion of hypertree decomposition has been shown to provide a powerful tool to address issue *(ii)*, whereas the reasoning engine DLV^{DB} presented in [62] has been used to transparently evaluate logic programs directly on generic database systems. In order to describe the optimization process discussed by [6], let us first consider programs composed of one rule.

A single logic rule r can be seen as a conjunctive query (possibly with negation), whose result must be stored in the head predicate. The optimized evaluation of r starts from the computation of a hypertree decomposition *HD* for this query, which is interpreted as a distributed query plan. To this end and, in particular, in order to take into account data distribution, each node p of the hypertree *HD* is labelled with the database where the partial data associated with it reside. Formally, let $Site(p)$ denote the site associated with the node p in *HD* and let $net(Site(p), Site(p'))$ be the unitary data transfer cost from $Site(p)$ to $Site(p')$. Let $\lambda(p)$ be the set of atoms referred by p, and $\chi(p)$ the variables covered by p. Then, $Site(p)$ is chosen among the databases where the relations in $\lambda(p)$ reside by computing:

$h_m = arg\ min_{h_i \in \lambda(p)} \{\Sigma_{h_j \in \lambda(p)} |rel(h_j)| \times net(Site(h_j), Site(h_i))\}$.

Thus, $Site(p)$ is chosen to be the site where the database with h_m is stored, and this choice minimizes the cost of transferring all data required to evaluate the subproblem associated with $\lambda(p)$.

Eventually, in order to optimize the distribution of rule evaluation, the adopted cost function is further refined in [6] by taking into account other aspects, such as the estimated cardinality of join results. The query decomposition strategy, in addition, includes options like moving data from one site to another, and changing the order of joins based on data localization. Once the hypergraph $\mathcal{H}_r$ for r is obtained and the above cost function is defined, a hypertree decomposition HD_r minimizing such function is computed via the algorithm discussed in [60]. Then, a distributed plan for r is composed accordingly. Intuitively, the plan is built in such a way that it evaluates joins bottom-up, from the leaves to the root, by suitably transferring data if the sites of a child node and its father are different. Independent sub-trees are executed in parallel processes.

When the program is composed of several rules, three further kind of optimizations can be applied: *(i)* rule unfolding optimization, *(ii)* inter-components optimization, *(iii)* intra-component optimization. Since the rule optimization strategy is particularly suited for rules having long bodies, in presence of filters (or queries) in the program, an unfolding [55] optimization step is carried out to make rule bodies as long as possible, and to reduce their number. In particular, the inter-components optimization [11] is in charge of dividing the input (possibly unfolded) program $\mathcal{P}$ into subprograms, according to the dependencies among the predicates occurring in it, and by identifying which of them can be evaluated in parallel. This allows for concurrently evaluating rules involved within the same component.

5 Further Methods and Conclusions

The chapter describes structural decomposition methods aimed at defining suitable generalizations of hypergraph acyclicity, by focusing in particular on hypertree decompositions. The natural research question to be addressed is whether it is possible to go beyond this method, by defining more general notions the still retain the desirable conditions pointed out in the Introduction.

Observe that in the λ labelling of hypertree decompositions one could use as available resources all subsets of the hyperedges of a hypergraph $\mathcal{H}$, instead of using just the original hyperedges. This way, it is possible to recover the full power of generalized hypertree decompositions, but with an exponential blow-up. If, on the other hand, only polynomially many subsets are used, it is possible to obtain a number of variants of hypertree decompositions that still fulfil conditions *(i)*, *(ii)*, and *(iii)*. This is precisely the idea of subset-based decompositions as defined in [34]. Specific subset-based decompositions are the *component decompositions* defined in [34] and the *spread-cut* decompositions defined in [13], while an interesting way of computing a subset-based decomposition *dynamically* provides the new notion of *greedy hypertree decomposition* [35, 38].

A more radical approach to improve HDs, and even GHDs, was taken by Grohe and Marx [40], who defined *fractional hypertree decompositions (FHDs)*, giving rise to the notion of *fractional hypertree width* $fhw(\mathcal{H})$ of a hypergraph $\mathcal{H}$ (or of a query). The fractional hypertree width width is more general than the generalized hypertree width, and hence $fhw(\mathcal{H}) \leq ghw(\mathcal{H})$. Moreover, there are classes of hypergraphs having unbounded ghw, but bounded fhw. A problem with fractional hypertree decompositions is that they are intractable [19]. However, an approximate decomposition whose width is bounded by a cubic function of the fractional hypertree width can be computed in polynomial time [53]. A yet more general width-concept is the *submodular width* [54]. Unfortunately, recognizing queries of bounded submodular width is not known to be tractable. Moreover, having bounded submodular width does not guarantee a polynomial-time combined complexity as for hypertree width, but just fixed-parameter tractability where the query hypergraph is used as a parameter.

Acknowledgements The work was supported by project "Ba2Know (Business Analytics to Know) Service Innovation - LAB", No. PON03PE_00001_1 funded by the Italian Ministry of University and Research (MIUR), and by project "Smarter Solutions in the Big Data World (S2BDW)", funded by the Italian Ministry for Economic Development (MISE) within the programme PON "Imprese e competitivitá" 2014–2020.

References

1. C.R. Aberger, S. Tu, K. Olukotun, C. Ré, EmptyHeaded: A relational engine for graph processing, in *Proceedings of SIGMOD 2016* (2016)
2. C.R. Aberger, S. Tu, K. Olukotun, C. Ré, Old techniques for new join algorithms: A case study in RDF processing, in *CoRR*, arXiv:abs/1602.03557 (2016)
3. I. Adler, G. Gottlob, M. Grohe, Hypertree width and related hypergraph invariants. Eur. J. Comb. **28**(8), 2167–2181 (2007)
4. F.N. Afrati, M. Joglekar, C. Ré, S. Salihoglu, J.D. Ullman, GYM: A multiround join algorithm in mapreduce, in *CoRR*, arXiv:abs/1410.4156 (2014)
5. M. Aref, B. ten Cate, T.J. Green, B. Kimelfeld, D. Olteanu, E. Pasalic, T.L. Veldhuizen, G. Washburn, Design and implementation of the logicblox system, in *Proceedings of SIGMOD 2015* (2015), pp. 1371–1382
6. R. Barilaro, F. Ricca, G. Terracina, Optimizing the distributed evaluation of stratified programs via structural analysis, in *Proceeding of 11th International Conference on Logic Programming and Nonmonotonic Reasoning (LPNMR 2011), Vancouver, Canada, 2011*, Lecture Notes in Computer Science (Springer, Heidelberg, 2011), pp. 217–222
7. P.A. Bernstein, N. Goodman, Power of natural semijoins. SIAM J. Comput. **10**(4), 751–771 (1981)
8. H.L. Bodlaender, A linear time algorithm for finding tree-decompositions of small treewidth, in *Proceeding of STOC 1993* (1993), pp. 226–234
9. A.A. Bulatov, The complexity of the counting constraint satisfaction problem. J. ACM **60**(5), 34:1–34:41 (2013)
10. A.A. Bulatov, M. Dyer, L.A. Goldberg, M. Jerrum, C. Mcquillan, The expressibility of functions on the boolean domain, with applications to counting CSPs. J. ACM **60**(5), 32:1–32:36 (2013)
11. F. Calimeri, S. Perri, F. Ricca, Experimenting with parallelism for the instantiation of ASP programs. J. Algorithms Cogn. Inf. Log. **63**(1–3), 34–54 (2008)
12. H. Chen, S. Mengel, A trichotomy in the complexity of counting answers to conjunctive queries, in *Proceeding of ICDT 2015* (2015), pp. 110–126
13. D.A. Cohen, P. Jeavons, M. Gyssens, A unified theory of structural tractability for constraint satisfaction problems. J. Comput. Syst. Sci. **74**(5), 721–743 (2008)
14. V. Dalmau, P. Jonsson, The complexity of counting homomorphisms seen from the other side. Theory Comput. Syst. **329**(1–3), 315–323 (2004)
15. J. Dean, S. Ghemawat, Mapreduce: A flexible data processing tool. Commun. ACM **53**(1), 72–77 (2010)
16. R. Dechter, *Constraint Processing* (Morgan Kaufmann Publishers Inc., 2003)
17. R. Dechter, N. Flerova, R. Marinescu, Search algorithms for M Best solutions for graphical models, in *Proceeding of AAAI 2012* (2012), pp. 1895–1901

18. R. Fagin, Degrees of acyclicity for hypergraphs and relational database schemes. J. ACM **30**(3), 514–550 (1983)
19. W. Fischl, G. Gottlob, R. Pichler, General and fractional hypertree decompositions: Hard and easy cases, in *CoRR*, arXiv:abs/1611.01090 (2016)
20. L. Ghionna, L. Granata, G. Greco, F. Scarcello, Hypertree decompositions for query optimization, in *Proceeding of ICDE 2007* (2007), pp. 36–45
21. L. Ghionna, G. Greco, F. Scarcello, H-DB: A hybrid quantitative-structural sql optimizer, in *Proceeding of CIKM 2011* (2011), pp. 2573–2576
22. G. Gottlob, G. Greco, Decomposing combinatorial auctions and set packing problems. J. ACM **60**(4), 24:1–24:39 (2013)
23. G. Gottlob, N. Greco, N. Leone, F. Scarcello, Hypertree decompositions: Questions and answers, in *Proceeding of PODS 2016* (2016), pp. 57–74
24. G. Greco, F. Scarcello, Tractable optimization problems through hypergraph-based structural restrictions, in *Proceeding of ICALP 2009* (2009), pp. 16–30
25. G. Gottlob, G. Greco, F. Scarcello, Treewidth and hypertree width, in *Tractability: Practical Approaches to Hard Problems*, ed. by L. Bordeaux, Y. Hamadi, P. Kohli (2012)
26. G. Gottlob, M. Grohe, N. Musliu, M. Samer, F. Scarcello, Hypertree decompositions: Structure, algorithms, and applications, in *Proceeding of WG 2005* (2005), pp. 1–15
27. G. Gottlob, S.T. Lee, G. Valiant, P. Valiant, Size and treewidth bounds for conjunctive queries. J. ACM **59**(3), 1–35 (2012)
28. G. Gottlob, N. Leone, F. Scarcello, Advanced parallel algorithms for acyclic conjunctive queries. Technical Report DBAI-TR-98/18, Technical University of Vienna (1998)
29. G. Gottlob, N. Leone, F. Scarcello, On tractable queries and constraints, in *Proceeding of DEXA 1999* (1999), pp. 1–15
30. G. Gottlob, N. Leone, F. Scarcello, The complexity of acyclic conjunctive queries. J. ACM **48**(3), 431–498 (2001)
31. G. Gottlob, N. Leone, F. Scarcello, Computing LOGCFL certificates. Theor. Comput. Sci. **270**(1–2), 761–777 (2002)
32. G. Gottlob, N. Leone, F. Scarcello, Hypertree decompositions and tractable queries. J. Comput. Syst. Sci. (Conference Version has Appeared in PODS 1999) **64**(3), 579–627 (2002)
33. G. Gottlob, N. Leone, F. Scarcello, Robbers, marshals, and guards: Game theoretic and logical characterizations of hypertree width. J. Comput. Syst. Sci. **66**(4), 775–808 (2003)
34. G. Gottlob, Z. Miklós, T. Schwentick, Generalized hypertree decompositions: NP-hardness and tractable variants. J. ACM **56**(6), 30:1–30:32 (2009)
35. G. Greco, F. Scarcello, The power of tree projections: Local consistency, greedy algorithms, and larger islands of tractability, in *Proceeding of PODS 2010* (2010), pp. 327–338
36. G. Greco, F. Scarcello, Structural tractability of constraint optimization, in *Proceeding of CP 2011* (2011), pp. 340–355
37. G. Greco, F. Scarcello, Counting solutions to conjunctive queries: Structural and hybrid tractability, in *Proceeding of PODS 2014* (2014), pp. 132–143
38. G. Greco, F. Scarcello, Greedy strategies and larger islands of tractability for conjunctive queries and constraint satisfaction problems. Inf. Comput. **252**, 201–220 (2017)
39. G. Greco, F. Scarcello, The power of local consistency in conjunctive queries and constraint satisfaction problems. SIAM J. Comput. (2017)
40. M. Grohe, D. Marx, Constraint solving via fractional edge covers. ACM Trans. Algorithms **11**(1), 4:1–4:20 (2014)
41. I.F. Ilyas, G. Beskales, M.A. Soliman, A survey of top-k query processing techniques in relational database systems. ACM Comput. Surv. **40**(4), 11:1–11:58 (2008)
42. M. Joglekar, R. Puttagunta, C. Ré, Aggregations over generalized hypertree decompositions, in *Proceeding of PODS 2016* (2016)
43. M.R. Joglekar, C.M. Ré, It's all a matter of degree: Using degree information to optimize multiway joins, in *Proceeding of ICDT 2016* (2016), pp. 11:1–11:17
44. O. Kalinsky, Y. Etsion, B. Kimelfeld, Flexible caching in trie joins, in *CoRR*, arXiv:abs/1602.08721 (2016)

45. R.M. Karp, V. Ramachandran, Parallel algorithms for shared-memory machines, in *Handbook of Theoretical Computer Science*, vol. A (MIT Press, 1990), pp. 869–941
46. K. Kask, R. Dechter, J. Larrosa, A. Dechter, Unifying tree decompositions for reasoning in graphical models. Artif. Intell. **166**(1–2), 165–193 (2005)
47. M.A. Khamis, H. Ngo, D. Suciu, Worst-case optimal algorithms for conjunctive queries with functional dependencies, in *Proceeding of PODS 2016* (2016)
48. M.A. Khamis, H.Q. Ngo, C. Ré, A. Rudra, Joins via geometric resolutions: Worst-case and beyond, in *Proceeding of PODS 2015* (2015), pp. 213–228
49. M.A. Khamis, H.Q. Ngo, A. Rudra. FAQ: Questions asked frequently, in *Proceedings of PODS 2016* (2016)
50. B. Kimelfeld, Y. Sagiv, Incrementally computing ordered answers of acyclic conjunctive queries, in *Proceedings of NGITS 2006* (2006), pp. 141–152
51. P.G. Kolaitis, Constraint satisfaction, databases, and logic, in *Proceedings of IJCAI 2003* (2003), pp. 1587–1595
52. E.L. Lawler, A procedure for computing the k best solutions to discrete optimization problems and its application to the shortest path problem. Manag. Sci. **18**(7), 401–405 (1972)
53. D. Marx, Approximating fractional hypertree width. ACM Trans. Algorithms **6**(2), 29:1–29:17 (2010)
54. D. Marx, Tractable hypergraph properties for constraint satisfaction and conjunctive queries. J. ACM **60**(6), 42:1–42:51 (2013)
55. J. Minker (ed.), *Foundations of Deductive Databases and Logic Programming* (Morgan Kaufmann Publishers Inc., Washington DC, 1988)
56. H.Q. Ngo, C. Ré, A. Rudra, Skew strikes back: New developments in the theory of join algorithms. SIGMOD Rec. **42**(4), 5–16 (2013)
57. R. Pichler, S. Skritek, Tractable counting of the answers to conjunctive queries. J. Comput. Syst. Sci. **79**(6), 984–1001 (2013)
58. O. Reingold, Undirected connectivity in log-space. J. ACM **55**(4), 17:1–17:24 (2008)
59. N. Robertson, P. Seymour, Graph minors. II. Algorithmic aspects of tree-width. J. Algorithms **7**(3), 309–322 (1986)
60. F. Scarcello, G. Greco, N. Leone, Weighted hypertree decompositions and optimal query plans. J. Comput. Syst. Sci. **73**(3), 475–506 (2007)
61. R.E. Tarjan, M. Yannakakis, Simple linear-time algorithms to test chordality of graphs, test acyclicity of hypergraphs, and selectively reduce acyclic hypergraphs. SIAM J. Comput. **13**(3), 566–579 (1984)
62. G. Terracina, N. Leone, V. Lio, C. Panetta, Experimenting with recursive queries in database and logic programming systems. Theory Pract. Log. Program. (TPLP) **8**(2), 129–165 (2008)
63. L.G. Valiant, A bridging model for parallel computation. Commun. ACM **33**(8), 103–111 (1990)
64. T.L. Veldhuizen, Triejoin: A simple, worst-case optimal join algorithm, in *Proceedings of ICDT 2014* (2014), pp. 96–106
65. M. Yannakakis, Algorithms for acyclic database schemes, in *Proceedings of VLDB 1981* (1981), pp. 82–94

Multimedia Data Modeling and Management

Vincenzo Moscato and Antonio Picariello

Abstract Nowadays, multimedia data is surely one of the most popular and pervasive information and communication media that accompanies us in almost every walk of lives. They allow fast and effective communication and sharing of information about peoples' lives, their behaviors, works, interests, and they are also the digital testimony of facts, objects, and locations and have become an essential component of social media networks. Technically speaking, how to organize and structure this huge amount of data using different paradigms, so that we can easily get useful information, has been a challenging research field for decades. In this chapter we will describe the main results produced by the Multimedia Database Research Group of University of Naples in this area: models for representing multimedia data and the related knowledge and techniques for their storage, indexing and retrieval. In addition, we also point out several applications, with a particular emphasis on social media networks.

1 Introduction

Multimedia field has dramatically grown in the last two decades. From the first ACM Multimedia Conference in Anaheim, CA in 1993, at which most participants even did not know the meaning of the word "multimedia" to current days where multimedia is a solid theory and used in tons of applications [1, 2]. If we take a look at the indexed topics, we realize that at the very beginning multimedia meant to be an integration of different research fields such as: video and image processing (coming from Computer Vision and Image Processing research communities), compression,

V. Moscato (✉) · A. Picariello
Universitá di Napoli "Federico II", DIETI, via Claudio 21, 80125 Napoli, Italy
e-mail: vmoscato@unina.it

V. Moscato · A. Picariello
CINI – ITEM National Lab, via Cinzia, 80125 Napoli, Italy

S. Flesca et al. (eds.), *A Comprehensive Guide Through the Italian Database Research Over the Last 25 Years*, Studies in Big Data 31, DOI 10.1007/978-3-319-61893-7_16

communication and synchronization (from Signal Processing and Network communities), data modeling, storage and retrieval (from Database groups), and other multimedia applications.

Currently, multimedia still remains an interdisciplinary field, which contains elegant theories and practical applications ranging from all parts of our lives: from education, engineering, scientific research to art, advertisement, entertainment, and many others. However, the main characteristic of multimedia information is that of having heterogeneous data to store and organize in an "integrated view".

Actually, how to organize and structure this huge amount of data using different paradigms has been a challenging research field for decades, especially for building effective multimedia applications. Indeed, archiving, organizing, and searching multimedia data is a very complex activity. From a certain perspective, multimedia data are inherently "subjective" [3]: for example, the association of a meaning and the corresponding content description of an image as well as the evaluation of the differences between two images or two pieces of music usually depend on the user who is involved in the evaluation process. For retrieval, such subjective information needs to be combined with objective information, such as image color histograms or sound frequencies, that is obtained through (generally imprecise) data analysis processes. Therefore, the inherently nature of multimedia data, both at subjective and objective levels, may lead to multiple, possibly inconsistent, interpretations of data. The various properties of these objects cannot be captured properly by relational or object-oriented models: multimedia systems should so provide new functionalities, depending on the type of multimedia data being stored. Within this context, new challenges ranging from problems related to data representation to indexing and retrieval of such complex information, have to be addressed.

In this chapter we will describe the main models that has been adopted for representing multimedia data and the related knowledge and the most diffused techniques for their indexing, retrieval and recommendation, at the same time, describing several applications that took some advantages from the introduced techniques.

In particular, we will provide an overview of the most representative research results obtained in the management of multimedia data proposed by the Multimedia Database Group of the University of Naples "Federico II", from the first attempts to extend relational data model to the recent definition of *Multimedia Social Networks* following a Big Data perspective.

2 Multimedia Database: Data Modeling, Indexing, Retrieval

In analogy with standard database, a *multimedia database* may be viewed as a "controlled" collection of multimedia data items, such as text, images, graphic objects, sketches, video, and audio.

In this context, a *MultiMedia DataBase Management System* (MMDBMS) is the system that provides support for multimedia data types, adding the usual facilities for the creation, storage, access, query, and control of the multimedia database.

Consequently, the different data types involved in multimedia databases require special methods for optimal storage, access, indexing, and retrieval. The MMDBMS should accommodate these special requirements by providing high-level abstractions to manage the different data types, along with a suitable interface for their presentation. In our research group, we exploited different theoretical lines: first, we developed a model and an algebra for multimedia data, based on an extension of the relational model; successively, in order to capture the complex semantics attached to multimedia data, we also provided an ontological model that can be used for storing and managing different levels of multimedia data in terms of rough data, intermediate and high level semantic concepts, as well as some abstractions that can be observed over them.

2.1 A NF^2 Image Database

We started working on the Image DataBase (IDB) project at the early 2000, in a joint research with the University of Maryland at College Park[1] and the University of Turin.[2]

The basic motivation of the project was the lack in the scientific world of a formal theory on how to model and represent data within a IDB. For this reason, we decided to build a new theory over the top of standard object relational databases.

Typical descriptions of images are given by means of colors, textures, shapes and spatial descriptors. The task of associating a high level textual feature to a low level visual descriptor is usually performed by means of a set of image processing and analysis algorithms, that put into evidence high level features, each one associated with a certain grade of *confidence*. Thus, it is clear that information in image database is inherently both complex and uncertain, and managing uncertainty becomes a fundamental topic.

The relational model has been extended in many ways, depending on "where" *fuzziness* is introduced and "what" one means with uncertainty. Specifically, the relational model of data has been extended to incorporate uncertainty either at the tuple level or at the attribute level. In the "tuple-level approach", each tuple may have one or more uncertainty attributes; uncertainty attributes are usually real numbers, or intervals on real numbers. In the "attribute-level approach", the information about the certainty degree is associated directly to the single attribute values. In image databases, attribute level approaches seem preferable, since in content based image

[1] V.S. Subrahmanian's research group.

[2] M.L. Sapino's research group.

Fig. 1 Lake image

description each feature is produced by a distinct image processing technique, that returns multiple possible answers, with different grades of certainty.

Several aspects of the problem of managing uncertainty in Database and Knowledge Base have been discussed by [4], where an algebra has been developed to deal with uncertainty in the tuple level approach, and in an annotated relations scenario; the meaning of annotated relations has been deeply investigated, and an extension of classical relational algebra operators has been introduced.

Just to make clear the main idea of the proposed model, let us consider the image in Fig. 1.

Leveraging an image processing and analysis system, it is possible to obtain a description of the content of this image in terms of color and shape, together with the grade of uncertainty that we expect each image processing algorithm to produce. For this picture we obtain, for example, the following values of color features with their degrees of uncertainty:

$$\{\langle \text{Green}, 0.6 \rangle, \langle \text{White}, 0.7 \rangle, \langle \text{Gray}, 0.89 \rangle, \langle \text{Blue}, 0.65 \rangle\}$$

and the following values of shape features, with their uncertainty degrees:

$$\{\langle \text{Rectangle}, 0.8 \rangle, \langle \text{Ellipse}, 0.7 \rangle\}.$$

An intelligent system, using a classifier, might associate to the previous features the description of the content of the image itself, e.g.:

$$\{\langle \text{A lake with a snowy mountain}, 0.7 \rangle\}.$$

From the above described example, it is clear that the representation requirements of image data and, more in general, of multimedia data, cannot be satisfied by the classical, flat relational model: besides, a probabilistic extension of the relational model has been proposed [3, 4] and an algorithm to reduce the probabilistic model to the annotated relational model has been provided. The proposed reduction from the probabilistic relations to annotated relations, while providing an easy way to manage the probabilistic relations (relying on the existing algebra for annotated relations) has as a drawback a dramatic increase in the data volume in the database. This problem would be overcome if a suitable algebra on an attribute level data model was used. These are the motivations for this project: to define a data model for image databases which is at least as expressive as annotated relations, while allowing a compact representation of data. Specifically, we defined a fuzzy data model, dealing with fuzziness at the attribute level, and we developed an algebra for this model, introducing new algebraic operators in order to take into account the fuzziness implicitly related to image descriptors, both at a low level and at high level.

The resulting framework is a flexible, compact and powerful model that may be used as the basis for a query by example in image database systems.

2.2 *Multimedia Ontologies*

The need for an high level representation that captures the semantics of a multimedia object led at the beginning of the 90's to the development of the MPEG-7 standard for describing multimedia documents. This standard provides metadata descriptors for structural and low level aspects of multimedia documents, as well as metadata for information about their creators and their format, and other features. Since MPEG-7 is defined in terms of an XML schema, the semantics of its elements has no formal grounding. Thus, the representation and understanding of concepts and relationships among them is only possible through formal languages and ontologies [5]. In the last decade, several projects have been presented about multimedia systems based on knowledge models, image ontologies, fuzzy extension of ontology theories. In almost all the works, multimedia ontologies are effectively used to perform semantic annotation of the media content by associating the terms of the ontology with the individual elements of the image or video domains, thus demonstrating that the use of ontologies can enhance classification precision and image retrieval performance.

For this reason, many efforts to build ontologies that can bridge the *semantic gap* have been done. However, till now there is not yet an accepted solution to the problem of how to represent, organize, and manage multimedia data and the related semantics by means of a formal framework.

In particular, our research group tried to give an answer to the following big questions:

- what a multimedia ontology is: is it a taxonomy, or a semantic network of metadata (tags, annotations)?
- does a multimedia ontology support concrete data: what is the role of rough data (image, video, audio data), if any?
- what a multimedia semantics is: how to define and capture the semantics of multimedia data?
- how to build extensional ontologies: once defined a suitable formal framework, can we automatically build the defined multimedia ontologies?

We started at the beginning of 2005 working on a project in cooperation with the Polytechnic of Milan[3] aiming to define multimedia ontologies for images. We first proposed an extension of the NF^2 IDB model with an ontological model, also

[3]L. Tanca's research group.

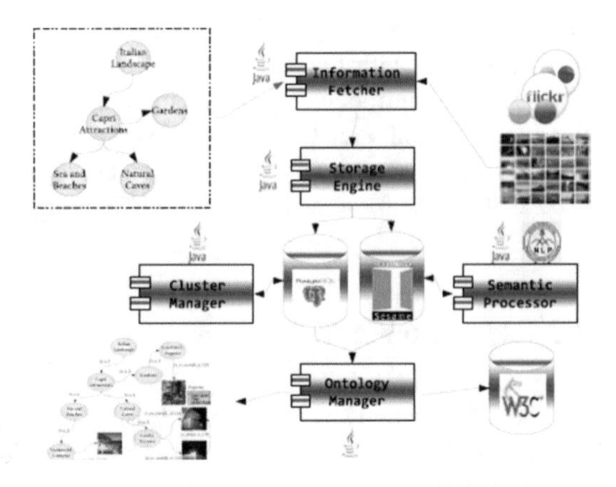

Fig. 2 MOWIS system architecture

describing a general architecture for supporting the creation and management of multimedia objects. In order to include the entire semantics of multimedia data, ontology concepts and languages, such as OWL, and the DL-based reasoning services that complement such languages should be extended. An image ontology [6], in fact, should allow specifications for:

- Spatial relationships that exist between the different media entities characterizing a media object or an event.
- Uncertainty of the concepts and their media properties.
- Association between low-level properties and semantic properties of images.
- An associated reasoning service which can use the available feature observations and concept likelihoods to infer probabilities for other concepts in presence of uncertain relationships.

An effective framework for supporting such theories was then developed within the Italian PRIN Project COOPERARE.[4]

We implemented the MOWIS (Multimedia Ontologies from Web Information Sources) system [7–9] for automatic construction of multimedia ontologies using the Flickr web services that provide images, tags, keywords and sometimes useful annotation describing both the image content and personal interesting information.

Figure 2 shows at a glance the implemented MOWIS system. In particular, the MOWIS ontology building process is made of:

1. definition of an initial taxonomy containing few high level nodes (related to the main concepts of a specific domain),
2. extraction of useful information (images and annotations related to the taxonomy concepts) from several annotated web repositories,

[4]Content Organization, Propagation, Evaluation and Reuse through Active Repositories, in collaboration with the University of Turin – M.L. Sapino– and University of Bologna – I. Bartolini and M. Patella.

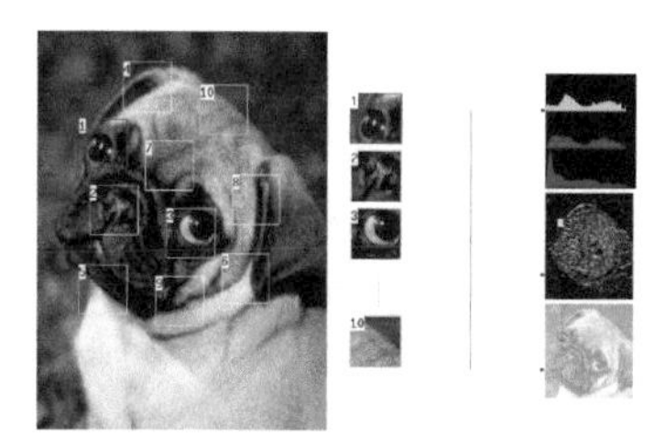

Fig. 3 Image information path and extracted features

3. content-based analysis on the row-data and a semantic processing on the related textual annotations,
4. semantic processing of textual tagged data, and
5. the ontology construction.

The image ontology is thus generated in an incremental way and in correspondence of pick-up operations from the Flickr repository.

2.3 Image and Video Indexing Based on Animate Vision

At the beginning of 2000, we started to study some novel indexing techniques for video and image databases based on the *Animate Vision* paradigm [10, 11].[5]

Leveraging Animate Vision theory, from one hand, it has been shown how, by embedding within image inspection algorithms active mechanisms of biological vision such as saccadic eye movements and fixations, a more effective content-based retrieval in image database can be achieved. From the other one, it has been shown by taking advantage of such "foveated" representation of an image, it is possible the partitioning of a video into shots.

The first goal was to obtain a low-level description of images in terms of content features, using classical Computer Vision techniques.

We decided to use a *salient points* based technique - exploiting the Animate Vision model for image analysis - that uses color, texture and shape information associated with those regions of the image that are "relevant" for human attention (*Focus of Attention*), in order to obtain a compact characterization (namely *Information Path*) that could be also used to evaluate the similarity between images, and for indexing issues (see Fig. 3).

An Information Path $\mathcal{IP} = \langle F_s(p_s; \tau_s), h_b(F_s), \Sigma_{F_s} \rangle$ can be seen as a particular data structure that contains, for each region $F(p_s; \tau_s)$ surrounding a given salient point (where p_s is the center of the region and τ_s is the observation time spent by a human to detect the point), the color features in terms of HSV histogram $h_b(F_s)$, and the texture and shape features in terms of wavelet covariance signatures Σ_{F_s} [12].

[5]In cooperation with the University of Salerno – G. Boccignone.

With respect to indexing and retrieval of images in large databases, the greatest difficulty was to find features that effectively represent image content, while adopting image data structures that organize efficiently the feature space. In the framework of *Content Based Image Retrieval* (CBIR), *Query By Example* (QBE) was considered a suitable approach because the user handles an intuitive query representation: the form of the query, namely an image, is that of the data to be evaluated. However, a hallmark all too easily overlooked is that when the user is performing a query, she/he is likely to have some semantic specification in mind, e.g. "I want to see a portrait", and the portrait example provided to the query engine is chosen to best represent the semantics. Unfortunately, traditional image databases were not able to express either such semantics or similarity rules consistent with semantics; this problem is known as "semantic gap".

To overcome such problem, we proposed a "context-sensitive" methodology for image retrieval based on the QBE approach and leveraging images' description in the shape of Information Path [13].

Images - represented in the Information Path features space - are recursively clustered using a balanced version of *Expectation Maximization* (EM) algorithm and a tree indexing structure, whose high-level nodes correspond to database categories and subcategories (high-level concepts), is then generated. Given a query image I_q and the dimension of the desired results set, the most similar images are retrieved in the following steps (see Fig. 4):

1. map the image in the features space by computing the Image Path under free viewing conditions, $\mathcal{IP}_q$;
2. discover the best K categories that may describe the image on the base of a distance between I_q and the centroid of the high-level nodes of the tree indexing structure;

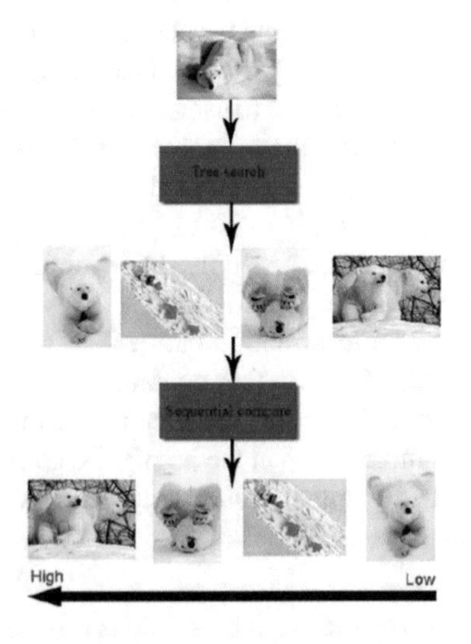

Fig. 4 Animate query process

3. for each selected category among the best K discovered, by traversing the associated subtrees, retrieve the target images within the category at minimum distance from the query image;
4. refine results by choosing the images most similar to the query image by performing a sequential scanning of the previous set of retrieved images and evaluating the similarities using the Information Path Matching algorithm [12].

In addition, we proposed a novel approach for video partitioning into shots. In particular, we used the Information Path Matching algorithm - applied for each couple of consecutive frames in a video stream - to determine possible decreases of attention consistency that can correspond to abrupt or gradual transitions [14].

3 Application of Multimedia DataBase Systems

The described theories are at the basis of a number of applications to real scenarios. In the following, we will show the most important implementations in Cultural Heritage and Video Surveillance domains.

3.1 Video Surveillance: The Activity Detection Project

The idea beyond the Activity Detection Project[6] is that of providing a flexible and intelligent framework for video surveillance applications [15, 16]. Let us consider a large surveillance videos database, a user may be interested in executing queries, such as *find all videos or part of a video where*:

- a package is transferred by one person to another;
- "Antonio Picariello" transfers a package to another person;
- "Antonio Picariello" transfers a package to "Vincenzo Moscato".

In this project,[7] we made the following contributions. We first provided a *Probabilistic Activity Description Language* (PADL) that allows activities to be described on top of some basic suite of image processing algorithms. We thus presented a new probabilistic logic model for representing events and efficient algorithms to accurately identify those events. In short, the key characteristics of PADL are the following:

1. it uses a full first-order logic formulation of activities;
2. it is the first framework where users can articulate any definition of an activity and have instances of those activities found whereas past work seems to mostly

[6]http://www.umiacs.umd.edu/research/LCCD/projects/Activity_Detection.jsp.

[7]Project in cooperation with V.S. Subrahmanian – University of Maryland at College Park and M. Albanese – George Mason University.

assume that activities were defined by vision experts or somehow learned from data;
3. it integrates boolean and probabilistic features and predicates and is the first activity detection that merges the power of logic and probability;
4. it introduces efficient offline algorithms as well as the first online algorithm to detect the smallest video subsequences matching a given activity definition.

The system has been proved in real video surveillance system at the University of Naples parking lots.

In addition, we presented a computational framework for human activity representation based on Petri Nets. In particular, we proposes a Petri Nets extension - *Probabilistic Petri Nets* (PPN) - and show how this model is well suited to address Activity Detection problem. We then focused on answering two types of questions: (i) what are the minimal sub-videos in which a given activity is identified with a probability above a certain threshold and (ii) for a given video, which activity from a given set occurred with the highest probability? We provided the PPN-MPS algorithm for the first problem, as well as two different algorithms (naivePPNMPA and PPN-MPA) to solve the second one.

Finally, more recently, we developed two systems for the detection of "unexplained" activities in a video based on the theory of possible worlds [17] and on probabilistic graphs [18].

3.2 Multimedia Recommendation

Another interesting application was the adoption of some recommendation techniques [19] in the Multimedia realm, especially for Cultural Heritage applications.[8]

In this project, we empowered standard algorithms and tools, using all the facilities provided by our multimedia database system [20]. A typical scenario where an effective multimedia recommender system would be desirable is a virtual museum offering web-based access to a multimedia collection of digital reproductions of paintings: let us consider the Uffizi Gallery in Florence, Italy. For instance, consider a user visiting this virtual museum, and suppose a user initially requests some paintings depicting the "Holy Mary". While observing such paintings, she is attracted by an Albrecht Dürer's painting entitled *Madonna col Bambino* (Fig. 5(a)). It would be helpful if the system – based on these first interactions – could learn her preferences and suggest other paintings representing the same or related subjects, depicted by the same or other related authors or items that have been requested by users with similar preferences. As an example, a user who is currently watching the Dürer's painting in Fig. 5(a) might be recommended to see a Jacopo Carucci's painting entitled *Madonna col Bambino e San Giovannino* (Fig. 5(b)), that is quite similar to the current picture in terms of color and texture, and *Madonna col Bambino* by

[8]Research in cooperation with M. Albanese, GMU and A. d'Acierno, Italian CNR.

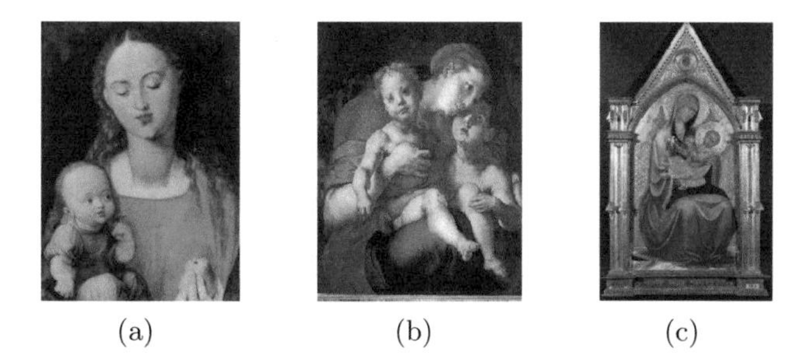

Fig. 5 Paintings depicting the "Holy Mary"

Andrea Vanni (Fig. 5(c)), that is not similar in terms of low level features but similar in terms of semantic content.

Starting from these considerations, a multimedia recommender system has to deal with a set of *users* $U = \{u_1 \ldots, u_m\}$ and a set of multimedia *items* $O = \{o_1, \ldots, o_n\}$. For each pair (u_i, o_j), a recommender can compute a *score* (or a *rank*) $r_{i,j}$ that measures the expected interest of user u_i in item o_j (or the expected utility of item o_j for user u_i), using a *knowledge base* and a *ranking* algorithm that generally could consider different combinations of the following characteristics: (i) user preferences and past behavior, (ii) preferences and behavior of the user community, (iii) multimedia items' features and how they can match user preferences, (iv) user feedbacks, (v) context information and how recommendations can change together with the context.

Our multimedia recommender system is formed by the following components [21].

- A *pre-filtering* module that selects for each user u_i a subset $O_i^c \subset O$ containing items that are good candidates to be recommended; such items usually match user preferences and needs. We employ *high-order star-structured co-clustering* techniques to address the problem of heterogeneous data filtering, where a user is represented as a set of vectors.
- A *ranking* module that assigns w.r.t. user u_i a rank $r_{i,j}$ to each candidate item o_j in O_i^c using a *hybrid* approach based on the PageRank algorithm that exploits in several ways multimedia items' features (with the related high and low level similarity) and users' preferences, feedbacks (in the majority of cases in terms of *ratings*), moods (by opinion or sentiment analysis) and behavior.
- A *post-filtering* module that dynamically excludes, for each user u_i, some items from the recommendations' list; in this way, a new set $O_i^f \subseteq O_i^c$ is obtained on the base of user feedbacks and other contextual information (such as data coming from the interactions between the user and the application).

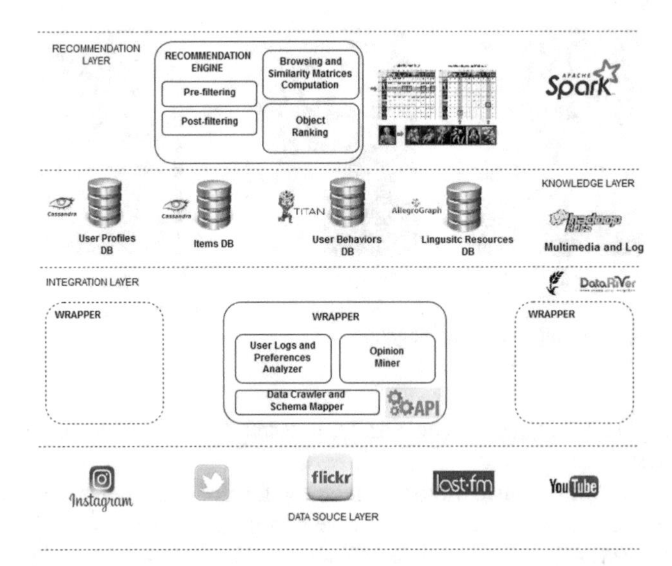

Fig. 6 Big data infrastructure for multimedia recommendation

We recently developed a big data infrastructure to support the multimedia recommendation strategy (see Fig. 6) within the research activity of DATABENC project.[9]

4 Multimedia Social Networks

In the last few years, the use of *Online Social Networks* (OSNs) has been rapidly growing allowing people, that lives in different places, to make friends and to share, comment and observe different types of multimedia content, producing a large amount of data showing *Big Data* features, mainly due to their high change rate, their large volume and intrinsic heterogeneity.

Indeed, OSNs represent a new kind of information network that differs significantly from existing networks like the Web, where known hyperlinks between content objects constitute a graph that is used to organize, navigate, and rank information. Instead, OSNs exhibit a particular structure characterized by few links between content objects, in turn a certain number of links exist between content objects and users and between the users themselves. *Social Network Analysis* (SNA) methodologies have been recently introduced to study the properties of such kind of information net-

[9]DATABENC is the High Technology District for Cultural Heritage management of the Campania Region, in Italy (www.databenc.it) and in collaboration with the Polytechnic of Milan, University of Turin, the University of Salerno – F. Colace and M. De Santo, and the University of Bologna [22, 23].

works with the aim of supporting a wide range of applications: information retrieval, influence analysis, recommendation, marketing, event recognition, user profiling, and so on.

In our vision, an additional challenge in the management of OSNs derive from the presence of multimedia information and several questions arise, if we consider the important role that multimedia data can assume in a social network:

- Is it possible to exploit multimedia features and notion of *similarity* to discover more links? Are such links effectively useful for analytics?
- Can the different types of user annotations (e.g. tag, comment, keywords, status, etc.) and interactions with multimedia objects provide a further support for an advanced network analysis?
- Is it possible to integrate and efficiently manage in a unique network the information coming from OSNs and that are related to multimedia sharing systems (for example, a Facebook user has usually an account also on Instagram or Flickr)? How we can deal with a very large volume of data?
- In this context how is possible to model all the various relationships between users and multimedia objects? Are the graph-based strategies still the most suitable solutions?

The preliminary step to provide an answer for the above questions lies in the introduction of a model for *Multimedia Social Networks* (MSNs): integrated networks that combine the information on users belonging to one or more social communities, with all the multimedia contents that can be generated and used within the related environments.

We presented a novel data model for MSNs based on the *hypergraph* structure that allows us to represent in a simple way all the different kinds of relationships that are typical of a social network, in particular among multimedia contents, among users and multimedia contents and among users themselves, at the same time supporting several kinds of applications by means of the introduction of several *ranking* functions [24].

In our vision, a MSN (see Fig. 7) is basically composed by three different entities:

- **Users** - the set of persons and organizations constituting the particular social community, with the information concerning their profile, interests, preferences, etc.
- **Multimedia Objects** - the set of multimedia resources (i.e. texts, images, video, audio, etc.) that can be shared within a MSN community, together with High level (*metadata*) and low level information (*features*).
- **Annotation Assets** - the most significant terms or named entities of a given domain, namely *topics*, exploited by users to annotate multimedia data and derived from the analysis of textual information such as keywords, labels, tags, comments etc.

Several types of relationships can be established among the described entities: a user can annotate an image with a particular tag, two friends can comment the same post, a user can tag another user in a photo, a user can share some videos within a

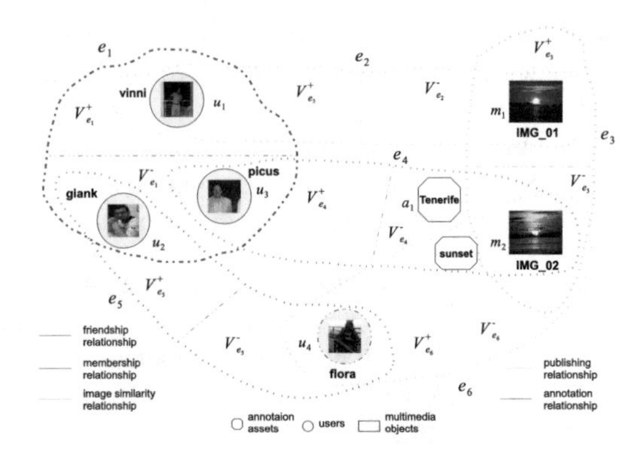

Fig. 7 Representation of a MSN via hypergraph

group, etc. Analyzing the different types of relationships that can be established in the main social media networks (see Fig. 7), we have identified three categories:

- **User to User** relationships, describing user actions towards other users;
- **Similarity** relationships, describing a relatedness between two multimedia objects, users or annotation assets;
- **User to Multimedia** relationships, describing user actions on multimedia objects, eventually involving some annotation assets or other users.

Our model is thus quite general to model not only the most diffused OSNs (such as Facebook, Twitter, etc.) and the main social networks for multimedia sharing (such as Panoramio, Flickr, YouTube, Instagram, etc.), but also other kinds of networks (e.g., LinkedIn, ResearchGate, etc.).

5 Conclusion

In this chapter we described some of the main results produced by the Multimedia Database Research Group of University of Naples in the multimedia database field. In particular, we presented models for representing multimedia data and the related knowledge and advanced techniques for their storage, indexing and retrieval. In addition, we also pointed out several applications, with a particular emphasis on recommender systems, video surveillance, social media networks and Cultural Heritage domain.

References

1. R. Jain, A. Del Bimbo, T.-S. Chua, B. Furht, Survey papers in multimedia-guest editorial. Multimed. Tools Appl. **51**(1), 1–4 (2011)

2. X. Li, T. Uricchio, L. Ballan, M. Bertini, C.G. Snoek, A.D. Bimbo, Socializing the semantic gap: A comparative survey on image tag assignment, refinement, and retrieval. ACM Comput. Surv. (CSUR) **49**(1), 14 (2016)
3. A. Picariello, M.L. Sapino, Managing uncertainties in image databases, in *Semantic-Based Visual Information Retrieval* (2007), pp. 292–310
4. A. Chianese, A. Picariello, L. Sansone, M.L. Sapino, Managing uncertainties in image databases: A fuzzy approach. Multimed. Tools Appl. **23**(3), 237–252 (2004)
5. M.C. Suárez-Figueroa, G.A. Atemezing, O. Corcho, The landscape of multimedia ontologies in the last decade. Multimed. Tools Appl. **62**(2), 377–399 (2013)
6. A. Penta, A. Picariello, L. Tanca, Multimedia knowledge management using ontologies, in *Proceedings of the 2nd ACM Workshop on Multimedia Semantics* (ACM, 2008), pp. 24–31
7. V. Moscato, A. Penta, F. Persia, A. Picariello, Mowis: A system for building multimedia ontologies from web information sources, in *IIR* (2010), pp. 89–93
8. A. Chianese, V. Moscato, F. Persia, A. Picariello, C. Sansone, A framework for building multimedia ontologies from web information sources, in *SEBD* (2012), pp. 83–90
9. A. Chianese, V. Moscato, A. Picariello, A system for building multimedia ontologies from web information sources, in *New Trends in Software Methodologies, Tools and Techniques - Proceedings of the Eleventh SoMeT 2012, Genoa, Italy, 26–28 September 2012* (2012), pp. 379–394
10. D.H. Ballard, Animate vision. Artif. Intell. **48**(1), 57–86 (1991)
11. L. Itti, C. Koch, E. Niebur, A model of saliency-based visual attention for rapid scene analysis. IEEE Trans. Pattern Anal. Mach. Intell. **20**(11), 1254–1259 (1998)
12. G. Boccignone, A. Picariello, V. Moscato, M. Albanese, Image similarity based on animate vision: Information path matching, in *Multimedia Information Systems* (2002), pp. 66–75
13. G. Boccignone, A. Chianese, V. Moscato, A. Picariello, Context-sensitive queries for image retrieval in digital libraries. J. Intell. Inf. Syst. **31**(1), 53–84 (2008)
14. G. Boccignone, A. Chianese, V. Moscato, A. Picariello, Foveated shot detection for video segmentation. IEEE Trans. Circuits Syst. Video Technol. **15**(3), 365–377 (2005)
15. M. Albanese, R. Chellappa, V. Moscato, A. Picariello, V. Subrahmanian, P. Turaga, O. Udrea, A constrained probabilistic petri net framework for human activity detection in video. IEEE Trans. Multimed. **10**(8), 1429–1443 (2008)
16. M. Albanese, R. Chellappa, N. Cuntoor, V. Moscato, A. Picariello, V. Subrahmanian, O. Udrea, Pads: A probabilistic activity detection framework for video data. IEEE Trans. Pattern Anal. Mach. Intell. **32**(12), 2246–2261 (2010)
17. M. Albanese, C. Molinaro, F. Persia, A. Picariello, V. Subrahmanian, Discovering the top-k unexplained sequences in time-stamped observation data. IEEE Trans. Knowl. Data Eng. **26**(3), 577–594 (2014)
18. C. Molinaro, V. Moscato, A. Picariello, A. Pugliese, A. Rullo, V. Subrahmanian, Padua: Parallel architecture to detect unexplained activities. ACM Trans. Internet Technol. (TOIT) **14**(1), 3 (2014)
19. F. Ricci, L. Rokach, B. Shapira, P.B. Kantor (eds.), *Recommender Systems Handbook* (Springer, Berlin, 2011)
20. M. Albanese, A. d'Acierno, V. Moscato, F. Persia, A. Picariello, A multimedia recommender system. ACM Trans. Internet Technol. **13**(1), 3 (2013)
21. F. Colace, M.D. Santo, L. Greco, V. Moscato, A. Picariello, A collaborative user-centered framework for recommending items in online social networks. Comput. Hum. Behav. **51**, 694–704 (2015)
22. I. Bartolini, V. Moscato, R.G. Pensa, A. Penta, A. Picariello, C. Sansone, M.L. Sapino, Recommending multimedia visiting paths in cultural heritage applications. Multimed. Tools Appl. **75**(7), 3813–3842 (2016)

23. F. Colace, M.D.E. Santo, V. Moscato, A. Picariello, F.A. Schreiber, L. Tanca, Patch: A portable context-aware atlas for browsing cultural heritage, in *Data Management in Pervasive Systems* (Springer, Berlin, 2015), pp. 345–361
24. F. Amato, V. Moscato, A. Picariello, G. Sperlí, Multimedia social network modeling: a proposal, in *2016 IEEE Tenth International Conference on Semantic Computing (ICSC)* (IEEE, 2016), pp. 448–453

Part IV
Knowledge Discovery and Data Mining

How Data Mining and Machine Learning Evolved from Relational Data Base to Data Science

G. Amato, L. Candela, D. Castelli, A. Esuli, F. Falchi, C. Gennaro, F. Giannotti, A. Monreale, M. Nanni, P. Pagano, L. Pappalardo, D. Pedreschi, F. Pratesi, F. Rabitti, S. Rinzivillo, G. Rossetti, S. Ruggieri, F. Sebastiani and M. Tesconi

1 Introduction

During the last 35 years, data management principles such as physical and logical independence, declarative querying and cost-based optimization have led to profound pervasiveness of relational databases in any kind of organization. More importantly, these technical advances have enabled the first round of business intelligence applications and laid the foundation for managing and analyzing Big Data today. The 90's have been exceptional years for the invention and development of solid data mining and machine learning algorithms [1, 2, 56] building on existing statistical and artificial intelligence theories. Open and proprietary software libraries and analytical platforms have bloomed in parallel with the development of a robust methodological approach to the development of analytical processes capable of extracting valuable knowledge out of large masses of data: the Knowledge Discovery in Databases (KDD) [33]. When the data deluge began, the KDD technologies were well prepared so that the new advances stimulated by the many novel challenges and opportunities associated with Big Data took place in parallel very effective field demonstrations in a wide array of domains, thus activating a virtuous cycle between innovation and research.

G. Amato · L. Candela · D. Castelli · A. Esuli · F. Falchi · C. Gennaro ·
F. Giannotti (✉) · M. Nanni · P. Pagano · F. Pratesi · F. Rabitti · S. Rinzivillo · F. Sebastiani
ISTI - CNR, Pisa, Italy
e-mail: fosca.giannotti@isti.cnr.it

A. Monreale · L. Pappalardo · D. Pedreschi · G. Rossetti · S. Ruggieri
University of Pisa, Pisa, Italy

M. Tesconi
IIT-CNR, Pisa, Italy

S. Flesca et al. (eds.), *A Comprehensive Guide Through the Italian Database Research Over the Last 25 Years*, Studies in Big Data 31, DOI 10.1007/978-3-319-61893-7_17

The data deluge has been really impressive: digital technology has become ubiquitous and very much part of public and private organizations and individuals. People and things have become increasingly interconnected. Smartphones, buildings, cities, vehicles and other environments and devices have been filled with digital sensors, all of them creating evermore data. New high-throughput scientific instruments, telescopes, satellites, accelerators, supercomputers, sensor networks, and running simulations have generated and are generating massive amounts of data.

Big data have been blossoming together with the hope to harness the knowledge they hide to solve the key problems of society, business and science. However, turning an ocean of messy data into knowledge and wisdom is an extremely challenging task. Heterogeneity, scale, timeliness, complexity, and privacy problems with Big Data are the key issues to be addressed at all phases of the pipeline that can create value from data.

Big Data is not natively in structured format; for example, tweets and blogs are weakly structured pieces of text, while images and video are structured for storage and display, but not according its semantic content to enable search: transforming such content into a structured format for later analysis has been and is a major challenge. The value of data explodes when it can be linked with other data, thus data integration is a major creator of value. Since most data is directly generated in digital format today, we have the opportunity and the challenge both to influence the creation to facilitate later linkage and to automatically link previously created data.

In this context, at the end of the 90s a new analytical trend joined data mining and machine learning: the emergence of *network science* [13]. Once again, the availability of large graph data emerging from the web has allowed to discover general patterns and statistical laws regulating statics and dynamics of complex networks.

Another relevant impact of big data is the opportunity to observe and measure how our society intimately works: the digital breadcrumbs of human activities carried the capacity to scrutinize the ground truth of individual and collective behaviour at an unprecedented detail. Multiple dimensions of our social life have been increasingly "proxied" by big data: automated payment systems record the tracks of our purchases; search engines record the logs of our queries on the web; wireless networks and mobile devices record the traces of our movements; social media record the traces of our opinions and emotions; social networks record the traces of our interactions. This new scenario took the name of *social mining* and it is clear that such challenge requires high-level analytics, modeling and reasoning across all the social dimensions above.

This chapter proposes an account of the scientific and technical evolution of data mining and machine learning from relational data bases to data science, focusing on "making sense" of data generated as by-product of ICT mediated human activities, i.e. on the analytical methods and process, intentionally neglecting the amazing advances on the efficiency and scalability of the algorithms as well as their ability to deal with massive streaming data. The chapter tells the story of this evolution through the research achievements of a network of research labs in Pisa across the CNR and the University, which contribute to the birth and life of the Italian database community of SEBD. The next sections discuss are dedicated to the main trends and results of these groups in the following areas. **Mobility data analysis** leverages the spatio-temporal dimensions of big data to the purpose of understanding human mobility behavior, evolutionary patterns, daily activity patterns, geographic patterns. **Social network analysis** studies the architecture of interpersonal relationships, with the purpose of understanding the structure and the dynamics of the fabric of human society. **Multimedia media mining** methods for making sense of heterogeneous data, sensed from different on line sources: tweets, mails, blogs, web pages, link structures, videos etc., to the purpose of extracting the hidden semantics from them. **Sentiment mining**. At the crossroads of natural language processing and information retrieval, a key topic is opinion/sentiment mining, aimed at harnessing the emotional content of user generated texts. The last two sections present two activities that have a transversal impact. First, privacy aware analytics to prevent "by-design" the risks of invading the sphere of personal information and the reflection on the ethical consequences of predictive analytics. Second, the new opportunities that data infrastructures and virtual research environments bring to data scientists.

2 Mobility Data Analysis

In the last decades, the wide-spread availability of geo-localization devices and the technologies to store and analyze the data they generate had a huge impact on the research areas dealing with spatial and spatio-temporal data. GPS traces and other forms of mobility data quickly became a focus for researchers from several disciplines, especially in the domain of data mining. These mobility data provide a new powerful social microscope, which may help us understand human mobility, and discover the hidden patterns and models that characterize the trajectories humans follow during their daily activity.

Revisitations of standard data mining problems and techniques quickly appeared, many of them coming from the important contributions provided by the Italian community in the last 15 years: trajectory patterns, i.e. frequent patterns describing sequences of places and possibly timings that are common to a large number of input trajectories; trajectory clustering, i.e. grouping similar trajectories [9] or trajectory segments [45] into homogeneous groups; location prediction, i.e. forecasting the position that a moving object will have in the near future [48]; recognition of movement activity, i.e. associating to a trajectory or to a stop area the activity it was aimed to [58], such as going to work, leisure, shopping, etc.

In the rest of this section we will mention two examples that had a great impact on the research community, in terms of references and applications that stemmed from them: Trajectory Patterns and Mobility Profiles.

2.1 Trajectory Patterns

In some contexts, the moving objects we are examining might act in a similar way, even if they are not spatially located together. For instance, similar daily routines might lead several individuals to drive their car along the same routes, even if they leave home at very different hours of the day. Or, tourists that visit a city on different days of the year might actually visit it in the same way – for instance by visiting the same places in the same order and spending there approximately the same amount of time – because they simply share interests and attitude. The kind of questions we might try to answer in this cases is: *Are there groups of objects that perform a sequence of movements, with similar timings though possibly during completely different moments?* Accordingly, ***T-Patterns*** [39] (abbreviation of *Trajectory patterns*) provide sequences of spatial locations with typical transition times, such as Railway Station $\xrightarrow{15min}$ Museum $\xrightarrow{2h15min}$ Castle Square. This might represent the typical behavior of tourists that rapidly reach a museum from the railway station and spend there about two hours before getting to the adjacent square.

The set of spatial regions to be used to form patterns is a major parameter of the method, i.e., the spatial extension of "Railway Station" and any other place considered relevant for the analysis. However, the algorithm proposed also contains heuristics to automatically define such regions, based on coverage of dense areas, in case there is no domain expert to provide them.

T-Patterns represented the first attempt to automatically infer from raw GPS traces an higher abstraction of movement, capturing key places (the regions in the patterns) and temporal evolution. This information has been later exploited as building block of various applications, such as prediction (the WhereNext method [48], which predicts the next most likely area a moving object will visit), or the identification of hot routes in the city [40].

2.2 Mobility Profiles

Despite the great attention that the analysis of individual trajectories attracted, for a very long time the individuals themselves have not been considered as a relevant subject of analysis. **Mobility profiles** [68] represent the first clear step on the opposite direction, by analysing individuals (rather than just large groups) with the purpose of understanding systematic mobility, as opposed to occasional movements, which is fundamental in some mobility planning applications, e.g. public transport.

The objective is to use the set of trips of an individual user to find his/her routine behaviors. That is realized by grouping together similar trips based on concepts of spatial distance and temporal alignment, with corresponding thresholds for both the spatial and temporal components of the trips. In order to be defined as *routine*, a behavior needs to be supported by a significant number of similar trips of the user. The technology adopted to achieve that is a clustering algorithm that groups together the similar trajectories, each cluster representing a routine [9]. In particular, the algorithm combines density-based methods (known to deal well with non-spherical clusters and noisy data, both being typical features of trajectory data) with bisecting k-means (used to obtain compact clusters). Each group obtained is then summarized by its central element.

Individual Mobility profiles enable several applications, ranging from deeper traffic analyses (indeed, the traffic traversing a given area can now be described also by a systematicity index, measuring the percentage of trips that are routinaries for the individuals involved) and the creation of predictive models (as in the case of MyWay [69], where ongoing trips are compared against the user's routines, and in case of match they are used to predict how the trip will continue) to services like carpooling [42].

2.3 *The Borders of Human Mobility*

The problem of discovering the geographic borders from human mobility at the low spatial resolution of municipalities or counties is a far reaching problem, motivated by providing policy makers with suggestions about the best administrative partitions for the government of the territory. In [23, 57], we adopt a social network analysis view to mobility data to reach a better understanding of human mobility patterns, leveraging the underlying, hidden connections that human mobility establishes among different places. Starting from a given zoning of the territory, tessellated into census zones, we construct a network whose nodes are the zones and the weighted edges between any two zones represent the number of travels originating in the first and ending in the second. The analysis phase consists in discovering densely connected sub-graphs in this network by means of a community detection method, thus highlighting groups of zones that are highly connected by many travels compared to the lower connectivity among different modules. This an example how a network mining method can be adopted to reveal the hierarchical structure of a complex phenomenon, and highlight the thresholds at which we separate*macro-, meso- and micro-levels* of the system.

2.4 *Returners and Explorers*

Another interesting line of research has been at the crossroad of mobility data mining and network science. Network science is aimed at discovering the global models of complex social phenomena, by means of statistical macro-laws governing basic quantities, which show the behavioral diversity in society at large. Data mining is aimed at discovering local patterns of complex social phenomena, by means of micro-laws governing behavioral similarity or regularities in sub-populations. The objective of combining micro and macro laws has been pursued in [54] where taking advantage of massive digital traces of human whereabouts a series of novel insights on the quantitative patterns characterizing human mobility have been discovered and used to anchor to reality the abstract models of human mobility. Our work starts from the recent consensus on the fact that the considerable variability in the characteristic travelled distance of individuals coexists with a high degree of predictability of their future locations. Here we shed light on this surprising coexistence by systematically investigating the impact of recurrent mobility on the characteristic distance travelled by individuals. Using both mobile phone and GPS data, we discover the existence of two distinct classes of individuals: returners and explorers. As existing models of human mobility cannot explain the existence of these two classes, we develop more realistic models able to capture the empirical findings. Finally, we show that returners and explorers play a distinct quantifiable role in spreading phenomena and that a correlation exists between their mobility patterns and social interactions.

2.5 Sociometer: Classification of City Users and Flows

One very promising source of mobility information are mobile phones traces, most commonly collected in the form of Call Detail Records (CDRs), i.e. records of the phone calls performed that describe the starting time and location (in terms of antenna connected in that moment) of the call. In this context, an analysis method named *Sociometer* was developed, aimed to associate to each user her role w.r.t. a specific area, such as *resident*, *visitor*, etc.

Adopting a vision similar to the mobility profiles described above, our work started from the analysis of the single users' behavior. The approach summarizes the CDR data of each user through a temporal distribution of her calls within the spatial area under consideration, measuring the percentage of days that her was seen at different hours of the day (grouped into three intervals of around 8 h each), in different days of the week (grouped into *week-days* and *week-ends*), in different weeks. The basic idea is that different city users will produce different kinds of temporal distribution, for instance residents will most likely be present on all the time slots, while commuters will be seen only during working hours/days.

The personal *fingerprints* are clustered to identify the most relevant calling patterns, which are then classified through a standard K-NN classification schema. Earlier versions of the solution were based on a manual labelling of the relevant calling patterns found [35], while most recent ones compare them against a pre-defined set of representative distributions, called *archetypes* [36].

3 Social Network Analysis

Nowadays Complex Networks are pervasively used to model and describe the behaviors of a wide range of real world phenomena. Social relationships, biological interactions, transportation, commercial exchanges are only few of the several scenarios usually studied with the support of instruments borrowed by graph theory. Countless problems are formulated, or can be formulated, upon such structures: Community Discovery, Link Prediction, Tie Strength estimation are only few of them. Among all the fields that emerged in the last decades Social Network Analysis, SNA, is the one that makes use of graph mining techniques to understand human behaviors. SNA research has certainly be facilitated by the ever-growing popularity of online social network platforms data available. Such unprecedented sources of human generated data naturally modelled by the tools and theories offered by graph theory have lead to the rising of this novel field of research. Among the vast SNA literature, our research group has effectively contributed to the following themes: Multidimensional network analysis, Community Discovery, Network Analytics & Mobility and analysis of diffusive patterns.

3.1 *Multidimensional Networks*

Most real life networks are intrinsically multidimensional, and some of their properties may be lost if the different dimensions are not taken into account. In other cases, it is natural to derive multiple dimensions connecting a set of nodes from the available data to the end of analyzing some phenomena. In order to study this complex scenario a framework that extends the classical graph theory is needed. Reasoning on multidimensional networks seems clear that the usual graph model is not enough to represent all the available information. In our work *"Foundations of Multidimensional Network Analysis"* [14], using a multigraph representation, we proposed and evaluated on real datasets a multidimensional framework able to capture the interplay among dimensions and to overcome some limits that made the classical monodimensional measures unsuitable in this complex scenario. Such framework was then extended, in [53] where we formulate an approach to estimate tie strength on multidimensional networks and validate it on a multigraph built upon the social relationships of users interacting on three different online platform, namely Facebook, Twitter and Foursquare. Moreover, in [60] we proposed and evaluated a set of Link Prediction approach specifically tailored for multidimensional networks.

3.2 *Community Discovery*

The problem of identifying communities in complex networks is very popular among network scientists, as witnessed by an impressive number of valid works in this field.

Traditionally, a community is defined as a dense subgraph, in which the number of edges among the members of the community is significantly higher than the outgoing edges. Our survey [22] explores all the most popular techniques to find communities in complex networks and categorize them into eight main categories: Feature Distance, Internal Density, Bridge Detection, Closeness, Structure Definition, Link Clustering, Meta Clustering and Diffusion. In [24] we propose a bottom-up approach to efficiently extract overlapping communities: DEMON. DEMON leverages the nodes perspective to identify meaningful network substructures: it works by identify local-communities at the ego-network level exploiting label propagation and then merging them in an incremental fashion. Our approach has been used as a proxy for users homophily to support network quantification tasks [47]; as filter to reduce the computational cost of Link Prediction approaches [61]; as well as to bound set of Skype users while searching a network driven methodology to relate service usage to network position [62]. Moreover, in order to cope with the evolving nature of interaction networks, we proposed an online dynamic community discovery algorithm, TILES [63], able to track community life cycles as new perturbations appears in the network (i.e. appearance/ vanishing of nodes as well as edges).

4 Sentiment Analysis

A large proportion of the data that is generated daily, and that needs to be processed by search and mining algorithms, is of a textual, non-structured nature; these data have traditionally been the domain of information retrieval and text mining. After the advent of the so-called "Web 2.0", a lot of textual content is user-generated, and its nature is not purely descriptive: that is, it is not confined to describing facts or states of affairs in an objective, detached way, but is instead rich in subjective, opinionated content. Harnessing the opinions and emotions expressed by the authors of these textual contents is the object of *sentiment analysis* (also known as *opinion mining*), an area at the crossroads of natural language processing and information retrieval that has blossomed in the mid years of the past decade, and that has been receiving increased attention, from industry and the scientific community alike, ever since.

4.1 Automatically Expanding Sentiment Lexicons

Possibly the most important task underlying attempts to tap into this kind of data is *sentiment classification*, the task of classifying an item of user-generated content (UGC – e.g., a tweet, a product review, a post on a social networking service) according to the sentiment it conveys (or opinion it expresses) about a certain entity. While this shares many characteristics with the task of classifying text by topic, the traditional "bag of words" (BoW) approach to representing textual content cannot be used for classifying text by sentiment: to see why, simply consider the fact that two sentences such as "A horrible hotel in a beautiful town" and "A beautiful hotel in a horrible town" would be assigned the same class if relying on a BoW representation, while they convey radically different sentiment. As a result, classification by sentiment fundamentally relies on the availability of a *sentiment dictionary*, i.e., an online dictionary where lexical entries (e.g., words, or word senses) are tagged in terms of whether they convey a sense of positivity (e.g., "truthful", "sublime") or negativity (e.g., "inaccurate", "pathetic"). However, manually curated sentiment dictionaries characterised by a high coverage of the language rarely exist in practice, especially for less resourced languages. As a result, our group investigated a number of language-independent methods for automatically tagging by sentiment existing online dictionaries.

A first method we developed was based on gloss classification, i.e., on classifying a lexical entry as positive or negative by classifying the textual definition of the entry ("gloss"); the method was first applied to classifying words according to the positive vs. negative dichotomy [30], and later extended to also identify neutral words (i.e., words that convey no sentiment; e.g., "inorganic", "quadratic") [29]. This method was deployed in practice in order to tag the English-language version of WordNet; the result was SentiWordNet [31], a sentiment lexicon now routinely used by hundreds of research groups worldwide.

A second method we later developed was based on random walks, and assumed that a positive (resp., negative) word being defined (the *definiendum*) is defined by mostly using positive (resp., negative) words in the gloss (the *definiens*). By assuming that positivity and negativity "flow" along the links connecting the definiendum with the words contained in the definiens, random walks on the word graph can be used for performing fine-grained computations of how positive/negative a word in a dictionary is. This method was applied to refining SentiWordNet; this led to a more accurately tagged version, called SentiWordNet 3.0 [12], which is the version now currently available.

4.2 Cross-Lingual and Cross-Domain Sentiment Classification

Cross-lingual sentiment classification is the task of classifying by sentiment text expressed in a target language (e.g., Urdu) when training data are available only for a source language (e.g., English). *Cross-domain* sentiment classification instead refers to sentiment classification of texts about a target domain (e.g., reviews of books) when training data are available only for a source domain (e.g., reviews about CDs). In [52] we have developed a technique that can tackle cross-language *and* cross-domain sentiment classification at the same time (e.g., classifying reviews of books in Urdu when only training reviews of CDs in English are available). The technique, called *Distributional Correspondence Indexing* (DCI), leverages the "distributional hypothesis", i.e., the hypothesis that words with similar meanings tend to occur in the same contexts. DCI derives term representations in a vector space common to both languages/domains where each dimension reflects its distributional correspondence to a pivot, i.e., to a highly predictive term that behaves similarly across languages/domains. Experiments show that DCI obtains better performance than current state-of-the-art techniques for cross-lingual and/or cross-domain sentiment classification.

4.3 Sentiment Quantification

While sentiment classification is important, in [32] we argued that, in many cases of applicative interest (e.g., when analysing tweets or product reviews), the final goal is often not the classification of individual items, but the estimation of the percentage of items that belong to a certain class; in other words, in these cases we are interested not in sentiment classification, but in sentiment *quantification*.

Research has shown that quantification is best tackled by quantification-specific algorithms, and not by using standard classification algorithms followed by counting the number of items that have been assigned the class. In [37, 38] we conducted an extensive analysis of existing quantification algorithms as applied to analysing tweets by sentiment; the results confirmed that applying standard classification technology when quantification is the real goal, is suboptimal. Similar conclusions were reached when, instead of standard multi-class quantification, we tackled *ordinal* quantification [25], i.e., the task characterized by a set of classes on which a total order is defined.

5 Multimedia Analysis

Content-based Multimedia Information Retrieval (CBMIR) on a very large scale has been a very active multidisciplinary research field during the last 25 years. Multimedia retrieval involves topics ranging from similarity search, metric access methods and big data, to features extraction, deep learning and smart cameras. The explosion of multimedia data caused by the diffusion of mobile devices and social media, has increased the relevance of this topic for both industries and governments. We show that the combination of state-of-the-art data structures and deep neural networks allows multimedia analysis that have been considered unachievable for many years because of issues such as semantic gap and curse of dimensionality.

In 2016, a benchmark consisting of 97M deep features[1] extracted from the Yahoo Creative Commons 100M (YFCC100M) dataset was presented and two approximate similarity search techniques were tested on it [7]. In this Section we start from this

[1] http://www.deepfeatures.org.

recent result, to discuss the most relevant research results of the last 25th years that made this possible.

Starting from the CoPhIR dataset dating back to 2009, large datasets have been created using the multimedia shared by users on social media [15]. The proliferation of easily and quickly accessible social media data can be used by researchers for many different purposes. For example, such data has already proved useful for many scenarios such as that of emergency management [11], intelligence [3], eHealth [26], and social networks security. In recent years, we have observed the explosion of image-sharing services such as Flickr and Instagram. For instance, Instagram has 600 Million Monthly Users and it was estimated that about 85 million photos are shared everyday Since by sharing photos, users could also express opinions or sentiments, social media images provide a potentially rich source for understanding public opinions.

The features extracted from the images in [7] are the activations of an hidden layer of a Convolutional Neural Networks. This information, automatically extracted from pre-trained deep neural networks, has recently show outstanding results, rapidly becoming state of the art in many computer vision applications that have used global (e.g., MPEG-7 Visual Descriptors) and local features (e.g., SIFT, SURF, BRIEF) for decades. Moreover, deep learning is allowing tasks that were not even considered before. In [18], as an example, the authors presented a deep learning based method for searching in a visual feature space, by learning to translate a textual query into a visual representation allowing text searching in non-annotated (not even automatically) image datasets. Deep Learning is also substituting local features based techniques in smart cameras applications such as parking occupancy detection [8].

While Computer Vision is significantly contributing to make multimedia analysis more effective, the large and increasing amount of multimedia available through social media requires large scale and big data algorithms. Among several approaches to address the problem of efficient search in large archives of image features, one that is very promising is the use of inverted indices. Reference [5] introduced MI-File, an approach that allows using inverted files to perform similarity search with an arbitrary similarity function. In [4] a Surrogate Text Representation (STR) derived from the MI-File has been proposed. The conversion of the permutations in a textual form allows using off-the-shelf text search engines for similarity search. Another solution that exploits a text retrieval engine to perform image similarity search, introduced in [6], uses a straightforward quantization of the vector components of the DCNN features.

6 Social Mining and Ethics

In a world more and more connected, we are witnessing an incredible growth in the generation and sharing of data originating from the digital breadcrumbs of human activities and sensed as a by-product of the ICT systems that we use everyday. Thanks to the massive availability of this data, human behavior can be observed at large scale.

New powerful data-driven tools may be designed and developed to exploit this data for improving the world in many different ways. We can use GPS/GSM data to observe and measure the behavior of a population, to build better cities tailored to the movement of the population, with lower commuting times and lower pollution. We can exploit medical data to build classifiers able to help in diagnosing and curing diseases. We can use industrial data to improve the production processes, and create smarter and more secure factories. We can do a lot of other incredible and useful things with the support of data and analytical tools able to extract useful knowledge from raw data.

These data describing human activities are at the heart of the idea of a *knowledge society*, where the understanding of social phenomena is sustained by the knowledge extracted from the miners of big data across the various social dimensions by using social mining technologies. However, the opportunities of discovering interesting patterns from human data can be outweighed due to the high risks of ethical issues in data processing and analysis and ethical consequences of their suggestions and predictions. Important ethical risks are: (i) *privacy violations*, when uncontrolled intrusion into the personal data of the subjects occurs, and (ii) *discrimination*, when the discovered knowledge is unfairly used in making discriminatory decisions about the (possibly unaware) people who are classified, or profiled.

In the literature some works have shown that data analytics and ethics are not necessary enemies: practical and impactful *data-driven and knowledge-based* services can be designed obtaining data and service quality while enforcing ethical requirements. The key factor is to develop data analytics technologies that *by-design* enforce ethical value requirements to provide safeguards of fairness. This vision is fully compliant with the European General Data Protection Regulation which will be applied on 25 May 2018 and that especially encourages the application of the *privacy by-design* principle.

In the context of privacy protection in big data analytics, Monreale et al. [51] propose the instantiation of the *privacy-by-design* paradigm [20], introduced by Ann Cavoukian, in the 1990s, to the designing of big data analytical services. This methodology was applied to guarantee privacy in the following fields.

Privacy in Data Mining Outsourcing. Giannotti et al. in [41] propose a method for the outsourcing of the association rule mining task while ensuring privacy protection. The results show how an organization can outsource transactional data to an untrusted third party, such as a cloud provider, and obtain a data mining service in a privacy-preserving manner. In this particular scenario, not only the underlying data but also the mined results are not intended for sharing and must remain private because they are considered valuable strategic information. The proposed schema, before sending the transactional data to the third party, applies an encryption based on the addition of fake transactions to the original data in such a way that each item (itemset) becomes indistinguishable with at least $(k - 1)$ other items (itemsets). This framework guarantees that not only individual items, but also any group of items, have the property of being indistinguishable from at least k other groups in the worst case, and actually many more in the average case. The consequence is that a possible attack has a very limited probability of success in guessing actual items contained either in the transaction data or in the mining results. However, the data owner any time queries the third party can efficiently decrypt correct mining results.

Privacy in Mobility Data Publishing. Monreale et al. [49] present a framework offering an instance of the privacy by-design paradigm concerning personal mobility trajectories, obtained from GPS devices or cell phones. The designed method enforces privacy protection while enabling clustering analysis useful for understanding human mobility behavior in specific urban areas. The released movement data are made anonymous by a process that applies a data-driven spatial generalization of the trajectories. This data-driven approach allows to generalize more areas with high traffic density with respect to urban areas with lower level of traffic. The results obtained with the application of this frame- work show how trajectories can be anonymized to a high level of protection against re-identification while preserving the possibility of mining clusters of trajectories, which enables novel powerful analytic services for info-mobility or location-based services.

Privacy in Distributed Analytical Systems. Monreale et al. in [50] apply the privacy-by-design methodology also in a distributed setting where an untrusted central station is able to collect some aggregate statistics computed by each individual node that observes a stream of mobility data. The central station stores the received statistical information and computes a summary of the traffic conditions of the whole territory, based on the information collected from data collectors. The proposed methodology guarantees for each node of the system privacy protection at individual level by applying a data transformation based on the well-known differential privacy model [27].

In the context of discrimination data analysis, two main lines of research are being pursued (see [59] for a survey).

Discrimination discovery from data consists in the actual discovery of an unjustified difference in treatment of individuals in a large amount of historical decision records. A process for direct and indirect discrimination discovery on social groups using classification rule mining and filtering was originally proposed [66]. The process is guided by legally grounded measures of discrimination, possibly including statistical tests of confidence. An alternative view of "discovery as attack" is investigated in [67], in which attack strategies of privacy models are used to unveil discrimination hidden behind redlining practices. Discrimination against individuals has been instead modeled with a k-NN approach, following the legal methodology of situation testing, and applied to a real case study in research project funding [65].

Discrimination prevention consists of removing bias in the machine learning and data mining process. Bias can be present in the training data and in the learning algorithm. Data sanitization for discrimination prevention has been investigated in [64], by first reducing the t-closeness model of privacy to a model for non-discrimination, and then adapting state-of-the- art data sanitization methods for t-closeness. An approach dealing with both privacy and discrimination sanitization is in [43]. Regarding learning algorithms, a modified voting mechanism of rule-based classifiers in order to reduce the weight of possibly discriminatory rules has been proposed [55].

7 Data Infrastructure and Virtual Research Environments as Data Science Enablers

Data Infrastructures [28] open new opportunities to data mining and machine learning activities by facilitating faster and cheaper investigations and enabling a more rapid expansion of their volume. In fact, they are conceived to realise large scale software ecosystems suitable for the big data challenges including analytics [44]. They offer the entire spectrum of resources (data, software, methods, services, computing) needed to carry out a certain investigation "as-a-Service" thus relieving researchers to operate and maintain them. Moreover, they are progressively introducing mechanisms that limit as much as possible the exposure of the researcher to technicalities and challenges related with access to the necessary distributed and heterogeneous set of resources.

Novel data infrastructures support the entire data processing chain, from the collection and preparation of the necessary datasets, to the analytics steps till the publication of the produced outcomes. Along this chain, unifying and open capabilities are provided thus to make it possible, for example, to access uniformly different datasets or to simply plug-in new tools/methods and data whenever needed. The resource space offered can also be made available to researchers through tailored *views*, i.e. web-based working environments known as *Virtual Research Environments* [16], where (*a*) researchers can focus on a specific investigation by having at their fingerprint what is needed; (*b*) what researchers produce is equipped with rich enough metadata thus to become a new resource compliant with Open Science [34] practices; (*c*) researchers are also provided with state-of-the-art facilities promoting collaboration and cooperation.

Driven by this rationale a software system named *gCube* [10] has been designed and developed. This technology is enacting the *D4Science Infrastructure* [17] and exploited to create and operate more than 70 diverse VREs. Overall, these VREs are serving more than 3100 (returning) scientists in 44 countries across a rich array of diverse communities including the KDD community via the SoBigData RI.

Along the years, gCube has been progressively endowed with (*a*) a rich array of *mediators* for interfacing with existing *systems* and their enabling technologies including distributed computing infrastructures (e.g. EGI [19]) and data providers (e.g. by relying on standards like OAI-PMH, SDMX, OGC W*S) as well as for making it possible for *third-party* service providers to easily exploit gCube facilities (e.g. OAuth, OGC W*S, REST APIs), (*b*) a set of basic services including a shared workspace where the objects used and resulting from VRE activity (beyond simple files) can be stored, organised and accessed as if they were in a "standard" file-manager; a social networking area where the member of each VRE can have discussions, share news and other material of interest, rate each item of a discussion, classify the discussion items by hashtags, refer to people or groups thus to call for actions from them, etc.; a user management area where authorized people are allowed to manage VRE membership, to create groups, assign members to groups, assign roles to member, invite new members, etc.; an open, customizable and extensible set of facilities made available for the needs of the specific community context. These include a project management and issue-tracking system with a wiki, a rich and extensible data analytics platform [10, 21], a flexible "products" catalogue where any (research) artefact produced in the VRE and worth being "published" can be easily made available by equipping it with rich metadata including license and provenance-related ones, a rich array of domain data management facilities.

VREs are created by using a wizard-based approach where a VRE designer is simply requested to select (among the available ones) the facilities and resources he/she is willing to have in the VRE, and then upon approval the VRE is automatically provisioned and made available by a web-based portal.

Overall, the data analytics platform resulting from gCube is characterised by the following key principles:

- *Extensibility*: the platform is "open" with respect to (*i*) the analytics techniques and methods it offers and supports and (*ii*) the computing infrastructures and solutions it relies on to enact the processing tasks. It is based on a plug-in architecture to support adding new algorithms and methods as well as new computing platforms;
- *Distributed processing*: the platform is conceived to execute processing tasks by relying on "local engines"/"workers" that can be deployed in multiple instances and execute tasks in parallel and seamlessly. The platform is able to rely on computing resources offered by both well-known e-Infrastructures (e.g. EGI) as well as resources made available by the Research Infrastructures or communities to deploy instances of the "local engines"/"workers". This is key to make it possible to "move" the computation close to the data;
- *Multiple interfaces*: the platform offers its services via both a (web-based) graphical user interface and a (web-based) programmatic interface thus to enlarge the possible application contexts. For instance, having a proper API facilitates the development of components capable to execute processing tasks from well-known applications (e.g. R, KNIME);
- *Cater for scientific workflows* [46]: the platform is both exploitable by existing WFMS (e.g. a node of a workflow can be the execution of a task/method offered by the platform) and support the execution of a workflow specification (e.g. by relying on one or more instances of WFMSs);
- *Easy to use*: the platform should is easy to use for both (*a*) algorithms/methods providers, i.e. scientists and practitioners called to realise processing methods of interest for the specific community, and (*b*) algorithms/methods users, i.e. scientists and practitioners called to exploit existing methods to analyse certain datasets;
- *Open science friendly*: the platform is transparently injecting open science practices in the processing tasks executed through it. This includes mechanisms for capturing and producing "provenance records" out of any computing task, mechanisms aiming at producing "research objects" so as to make it possible for others to repeat the task and reproduce the experiment.

These key principles make this analytics platform suitable for the challenges KDD community is facing.

8 Conclusions

Twenty-five years ago, most statisticians and computer scientists looked with skepticism at the novel community of KDD scientists, trying to reformulate the analytical process as data driven discovery. Indeed, such visionary endeavor, combined with the advent of big data and spectacular advances in high performance computing, has brought what we call today *data science*: a disruptive paradigm shift impacting all disciplines that pushes towards novel scientific methods where "top down" modelling of phenomena coexists with "bottom up" discoveries from data.

Data abundance combined with powerful data science techniques has the potential to dramatically improve our lives by enabling new services and products, while improving their efficiency and quality.

Many of today's scientific discoveries are already fueled by developments in statistics, data mining, machine learning, network science, databases, and visualization, and we can expect advances in any field related to the comprehension of complex phenomena as in medicine and health (network/personalized medicine), manufacturing (industry 4.0), social dynamics, urban planning, sustainable development.

The importance of data science is widely acknowledged, but there are also great concerns about the irresponsible use of data and models. Automated data driven decisions may be unfair or non-transparent. Confidential data may be shared unintentionally or abused by third parties. Each step in the data science pipeline (from raw data to conclusions) may create inaccuracies, e.g., if the data used to learn a model reflects existing social biases, the algorithm is likely to incorporate these biases. The ethics of data science is a challenging research topic where computer scientists play a central role, we contributed to change society and we cannot escape the responsibility to understand the impact of the digital transformation helping in catching the opportunities mitigating the risks.

Finally, there is an urgent need to start harnessing these opportunities for scientific advancement and for the social good, compared to the currently prevalent exploitation of big data for commercial purposes (e.g. user profiling and behavioral advertising) or, worse, social control and surveillance. The main obstacle to this accomplishment, besides the scarcity of data scientists, is the lack of a large-scale open ecosystem where big data and social mining research can be carried out.

This is why we propose to establish SoBigData, the Social Mining & Big Data Ecosystem: a research infrastructure (RI) providing an integrated ecosystem for ethic-sensitive scientific discoveries and advanced applications of social data mining on the various dimensions of social life, as recorded by "big data". The research community will use the SoBigData RI facilities as a "secure digital wind-tunnel" for large-scale social data analysis and simulation experiments. SoBigData will serve the wide cross-disciplinary community of data scientists, i.e., researchers studying all aspects of societal complexity from a data- and model-driven perspective, including data and text miners, computer scientists, socio-economic scientists, network scientists, political scientists, humanities researchers, and more.

References

1. R. Agrawal, T. Imieliński, A. Swami, Mining association rules between sets of items in large databases, in *Acm Sigmod Record*, vol. 22 (ACM, 1993), pp. 207–216
2. R. Agrawal, R. Srikant, Algorithms for mining association rules in large databases, in *Proceedings of the 20th VLDB Conference*, vol. 2 (1994), pp. 141–182
3. C. Aliprandi, A.E. De Luca, G. Di Pietro, M. Raffaelli, D. Gazzè, M.N. La Polla, A. Marchetti, M. Tesconi, Caper: crawling and analysing facebook for intelligence purposes, in *2014 IEEE/ACM International Conference on Advances in Social Networks Analysis and Mining (ASONAM)* (IEEE, 2014), pp. 665–669
4. G. Amato, P. Bolettieri, F. Falchi, C. Gennaro, F. Rabitti, Combining local and global visual feature similarity using a text search engine, in *International Workshop on Content-Based Multimedia Indexing (CBMI)* (IEEE, 2011), pp. 49–54
5. G. Amato, C. Gennaro, P. Savino, Mi-file: using inverted files for scalable approximate similarity search. Multimed. Tools Appl. **71**(3), 1333–1362 (2014)
6. G. Amato, F. Debole, F. Falchi, C. Gennaro, F. Rabitti, Large scale indexing and searching deep convolutional neural network features, in *International Conference on Big Data Analytics and Knowledge Discovery* (Springer, Berlin, 2016), pp. 213–224
7. G. Amato, F. Falchi, C. Gennaro, F. Rabitti, YFCC100M-HNfc6: a large-scale deep features benchmark for similarity search, in *International Conference on Similarity Search and Applications* (Springer, Berlin, 2016), pp. 196–209
8. G. Amato, F. Carrara, F. Falchi, C. Gennaro, C. Meghini, C. Vairo, Deep learning for decentralized parking lot occupancy detection. Exp. Syst. Appl. **72**, 327–334 (2017)
9. G. Andrienko, N. Andrienko, S. Rinzivillo, M. Nanni, D. Pedreschi, F. Giannotti, *Interactive Visual Clustering of Large Collections of Trajectories. VAST: Symposium on Visual Analytics Science and Technology* (2009)
10. M. Assante, L. Candela, D. Castelli, G. Coro, L. Lelii, P. Pagano, Virtual research environments as-a-service by gCube. PeerJ Preprints (2016)
11. M. Avvenuti, S. Cresci, F. Del Vigna, M. Tesconi, Impromptu crisis mapping to prioritize emergency response. Computer **49**(5), 28–37 (2016)
12. S. Baccianella, A. Esuli, F. Sebastiani, Sentiwordnet 3.0: an enhanced lexical resource for sentiment analysis and opinion mining, in *Proceedings of the 7th Conference on Language Resources and Evaluation (LREC 2010)* (2010)
13. A.L. Barabási, R. Albert, Emergence of scaling in random networks. Science **286**(5439), 509–512 (1999)
14. M. Berlingerio, M. Coscia, F. Giannotti, A. Monreale, D. Pedreschi, Multidimensional networks: foundations of structural analysis. World Wide Web **16**(5–6), 567–593 (2013)
15. P. Bolettieri, A. Esuli, F. Falchi, C. Lucchese, R. Perego, T. Piccioli, F. Rabitti, CoPhIR: a test collection for content-based image retrieval (2009), arXiv:0905.4627
16. L. Candela, D. Castelli, P. Pagano, Virtual research environments: an overview and a research agenda. Data Sci. J. **12**, GRDI75–GRDI81 (2013)
17. L. Candela, D. Castelli, A. Manzi, P. Pagano, Realising virtual research environments by hybrid data infrastructures: the D4 science experience, in *International Symposium on Grids and Clouds (ISGC) 2014 23–28 March 2014, Academia Sinica, Taipei, Taiwan, PoS(ISGC2014)022. Proceedings of Science* (2014)
18. F. Carrara, A. Esuli, T. Fagni, F. Falchi, A.M. Fernández, Picture it in your mind: generating high level visual representations from textual descriptions (2016), arXiv:1606.07287
19. E. Fernández-del Castillo, D. Scardaci, Á.L. García, The EGI federated cloud e-infrastructure, in *Procedia Computer Science - 1st International Conference on Cloud Forward: From Distributed to Complete Computing*, vol. 68 (2015)
20. A. Cavoukian, Privacy design principles for an integrated justice system - working paper (2000), https://www.ipc.on.ca/index.asp?layid=86&fid1=318

21. G. Coro, L. Candela, P. Pagano, A. Italiano, L. Liccardo, Parallelizing the execution of native data mining algorithms for computational biology. Concurr. Comput.: Pract. Exp. **27**(17), 4630–4644 (2015)
22. M. Coscia, F. Giannotti, D. Pedreschi, A classification for community discovery methods in complex networks. Stat. Anal. Data Min. **4**(5), 512–546 (2011)
23. M. Coscia, S. Rinzivillo, F. Giannotti, D. Pedreschi, Optimal spatial resolution for the analysis of human mobility, in *Proceedings of the International Conference on Advances in Social Networks Analysis and Mining (ASONAM)* (IEEE, 2012), pp. 248–252
24. M. Coscia, G. Rossetti, F. Giannotti, D. Pedreschi, Demon: a local-first discovery method for overlapping communities, in *Proceedings of SIGKDD International Conference on Knowledge Discovery and Data Mining* (ACM, 2012), pp. 615–623
25. G. Da San Martino, W. Gao, F. Sebastiani, Ordinal text quantification, in *Proceedings of the 39th ACM Conference on Research and Development in Information Retrieval (SIGIR 2016)* (2016), pp. 937–940
26. F. Del Vigna, M. Petrocchi, A. Tommasi, C. Zavattari, M. Tesconi, Semi-supervised knowledge extraction for detection of drugs and their effects, in *International Conference on Social Informatics* (Springer, Berlin, 2016), pp. 494–509
27. C. Dwork, Differential privacy, in *Automata, Languages and Programming*, ed. by M. Bugliesi, B. Preneel, V. Sassone, I. Wegener. Lecture Notes in Computer Science, vol. 4052 (Springer, Berlin, 2006), pp. 1–12. doi:10.1007/11787006_1
28. P.N. Edwards, S.J. Jackson, G.C. Bowker, C.P. Knobel, Understanding infrastructure: dynamics, tensions, and design. Working paper, National Science Foundation (2007), http://hdl.handle.net/2027.42/49353
29. A. Esuli, F. Sebastiani, Determining term subjectivity and term orientation for opinion mining, in *Proceedings of the 11th Conference of the European Chapter of the Association for Computational Linguistics (EACL)*, pp. 193–200
30. A. Esuli, F. Sebastiani, Determining the semantic orientation of terms through gloss analysis, in *Proceedings of the 14th ACM International Conference on Information and Knowledge Management (CIKM 2005)* (2005), pp. 617–624
31. A. Esuli, F. Sebastiani, Sentiwordnet: a publicly available lexical resource for opinion mining, in *Proceedings of the Conference on Language Resources and Evaluation (LREC)* (2006), pp. 417–422
32. A. Esuli, F. Sebastiani, Sentiment quantification. IEEE Intell. Syst. **25**(4), 72–75 (2010)
33. U.M. Fayyad, G. Piatetsky-Shapiro, P. Smyth, R. Uthurusamy, *Advances in Knowledge Discovery and Data Mining*, vol. 21 (AAAI Press, Menlo Park, 1996)
34. B. Fecher, S. Friesike, Open science: one term, five schools of thought, in *Opening Science*, ed. by S. Bartling, S. Friesike (Springer, Berlin, 2014), pp. 17–47
35. B. Furletti, L. Gabrielli, C. Renso, S. Rinzivillo, Analysis of GSM calls data for understanding user mobility behavior (2013)
36. L. Gabrielli, B. Furletti, R. Trasarti, F. Giannotti, D. Pedreschi, City users' classification with mobile phone data, in *IEEE Big Data* (2015)
37. W. Gao, F. Sebastiani, Tweet sentiment: from classification to quantification, in *Proceedings of the 7th International Conference on Advances in Social Network Analysis and Mining (ASONAM 2015)* (Paris, FR, 2015), pp. 97–104
38. W. Gao, F. Sebastiani, From classification to quantification in tweet sentiment analysis. Soc. Netw. Anal. Min. **6**(19), 1–22 (2016)
39. F. Giannotti, M. Nanni, F. Pinelli, D. Pedreschi, Trajectory pattern mining, in *Proceedings of the 13th ACM SIGKDD International Conference on Knowledge Discovery and Data Mining* (KDD, ACM, 2007), pp. 330–339
40. F. Giannotti, M. Nanni, D. Pedreschi, F. Pinelli, C. Renso, S. Rinzivillo, R. Trasarti, Unveiling the complexity of human mobility by querying and mining massive trajectory data. VLDB J. **20**(5), 695–719 (2011)
41. F. Giannotti, L.V.S. Lakshmanan, A. Monreale, D. Pedreschi, W.H. Wang, Privacy-preserving mining of association rules from outsourced transaction databases. IEEE Syst. J. **7**(3), 385–395 (2013)

42. R. Guidotti, M. Nanni, S. Rinzivillo, D. Pedreschi, F. Giannotti, Never drive alone: boosting carpooling with network analysis. Inf. Syst. **64**, 237–257 (2016)
43. S. Hajian, J. Domingo-Ferrer, A. Monreale, D. Pedreschi, F. Giannotti, Discrimination- and privacy-aware patterns. Data Min. Knowl. Discov. **29**(6), 1733–1782 (2015)
44. S. Khalifa, Y. Elshater, K. Sundaravarathan, A. Bhat, P. Martin, F. Imam, D. Rope, M. Mcroberts, C. Statchuk, The six pillars for building big data analytics ecosystems. ACM Comput. Surv. **49**(2), 33 (2016)
45. J.G. Lee, J. Han, Trajectory clustering: a partition-and-group framework, in *In SIGMOD* (2007), pp. 593–604
46. C.S. Liew, M.P. Atkinson, M. Galea, T.F. Ang, P. Martin, J.I.V. Hemert, Scientific workflows: moving across paradigms. ACM Comput. Surv. **49**(4) 66 (2016)
47. L. Milli, A. Monreale, G. Rossetti, D. Pedreschi, F. Giannotti, F. Sebastiani, Quantification in social networks, in *2015 IEEE International Conference on Data Science and Advanced Analytics (DSAA)*, vol. 36678 (IEEE, 2015), pp. 1–10
48. A. Monreale, F. Pinelli, R. Trasarti, F. Giannotti, Wherenext: a location predictor on trajectory pattern mining, in *ACM SIGKDD Conference on Knoledge Discovery and Data Mining (KDD)* (2009)
49. A. Monreale, G.L. Andrienko, N.V. Andrienko, F. Giannotti, D. Pedreschi, S. Rinzivillo, S. Wrobel, Movement data anonymity through generalization. TDP **3**(2), 91–121 (2010)
50. A. Monreale, W.H. Wang, F. Pratesi, S. Rinzivillo, D. Pedreschi, G. Andrienko, N. Andrienko, Privacy-preserving distributed movement data aggregation, in *AGILE* (Springer, Berlin, 2013)
51. A. Monreale, S. Rinzivillo, F. Pratesi, F. Giannotti, D. Pedreschi, Privacy-by-design in big data analytics and social mining. EPJ Data Sci. **3**(1), 10 (2014). doi:10.1140/epjds/s13688-014-0010-4
52. A. Moreo Fernández, A. Esuli, F. Sebastiani, Distributional correspondence indexing for cross-lingual and cross-domain sentiment classification. J. Artif. Intell. Res. **55**, 131–163 (2016)
53. L. Pappalardo, G. Rossetti, D. Pedreschi, "How well do we know each other?" detecting tie strength in multidimensional social networks, in *2012 IEEE/ACM International Conference on Advances in Social Networks Analysis and Mining (ASONAM)* (IEEE, 2012), pp. 1040–1045
54. L. Pappalardo, F. Simini, S. Rinzivillo, D. Pedreschi, F. Giannotti, A.L. Barabasi, Returners and explorers dichotomy in human mobility. Nat. Commun. **6**, 8166 (2015). doi:10.1038/ncomms9166
55. D. Pedreschi, S. Ruggieri, F. Turini, Measuring discrimination in socially-sensitive decision records, in *Proceedings of the SIAM International Conference on Data Mining (SDM 2009)* (SIAM, 2009), pp. 581–592
56. J.R. Quinlan, *C4. 5: Programs for Machine Learning* (Elsevier, San Francisco, 2014)
57. S. Rinzivillo, S. Mainardi, F. Pezzoni, M. Coscia, D. Pedreschi, F. Giannotti, Discovering the geographical borders of human mobility. KI-Künstl. Intell. **26**(3), 253–260 (2012)
58. S. Rinzivillo, L. Gabrielli, M. Nanni, L. Pappalardo, D. Pedreschi, F. Giannotti, The purpose of motion: learning activities from individual mobility networks, in *International Conference on Data Science and Advanced Analytics, DSAA* (2014). doi:10.1109/DSAA.2014.7058090
59. A. Romei, S. Ruggieri, A multidisciplinary survey on discrimination analysis. Knowl. Eng. Rev. **29**(5), 582–638 (2014)
60. G. Rossetti, M. Berlingerio, F. Giannotti, Scalable link prediction on multidimensional networks, in *International Conference on Data Mining Workshops (ICDMW)* (IEEE, 2011), pp. 979–986
61. G. Rossetti, R. Guidotti, I. Miliou, D. Pedreschi, F. Giannotti, A supervised approach for intra-/inter-community interaction prediction in dynamic social networks. Soc. Netw. Anal. Min. **6**, 86 (2016)
62. G. Rossetti, L. Pappalardo, R. Kikas, D. Pedreschi, F. Giannotti, M. Dumas, Homophilic network decomposition: a community-centric analysis of online social services. Soc. Netw. Anal. Min. J. **6**, 103 (2016)
63. G. Rossetti, L. Pappalardo, D. Pedreschi, F. Giannotti, Tiles: an online algorithm for community discovery in dynamic social networks, in *Machine Learning* (2016), pp. 1–29

64. S. Ruggieri, Using t-closeness anonymity to control for non-discrimination. Trans. Data Priv. **7**(2), 99–129 (2014)
65. S. Ruggieri, F. Turini, A KDD process for discrimination discovery, in *Proceedings of Machine Learning and Knowledge Discovery in Databases (ECML-PKDD 2016) Part III*. LNCS, vol. 9853 (Springer, Berlin, 2016), pp. 249–253
66. S. Ruggieri, D. Pedreschi, F. Turini, Data mining for discrimination discovery. ACM Trans. Knowl. Discov. Data **4**(2), Article 9 (2010)
67. S. Ruggieri, S. Hajian, F. Kamiran, X. Zhang, Anti-discrimination analysis using privacy attack strategies, in *Proceedings of Machine Learning and Knowledge Discovery in Databases (ECML-PKDD) Part II*. LNCS, vol. 8725 (2014), pp. 694–710
68. R. Trasarti, F. Pinelli, M. Nanni, F. Giannotti, Mining mobility user profiles for car pooling, in *Proceedings of the 17th ACM SIGKDD International Conference on Knowledge Discovery and Data Mining* (KDD '11, ACM, New York, 2011), pp. 1190–1198
69. R. Trasarti, R. Guidotti, A. Monreale, F. Giannotti, Myway: location prediction via mobility profiling, in *Information Systems* (2015)

A Tour from Regularities to Exceptions

Fabrizio Angiulli, Fabio Fassetti, Luigi Palopoli and Domenico Ursino

1 Overview

The enormous growth of information available in database systems has led to a significant development of techniques for knowledge discovery. At the heart of the knowledge discovery process is the application of data mining algorithms in charge of extracting hidden relationships among pieces of stored information. Information thus extracted from databases have widespread use in great many application domains and contexts.

In this paper, we focus on traditional, structured databases as data sources. In this realm information is available at two different levels, that of the database instance (known as *extensional level*) and that of the database schema (often called *intensional level*). For both of them, it is interesting to devise methods that automatically extract useful information to be used, for instance, in re-engineering applications. Also, it is sometimes useful to search databases for hidden regularities or, viceversa, to look for exceptional data bunches. These two dichotomies sketch, in a nutshell, the structure of this paper in which we summarize our long-term contributions in the area of data mining and data integration.

F. Angiulli (✉) · F. Fassetti · L. Palopoli
DIMES, University of Calabria, Calabria, Italy
e-mail: fabrizio.angiulli@unical.it

F. Fassetti
e-mail: fabio.fassetti@unical.it

L. Palopoli
e-mail: luigi.palopoli@unical.it

D. Ursino
DICEAM, University "Mediterranea" of Reggio Calabria, Calabria, Italy
e-mail: ursino@unirc.it

S. Flesca et al. (eds.), *A Comprehensive Guide Through the Italian Database Research Over the Last 25 Years*, Studies in Big Data 31, DOI 10.1007/978-3-319-61893-7_18

To illustrate, in the first part of this paper, we deal with discovering regularities and exceptions in data. Searching for regularities in data instances is, actually, at the heart of many data mining tasks, including classification, regression, clustering and rule discovery. This latter is the specific subject we shall focus on first, by considering two forms of it, namely, associative rules induction and metaquerying. After that, we shall discuss the problem of looking for exceptions, aka outliers in database instances. In this respect, it must be noted that searching regularities and exceptions does not complement one another, and specific and often unrelated techniques are needed in order to look for them. We shall discuss about outlier detection in ordinary data instances and in data streams and about the related problems of constructing explanations justifying for data bunches to be considered as exceptional.

Finally, we tackle the context of extracting regularities and exceptions in data schemas and we illustrate their exploitation to the problem of data integration and, ultimately, of information system cooperation.

2 Extensional Data Mining

2.1 Dependency Discovery

This task aims at revealing types of correlations holding among pieces of information stored in a database, which can be somehow expressed in the form of logical-like rules. For instance, with boolean association rules, one describes co-occurrences of items within databases of itemsets.

Not all rules induced from a database are, however, interesting: this will be the case only if it describes a relationship that is "mostly valid". To state such a validity, *indices* are used, that is, functions with values usually in [0, 1]. Support and confidence are classical indices employed in the data mining field. Intuitively, when a rule scores a high *support*, it is worth to further consider it, since there exist a significant fraction of the database tuples that satisfy the conjunction of the atoms in the rule. *Confidence* shows to what extent a given rule is true within the database at hand. Just for the way of example, a *confidence* value of 0.7, associated to the boolean association rule of the form $I_1, I_2, \ldots, I_n \rightarrow I$, tells that 70 percent of the stored itemsets where $I_1, I_2, \ldots, I_n$ are true (that is, occur) also have I true.

2.1.1 Association Rules Induction

Above we have made reference to the simplest form of an association rule, that is, the boolean one. In several application domains, though, they are not enough to encode interesting data dependencies.

Quantitative association rules use conditions of the form: (i) $A = c$; (ii) $A \neq c$; (iii) $A \in [l, u]$; (iv) $A \notin [l, u]$; where A is an attribute, l and u are values and $[l, u]$

denotes the set of numbers x s.t. $l \leq x \leq u$. In either forms, inducing association rules is a quite widely used data mining technique. Despite their widespread utilization, a thorough analysis of the complexity of the associated tasks was not been developed when we begun working on the subject.

We define a form of association rules that generalizes over the quantitative, categorical and the boolean attributes. We allow the null values (indicated by ϵ) to occur in the database, denoting the absence of information, for which it is forbidden to specify conditions. We capture [13, 15] the formal framework of boolean rules by calling *boolean* a database defined on a set of attributes taking value over $\{c, \epsilon\}$, where c is an arbitrary constant. In this setting, an association rule $(B_1 = c) \wedge \ldots \wedge (B_p = c) \Rightarrow (H_1 = c) \wedge \ldots \wedge (H_q = c)$ will encode the boolean association rule $B_1, \ldots, B_p \Rightarrow H_1, \ldots, H_q$.

Given a database T, support and confidence for association rule $B \Rightarrow H$ are:

- the *support* of $B \Rightarrow H$ in T, written $sup(B \Rightarrow H, T)$, is $|T_{B \wedge H}|/|T|$, and
- the *confidence* of $B \Rightarrow H$ in T, written $cnf(B \Rightarrow H, T)$, is $|T_{B \wedge H}|/|T_B|$,

where T_C denotes the set of tuples of T which satisfy the condition C. In our investigation, we considered two additional indexes, namely *gain* and *laplace*, that we shall not discuss further here.

Association rule induction problem: Let I be a set of attributes, let T be a database on I, let k, $1 \leq k \leq |I|$, be a natural number, and let s, $0 < s \leq 1$, be a rational number. Furthermore, let ρ be an index. The association rule induction problem $\langle I, T, \rho, k, s \rangle$ is as follows: Is there a non-trivial association rule R such that $|R| \geq k$ and $\rho(R, T) \geq s$?

Complexity results are summarized in Table 1. Before commenting on those, we recall that problems belonging to classes as AC^0, TC^0, L, and LOGCFL are very efficiently parallelizable (indeed $AC^0 \subseteq TC^0 \subseteq NC_1 \subseteq L \subseteq LOGCFL \subseteq NC_2 \subseteq P$; the reader is referred to [24] for basics on complexity classes), so that the algorithm design effort could be addressed accordingly:

Table 1 Complexity results for the *Association rule induction problem*

	Index	Database type	Constraints	Complexity
1	$sup, gain, laplace$	No nulls	No	NP-complete
2	cnf	No nulls	No	TC^0
3	All	With nulls	No	NP-complete
4	All	Sparse	No	L
5	All	Any	$\|I\|$ fixed	L
6	sup	With nulls	s fixed	NP-complete
7	sup	Boolean	$s\|T\|$ or k fixed	TC^0
8	sup	Boolean	$s\|T\|$ and k fixed	AC^0_2

- All the problems are NP-complete in the presence of null values (row 3 of Table 1); therefore, dealing with nulls makes "per se" the task of rule induction very demanding;
- The problem $\langle I, T, cnf, k, s\rangle$ becomes tractable when databases without nulls are considered (row 2), while the other problems remain intractable (row 1); this means that generating rules with high-confidence is easier than the case of other indices;
- In the presence of a bound on the length of the tuples (row 4) or of a bound on the number of attributes (row 5), then all the problems become highly parallelizable; this result can be understood if one considers that, when such bounds are imposed, the number of candidate rules is polynomial and different rules can be generated independently from one another.

2.1.2 Metaquerying

Metaquerying is a task aiming at extracting first order logical rules from a relational and a deductive database. This is a semi-supervised task in that the metaquery serve as the description of a class of patterns that the user is willing to discover. Differently from canonic association rules induction, patterns discovered using metaqueries can link information from several tables in databases. To illustrate, a metaquery **MQ** has the form $T \leftarrow L_1, \ldots, L_m$ where T and L_i are literal schemes $Q(Y_1, \ldots, Y_n)$, and Q is either an ordinary predicate name or a predicate variable. In the latter case, $Q(Y_1, \ldots, Y_n)$ can be instantiated to an atom with a predicate name denoting a relation in the database.

Answering a metaquery **MQ** on a database instance **DB** amounts to finding all substitutions σ, also called *instantiations*, of relational patterns appearing in **MQ** by atoms having as predicate names relations in **DB**, such that the Horn rule σ(**MQ**) holds in **DB** with a certain degree of plausibility, defined in terms of indices such as *support* and *confidence*.

We worked in the context outlined above with the two-sided aim of defining a clear and well-defined semantics and of studying the complexity of the underlying computational problems [11, 12, 14].

Metaquery semantics: *Let* **MQ** *be a metaquery,* **DB** *a database, and σ an instantiation. According to the type of the instantiation (namely, 0, 1, or 2), for any relational pattern L and atom A, $\sigma(L) = A$ implies the following:*

- ***type–0****: L and A have the same list of arguments;*
- ***type–1****: the arguments of A are obtained from those of L by permutation.*
- ***type–2****: the arguments of A are a superset of those of L, and the ones not appearing in L do not occur elsewhere in the instantiated rule.*

Metaquery induction problem: *Let $T \in \{0, 1, 2\}$ be an instantiation type and let I be a plausibility index. Let* **DB** *denote a database instance,* **MQ** *a metaquery, and k a threshold, $0 \leq k < 1$. Then, the combined (data, resp.) complexity of* $\langle$**DB**, **MQ**, $I, k, T\rangle$ *is the complexity, measured in the size of* **DB**, **MQ** *and k* (**DB**,

Table 2 Complexity results for the *Metaquery induction problem*

	Complexity measure	Problem type	Instantiation type	Indices	Threshold	Complexity
1	Comb. Compl.	General	$0, 1, 2$	$\mathbf{I}$	$k = 0$	NP-complete
2	Comb. Compl.	General	$0, 1, 2$	cvr, sup	$0 \leq k < 1$	NP-complete
3	Comb. Compl.	General	$0, 1, 2$	cnf	$0 \leq k < 1$	$\mathrm{NP}^{\mathrm{PP}}$-complete
4	Comb. Compl.	Acyclic	0	$\mathbf{I}$	$k = 0$	LOGCFL-complete
5	Comb. Compl.	Acyclic	$1, 2$	$\mathbf{I}$	$k = 0$	NP-complete
6	Comb. Compl.	Acyclic	$1, 2$	cvr, sup	$0 \leq k < 1$	NP-complete
7	Comb. Compl.	Semi-acyclic	$0, 1, 2$	$\mathbf{I}$	$k = 0$	NP-complete
8	Data Compl.	General	$0, 1, 2$	$\mathbf{I}$	$k = 0$	AC_0
9	Data Compl.	General	$0, 1, 2$	$\mathbf{I}$	$0 \leq k < 1$	TC_0

resp.), of deciding if there exists a type-T instantiation σ *such that* $I(\sigma(\mathbf{MQ})) > k$, *where* **DB** *has variable schema (fixed schema, resp.).*

Complexity results are summarized in Table 2, where it can be read that:

- The combined complexity of: (1) $\langle \mathbf{DB}, \mathbf{MQ}, I, 0, T \rangle$ is NP-complete for any instantiation type T; (2) $\langle \mathbf{DB}, \mathbf{MQ}, I, k, T \rangle$ is NP-hard for any index I; (3) $\langle \mathbf{DB}, \mathbf{MQ}, I, k, T \rangle$ is in NP for $0 \leq k < 1$, any T, and $I \in \{cvr, sup\}$.
- Instantiating metaqueries complying with a given bound on the confidence value turns out to be more complex than for other indices. This is due to the need of computing the *exact* count of tuples satisfying the body of an instantiation, whereas for other indices this is not required. In fact, this problem is related to the $\#P$-complete problem #SAT, where the question concerns *counting* the exact number of solutions of a given boolean formula. The class $\#P$ employed as an oracle is equivalent to another class related to counting, namely PP. Specifically, the combined complexity of $\langle \mathbf{DB}, \mathbf{MQ}, cnf, k, T \rangle$ turns out to be $\mathrm{NP}^{\mathrm{PP}}$-complete, a class "close" to PSPACE. Interestingly enough, the above result states the first natural problem known for the complexity class $\mathrm{NP}^{\mathrm{PP}}$.
- In order to single out tractable cases in the context of the combined complexity analysis, we individuate the classes of *(semi–)acyclic* metaqueries on the basis of the acyclicity of the *(semi–)hypergraph* associated with a metaquery: nodes are associated to both predicate and ordinary (only ordinary, in the case of the semi-hypergraph) variables occurring in the metaquery, while edges ecompass all variables appearing together in a certain relational pattern. Interestingly, the combined complexity of $\langle \mathbf{DB}, \mathbf{MQ}, I, 0, 0 \rangle$ for acyclic metaqueries is LOGCFL-complete, hence, highly parallelizable. However, acyclicity is not sufficient to guarantee tractability in general.
- As for the data complexity, depending on the threshold k, its membership ranges from AC_0 $(k = 0)$ to TC_0 $(0 \leq k < 1)$, hence, highly parallelizable.

2.2 Mining Exceptions

Outlier mining approaches tackle the knowledge discovery problem from a perspective which is reversed with respect to that of regularity mining ones. Specifically, the goal of outlier detection techniques is to isolate a few individuals deemed as exceptional, also referred to as *outliers*.

While in many contexts outliers are considered as noise that must be eliminated, as pointed out elsewhere "one persons noise could be another persons signal", and thus outliers themselves can represent the knowledge of interest in many applications, as in medical analysis, intrusion detection, surveillance systems, data cleaning, to cite a few.

Actually, exception mining techniques can be grouped in two main categories, that are *outlier detection* and *outlier explanation*.

2.2.1 Outlier Detection

Outlier identification has its roots in statistics: "An outlier is an observation that deviates so much from other observations as to arouse suspicions that it was generated by a different mechanism". According to most statistical approaches, outliers are those points that satisfy a discordancy test, that is, that are significantly far from what would be their expected position given the hypothesized distribution [1].

Approaches to outlier detection pertaining to the data mining field can be classified in supervised, semi-supervised, and unsupervised.

Unsupervised outlier detection. Unsupervised methods search for outliers in an unlabelled data set by assigning to each object a score which reflects its degree of abnormality. Most of the unsupervised approaches proposed in the data mining literature can be classified as deviation-based, distance-based, density-based, isolation-based, and others [1].

Among the unsupervised outlier detection methods, distance-based outlier ones occupy a prominent position. Distance-based outlier detection has been introduced by [26] to overcome the limitations of model-based statistical methods, that are the methods requiring that the data fits an hypothesized distribution. According to the original definition, an object obj is a distance-based outlier in a dataset with respect to parameters k and R if less than k objects in the data set lie within distance R from obj. Subsequently, some variants of the basic distance-definition have been introduced in the literature that have become popular during the years. Specifically, in order to provide a ranking of the outliers and with the aim of taking into account the whole neighborhood of the objects, [9] proposed to rank objects on the basis of the sum of the distances $\omega_k(\cdot)$ (or, equivalently, the average distance) from their k nearest neighbors, a measure called *weight* and also referred to as aKNN score (for average KNN).

The first two algorithms for mining distance-based outliers in large data sets were presented in [26]. However, none of these methods scales well for both large and high-dimensional data, and this originates efforts for developing scalable algorithms that are scalable in both the size and the dimensionality of the data.

HilOut algorithm. Within this scenario [9, 10] proposed an algorithm, called *HilOut*, for detecting the top-n distance-based outliers according to the weight score. The major contributions of this research have been a novel distance-based outlier definition and the first distance-based outlier algorithm guaranteeing an approximate solution within a deterministic factor in time linear with respect to the dataset size.

The algorithm relies on the definition of approximate set $\{a_1, \ldots, a_n\}$ of outliers: elements a_i of this set have a weight greater than the weight of the true outliers within a pre-defined factor $\epsilon \geq 1$ (that is $\epsilon \omega_k(a_i) \geq \omega_k(o_i)$, where o_i denotes the true i-th top outlier, $i = 1, \ldots, n$).

Let n and d be the dataset size and dimensionality, respectively. The algorithm consists of two phases. The first provides an approximate solution, within a rough deterministic factor, after executing at most $d + 1$ sorts and scans of the data set, with temporal cost $O(d^2 nk)$ and spatial cost $O(nd)$,

Specifically, the algorithm avoids the distance computation between each pair of points by making use of the space-filling curves (and, specifically, the Hilbert curve), that are mappings of an hypercube $D = [0, 1]^d$ into the interval $I = [0, 1]$, to linearize the dataset. The mapping assures that if two points are close in I, they are close in D too, although the reverse in not always true.

During the first phase the algorithm calculates a lower and an upper bound to the weight of each point by exploiting Hilbert curves, and determines the points candidate to belong to the solution set, or candidate outliers. If the number of candidates n^* is n, then the algorithm stops reporting the exact solution. Otherwise, the second phase is needed, which calculates the exact solution with a final scan of temporal cost $\mathcal{O}(n^* nd)$. Experimental results highlighted that in practice the algorithm always stops, reporting the exact solution, during the first phase after much less than $d + 1$ steps.

DOLPHIN algorithm. The DOLPHIN algorithm [6, 7] detects distance-based outliers according to the definition of [26]. The algorithm, designed to work on disk-resident datasets, maintains an in-memory data structure, called INDEX. DOLPHIN performs two sequential scans of the dataset file. During the first scan INDEX is employed to maintain a summary of the portion of the dataset already examined. In particular, for each incoming dataset object obj, the objects stored in INDEX are exploited in order to determine if obj is an inlier. The object obj will be inserted into INDEX if it is not recognized as an inlier. By adopting this strategy it is guaranteed that INDEX contains all the outliers occurring in the portion of the dataset already scanned. Moreover, some of the objects stored in INDEX, called *proved inliers*, can be recognized as inliers on the basis of the objects read after them. When the first dataset scan finishes, INDEX contains a superset of the dataset outliers. During the second scan, the candidate outliers stored in INDEX are compared with all the dataset objects and at the end of the scan, INDEX contains all and only the outliers of the dataset.

It is proved that the size of INDEX is $O(k/p)$, where p denotes the probability that two randomly picked objects, one from INDEX and the other from the dataset, are neighbors. As for the value of k/p, for meaningful combinations of the parameters

k and R, that are those associated with a pre-determined fraction α (usually of the order of 1%) of objects to be recognized as outliers, provably corresponds to a small fraction (of the order of 1%) of the dataset size. Importantly, probably approximately correct values for k and R associated with a given α can be quickly determined by means of a fixed-size sampling procedure.

Summarizing, the spatial cost is $O(k/p)$, the temporal cost is $O(nk/p)$, which for k fixed is linear in the dataset size, and the I/O cost is linear, since it corresponds to the cost of sequentially reading the input dataset file twice. DOLPHIN has been compared with state of the art distance-based outlier detection algorithms showing that it is much more efficient.

Outliers in data streams. In many emerging applications, such as fraud detection, network flow monitoring, telecommunications, data management and others, data arrive continuously and it is either unnecessary or impractical to store all incoming objects. In this context, a challenge is to find the most exceptional objects among the flow of incoming data, also called a *data stream*. Data mining on data streams is often performed based on certain time intervals, called windows.

Finding outliers in data streams is a relatively novel and challenging research area [25]. The method proposed in [8], called STORM (for STream OutlieR Miner), introduces a novel concept of querying for outliers. Specifically, previous work deals with continuous queries, that are evaluated as objects arrive. Conversely, one-time queries are evaluated once over a point-in-time. The underlying intuition is that, due to evolution, stream characteristics can change over time and, hence, by classifying single objects when a data analysis is required, the concept drift, a challenging characteristics of data streams, can be captured.

Semi-supervised outlier detection. Semi-supervised methods assume that only normal examples are given. The goal is to find a description of the data, that is a rule partitioning the object space into an accepting region, containing the normal objects, and a rejecting region, containing all the other objects. These methods are also called one-class classifiers or domain description techniques.

In [16] the concept of *outlier detection solving set* is defined, a subset of the input data set representing a model that can be used to predict distance-based outliers according to the weight score. The computational complexity of computing a minimum cardinality solving set is analyzed, showing that it is in general an intractable problem. An algorithm, called *Solving Set*, that computes with sub-quadratic time requirements a solving set and the top-n outliers is described, and experimental evidence that the solving set is a fraction of the overall data set and that the false positive rate obtained using it is negligible is given.

Parallel/distributed and GPU based strategies for solving set computation are described in [19, 21]. Other compressed representation for novelty detection based on distance-based definitions are introduced in [4, 5] and compared with well-established one-class classification methods.

Outlier Explanation. In many real situations, one is given a data population characterized by a certain number of attributes, and information are provided that one of the individuals in that data population is abnormal, but no reason whatsoever is given as to why this particular individual is to be considered abnormal. The problem we deal with next is precisely to single out such reasons.

This problem has many practical applications. For instance, in analyzing health parameters of a sick patient it is relevant to single out parameters mostly differentiating them from those of the healthy population. As another example, a data history associated with an athlete that has established an exceptional performance can be analyzed to detect characteristics determining that performance.

Despite its wider applicability, a limited attention has been paid to the subject at the time we started working on it. We tackled the problem under several directions, since, due to its peculiarities, it needed the designing of specific techniques on the basis of one or more outliers in input and on the basis of the presence of numerical or categorical attributes. The following table reports the techniques we develop for the different contexts, together with the referred work.

	categorical data	numerical data
one outlier	FOP [17, 18]	OPD [23]
more outlier	EXPREX [20]	EXPREX [20]

In order to illustrate the scenario, we refer to the first case, namely a categorical dataset and one outlier provided in input. Nevertheless, the underlying ideas are shared by the different scenarios.

Consider the example consisting in a portion of the Zoo dataset (Fig. 1a). This database consists of 15 boolean-valued and two numerical attributes representing animal features. The table reports some of the attributes collected for some animals. It is known that the platypus is an exceptional animal being it a mammal, but laying eggs. This intuitive notion can be formally illustrated on the database by noticing that among dataset objects having value "y" for the attribute *eggs*, the platypus is the only animal having value "y" for the attribute *milk*. Obviously the value "y" for the

name	eggs	milk	legs
⋮	⋮	⋮	⋮
flea	y	n	6
gnat	y	n	6
gull	y	n	2
octopus	y	n	8
ostrich	y	n	2
parakeet	y	n	2
penguin	y	n	2
platypus	y	y	4
polecat	n	y	4
pony	n	y	4
wasp	y	n	6
⋮	⋮	⋮	⋮

(a)

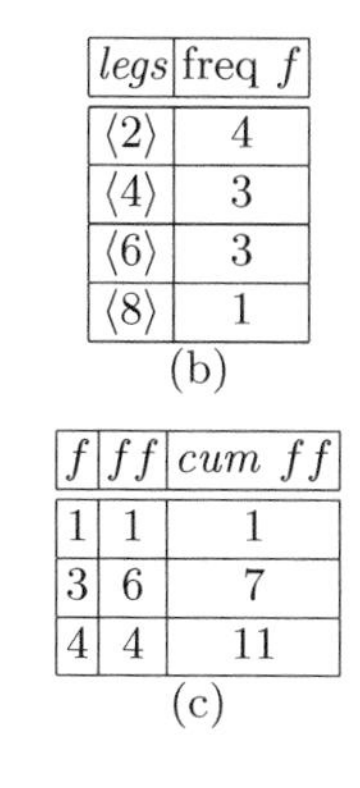

legs	freq *f*
⟨2⟩	4
⟨4⟩	3
⟨6⟩	3
⟨8⟩	1

(b)

f	*ff*	*cum ff*
1	1	1
3	6	7
4	4	11

(c)

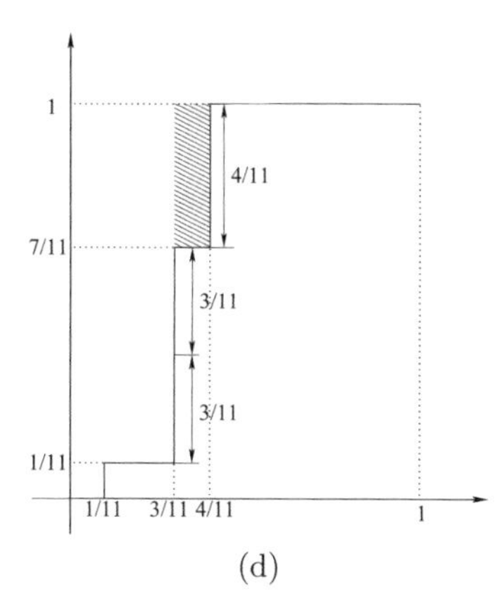

(d)

Fig. 1 Illustration of the outlier explanation problem

attribute *milk* is not an exceptional feature "per se", but it is surprising when attention is restricted to the animals which lay eggs. This is a case where a *local* property is individuated, where the attribute *eggs* plays the role of *explanation* for the *outlying property*, *milk*, of the platypus.

The intuition underlying our definition of exceptionality is that a set of attributes, or *property*, makes an object exceptional if the frequency of the combination of values assumed by that object on those attributes is rare if compared to the frequencies associated with the other combinations of values assumed on the same attributes by the other objects of the database.

Indeed, considering frequency values may lead to incorrect conclusions. As an example, consider a key-attribute. Obviously, the value assumed by the outlier object on that attribute occurs just once on the dataset, but this cannot be considered exceptional since all the values occur once.

Our approaches, then, try to capture this intuition in the various scenarios where we search for *property-explanation pairs*.

One outlier – Categorical data. The technique we propose to deal with the case of one user-provided outlier in a categorical dataset is based on (1) the construction of the histogram of frequency values (Fig. 1b), (2) the construction of the cumulated histogram of the frequency of frequencies (Fig. 1b) and, (3) the quantification of the "degree of unbalanceness" between the frequency of the object o under consideration and the frequencies of the rest of the database (Fig. 1b). This latter step is performed by measuring the area above the cumulated frequency of frequencies histogram of the database w.r.t. the set of attributes of interest, starting from the frequency of o. Indeed, the larger this area is, the smaller the frequency of o is w.r.t. the frequencies associated with the other data set objects. To illustrate, Fig. 1 reports the computation of the outlierness of the property *legs* for the platypus.

One outlier – Numerical data. When dealing with numerical attributes, in order to extend the previously stated intuition, a key aspect is being able to efficiently estimate both the *cumulated density frequency* (*cdf*) and the related *probability density frequency* (*pdf*), as well as to exploit them to measure the associated imbalance. This is in fact our main contribution in this scenario.

The *Outlying Property Factor* (OPF) of a numerical attribute a in o w.r.t. the dataset is defined as follows:

$$OPF_a(o) = \Omega\left(\int_{f_a(o[a])}^{\sup(f_a)} (1 - G_a(f))\mathrm{d}f - \int_0^{f_a(o[a])} G_a(f)\mathrm{d}f\right). \quad (1)$$

where, Ω denotes a function from the set of real numbers to [0, 1], and

$$G_a(\varphi) = Pr(X_a^f \leq \varphi) = \int_0^{\varphi} Pr(X_a^f = v)\mathrm{d}v, \quad (2)$$

with X_a^f denoting the random variable whose pdf represents the relative likelihood for the pdf f_a, representing the density of the active domain of a, to assume a certain value. Thus, $G_a(\varphi)$ measures the probability to observe in a a density value not exceeding φ. To maintain the analogy with the previous definition, the first integral in OPF measures the *area above* the cdf $G_a(f)$ for $f > f_a(o[a])$, while the second integral in OPF measures the *area below* the cdf G_a for $f \leq f_a(o[a])$.

Two problems are related to the applicability above definition: (1) to estimate the empirical pdfs from dataset values, and (2) to determine the "natural" intervals of homogeneous values to be employed to form explanations. The strategies we have designed to solve these two problems exploit a common framework, which is based on Kernel Density Estimation (KDE).

More outliers. The work [20] extends the perspective of previously described approaches in order to be able to deal with *groups*, or *sub-populations*, of anomalous individuals. As an example, consider a restrict group of longevous individuals; it would be very useful to single out properties, namely genetic traits, differentiating them as a whole from normal individuals.

We designed exceptionality scores well-tailored for comparing a rare population with a large one, which exploit the notion of *randomization test* based on the *Pearson chi-square* criterion, for categorical properties, and on the *Cramér-von-Mises* criterion, for numerical properties. These criteria evaluate the badness of fit of a probability distribution F compared to a sample set. In particular, we employ as reference distribution F the empirical distribution function associated with the population of inliers and, as the sample set, the population of outliers.

3 Intensional Data Mining

Analogously to what happens for the extensional level [36], also for the intensional one it is possible to define regularities and exceptions. In particular, both of them, as a whole, form the so-called *interschema properties* [29, 32].

These can be partitioned into *nominal properties* [30], which involve lexicon, and *structural properties* [37], which involve the structure of sources to integrate.

Nominal properties, in turn, can be divided in synonymies, homonymies and hyponymies. A *synonymy* between two concepts C_1, belonging to a schema S_1, and C_2, belonging to a schema S_2, indicates that they have the same meaning but different names. A *homonymy* between $C_1 \in S_1$ and $C_2 \in S_2$ denotes that they have the same name but different meaning [29, 32]. A concept $C_1 \in S_1$ is a hyponym of a concept $C_2 \in S_2$ (which, in turn, is the hypernym of C_1) if C_1 has a more specific meaning than C_2 [31]. Synonymies and homonymies are examples of regularities, whereas homonymies represent a case of exceptions.

The main *structural properties* [40] are type conflicts, subschema similarities and complex knowledge patterns. A *type conflict* denotes that the same concept is represented by means of different schema structures in the involved databases (for instance, by means of an attribute in S_1 and an entity in S_2). A *sub-schema similarity* indicates that a subschema of S_1 and another of S_2 could represent the same concept although they seem to have a different structure. A *complex knowledge pattern* represents a (generally complex) relationship involving several concepts possibly belonging to several different schemas [28, 35]. Also for structural properties we can recognize regularities (in this case, subschema similarities) and exceptions (in this case, type conflicts). Interestingly, complex knowledge patterns can encompass both regularities and exceptions. In the literature, different techniques for the extraction of interschema properties have been proposed. The interested reader can find a survey in [38].

In this chapter, we shall illustrate one of them based on the assumption that two concepts are similar if the concepts belonging to their neighborhoods are similar. Now, the question is: given a concept C, which are its neighbors? To answer this question, it is necessary to define a metric aiming at detecting the semantic relationship degree between two concepts. Thanks to this metric, it is possible first to detect the closest neighbors of C, then to determine the neighbors of the closest neighbors, and so forth, until to a certain neighbor distance has been covered. Clearly, in the definition of the semantics of C, the contribution of its closest neighbors must be higher than the one of its farthest neighbors. In other words, as this process moves away from C, the weight of neighbors in determining the semantics of C decreases.

To define the metric of our interest, some support information is needed. To determine this information, it is necessary to associate some suitable graphs (called Semantic Distance Graphs - SD Graphs) with the involved E/R schemas. Specifically, given an E/R schema S, the corresponding SD-Graph can be represented as $G(S) = \langle N(S), D(S) \rangle$, where $N(S)$ indicates the set of the nodes of $G(S)$, whereas $A(S)$ denotes the set of its arcs. There is a node in $G(S)$ for each entity, relationship or attribute of S. $A(S)$ consists of two subsets, namely: (i) $SA(S)$, i.e., the set of *solid arcs*, denoting a strong relationship between two concepts; (ii) $DA(S)$, i.e., the set of *dashed arcs*, indicating a weak relationship between two concepts.

There is a solid arc: (i) from an entity or a relationship to each of its attributes; (ii) from a relationship to each of the entities linked by it; (iii) from a key attribute to the corresponding entity; (iv) from the "child" entity to the "father" one of an *isa* relationship.

There is a dashed arc: (i) from a non-key attribute to the corresponding entity or relationship; (ii) from an entity to each of the relationships it participates to; (iii) from a "father" entity to each of its "children entities" of an *isa* relationship.

The more the shortest path connecting two nodes encompasses dashed arcs, the more they must be considered semantically distinct. We define $D\text{-}path_n$ in $G(S)$ a path with n dashed arcs and any number of solid arcs. Given two nodes x and y in $G(S)$, the *D-shortest* path from x to y, denoted by $\langle x, y \rangle$, is the path from x to y characterized by the minimum number of D-arcs.

Given a node x, the neighborhood, or *context* of level $i \geq 0$ of x is defined as: $cnt(x, i) = \{y \mid y \in N(S), y \neq x, \langle x, y \rangle \text{is a } D - path_i \text{ in G(S)}\}$.

The *A_cnt* of level i $(i \geq 0)$ of a node x is defined as as the set of the attributes belonging to $cnt(x, i)$. The *E_cnt* of level i $(i \geq 0)$ of a node x encompasses all the entities belonging to all the contexts of level j $(0 \leq j \leq i)$ of x.

In order to determine if two concepts, belonging to two different schemas, are similar, it is necessary to first examine their contexts of level 0, then the ones of level 1, and so forth.

Let $x \in G(S_1)$ and $y \in G(S_2)$ be two nodes. In order to compute the similarity degree of their A_cnts, a full bipartite graph is constructed. In this graph, there is a node for each attribute of x and y. Each edge has associated a weight on the basis of the similarity degree of the corresponding attributes. Given two attributes A_1 of x and A_2 of y, the weight of the edges linking them is computed by means of the following formula: $\gamma(A_1, A_2) = w_l \times L(A_1, A_2) + w_d \times D(A_1, A_2) + w_k \times K(A_1, A_2)$, where w_n, w_d and w_k are weighting factors whose sum is equal to 1, and $L(A_1, A_2)$, $D(A_1, A_2)$ and $K(A_1, A_2)$ consider the lexicographic, domain and "key characterization" similarities of A_1 and A_2, respectively. After the bipartite graph has been constructed, the computation of the similarity degree of the two A_cnts is performed by computing a suitable objective function associated with the maximum weight matching related to this bipartite graph. This objective function is given by the sum of the weights of the selected arcs multiplied by 2 and divided by the number of the nodes of the bipartite graph. The corresponding value belongs to the real interval [0, 1]. In an analogous way, it is possible to compute the similarity degree of two E_cnts. Finally, the similarity degree $c - sim(x, y, i)$ of two contexts $cnt(x, i)$ and $cnt(y, i)$ is a weighted mean of the similarity degree of their A_cnts and their E_cnts.

To compute the similarity $sim(x, y)$ between x and y, it is necessary to apply an iterative procedure. First $c - sim(x, y, 0)$ is computed. Then, $c - sim(x, y, 1)$ is determined and exploited to refine the previously computed value. This way of proceeding is performed until to a user predefined level n is reached, or until to the value of $sim(x, y)$ reaches a fixed point, with farthest levels influencing the computed value less than closer ones. For this latter purpose, we exploit a quadratic decrease function.

Once the similarity coefficients of all the objects of two schemas S_1 and S_2 have been determined, it is possible to compute interschema properties:

- There exists a *synonymy* between two objects $C_1 \in S_1$ and $C_2 \in S_2$ if they have the same type (i.e., both of them are entities or relationships or attributes) and their similarity coefficient $sim(C_1, C_2)$ is higher than a certain threshold th.
- There exists a *homonymy* between two objects $C_1 \in S_1$ and $C_2 \in S_2$ if they have the same name and the same type but their similarity coefficient is lower than (1-th).
- There exists a *type conflict* between two objects $C_1 \in S_1$ and $C_2 \in S_2$ if they have different types but their coefficient is higher than a threshold th.

The interested reader can find all details about this approach in [32].

In order to compute *subschema similarities*, a very similar approach can be adopted. In fact, the neighborhood of a subschema is nothing more than the union of the neighborhoods of the objects forming the subschema [40]. The approach to computing *hyponymies* is very similar to the one described above [31]. Finally, in order to determine complex knowledge patterns, a variant of description logics can be exploited [28]. As previously pointed out, interschema properties play a key role in schema integration and, ultimately, in the construction of Cooperative Information Systems and Data Warehouses [39].

The interested reader can find the description of a system performing interschema property computation by means of the approach described above, as well as schema integration for the construction of a Cooperative Information Systems or a Data Warehouse in [33, 34].

4 Future Trends

As current trends, a renewed interest is witnessed for the field of metaquerying due to the success of the semantic web and to its strict relationship with ontological, that carries in the spotlight the need of querying classes and concepts [27]. Furthermore, notable research directions pertaining outliers are devoted to the needs of certain applications, e.g., in terms of interpretability or capability to describe the reasons underlying unusual behaviours of users, to provide user interfaces to navigate within data and to visualize outliers, and towards tailoring the discovery process on users' goals rather than on some pre-defined notion of abnormality [3].

As another interesting challenge for researchers, it is worth to emphasize the relevance of designing techniques able to work in domain so complex to necessarily require new and specialized methodologies. Among these, relevant contributions could be provided to fields like physics or biology because of the huge amount of data already produced in these areas characterized by high variability and uncertainty.

Moreover, of particular interest is the analysis of networks and of their dynamics related to the formidable popularity of online social networks and availability of huge amount of user-generated content. This has lead to a significant increase in the number of studies about social network in the area of the social sciences network modeling and analysis in the area of machine learning and data mining. All that is aimed to extract knowledge about relations and diffusion processes in order to shed lights on social behaviors and interactions among individuals or discover anomalous, malicious individuals who attempt to perform illegal activities and cause harm to other users [2, 22].

References

1. C. Aggarwal, *Outlier Analysis* (Springer, Berlin, 2013)
2. L. Akoglu, H. Tong, D. Koutra, Graph based anomaly detection and description: a survey. Data Min. Knowl. Discov. **29**(3), 626–688 (2015)
3. L. Akoglu, F. Bell, E. Müller, T.E. Senator (eds.), *ACM KDD Workshop on Outlier Definition, Detection and Description on Demand, San Francisco, USA* (2016)
4. F. Angiulli, Condensed nearest neighbor data domain description. IEEE Trans. Pattern Anal. Mach. Intell. **29**(10), 1746–1758 (2007)
5. F. Angiulli, Prototype-based domain description for one-class classification. IEEE Trans. Pattern Anal. Mach. Intell. **34**(6), 1131–1144 (2012)
6. F. Angiulli, F. Fassetti, An efficient method for outlier detection, in *SEBD* (2008), pp. 326–333

7. F. Angiulli, F. Fassetti, DOLPHIN: an efficient algorithm for mining distance-based outliers in very large datasets. ACM Trans. Know. Discov. Data **3**(1), 4:1–4:57 (2009)
8. F. Angiulli, F. Fassetti, Distance-based outlier queries in data streams: the novel task and algorithms. Data Min. Knowl. Discov. **20**(2), 290–324 (2010)
9. F. Angiulli, C. Pizzuti, Fast outlier detection in high dimensional spaces, in *PKDD* (2002), pp. 15–26
10. F. Angiulli, C. Pizzuti, Outlier mining in large high-dimensional data sets. IEEE Trans. Knowl. Data Eng. **17**(2), 203–215 (2005)
11. F. Angiulli, R. Ben-Eliyahu-Zohary, G. Ianni, L. Palopoli, Computational properties of metaquerying problems, in *PODS* (2000), pp. 237–244
12. F. Angiulli, G. Ianni, L. Palopoli, Metaquerying: proprietà e tecniche di implementazione, in *SEBD* (2000), pp. 317–330
13. F. Angiulli, G. Ianni, L. Palopoli, On the complexity of mining association rules, in *SEBD* (2001), pp. 177–184
14. F. Angiulli, R. Ben-Eliyahu-Zohary, G. Ianni, L. Palopoli, Computational properties of metaquerying problems. ACM Trans. Comput. Log. **4**(2), 149–180 (2003)
15. F. Angiulli, G. Ianni, L. Palopoli, On the complexity of inducing categorical and quantitative association rules. Theor. Comput. Sci. **314**(1–2), 217–249 (2004)
16. F. Angiulli, S. Basta, C. Pizzuti, Distance-based detection and prediction of outliers. IEEE Trans. Knowl. Data Eng. **18**(2), 145–160 (2006)
17. F. Angiulli, F. Fassetti, L. Palopoli, Un metodo per la scoperta di proprietà inattese, in *SEBD* (2006), pp. 321–328
18. F. Angiulli, F. Fassetti, L. Palopoli, Detecting outlying properties of exceptional objects. ACM Trans. Database Syst. **34**(1), 1–62 (2009)
19. F. Angiulli, S. Basta, S. Lodi, C. Sartori, Distributed strategies for mining outliers in large data sets. IEEE Trans. Knowl. Data Eng. **25**(7), 1520–1532 (2013)
20. F. Angiulli, F. Fassetti, L. Palopoli, Discovering characterizations of the behavior of anomalous subpopulations. IEEE Trans. Knowl. Data Eng. **25**(6), 1280–1292 (2013)
21. F. Angiulli, S. Basta, S. Lodi, C. Sartori, GPU strategies for distance-based outlier detection. IEEE Trans. Parallel Distrib. Syst. **27**(11), 3256–3268 (2016)
22. F. Angiulli, F. Fassetti, E. Narvaez, Anomaly detection in networks with temporal information, in *Discovery Science* (Italy, Bari, 2016), pp. 359–375
23. F. Angiulli, F. Fassetti, G. Manco, L. Palopoli, Outlying property detection with numerical attributes. Data Min. Knowl. Discov. **31**(1), 134–163 (2017)
24. M.R. Garey, D.S. Johnson, *Computers and Intractability: A Guide to the Theory of NP-Completeness* (W.H. Freeman, New York, 1979)
25. M. Gupta, J. Gao, C. Aggarwal, J. Han, Outlier detection for temporal data: a survey. IEEE Trans. Knowl. Data Eng. **26**(9), 2250–2267 (2014)
26. E. Knorr, R. Ng, A unified notion of outliers: properties and computation, in *KDD* (1997), pp. 219–222
27. M. Lenzerini, L. Lepore, A. Poggi, Answering metaqueries over hi (OWL 2 QL) ontologies, in *IJCAI, New York, USA* (2016), pp. 1174–1180
28. L. Palopoli, D. Saccà, D. Ursino, DL_P: a description logic for extracting and managing complex terminological and structural properties from database schemes. Inf. Syst. **24**(5), 410–424 (1999)
29. L. Palopoli, D. Saccà, D. Ursino, Semi-automatic techniques for deriving interscheme properties from database schemes. Data Knowl. Eng. **30**(4), 239–273 (1999)
30. L. Palopoli, G. Terracina, D. Ursino, A graph-based approach for extracting terminological properties of elements of XML documents, in *ICDE* (IEEE Computer Society, Heidelberg, Germany, 2001), pp. 330–337
31. L. Palopoli, D. Saccà, G. Terracina, D. Ursino, A technique for deriving hyponymies and overlappings from database schemes. Data Knowl. Eng. **40**(3), 285–314 (2002)
32. L. Palopoli, D. Saccà, G. Terracina, D. Ursino, Uniform techniques for deriving similarities of objects and subschemes in heterogeneous databases. IEEE Trans. Knowl. Data Eng. **15**(2), 271–294 (2003)

33. L. Palopoli, G. Terracina, D. Ursino, Dike: a system supporting the semi-automatic construction of cooperative information systems from heterogeneous databases. Softw. Pract. Exp. **33**(9), 847–884 (2003)
34. L. Palopoli, G. Terracina, D. Ursino, Experiences using DIKE, a system for supporting cooperative information system and data warehouse design. Inf. Syst. **28**(7), 835–865 (2003)
35. L. Palopoli, G. Terracina, D. Ursino, A plausibility description logic for handling information sources with heterogeneous data representation formats. Ann. Math. Artif. Intell. **39**(4), 385–430 (2003)
36. L. Pontieri, D. Ursino, E. Zumpano, An approach for the extensional integration of data sources with heterogeneous representation formats. Data Knowl. Eng. **45**(3), 291–331 (2003)
37. L. Palopoli, D. Rosaci, G. Terracina, D. Ursino, A graph-based approach for extracting terminological properties from information sources with heterogeneous formats. Knowl. Inf. Syst. **8**(4), 462–497 (2005)
38. E. Rahm, P. Bernstein, A survey of approaches to automatic schema matching. VLDB J. **10**(4), 334–350 (2001)
39. D. Rosaci, G. Terracina, D. Ursino, An approach for deriving a global representation of data sources having different formats and structures. Knowl. Inf. Syst. **6**(1), 42–82 (2004)
40. G. Terracina, D. Ursino, A uniform methodology for extracting type conflicts and subscheme similarities from heterogeneous databases. Inf. Syst. **25**(8), 527–552 (2000)

Relational Data Mining in the Era of Big Data

Annalisa Appice, Michelangelo Ceci and Donato Malerba

Abstract The aim of this article is to synthetically describe a sample of distinct approaches and applications of Relational Data Mining, which address the issue of managing complex, and possibly big, amounts of data. Specifically, we report a brief review of the literature on Relational Data Mining in the fields of Spatial Data Mining, Process Mining, Network Data Analysis and Stream Data Mining, with an emphasis on the Italian research. For each field, we describe the milestones that have been reached, as well as the future research trends that are fuelled by the emergent ubiquity of Big Data.

Keywords Relational data mining · Spatial data mining · Process mining · Stream data mining · Sensor network analysis · Big data analytics

1 Introduction

Relational Data Mining is a branch of Data Mining whose main goal is to overcome the problem of single table assumption, that is made in classical data mining, statistical and machine learning approaches [32]. The single table assumption assumes that the training set can be represented as a single relational table, where each row corresponds to an example and each column to a predictor variable or to the target variable. This assumption seems quite restrictive in several data mining applications,

A. Appice (✉) · M. Ceci · D. Malerba
Dipartimento di Informatica, Universita degli Studi di Bari Aldo Moro,
via Orabona 4, 70125 Bari, Italy
e-mail: Annalisa.Appice@uniba.it

M. Ceci
e-mail: Michelangelo.Ceci@uniba.it

D. Malerba
e-mail: Donato.Malerba@uniba.it

A. Appice · M. Ceci · D. Malerba
CINI, Consorzio Interuniversitario Nazionale per l'Informatica, Bari, Italy

A. Appice · M. Ceci · D. Malerba
CILA Centro Interdipartimentale di Logica e Applicazioni, Bari, Italy

S. Flesca et al. (eds.), *A Comprehensive Guide Through the Italian Database Research Over the Last 25 Years*, Studies in Big Data 31, DOI 10.1007/978-3-319-61893-7_19

where data to be represented are complex, are stored in a database and are organized into several tables for reasons of efficient storage and access. In this context, both predictor variables and the target variables (if any) are represented as attributes of distinct tables (relations) eventually related each other by means of foreign key constraints.

In the 2000s, we have witnessed to a proliferation of Relational Data Mining approaches due to the high number of application domains yielding relational data. Specifically, a variety of Relational Data Mining algorithms have been formulated to be directly applied to various relational representations of collections of interconnected entities, namely networked data. We note that this significant research effort has been mainly motivated by the ubiquity of networked data. In fact, text data, social data, spatio-temporal data and sequence data can be easily seen as a kind of network data, where entities are objects and connections are social/spatial/temporal relationships. By resorting to a network representation of this kind of data, the application of Relational Data Mining algorithms appears straightforward, at least in principle. In particular, Relational Data Mining algorithms allow us to naturally take into account the various forms of correlation, which may bias learning in network data. Furthermore, discovered relational patterns may reveal those relationships which correspond to the network dependencies. In addition, a wide plethora of Relational Data Mining algorithms have been formulated for descriptive and predictive tasks [32]. Applications include spatial, spatio-temporal data mining, process mining, stream data mining.

One of the main issues to be considered in Relational Data Mining is the inherent complexity of data and tasks, which has called for algorithms inherently more expensive computationally-wise: larger hypothesis spaces are searched and evaluation of a single hypothesis becomes more complex. Although various major accomplishments have been performed in the direction of efficiency and scalability [15], unfortunately existing Relational Data Mining systems cannot be commonly applied for dealing effectively with very large (big) data sets [87].

Big Data starts with large-volume, heterogeneous, autonomous sources having distributed and decentralized control, and seeks to explore complex and evolving relationships among data. Big data have often a complex relational structure. The inherent complexity of data is expressed into the heterogeneous and diverse dimensionality, as well as in the existence of complex and evolving relationships [88]. Therefore, despite the efficiency and scalability limits, Relational Data Mining appears a natural approach for representing and handling such large complex data. A simple way to use Relational Data Mining systems is to simplify Big Data by reducing the amount of information. We could compress several data points into one example, even if the resulting example would need to appropriately capture a proportion of the data distribution. Although a few researchers [87] have explored the viability of a data compression technique smoothly integrated into Relational Data Mining frameworks as a part of domain knowledge, this solution still neglects the specific challenges of Big Data.

First, Big Data are often stored at different locations and data volumes may continuously grow, so an effective computing platform will have to take distributed large-scale data storage into consideration for computing. For example, typical Relational Data Mining algorithms require all data to be loaded into the main memory. However, this is becoming a clear technical barrier for Big Data because moving data (and their relationships) across different locations is expensive. Second, depending on different domain applications, the data privacy and information sharing mechanisms between data producers and data consumers can be significantly different. In addition to the above privacy issues, the application domains can also provide additional information to benefit or guide the Big Data Mining algorithm design. While the application domain knowledge is already used in various high-level Relational Data Mining algorithms, in order to speed-up knowledge-driven Relational Data Mining discovery [15], the data privacy and information sharing mechanisms between data producers and data consumers can add unaddressed technical barriers to the Relational Big Data access. Finally, the traditional Relational Data Mining challenges concentrate on algorithm designs in tackling the difficulties raised by complex and dynamic data characteristics, while the Big Data volumes add the challenges of distributed data.

In this paper, we present and discuss a brief review of the literature reporting the state-of-the-art of research in Relational Data Mining in the field of Spatial Data Mining, Process Mining, Network Data Analysis and Stream Data Mining. This work gives emphasis to research done in Italy and somehow extends the paper [13], where relational data mining is not (at least explicitly) discussed. For each field, we report the most of Relational Data Mining chances that meet Big Data and the challenges carried out, showing the critical aspects and the advantages of different solutions.

2 Spatial Data Mining

One of the most popular applications of Relational Data Mining approaches is that of spatial data (i.e., in Spatial Data Mining) [60]. The motivations are manifold: First, spatial objects have a locational property, which implicitly defines several spatial relationships between objects such as topological relationships (e.g., intersection, adjacency), distance relationships, directional relationships (e.g., north-of) and hybrid relationships (e.g., parallel-to). Second, attributes of spatially interacting (i.e., related) units tend to be statistically correlated. Spatial cross-correlation refers to the correlation between two distinct attributes across space (e.g., the employment rate in a city depends on the business activities both in that city and in its neighborhood). Autocorrelation refers to the correlation of an attribute with itself across space (e.g., the price level for a good at a retail outlet in a city depends on the price for the same good in the nearby). In geography, spatial autocorrelation is justified by the Tobler's First law of geography, according to which "everything is related to everything else, but near things are more related than distant things" [55].

In the era of "Big Data", spatiotemporal data, whether captured through remote sensors (e.g., remote sensing imagery, Atmospheric Radiation Measurement (ARM) data) or large scale simulations (e.g., climate data) are always considered "Big" [85].

However, recent advances in instrumentation and computation make the spatiotemporal data even bigger, putting several constraints on data analytics capabilities. Though these improvements are leading to increase in volume, velocity, and variety of remote sensing data products and making it hard to manage and process, they are also enabling new applications. For example, improvements in temporal resolution allow monitoring biomass on a daily basis. Improvements in spatial resolution allow fine-grained classification (settlement types), damage assessments and critical infrastructure (e.g., Google generates about 25 PB of data per day, significant portion of which is spatiotemporal data (including images and videos)). The rate at which spatiotemporal data is being generated clearly exceeds our ability to organize and analyze them, in order to extract patters critical for understanding the dynamically changing world. Therefore, there is now interest in developing efficient management and analytical infrastructures for big spatial data [85].

Moreover, explicit modeling of spatial dependencies requires relational approaches and increases the computational complexity and further motivates efficient management and analytical infrastructures for Big Spatial Data. One of the best well-known approaches for taking autocorrelation into account is the spatial autoregressive (SAR) model [3]. Although there are some attempts in the literature to efficient solve SAR (see for instance [24] and see [52] for a comparison of several methods), they all suffer from the quadratic nature of the problem. Other ways to take autocorrelation into account are Markov Random Field (MRF) models [2] and Gaussian Process Learning and Mixture Models. In all cases, the complexity of the problems does not allow methods to scale well on Big Data. A recent platform to analyze Big Spatial Data has been proposed in [53] and in the project "Geospatial Big Data Management, Analysis and Service Platform Technology Development".[1] In both cases, however, the authors do not emphasize the role of the spatial autocorrelation phenomenon which, consequently, is not directly exploited in the Big Data analytics methods. A more recent and efficient solution, which embeds spatial autocorrelation in the induction of predictive models, has been presented in [77]. In this work, the authors propose to learn predictive clustering trees (PCTs) by embedding autocorrelation measures in the tree construction heuristics. This approach seems particularly promising because of the efficient way trees can be built and since PCTs can be used for the multi-target prediction.

In the literature there are several applications of spatial data mining approaches which take into account some form of (spatial) relations between objects. These include environmental modeling, ecology, agriculture, mobility, social sciences, energy etc. In this paper we will only focus on mobility and energy.

As for mobility, one of the main applications is in mining trajectories of moving objects. In this context, Monreale et al. [63] and Masciari et al. [61] have proposed methods for mining trajectory frequent patterns. These patterns can also be used for predicting the next position of the objects [63]. Other authors have also investigated related problems such as the definition of avoidance behavior between moving object trajectories [56] or mining (and querying) trajectories which contain gaps [64].

---

[1] Funded by the Ministry of Land, Infrastructure and Transport, South Korea.

As for social sciences, some studies have proposed to apply Relational Data Mining techniques to the spatially-aware analysis of social and economical phenomena. This is the case of the study of the level of deprivation of urban areas [16, 17]. Another example is the application in the humanities, in order to study the spatio-tempral characteristics of Roman Inscriptions [68] or to exploit the spatial position of layout components on printed historical documents [18].

As for the energy sector, Spatial Data Mining methods have been applied, in order to forecast the energy to be produced. For this problem, a recent solution is proposed in the project Vi-POC [20, 21]. The main goal of the Vi-POC is to support (renewable) energy providers with a framework for collecting, storing, analyzing, querying, and retrieving data coming from heterogeneous energy production plants (such as photovoltaic, wind, geothermal, Sterling engine, and running water) distributed over a wide territory. In Vi-POC both Artificial Neural Networks and Predictive Clustering Trees are used for forecasting purposes. In both cases, spatial autocorrelation is taken into account by specifically designed features [23].

3 Process Mining

Process Mining is used to extract process-related information from event data [83]. Event data are collected about any type of event, at any time and any place. They are the most important source of information [82]. They may concern "life events", "machine events", or both. They are associated with the execution of a certain business process and registered in a wide variety of data sources (i.e., databases, flat files, message logs, transaction logs, ERP systems and document management systems). Each event refers to a specific activity (i.e., a well-defined step in a business process). It may be recorded with additional information such as the resource (i.e., person or device) executing or initiating the activity, the timestamp of the event, or the data elements recorded with the event (e.g., the place where the activity is performed).

In the last decade, the omnipresence of event data has inspired the development of a large plethora of Process Mining algorithms, in order to explore event information in a meaningful way and learn knowledge related to the behavior of people, organizations, machines and systems [81].

In these algorithms, the wealth of information recorded with events is related to several perspectives such as the control-flow perspective (ordering of activities), the execution perspective (time performances and frequency of activities) and the organizational perspective (organization of resources) [6, 80]. We note that both the control-flow perspective and the organizational perspective may involve a relational representation of events. Specifically, in the control-flow perspective the activities are the entities, while the ordering of activities naturally defines the relationships between the activities. In the organizational perspective, the resources are the entities, while various social interactions (e.g., handover of work, working together and joint activities) naturally define the relationships between resources. Therefore, Relational Data Mining, as well its sub-fields like Inductive Logic Programming and Graph

Mining, are also an integral part of Process Mining fueled by the inherently relational structure of event data.

By bridging Relational Data Mining to Process Mining, Greco et al. [45] have initially explored the use of powerful relational representations in Process Mining. In particular, they have combined rich graph-based representations of workflow schemes with simple (i.e., stratified), yet powerful First-Order Logic rules (e.g., DATALOG rules), in order to express complex properties and constraints on the process executions.

Moreover, Greco et al. [46] have studied the problem of discovering correlations among general patterns of execution in a workflow by accounting for relational features expressing the order of executions.

Folino et al. [39] have investigated the usage of complex patterns for clustering complex objects and, in particular, sequential data, in order to identify all its hidden variants in process executions. They have extended existing approaches to accommodate more powerful structural, typically relational features, such as sets of activities, higher order k-grams and generic (non-contiguous) sub-sequences. They have used a special kind of pattern, based just on non-contiguous sub-sequences, also considered in [48].

Greco et al. [47] have also investigated the use of constrained graphs to describe the control flow of processes models. In such graphs, nodes represent the activities involved in the process, while edges describe the precedence relationship among such activities. Nodes and edges have some constraints to control the interaction among the activities. In another work, Greco et al. [49] have addressed the problem of discovering different process variants by clustering event data. They have focused on structural aspects by considering relational information describing which activities have been executed and in which order. They have also introduced an information-theoretic framework to simultaneously cluster the logged process executions, encode the structural information of event data, as well as a number of performance metrics associated with them.

Ghionna et al. [43] have defined an approach that combines the discovery of frequent structural patterns with a cluster-based anomaly detection procedure. They have defined the notion of a structural pattern, which effectively characterizes concurrent processes by accounting for typical routing relational constructs, in addition to the traditional order of execution constructs.

Greco et al. [44] have investigated the task of process modeling by exploring the benefit of accounting for the background knowledge that, in many cases, is available to the analysts taking care of the process (re-)design. They have proposed an approach that was based on encoding the information gathered from the log and the (possibly) given background knowledge in terms of precedence constraints. They have applied mining algorithms formulated in terms of reasoning problems over precedence constraints.

Diamantini et al. [30] have described the use of a graph representation, in order to express a business process. In addition, they have examined the use of graph-mining algorithms, in order to extract frequent process models. Subsequently, they have illustrated an algorithm to improve the quality of generated instance graph representation of a process modeling the presence of highly variable processes [31]. In particular, they have proposed to exploit causal relation inference rules, which are typical of filtering techniques.

Ferilli and Esposito [35] have described a framework, based on First-Order Logic, to standardize processes, in order to correctly carry out activities in an organization. They have proposed an incremental algorithm that is able to learn adapting models, as well as to express triggers and conditions on the tasks that make up the workflow. Ferilli et al. [36] have also showed an application of this First-Order Logic incremental algorithm to the user's daily routines, for predicting his/her needs and comparing the actual situation with the expected one. Subsequently, Ferilli et al. [37] have presented a new First-Order Logic-based approach to learn complex process models extended with complex, human-readable and interesting conditions.

More recently, Ferilli [33] have described a workflow management system, named WoMan, that was based on First-Order Logic. Its core is an automatic procedure that incrementally learns and refines workflow models from observed process executions. The system allows quick learning even in the presence of noise and changed behavior. An entire algorithmic apparatus has been presented, including translation and learning from a standard log format for case representation, import/export of workflow models from/into standard formalisms (Petri nets), and exploitation of the learned models for process simulation and monitoring. An extension of this apparatus is described in [38]. It allows us to address a predictive task, in order to predict the process behavior. Finally, in [34], Ferilli has deepened the representation aspects of the First-Order Logic-based formalism adopted by WoMan, highlighting its most outstanding strengths and comparing it to the current standard formalism (Petri nets).

Turi et al. [79] have described a Relational Data Mining system that represents the event data by resorting to a First-Order Logic formalism. They have used a deductive database, in order to store both the control flow and the organization of event data. They have improved the traditional logic formalism with hierarchical knowledge, in order to express activities and resources grouped in hierarchical categories, as well as with background knowledge, in order to express rules describing some independent knowledge. They have examined a multi-level, relational pattern discovery algorithm, in order to derive process relational patterns at multiple level of granularity. The discovery process has been performed according to a breadth-first search in the lattice of relational patterns spanned by a generality order between patterns. Multi-level relational patterns have been discovered so that, by descending/ascending through a hierarchy, it is possible to view the same activity or resource at different levels of abstraction (or granularity). Although this relational algorithm has resulted powerful, the relational frequent pattern discovery has appeared a very complex task, particularly in presence of massive event data. To deal with scalability issues, Turi et al. [79] have distributed discovery of multi-level approximate process patterns on a Grid platform by resorting to a partitioning procedure. This procedure divides event data in several data partitions by appropriately exploiting the relational structure of the data. They have adopted a computational framework to exploit the power of distributed computation and storage resources across the Internet. Finally, they have defined a schema to combine sets of local patterns and approximate global ones.

Appice et al. [8] have then extended this study by exploring a sampling procedure instead of a partitioning procedure to distribute the computation. Loglisci et al. [58] have described a further extension of this relational approach, which has integrated disjunctive forms into relational patterns discovered from event data. The discovery is performed via parallel, distributed computation of disjunctive forms, which enable relational patterns to express frequent variants of process models.

More recently, Ceci et al. [19] have resorted to a sequence- based represetnation a process execution. They have exploited sequential pattern mining and used the additional information about the activities matching a partial process model, in order to train nested prediction models from event logs.

Finally, Appice [4] has focused on a stream perspective of event data, introduced a graph-based formalism to represent the organization of a business process and described a stream-based graph mining algorithm, in order to discover time evolving organizational structures in event data streams. The proposed mining algorithm accounts for both the relational structure of the resources and the relational structure of the order of execution of activities.

To summarize the achievements of the state-of-the-art of research in Relational Data Mining and Process Mining, we note that the existing rich literature has definitely assessed that various relational formalisms (based on Inductive Logic Programming, Relational feature synthesis, Graph and/or Sequence structure representation) are very expressive in the event data scenario and ensure strict adherence to the observed practices. At the same time, the performed studies have mainly focused on both the expressiveness and the accuracy of Relational Data Mining in Process Mining, while the scalability is commonly neglected in the related literature. Although there are a few studies [8, 58, 79], which have examined procedures to distribute the process mining computation in the relational setting, they do not deal explicitly with the Big Event Data.

On the other hand, the widespread use of Big Data is heavily impacting organizations and individuals for which event data are collected. Organizations are heavily investing in Big Data technologies, but at the same time citizens, patients, customers, and employees are concerned about the use of their data [82]. Following the main stream of Data Mining research, van der Aalst and Damiani [84] have recently advocated the need of applying Big Data techniques, like MapReduce, to Process Mining as a means of enabling the computation of analytics on all available data points, as opposed to doing so on selected data samples. Accounting for this recommendation, Hernández et al. [50] have proposed various ways to compute the internal data abstractions used by various process discovery techniques within the MapReduce framework. Azzinini and Damiani [12] have also revised initial attempts of deploying Process Mining in the Big Data scenario. However, these initial attempts of deploying Big Data technologies in Process Mining do not pay attention to the Relational Data Mining. Performing the parallel computation of (relational) process patterns in Map Reduce is not a trivial task [84]. It requires extending partitioning or sampling procedures (such as those described in [8, 58, 79]), in order to run the Map phase by putting under the same key all event relational data, which must be considered together when performing the final computation (the Reduce phase). An additional open issue regards load balancing as the overall performance of MapReduce is highly dependent on the size of the lists to be handled at Reduce nodes. On the other hand, we cannot overlook that research in Relational Data Mining has recently focused on Big Data. For example, Srinivasan et al. [75] have started the investigation of the parallelization of general-purpose relational data and algorithm via MapReduce. In our opinion, this emerging research trend will fuel future developments of Relational Data Mining in Process Mining via Big Data technologies.

4 Network Data Mining

Networks have become ubiquitous in several social, economical and scientific fields, ranging from the Internet to social sciences, and including biology, epidemiology, geography, finance, and many others. Indeed, researchers in these fields have proven that systems of different nature can be represented as networks [65]. For instance, the Web can be considered as a network of web-pages, which may be connected with each other by edges representing various explicit relations, such as hyperlinks. Social networks can be seen as groups of members that can be connected by friendship relations or can follow other members because they are interested in similar topics of interests. Metabolic networks can provide insight about genes and their possible relations of co-regulation based on similarities in their expressions level. Finally, in epidemiology, networks can represent the spread of diseases and infections.

Due to this explosion of "big" network-structured data, many Data Mining techniques have been proposed in the last ten years to extract knowledge from them. Moreover, due to the fundamental importance of edges and links, many approaches work on relational data and take into account network autocorrelation, that is, a cross-correlation of an attribute with itself [26] in networked data. For instance, in social analysis, autocorrelation can be recognized in the homophily principle, that is, the tendency of nodes with similar values to be linked with each other [62], whereas in bioinformatics it is called *guilt-by-association* principle, which states that proteins sharing similar functional annotations tend to interact more frequently than proteins which do not share them or, in other terms, "when two proteins are found to interact in a high throughput assay, we also tend to use this as evidence of functional linkage" [51].

One of the first approaches has been presented in [1], where the authors have shown that collective classification in relational data often exhibits significant performance gains over conventional approaches, which classify instances individually. More recently [76] have proposed learning predictive clustering trees (PCTs) by embedding heuristics based on relational autocorrelation measures in the tree construction.

Just to mention some applications, in Bioinformatics, Relational Data Mining approaches have been used to inferring gene regulatory networks [22, 67, 69] and to predict the function of the genes by exploiting the guilt-by-association principle [78]. In Social Network analysis, they have been used for intra-/inter-community interaction prediction in dynamic social networks [72]. Finally, in mobility, Rinzivillo et al. [71] have addressed the issue of activity recognition by introducing Activity-Based Cascading (ABC) classification, which exploits Individual Mobility Networks (a model able to capture the salient aspects of individual mobility).

Data Mining solutions have been also proposed for more descriptive tasks, such as, mining changes in networks. This is done, for instance, in [14, 59], where the main idea is to describe how the network evolves. Specifically, in [14], the authors have introduced a hierarchical clustering methodology, which is able to detect clusters of temporal snapshots of a network, interpreted as eras of evolution. In [59], the main idea is to extract the evolution of relationships or properties in the network at consecutive time windows. Evolutions are expressed in the form of *change chains*, which are sequences of frequent patterns, that accommodate temporal information associated to the change.

5 Stream Data Mining

Stream Data Mining is concerned with extracting knowledge structures hidden in continuous streams of information (i.e., unbounded sequences of timestamped data records, which arrive on-line, at a high rate and at consecutive time points). In the last decade, the research in Stream Data Mining has gained a high attraction due to the importance of its applications and the increasing generation of streaming information. Applications of stream data analysis can vary from critical scientific and astronomical applications to important business, social and financial ones [41]. Stream Data Mining faces three principal challenges: volume, velocity, and volatility, which are the same key challenges of Big Data [54]. Volume and velocity require a high volume of data to be processed in limited time. Starting from the first arriving record, the amount of available data constantly increases from zero to potentially infinity. This requires incremental approaches, which incorporate information as it becomes available, and online processing if not all data can be kept. Volatility, on the other hand, corresponds to a dynamic environment with ever-changing patterns. Here, old data is of limited use, even if it could be saved and processed again later. This is due to change, that can affect the induced data mining models in multiple ways: change of the target variable, change in the available feature information, and drift.

Although, Stream Data Mining research has initially focused on upgrading classical Data Mining algorithms (assuming records represented as attribute-value vectors consistently with the single table assumption) to the stream scenario [42], it has also attracted some attention in the area of Relational Data Mining. This interest can be easily motivated by the existence of several real-world data stream applications, where data elements are complex data scattered in several database relations, as well as networked entities related by relationships.

To the best of our knowledge, Appice et al. [7] have authored the seminal Relational Data Mining study to mine data continuously stored in a relational database. Specifically, they have defined the first Relational Data Mining algorithm, that mines relational data streams (stored in a relational database), in order identify novelty patterns, which target new or unknown situations in the stream. They have proposed to mine relational data according to a data window model, that is, the stream is segmented in a sequence of data windows, where each data window consists of structured data elements (spanned in several relations) arriving in a user-defined period (e.g., daily or monthly). Relational patterns are mined from scratch each time a new data window arrives in the stream. The relational pattern discovery is performed by exploring level-by-level the lattice of relational patterns ordered according to a generality relation between patterns. A time window mechanism is defined, in order to filter out novelty patterns. This algorithm has been applied to the problem of detecting anomalies in the network traffic. By following this research direction, Fumarola et al. [40] have then defined an incremental algorithm to mine approximate relational patterns over a sliding time window of a relational data stream. They have described a false positive oriented algorithm, i.e., it does not discover any false negative frequent pattern. They have used a relational version of the SE-tree to efficiently store and retrieve relational patterns. They have defined a procedure to maintain an efficient and accurate approximation of the support of the frequent patterns over the sliding time window. More recently, Silva and Antunes [74] have proposed an algorithm to discover the set of relational frequent patterns in a large star schema, mining directly the data, in their original structure, and exploring the most efficient techniques for mining data streams. Finally, Deng et al. [29] have described a service-oriented manufacturing model to determine the activities of service-oriented manufacturing in the environment of relational data streams.

On the other hand, some interesting developments have been achieved by addressing the challenges of Stream Data Mining research into the subfield of Relational Data Mining, that is focused on the analysis of the network data. In particular, a significant research effort has been devoted to the design of Stream Data Mining algorithms to address various tasks, i.e., summarization, interpolation, forecasting, outlier and change detection, in geophysical data networks (i.e., geophysical data routinely sampled by remote sensor networks).

Mining geophysical data streams poses network-specific issues, like the presence of the spatial correlation, temporal correlation, as well as stream-specific issues like the concept drift according to the statistical properties of the data may change over the time, the presence of a huge volume of data that cannot be entirely recorded for future analysis, the need of processing data on-line, in (near) real time and with low latency. To deal with these challenges, Ciampi et al. [25] have defined a network pattern, namely trend cluster, for geophysical data stream processing. This pattern represents spatially related geosensors, which measure geophysical data evolving similarly over the time. An efficient data synopsis to deal with a huge volume of time-evolving geophysical data and speed-up the trend cluster discovery process has been described in [11]. An algorithm for the incremental trend cluster discovery with the sliding window mechanism has been illustrated in [10]. Finally, an in-network processing framework has been presented in [11], in order to strive for the decentralization of the trend cluster discovery in the an advanced tree-based WSN topology. Although, the trend discovery has been introduced to summarize geophysical data streams [11], it has been also considered, in order to compute stream interpolation models [9], as well as stream forecasting and outlier/change detection models [10]. Finally, Appice and Malerba [5] have also formulated an incremental algorithm to deal with streams of multi-variate networked data by computing a time-evolving network summarization and interpolation model.

Similarly, interesting results have been yielded in the context of social network data streams, as well as mobility data streams. For example, Rossetti et al. [73] have described an online algorithm to understand the relations between the homophily of individuals and the topological features expressed by specific network substructures. In [72], a supervised learning approach has been proposed, in order to exploit features computed by time-aware forecasts of topological measures calculated between social-aware node pairs. Loglisci and Malerba [57] have investigated the use of movement tracking technologies to enable the generation of mobility data in a streaming style. They have defined a stream data mining strategy, that enables the detection of two types of dense regions, one based on spatial closeness, the other one based on temporal proximity. Silva et al. [27] have described a novel structure, called micro-group, to represent the relationship among moving objects. They have formulated an incremental algorithm to maintain micro-groups and to capture their evolution on highly dynamic sub-trajectory data. Pappalardo et al. [66] have investigated the relations between human mobility patterns and socioeconomic development by opening an interesting perspective to study human behavior through multi-type (Big) data.

This partial overview of the literature shows that Stream Data Mining has reached important milestones also handling massive amounts of complex data, e.g., dealing with correlation properties, managing time-evolving data, using distributed processing, parallel processing and incremental learning. The future tendency of the research in this field is moving towards mining Big Data- oriented streams. This is proved by the emerging research trend [28, 86] investing in technological research using Kafka, Spark Flink, Storm, and Samza Streaming-based processing engines. Similarly the vast amount of networked data provides many opportunities for carrying on the development of data network applications building on big data technologies [70].

6 Conclusion

This article has presented some of the many research studies in Relational Data Mining, with an emphasis on the research performed in Italy. It covers various applications of Relational Data Mining, as well as fundamental challenges posed by the emerging era of Big Data. It shows a prolific academic research community, which has performed a challenging research in the field of Relational Data Mining by yielding relevant results in Spatial Data Mining, Process Mining, Network Data Analysis and Stream Data Mining. The next stage is a tighter cooperation with the Big Data technologies, in order to face together the most relevant challenges currently posed by the volume, velocity, variety and veracity of Big Data in a strictly relational scenario.

Acknowledgements The research described in this paper has been funded by the European project MAESTRA - Learning from Massive, Incompletely annotated, and Structured Data (Grant number ICT-2013-612944), the European project H2020 "TOREADOR - TrustwOrthy model-awaRE Analytics Data platform" (Grant number 988797).

References

1. P. Angin, J. Neville, A shrinkage approach for modeling non-stationary relational autocorrelation, in *Proceedings of 8th IEEE International Conference on Data Mining* (IEEE Computer Society, 2008), pp. 707–712
2. D. Anguelov, B. Taskar, V. Chatalbashev, D. Koller, D. Gupta, G. Heitz, A.Y. Ng, Discriminative learning of markov random fields for segmentation of 3d scan data, in *2005 IEEE Computer Society Conference on Computer Vision and Pattern Recognition (CVPR 2005), 20–26 June 2005, San Diego, CA, USA* (IEEE Computer Society, 2005), pp. 169–176
3. L. Anselin, *Spatial Econometrics: Methods and Models* (Kluwer, Dordrecht, 1988)
4. A. Appice, Towards mining the organizational structure of a dynamic event scenario. J. Intell. Inf. Syst. 1–29 (2017)
5. A. Appice, D. Malerba, Leveraging the power of local spatial autocorrelation in geophysical interpolative clustering. Data Min. Knowl. Discov. **28**(5–6), 1266–1313 (2014)
6. A. Appice, D. Malerba, A co-training strategy for multiple view clustering in process mining. IEEE Trans. Serv. Comput. **9**(6), 832–845 (2016)
7. A. Appice, M. Ceci, C. Loglisci, C. Caruso, F. Fumarola, M. Todaro, D. Malerba, A relational approach to novelty detection in data streams, in *Proceedings of the Seventeenth Italian Symposium on Advanced Database Systems, SEBD 2009, Camogli, Italy, June 21–24, 2009* (Edizioni Seneca, 2009), pp. 89–100
8. A. Appice, M. Ceci, A. Turi, D. Malerba, A parallel, distributed algorithm for relational frequent pattern discovery from very large data sets. Intell. Data Anal. **15**(1), 69–88 (2011)
9. A. Appice, A. Ciampi, D. Malerba, P. Guccione, Using trend clusters for spatiotemporal interpolation of missing data in a sensor network. J. Spat. Inf. Sci. **6**(1), 119–153 (2013)
10. A. Appice, P. Guccione, D. Malerba, A. Ciampi, Dealing with temporal and spatial correlations to classify outliers in geophysical data streams. Inf. Sci. **285**, 162–180 (2014)
11. A. Appice, A. Ciampi, D. Malerba, Summarizing numeric spatial data streams by trend cluster discovery. Data Min. Knowl. Discov. **29**(1), 84–136 (2015)
12. A. Azzini, E. Damiani, Process mining in big data scenario, in *Proceedings of the 5th International Symposium on Data-driven Process Discovery and Analysis (SIMPDA 2015), Vienna, Austria, December 9-11, 2015*, vol. 1527 of *CEUR Workshop Proceedings* (CEUR-WS.org, 2015), pp. 149–153
13. S. Bergamaschi, E. Carlini, M. Ceci, B. Furletti, F. Giannotti, D. Malerba, M. Mezzanzanica, A. Monreale, G. Pasi, D. Pedreschi, R. Perego, S. Ruggieri, Big data research in italy: a perspective. Engineering **2**(2), 163–170 (2016)
14. M. Berlingerio, M. Coscia, F. Giannotti, A. Monreale, D. Pedreschi, Evolving networks: eras and turning points. Intell. Data Anal. **17**(1), 27–48 (2013)
15. H. Blockeel, M. Sebag, Scalability and efficiency in multi-relational data mining. SIGKDD Explor. **5**(1), 17–30 (2003)
16. M. Ceci, A. Appice, D. Malerba, Spatial associative classification at different levels of granularity: a probabilistic approach, in *Knowledge Discovery in Databases: PKDD 2004, 8th European Conference on Principles and Practice of Knowledge Discovery in Databases, Pisa, Italy, September 20–24, 2004, Proceedings*, ed. By J. Boulicaut, F. Esposito, F. Giannotti, D. Pedreschi. Vol. 3202 of *Lecture Notes in Computer Science* (Springer, 2004), pp. 99–111

17. M. Ceci, A. Appice, D. Malerba, Discovering emerging patterns in spatial databases: a multi-relational approach, in *Knowledge Discovery in Databases: PKDD 2007, 11th European Conference on Principles and Practice of Knowledge Discovery in Databases, Warsaw, Poland, September 17-21, 2007, Proceedings*, ed. By J.N. Kok, J. Koronacki, R.L. de Mántaras, S. Matwin, D. Mladenic, A. Skowron. Vol. 4702 of *Lecture Notes in Computer Science* (Springer, 2007), pp. 390–397
18. M. Ceci, M. Berardi, D. Malerba, Relational data mining and ILP for document image understanding. Appl. Artif. Intell. **21**(4&5), 317–342 (2007)
19. M. Ceci, P.F. Lanotte, F. Fumarola, D.P. Cavallo, D. Malerba, Completion time and next activity prediction of processes using sequential pattern mining, in *Discovery Science - 17th International Conference, DS 2014, Bled, Slovenia, October 8-10, 2014. Proceedings*, vol. 8777 of *Lecture Notes in Computer Science* (Springer, 2014), pp. 49–61
20. M. Ceci, R. Corizzo, F. Fumarola, M. Ianni, D. Malerba, G. Maria, E. Masciari, M. Oliverio, A. Rashkovska, Big data techniques for supporting accurate predictions of energy production from renewable sources, in *Proceedings of the 19th International Database Engineering & Applications Symposium, Yokohama, Japan, July 13-15, 2015*, ed. By B.C. Desai, M. Toyama (ACM, 2015), pp. 62–71
21. M. Ceci, R. Corizzo, F. Fumarola, M. Ianni, D. Malerba, G. Maria, E. Masciari, M. Oliverio, A. Rashkovska. VIPOC project research summary (discussion paper), in *23rd Italian Symposium on Advanced Database Systems, SEBD 2015, Gaeta, Italy, June 14-17, 2015*, ed. By D. Lembo, R. Torlone, A. Marrella (Curran Associates, Inc., 2015), pp. 208–215
22. M. Ceci, G. Pio, V. Kuzmanovski, S. Dzeroski, Semi-supervised multi-view learning for gene network reconstruction. Plos One **10**(5), e0144031, 2015-12-07 00:00:00.0
23. M. Ceci, R. Corizzo, F. Fumarola, D. Malerba, A. Rashkovska, Predictive modeling of pv energy production: how to set up the learning task for a better prediction? IEEE Transactions on Industrial Informatics **PP**(99), 1–1 (2016)
24. M. Celik, B. Kazar, S. Shekhar, D. Boley, D.L. Northstar, A parameter estimation method for the spatial autoregression model. Technical Report Report No: 2005-00, AHPCRC, 2007
25. A. Ciampi, A. Appice, D. Malerba, G. Saponaro, D. Triglione, Clustering spatio-temporal data streams, in *Proceedings of the Eighteenth Italian Symposium on Advanced Database Systems, SEBD 2010, Rimini, Italy, June 20-23, 2010* (Esculapio Editore, 2010), pp. 230–241
26. N. Cressie, *Statistics for Spatial Data*, 1st edn. (Wiley, Chichester, 1993)
27. T.L.C.da Silva, K. Zeitouni, J.A.F.d. Macdo, M.A. Casanova, A framework for online mobility pattern discovery from trajectory data streams, in *2016 17th IEEE International Conference on Mobile Data Management (MDM)*, vol. 1 (2016), pp. 365–368
28. G. De Francisci Morales, A. Bifet, L. Khan, J. Gama, W. Fan, Iot big data stream mining, in *Proceedings of the 22Nd ACM SIGKDD International Conference on Knowledge Discovery and Data Mining, KDD '16* (ACM, 2016), pp. 2119–2120
29. H. Deng, Y.L. Wang, J. Yang, L.Q. Feng, Framework of service-oriented manufacturing based on multi-relational data stream mining, in *2012 International Conference on Computer Science and Service System* (2012), pp. 1427–1430
30. C. Diamantini, D. Potena, E. Storti, Clustering of process schemas by graph mining techniques (extended abstract), in *Sistemi Evoluti per Basi di Dati - SEBD 2011, Proceedings of the Nineteenth Italian Symposium on Advanced Database Systems, Maratea, Italy, June 26-29, 2011* (2011), p. 49
31. C. Diamantini, L. Genga, D. Potena, W.M.P. van der Aalst, Building instance graphs for highly variable processes. Expert Syst. Appl. **59**, 101–118 (2016)
32. S. Džeroski, N. Lavrač, *Relational Data Mining* (Springer, Berlin, 2001)
33. S. Ferilli, Woman: logic-based workflow learning and management. IEEE Trans. Syst. Man Cybern. Syst. **44**(6), 744–756 (2014)
34. S. Ferilli, The woman formalism for expressing process models, in *Advances in Data Mining. Applications and Theoretical Aspects - 16th Industrial Conference, ICDM 2016, New York, NY, USA, July 13-17, 2016. Proceedings*, vol. 9728 of *Lecture Notes in Computer Science* (Springer, 2016), pp. 363–378

35. S. Ferilli, F. Esposito, A logic framework for incremental learning of process models. Fundam. Inform. **128**(4), 413–443 (2013)
36. S. Ferilli, B.D. Carolis, D. Redavid, Logic-based incremental process mining in smart environments, in *Recent Trends in Applied Artificial Intelligence, 26th International Conference on Industrial, Engineering and Other Applications of Applied Intelligent Systems, IEA/AIE 2013, Amsterdam, The Netherlands, June 17-21, 2013. Proceedings*, vol. 7906 of *Lecture Notes in Computer Science* (Springer, 2013), pp. 392–401
37. S. Ferilli, B.D. Carolis, F. Esposito, Learning complex activity preconditions in process mining, in *New Frontiers in Mining Complex Patterns - Third International Workshop, NFMCP 2014, Held in Conjunction with ECML-PKDD 2014, Nancy, France, September 19, 2014, Revised Selected Papers*, vol. 8983 of *Lecture Notes in Computer Science* (Springer, 2014), pp. 164–178
38. S. Ferilli, F. Esposito, D. Redavid, S. Angelastro, Predicting process behavior in woman, in *AI*IA 2016: Advances in Artificial Intelligence - XVth International Conference of the Italian Association for Artificial Intelligence, Genova, Italy, November 29 - December 1, 2016, Proceedings*, vol. 10037 of *Lecture Notes in Computer Science* (Springer, 2016), pp. 308–320
39. F. Folino, G. Greco, A. Guzzo, L. Pontieri, Mining usage scenarios in business processes: outlier-aware discovery and run-time prediction. Data Knowl. Eng. **70**(12), 1005–1029 (2011)
40. F. Fumarola, A. Ciampi, A. Appice, D. Malerba, A sliding window algorithm for relational frequent patterns mining from data streams, in *Discovery Science, 12th International Conference, DS 2009, Porto, Portugal, October 3–5, 2009*, vol. 5808 of *Lecture Notes in Computer Science* (Springer, 2009), pp. 385–392
41. M.M. Gaber, A. Zaslavsky, S. Krishnaswamy, Mining data streams: a review. SIGMOD Rec. **34**(2), 18–26 (2005)
42. J. Gama, A.R. Ganguly, O.A. Omitaomu, R.R. Vatsavai, M.M. Gaber, Knowledge discovery from data streams. Intell. Data Anal. **13**(3), 403–404 (2009)
43. L. Ghionna, G. Greco, A. Guzzo, L. Pontieri, Outlier detection techniques for process mining applications, in *Proceedings of the Sixteenth Italian Symposium on Advanced Database Systems, SEBD 2008, 22–25 June 2008, Mondello, PA, Italy* (2008), pp. 263–270
44. G. Greco, A. Guzzo, F. Lupia, L. Pontieri, Process discovery under precedence constraints. ACM Trans. Knowl. Discov. Data **9**(4), 32:1–32:39
45. G. Greco, A. Guzzo, D. Saccà, A logic programming approach for planning workflows evolutions, in *2003 Joint Conference on Declarative Programming, AGP-2003, Reggio Calabria, Italy, September 3–5, 2003* (2003), pp. 75–85
46. G. Greco, A. Guzzo, G. Manco, D. Saccà, Mining correlations in workflows executions, in *Proceedings of the Thirteenth Italian Symposium on Advanced Database Systems, SEBD 2005, Brixen-Bressanone (near Bozen-Bolzano), Italy, June 19–22, 2005* (2005), pp. 137–148
47. G. Greco, A. Guzzo, G. Manco, L. Pontieri, D. Saccà, *Mining Constrained Graphs: The Case of Workflow Systems* (Springer, Berlin, 2006), pp. 155–171
48. G. Greco, A. Guzzo, L. Pontieri, D. Saccà, Discovering expressive process models by clustering log traces. IEEE Trans. Knowl. Data Eng. **18**(8), 1010–1027 (2006)
49. G. Greco, A. Guzzo, L. Pontieri, An information-theoretic framework for process structure and data mining. IJDWM **3**(4), 99–119 (2007)
50. S. Hernández, J. Ezpeleta, S.J. van Zelst, W.M.P. van der Aalst, Assessing process discovery scalability in data intensive environments, in *2nd IEEE/ACM International Symposium on Big Data Computing, BDC 2015, Limassol, Cyprus, December 7–10, 2015* (IEEE Computer Society, 2015), pp. 99–104
51. X. Jiang, N. Nariai, M. Steffen, S. Kasif, E. Kolaczyk, Integration of relational and hierarchical network information for protein function prediction. BMC Bioinform. **9**(1), 1–15 (2008)
52. B.M. Kazar, S. Shekhar, D.J. Lilja, R.R. Vatsavai, R.K. Pace, *Comparing Exact and Approximate Spatial Auto-regression Model Solutions for Spatial Data Analysis* (Springer, Berlin, 2004), pp. 140–161
53. L.J. Klein, F.J. Marianno, C.M. Albrecht, M. Freitag, S. Lu, N. Hinds, X. Shao, S. Bermudez Rodriguez, H.F. Hamann, Pairs: a scalable geo-spatial data analytics platform, in *Proceedings of the 2015 IEEE International Conference on Big Data (Big Data), BIG DATA'15* (IEEE Computer Society, Washington, DC, USA, 2015), pp. 1290–1298

54. G. Krempl, I. Zliobaite, D. Brzezinski, E. Hüllermeier, M. Last, V. Lemaire, T. Noack, A. Shaker, S. Sievi, M. Spiliopoulou, J. Stefanowski, Open challenges for data stream mining research. SIGKDD Explor. **16**(1), 1–10 (2014)
55. P. Legendre, Spatial autocorrelation: trouble or new paradigm? Ecology **74**(6), 1659–1673 (1993)
56. F. Lettich, L.O. Alvares, V. Bogorny, S. Orlando, A. Raffaetà, C. Silvestri, Detecting avoidance behaviors between moving object trajectories. Data Knowl. Eng. **102**, 22–41 (2016)
57. C. Loglisci, D. Malerba, *Mining Dense Regions from Vehicular Mobility in Streaming Setting* (Springer International Publishing, Cham, 2014), pp. 40–49
58. C. Loglisci, M. Ceci, A. Appice, D. Malerba, Relational disjunctive patterns mining for discovering frequent variants in process models, in *Sistemi Evoluti per Basi di Dati - SEBD 2011, Proceedings of the Nineteenth Italian Symposium on Advanced Database Systems, Maratea, Italy, June 26-29, 2011* (2011), pp. 227–238
59. C. Loglisci, M. Ceci, D. Malerba, Relational mining for discovering changes in evolving networks. Neurocomputing **150**, 265–288 (2015)
60. D. Malerba, A relational perspective on spatial data mining. IJDMMM **1**(1), 103–118 (2008)
61. E. Masciari, S. Gao, C. Zaniolo, Sequential pattern mining from trajectory data, in *17th International Database Engineering & Applications Symposium, IDEAS '13, Barcelona, Spain - October 09 - 11, 2013*, ed. By B.C. Desai, J. Larriba-Pey, J. Bernardinopages (ACM, 2013), pp. 162–167
62. M. McPherson, L. Smith-Lovin, J. Cook, Birds of a feather: homophily in social networks. Annu. Rev. Sociol. **27**, 415–444 (2001)
63. A. Monreale, F. Pinelli, R. Trasarti, F. Giannotti, Wherenext: a location predictor on trajectory pattern mining, in *Proceedings of the 15th ACM SIGKDD International Conference on Knowledge Discovery and Data Mining, KDD '09* (ACM, New York, NY, USA, 2009), pp. 637–646
64. M. Nanni, R. Trasarti, Querying and mining trajectories with gaps: a multi-path reconstruction approach (extended abstract), in *Proceedings of the Eighteenth Italian Symposium on Advanced Database Systems, SEBD 2010, Rimini, Italy, June 20–23, 2010*, ed. By S. Bergamaschi, S. Lodi, R. Martoglia, C. Sartori (Esculapio Editore, 2010), pp. 126–133
65. M.E.J. Newman, D.J. Watts, *The Structure and Dynamics of Networks* (Princeton University Press, Princeton, 2006)
66. L. Pappalardo, D. Pedreschi, Z. Smoreda, F. Giannotti, Using big data to study the link between human mobility and socio-economic development, in *2015 IEEE International Conference on Big Data (Big Data)* (2015), pp. 871–878
67. G. Pio, M. Ceci, D. D'Elia, C. Loglisci, D. Malerba, A novel biclustering algorithm for the discovery of meaningful biological correlations between micrornas and their target genes. BMC Bioinform. **14**(S-7), S8 (2013)
68. G. Pio, F. Fumarola, A.E. Felle, D. Malerba, M. Ceci, Discovering novelty patterns from the ancient christian inscriptions of rome. JOCCH **7**(4), 22:1–22:21 (2014)
69. G. Pio, M. Ceci, D. Malerba, D. D'Elia, Comirnet: a web-based system for the analysis of mirna-gene regulatory networks. BMC Bioinform. **16**(S-9), S7 (2015)
70. N. Pržulj, N. Malod-Dognin, Network analytics in the age of big data. Science **353**(6295), 123–124 (2016)
71. S. Rinzivillo, L. Gabrielli, M. Nanni, L. Pappalardo, D. Pedreschi, F. Giannotti, The purpose of motion: learning activities from individual mobility networks, in *International Conference on Data Science and Advanced Analytics, DSAA 2014, Shanghai, China, October 30 - November 1, 2014* (IEEE, 2014), pp. 312–318
72. G. Rossetti, R. Guidotti, I. Miliou, D. Pedreschi, F. Giannotti, A supervised approach for intra-/inter-community interaction prediction in dynamic social networks. Soc. Netw. Analys. Min. **6**(1), 86:1–86:20 (2016)
73. G. Rossetti, L. Pappalardo, R. Kikas, D. Pedreschi, F. Giannotti, M. Dumas, Homophilic network decomposition: a community-centric analysis of online social services. Soc. Netw. Analys. Min. **6**(1), 103:1–103:18 (2016)

74. A. Silva, C. Antunes, Multi-relational pattern mining over data streams. Data Min. Knowl. Disc. **29**(6), 1783–1814 (2015)
75. A. Srinivasan, T.A. Faruquie, S. Joshi, Data and task parallelism in ilp using mapreduce. Mach. Learn. **86**(1), 141–168 (2012)
76. D. Stojanova, M. Ceci, A. Appice, S. Dzeroski, Network regression with predictive clustering trees. Data Min. Knowl. Discov. **25**(2), 378–413 (2012)
77. D. Stojanova, M. Ceci, A. Appice, D. Malerba, S. Dzeroski, Dealing with spatial autocorrelation when learning predictive clustering trees. Ecol. Inf. **13**, 22–39 (2013)
78. D. Stojanova, M. Ceci, D. Malerba, S. Deroski, Using PPI network autocorrelation in hierarchical multi-label classification trees for gene function prediction. BMC Bioinform. **14**, 285 (2013)
79. A. Turi, A. Appice, M. Ceci, D. Malerba, Distributed discovery of multi-level approximate process patterns, in *Proceedings of the Sixteenth Italian Symposium on Advanced Database Systems, SEBD 2008, 22–25 June 2008, Italy, Mondello, PA*, ed. by S. Gaglio, I. Infantino, D. Saccà (2008), pp. 57–68
80. W.M.P. van der Aalst, *Process Mining - Discovery, Conformance and Enhancement of Business Processes* (Springer, Berlin, 2011)
81. W.M.P. van der Aalst, No knowledge without processes - process mining as a tool to find out what people and organizations really do, in *KEOD 2014 - Proceedings of the International Conference on Knowledge Engineering and Ontology Development, Rome, Italy, 21-24 October, 2014*, ed. By J. Filipe, J.L.G. Dietz, D. Aveiro (SciTePress, 2014), pp. IS–11
82. W.M.P. van der Aalst, Green data science - using big data in an "environmentally friendly" manner, in *ICEIS 2016 - Proceedings of the 18th International Conference on Enterprise Information Systems, Volume 1, Rome, Italy, April 25-28, 2016*, ed. By S. Hammoudi, L.A. Maciaszek, M. Missikoff, O. Camp, J. Cordeiro (SciTePress, 2016), pp. 9–21
83. W.M.P. van der Aalst, *Process Mining - Data Science in Action*, 2nd edn. (Springer, Berlin, 2016)
84. W.M.P. van der Aalst, E. Damiani, Processes meet big data: connecting data science with process science. IEEE Trans. Serv. Comput. **8**(6), 810–819 (2015)
85. R.R. Vatsavai, A. Ganguly, V. Chandola, A. Stefanidis, S. Klasky, S. Shekhar, Spatiotemporal data mining in the era of big spatial data: algorithms and applications, in *Proceedings of the 1st ACM SIGSPATIAL International Workshop on Analytics for Big Geospatial Data, BigSpatial '12* (ACM, New York, NY, USA, 2012), pp. 1–10
86. M. Wang, J. Liu, W. Zhou, Design and implementation of a high-performance stream-oriented big data processing system, in *2016 8th International Conference on Intelligent Human-Machine Systems and Cybernetics (IHMSC)*, vol. 01 (2016), pp. 363–368
87. H. Watanabe and S. Muggleton. Can ilp be applied to large datasets? in *Inductive Logic Programming: 19th International Conference, ILP 2009, Leuven, Belgium, July 02-04, 2009. Revised Papers*, ed. By L. De Raedt (Springer, Berlin, Heidelberg, 2010), pp. 249–256
88. X. Wu, X. Zhu, G.Q. Wu, W. Ding, Data mining with big data. IEEE Trans. Knowl. Data Eng. **26**(1), 97–107 (2014)

Data Mining in Databases: Languages and Indices

Elena Baralis, Tania Cerquitelli, Silvia Chiusano and Rosa Meo

Abstract Database systems methodologies and technology can provide a significant support to data mining processes. In this chapter we explore approaches which address the integration between data mining activities and DBMSs from different perspectives. More specifically, we focus on (i) specialized query languages which allow to define complex data mining tasks through the submission of query requests, and (ii) indices, i.e., physical data structures designed to improve the performance of mining algorithms.

Keywords Inductive databases · Data mining · Database indices · Association rules · Specialised query languages

1 Introduction

The topic of data mining is becoming every day more important since most any company and data center nowadays has accumulated in its history large volumes of data and is willing to analyse them and acquire knowledge in order to increase the competitive advantage over competitors or improve its business processes. The knowledge might be a model applicable to predict a target variable or descriptive, such that the user can use it in order to summarise the details of the data that cannot be analysed by a human given the large volumes of stored data. Of particular importance is the fact that very often the majority of these data, especially if it regards the past history of the business, such as clients information and past interaction, is stored in a

E. Baralis · T. Cerquitelli · S. Chiusano
Politecnico di Torino, corso Duca degli Abruzzi 24, Torino, Italy
e-mail: Elena.Baralis@polito.it

T. Cerquitelli
e-mail: Tania.Cerquitelli@polito.it

S. Chiusano
e-mail: Silvia.Chiusano@polito.it

R. Meo (✉)
Università di Torino, corso Svizzera 185, Torino, Italy
e-mail: Rosa.Meo@di.unito.it; meo@di.unito.it

S. Flesca et al. (eds.), *A Comprehensive Guide Through the Italian Database Research Over the Last 25 Years*, Studies in Big Data 31, DOI 10.1007/978-3-319-61893-7_20

persistent way in a DBMS which is almost always a relational database. An immediate conclusion arises: more technology and specialised systems should be developed by the database and data mining community in order to enhance and distribute facilities for the analysis of large volumes of data stored in relational databases.

Data mining is an interdisciplinary subfield of computer science aimed at discovering unexpected and potentially useful knowledge from large data collections. It includes a plethora of methods at the intersection of artificial intelligence, machine learning, statistics, and database systems, that can be profitably used to mine interesting patterns such as groups of similar data objects (cluster analysis), dependencies among data objects (association rule analysis), or a model describing data classes (classification). Traditionally data mining algorithms analyse large data collections stored into flat-files, even if possibly extracted from a DBMS. However this data extraction is not much practical, especially if a large volume of data is interested in the extraction. More effectively some research activity has addressed the application of data mining methods directly on relational data. Coupling data mining methods with the usual querying techniques can have a great potential in finally providing the end-user a richer, more diversified and interesting knowledge.

In this chapter we focus mainly on the knowledge that can be extracted from relational databases under the form of frequent patterns. In particular we focus on frequent itemsets and association rules that have been used with success in the past to solve predictive tasks such as classification [22] and to form a descriptive model of the dataset as well by the collection of frequent patterns extracted [3, 4].

Association rule mining, aiming at discovering correlations among data items, is the collection of data mining methods that can be more easily exploited for DBMS mining. Association rules are extracted from a transactional database $\mathcal{D}$. $\mathcal{D}$ is a collection of transactions, where each transaction is a set of data items. Association rules are usually represented in the form $A \rightarrow B$, where A and B are itemsets, i.e., sets of data items. Itemsets are characterized by their frequency of occurrence in $\mathcal{D}$, which is called support. Different constraints may be enforced to reduce the computational cost of itemset extraction, among which the most simple are support and item constraints [21, 37]. The support constraint enforces a threshold on the minimum support of the extracted itemsets. The item constraint enforces the extraction of the complete set of itemsets which include the required items.

Various efficient algorithms has been proposed for itemset extraction, which represents the most computationally intensive knowledge extraction task in association rule mining [2]. Ad-hoc main memory data structures are exploited to efficiently extract itemsets from data collections usually stored into binary files [2, 5, 17, 24, 28, 32, 34] Recently, disk-based extraction algorithms have been proposed to support the extraction from these large datasets [12, 15, 31].

In this chapter, we provide the following contributions: (i) an overview of the query languages and query optimisation techniques that can be used to interact with the database system and specify the type of patterns in which the analyst is interested in and (ii) a DBMS-based approach to support itemset mining queries. Relational DBMSs exploit indices, which are ad hoc data structures, to support the execution of complex queries. Following this approach, we describe the *IMine index*

(Itemset-Mine index), proposed in [6, 7], a compact data structure that provides a complete representation of transactional data supporting efficient itemset extraction from a relational DBMS.

2 An Overview of Specialised Query Languages for Data Mining

Inductive databases have been proposed to afford the problem of knowledge discovery from huge databases in [18]. This kind of databases integrates raw data with knowledge extracted from raw data, materialized under the form of patterns. With an inductive database the user/analyst performs a set of operations on data and on patterns using a specialized query language, powerful enough to perform all the required manipulations to support the KDD process, such as data preprocessing, pattern discovery and pattern post-processing.

- Selection of data to be mined.
- Specification of the type of patterns to be mined (descriptive or predictive).
- Specification of the background knowledge, for instance under the form of a concept hierarchy or ontology.
- Definition of constraints that the extracted patterns must satisfy in order to allow the user to specify the interesting patterns. This occurs usually by using measures like frequency, generality, coverage, similarity, novelty, etc.
- Satisfaction of the closure property (by storing the results in the database).
- Post-processing of results in order to allow the user to interact with the extracted patterns by browsing, apply selection templates, cross over patterns and data by selection of the data in which some patterns hold, or aggregating results.

A few query languages can be considered as candidates for inductive databases. We present a brief comparison between query languages that have been proposed for association rules extraction: MSQL, DMQL and MINE RULE. This allows us to compare the language design guidelines, with particular attention to the features of inductive databases for which they are designed.

2.1 MSQL

MSQL has been described in [19]. It comprises the following statements:

GetRules: it generates rules into a rule base;
SelectRules: it queries the rule base;
Create Encoding: it efficiently encodes discrete values into continuous valued attributes;

Satisfies and violates: they allow to cross-over data and rules, and that can be used in a data selection statement.

The main features of this language are the following. It has the ability to nest SQL expressions such as sorting and grouping in a unique MSQL statement. It satisfies the closure property and provides operators to manipulate results of previous queries. It can perform a cross-over between data and rules with operations allowing to identify subsets of data satisfying or violating a given set of rules. It distinguishes between rule generation and rule querying. Indeed, as the volume of generated rules might explode, rules might be extensively generated only at querying time, and not at generation time.

2.2 DMQL

DMQL has been presented in [16]. It consists of the specification of four major primitives for the management of:

1. the set of relevant data w.r.t. a data mining process; this is specified by means of conventional query.
2. the kind of knowledge to be discovered; it includes association rules, classification rules, characteristic descriptions that are a summarization of the common properties of the data, comparisons descriptions that discriminate the tuples belonging to a class with different classes, generalized relations obtained by generalizing a set of data according to the conceptual level described in a specified concept hierarchy.
3. the background knowledge by providing a set of primitives for the management of a set of concept hierarchies or generalization operators that assist the generalization processes.
4. the justification of the interestingness of the knowledge (i.e., by evaluation measure thresholds such as association rules measures like the classical support and confidence thresholds, the allowed noise and rule novelty).

2.3 MINE RULE

MINE RULE has been originally presented in [26]. This operator extracts a set of association rules from the database and stores them back in the database in a separate relation. This language is an extension of SQL. Its main features are the following.

1. Selection of the relevant set of data for a data mining process; this selection is applied at different granularity levels, that is at the row level (selection of a subset of the rows of a relation) or at the group level (group condition).

2. Definition of the structure of the rules to be mined and of constraints applied at different granularity levels; it might define either unidimensional association rule (i.e., rules elements are values of the same dimension or attribute), or multidimensional. Furthermore, rules constraints can be applied at the rule level (mining conditions) in order to filter single rules, or can be applied as cluster conditions, in order to build separately the two parts of the rules (body or head).
3. Definition of the grouping condition that determines which data of the relation can take part to an association rule;
4. Definition of rule evaluation measures (i.e., support and confidence thresholds).

2.4 Optimization of Mining Queries

In [25] we highlighted the relationships between two mining queries: equivalence, inclusion and dependence.

Equivalence: Two queries are equivalent if for all instances of the source data each rule r in the result set of the first query is also in the result set of the second query and vice versa with the same value of rules evaluation measures (support and confidence).

Inclusion: A first query is included in the second one if for all instances of the source data each rule r in the result set of the first query is also in the result set of the second query with the same value of the rules evaluation measures.

Dominance: A first query is dominated by a second one if for all instances of the source data each rule r in the result set of the first query is also in the result set of the second query. Furthermore, in the result of the second query the values of the rules evaluation measures are an upper bound of the values of the corresponding rules from the second query.

We showed the practical implications of the discussed principles with a set of algorithms designed for MINE RULE. These algorithms use also a new designed mining index called mining that allows to reduce the portion of database to be read in response to some classes of queries. In these cases the workload of the mining engine is greatly reduced or completely saved.

In [27] we proposed to optimize constraint-based queries on itemsets with the aim to reduce the overall computation time of a mining query. We introduced a very generic constraint-based language for the extraction of frequent itemsets and presented an optimization scheme that exploits the available materialization of previous queries. The optimization scheme proposed is based on query rewriting. We studied the conditions under which query rewriting is possible and suggested a way to find such a rewriting. For efficiency, we proposed a composition scheme of the materializations that makes usage of common and efficient operations in DBMSs, i.e., intersection and union of relations.

3 Using Indices to Mine Frequent Patterns

A cornerstone of efficient query processing in relational DBMSs is the exploitation of indices. An index is a specialized data structure that supports selective access to the subset of physical pages needed to process a query. A variety of different data structures have been proposed to support data access both for the relational and non-relational data representations.

While several disk-based data structures have been proposed to support itemset mining (e.g., [9, 12, 15, 34] which are further discussed in Sect. 3.3), the definition of index structures to support frequent pattern mining in relational DBMSs has been addressed only in [6, 7]. The *IMine index* (Itemset-Mine index) [6, 7] has been fully integrated into the PostgreSQL DBMS kernel [30]. It is a persistent data structure that provides efficient and effective data access to itemset mining algorithms. More specifically, the IMine index provides a compact and complete representation of transactional data and is characterized by several important properties.

Complete representation. The IMine index is created without enforcing any constraint (e.g., support or item constraint). Hence, it is a *covering index*, i.e., itemset mining can be performed by means of the index alone, without accessing the original database. The IMine index provides a *complete* representation of the transactional data, thus supporting itemset extraction with arbitrary support thresholds.

General structure. The structure of the IMine index is designed to support a variety of itemset extraction algorithms. These algorithms are typically characterized by different in-memory data representations (e.g., array list, prefix-tree) and techniques to explore the search space. The IMine index features efficient data access methods to load in memory the data needed by the considered extraction algorithm. The enforcement of item constraints is also supported by the IMine index access methods. Finally, the generality of the index structure allows it to gracefully adapt to both sparse and dense data distributions.

Efficient data access. The physical organization of the IMine index is designed to provide efficient access to physical data blocks during the mining process. Correlated data are stored in the same physical block, thus allowing a significant reduction of the number of block reads.

The IMine index exploits PostgreSQL open source DBMS [30] physical level access methods. The performance of the approach has been compared with state-of-the-art algorithms (i.e., Prefix-Tree [14] and LCM v.2 [35]) accessing binary data on a flat file. The IMine index always provides a better performance, which also scales linearly for large datasets.

3.1 IMine Index Structure

The IMine index is characterized by two levels of indexing. The first level stores the transactional data in a compact prefix-tree structure which provides a lossless representation of the data, the *Itemset-Tree* (*I-Tree*). The second level allows reading selected I-Tree portions during itemset mining. It is a B+Tree structure, the *Item-Btree* (*I-Btree*), that stores the physical location of all item occurrences in the I-Tree. Hence, it provides efficient access to the I-Tree data blocks to load in memory the transactions including a selected item.

IMine data access methods. Different itemset mining algorithms may exploit the IMine index structure to load data in memory. Three different data access methods are available, each one providing an in-memory representation appropriate for the selected mining algorithm (e.g., FP-tree for FP-growth [17], array-based structure for LCM [35]). These methods access different parts of the IMine index, depending on the adopted itemset mining algorithm and on the enforced support and/or item constraints. More specifically, the following data access methods have been designed. The *Frequent-item based projection,* method supports projection-based algorithms (e.g., FP-growth [17]), while the *Support-based projection,* supports level-based (e.g., APRIORI [2]), and array-based (e.g., LCM v.2 [35]) algorithms. Finally, the *Item-based projection,* loads in memory all transactions including a given item and is exploited for item constrained mining.

IMine is a covering index. Hence, the original transactional database is not accessed. During the mining process, only a small portion of the entire dataset is actually loaded in memory for the local search performed by the extraction algorithm. By accessing the IMine index, only the relevant index blocks are loaded in memory, thus significantly reducing the number of disk reads. Read disk blocks are stored in the buffer cache memory of PostgreSQL. Furthermore, a very limited data portion is actually loaded in memory at each step of the algorithm. Hence, more memory space becomes available for the mining process.

IMine physical organization. The design of the physical organization of the IMine index aims at minimizing the cost of reading the data needed by the current step of the mining algorithm. The selection of the blocks including the paths of interest is performed by means of the I-Btree. Thus, the number of disk blocks read to load the required I-Tree paths is the most important factor contributing to the I/O cost. The correlation between index parts, i.e., data paths accessed together during the current mining step, is exploited to reduce the I/O cost. More specifically, correlated index parts are stored together in the same disk block.

Furthermore, the I-Tree is partitioned in three distinct layers. The intuition driving the partitioning process is that items with very low support do not satisfy most support constraints. Hence, they are accessed only rarely during the mining process. The nodes corresponding to low-support items belong to the lower levels of the I-Tree. More specifically, the frequency of the node accesses performed by the mining process is considered to perform the partitioning. The interaction of the following three factors affects node access frequency: (a) the node support, which shows the number of paths including the node, (b) the global support of the item associated to the node, and (c) the distance of the considered node from the root, described by the node level in the tree.

3.2 *Itemset Mining*

The IMine index can support a variety of different itemset mining algorithms. The main difference among different approaches is in (a) the main memory data structure exploited to store the required data, and (b) the strategy adopted by the algorithm to visit the search space. IMine data access methods load in memory the data needed by the current step of the itemset mining algorithm. In each step of the mining process, data is read from the I-Tree and loaded in memory, in the appropriate structure for the selected mining algorithm. Next, mining takes place on the loaded data.

Enforcing constraints. The specification of constraints on the extraction process yields a subset of (more) interesting itemsets and may help the human analyst in focusing on relevant knowledge. Pushing constraint enforcement into the mining process would allow early pruning the search space, thus improving the efficiency of the mining process. Hence, a significant research activity has been focused on the definition of strategies for constrained itemset extraction [10, 11, 21, 29, 33]. A classification of constraints into anti-monotonic, monotonic, succinct, and convertible has been proposed in [29], which also addresses constraint enforcement into the FP-growth algorithm. The IMine index can directly support the access strategies described in [29]. More specifically, the items of interest can be straightforwardly selected by accessing the I-Btree.

3.3 *Disk-Based Strategies to Support Frequent Pattern Mining*

Several approaches have been proposed to support itemset extraction from flat file by means of disk-based data structures. These approaches do not support the tight integration of itemset extraction in a relational DBMS. However, they exploit some form of file indexing structure to support the mining process.

An interesting hybrid in-core/out-of-core approach has been presented by Lucchese et al. [23]. Even though the disk is exploited as an auxiliary means to extend scalability, the mining process is still mainly memory-based [23]. An index structure based on signature files is proposed in [20]. It supports the candidate frequent itemsets generation process, but the actual candidate frequency check requires further dataset access.

Fully disk-based mining algorithms have been proposed to support the extraction of knowledge from large datasets (e.g., B+tree-based indices [34], Inverted Matrix [12], Diskmine [15] I/O conscious optimizations [9], DRFP-tree [1], VLDB-Mine [8]). The Inverted Matrix is a disk-based data structure proposed by El-Hajj and Zaïane [12] to store the transactional dataset in an inverted matrix layout. The proposed data structure deals well with very sparse datasets, in which a large number of items are characterized by unitary support. B+tree-based indices to access data have been proposed by Ramesh et al. [34]. The adopted data representation is either vertical (e.g., ECLAT-Based [36]) or horizontal (e.g., APRIORI-Based [2]).

In [15] large databases are materialized on disk by storing different (recursively generated) projected databases whose size fits main memory. Each projection, represented as an FP-tree, is first materialized on disk, and then separately loaded in main memory for itemset extraction. Diskmine allows efficient memory saving and maximizes memory exploitation. However, storing all projections may require significant disk space, and subsequently cause a non-negligible I/O cost during the mining process.

The path tiling approach, proposed by Buehrer et al. [9, 13], is an efficient itemset mining techniques exploiting I/O conscious optimizations. More specifically, several data locality strategies are proposed to reduce the number of reads during the mining process. However, different data structures and mining algorithms may have different data locality requirements. Hence, different I/O conscious techniques should be defined for different mining approaches.

In [8] a persistent and hybrid structure, named VLDBMine, is proposed to compactly store huge transactional datasets characterized by a variable data distribution. VLDBMine has been designed to support existing in-core algorithms by enhancing memory usage, thus achieving scalability through different selective data retrieval methods.

4 Conclusions

The full integration of data mining techniques as DBMS services is a challenging goal yet to be achieved. In this chapter, we presented different integration attempts, which address both the query/mining specification language and the definition of physical data structures to improve the performance of mining algorithms. From one side, different SQL-like mining languages have been proposed to ease the continuous and exploratory interaction between the user and the DBMS. From the other side, the IMine index, discussed in this chapter, supports efficient itemset mining into a relational DBMS. It has been implemented into the PostgreSQL open source DBMS and it profitably exploits its services, among which physical level access methods and DBMS buffer management.

In the future we hope that new specialised query languages and index structures will be proposed in the same vein to facilitate the interaction of the user and the data analysis on the very big data stored in the so-called No-SQL databases.

References

1. M. Adnan, R. Alhajj, Drfp-tree: disk-resident frequent pattern tree. Appl. Intell. Springer **30**(2), 84–97 (2009)
2. R. Agrawal, R. Srikant, Fast algorithms for mining association rules in large databases, in *VLDB '94* (1994), pp. 487–499

3. R. Agrawal, R. Srikant, Mining sequential patterns, in *International Conference on Data Engineering*, Taipei, Taiwan, March 1995
4. R. Agrawal, T. Imielinski, A. Swami, Mining association rules between sets of items in large databases, in *Proc.ACM SIGMOD Conference on Management of Data*(British Columbia, Washington, D.C., 1993), pp. 207–216
5. R. Agrawal, T. Imilienski, A. Swami, Mining association rules between sets of items in large databases, in *SIGMOD'93*, Washington DC, May 1993
6. E. Baralis, T. Cerquitelli, S. Chiusano, Index support for frequent itemset mining in a relational dbms, in *ICDE* (2005), pp. 754–765
7. E. Baralis, T. Cerquitelli, S. Chiusano, Imine: index support for item set mining. IEEE Trans. Knowl. Data Eng. **21**(4), 493–506 (2009)
8. E. Baralis, T. Cerquitelli, S. Chiusano, A. Grand, Scalable out-of-core itemset mining. Inf. Sci. **293**, 146–162 (2015)
9. G. Buehrer, S. Parthasarathy, A. Ghoting, Out-of-core frequent pattern mining on a commodity pc, in *KDD '06* (2006), pp. 86–95
10. Y.-L. Cheung, Mining frequent itemsets without support threshold: with and without item constraints. IEEE Trans. Knowl. Data Eng. **16**(9), 1052–1069 (2004). Member-Ada Wai-Chee Fu
11. G. Cong, B. Liu, Speed-up iterative frequent itemset mining with constraint changes, in *ICDM* (2002), pp. 107–114
12. M. El-Hajj, O.R. Zaiane, Inverted matrix: Efficient discovery of frequent items in large datasets in the context of interactive mining. in *ACM SIGKDD* (2003)
13. A. Ghoting, G. Buehrer, S. Parthasarathy, D. Kim, A. Nguyen, Y.-K. Chen, P. Dubey, Cache-conscious frequent pattern mining on modern and emerging processors. VLDB J. **16**(1), 77–96 (2007)
14. G. Grahne, J. Zhu, Efficiently using prefix-trees in mining frequent itemsets, in *FIMI*, November 2003
15. G. Grahne, J. Zhu, Mining frequent itemsets from secondary memory, in *ICDM '04* (IEEE Computer Society, Washington, DC, USA, 2004), pp. 91–98
16. J. Han, Y. Fu, W. Wang, K. Koperski, O. Zaiane, DMQL: a data mining query language for relational databases, in *Proceedings of SIGMOD-96 Workshop on Research Issues on Data Mining and Knowledge Discovery* (1996)
17. J. Han, J. Pei, Y. Yin, Mining frequent patterns without candidate generation, in *SIGMOD '00* (2000), pp. 1–12
18. T. Imielinski, H. Mannila, A database perspective on knowledge discovery. Commun. ACM **39**(11), 58–64 (1996)
19. T. Imieliński, A. Virmani, Msql: a query language for database mining. Data Min. Knowl. Disc. **3**(4), 373–408 (1999)
20. B. Lan, B.C. Ooi, K.-L. Tan, Efficient indexing structures for mining frequently patterns, in *IEEE ICDE* (2002)
21. C.K.-S. Leung, L.V.S. Lakshmanan, R.T. Ng, Exploiting succinct constraints using fp-trees. SIGKDD Explor. Newsl. **4**(1), 40–49 (2002)
22. B. Liu, W. Hsu, Y. Ma, Integrating classification and association rule mining, in *Proceedings of the Fourth International Conference on Knowledge Discovery and Data Mining, KDD'98* (AAAI Press, 1998), pp. 80–86
23. C. Lucchese, S. Orlando, R. Perego, kdci: on using direct count up to the third iteration, in *FIMI* (2004)
24. H. Mannila, H. Toivonen, A. Inkeri Verkamo, Efficient algorithms for discovering association rules, in *KDD Workshop* (1994), pp. 181–192
25. R. Meo, Optimization of a language for data mining, in *Proceedings of the 2003 ACM Symposium on Applied Computing*, Melbourne, Florida, 2003
26. R. Meo, G. Psaila, S. Ceri, A new SQL-like operator for mining association rules, in *Proceedings of the 22st VLDB Conference*, Bombay, India, September 1996

27. R. Meo, M. Botta, R. Esposito, *Query Rewriting in Itemset Mining* (Springer, Berlin, 2004), pp. 111–124
28. S. Orlando, C. Lucchese, P. Palmerini, R. Perego, F. Silvestri, kDCI: a multi-strategy algorithm for mining frequent sets, in *FIMI* (2003)
29. J. Pei, J. Han, L.V.S. Lakshmanan, Pushing convertible constraints in frequent itemset mining. Data Min. Knowl. Discov. **8**(3), 227–252 (2004)
30. PostgreSQL. Postgresql, http://www.postgresql.org
31. G. Ramesh, W.A. Maniatty, M.J. Zaki, Indexing and data access methods for database mining, in *DMKD* (2002)
32. A. Savasere, E. Omiecinski, S.B. Navathe, An efficient algorithm for mining association rules in large databases, in *VLDB* (1995), pp. 432–444
33. R. Srikant, Q. Vu, R. Agrawal, Mining association rules with item constraints, in *KDD* (1997), pp. 67–73
34. H. Toivonen, Sampling large databases for association rules, in *VLDB* (1996), pp. 134–145
35. T. Uno, M. Kiyomi, H. Arimura, LCM ver. 2: efficient mining algorithms for frequent/closed/maximal itemsets, in *FIMI '04* (2004)
36. M.J. Zaki, Scalable algorithms for association mining. IEEE Trans. Knowl. Data Eng. **12**(3), 372–390 (2000)
37. L. Zhao, M.J. Zaki, N. Ramakrishnan, BLOSOM: a framework for mining arbitrary boolean expressions, in *KDD '06* (ACM Press, New York, NY, USA, 2006), pp. 827–832

20+ Years of Analytics on Complex Data: Impact, Issues, Challenges and Contributions

Stefano Basta, Giuseppe Manco, Elio Masciari and Luigi Pontieri

1 Introduction

Computer Science is a relatively young discipline, but in the last two decades the advances in hardware technology and software engineering has induced notable changes in the way users interact with computers. In particular, several processes involving data have changed in a radical manner. As a matter of fact, the amount of data stored in repositories has grown at impressive rates due to the rise of data sources, such as sensor networks, social networks or operational processes. Moreover, the heterogeneity of data has dramatically increased. In a word, data and their management have became more and more *complex*.

Besides the data issues, new challenges have emerged from an application viewpoint, enabled by the informatization and digitalization of even the most trivial processes in the society. New applications span from real-time monitoring of dependable systems by means of cyber-physical sensors, to information diffusion at all levels by means of persistently connected devices, to the capability of tracking and delivering ad-hoc services to individuals. We're living in a connected world and the new technologies allow us with unforeseen possibilities.

Along the recent years, new strategies (specifically, data loading, transformation and extraction, data management, and data mining approaches) have been proposed in order to deal with these unprecedented source of information. The advent of Big Data

S. Basta · G. Manco (✉) · E. Masciari · L. Pontieri
ICAR-CNR, Rende, Italy
e-mail: Giuseppe.manco@icar.cnr.it

S. Basta
e-mail: Stefano.Basta@icar.cnr.it

E. Masciari
e-mail: Elio.Masciari@icar.cnr.it

L. Pontieri
e-mail: Luigi.Pontieri@icar.cnr.it

S. Flesca et al. (eds.), *A Comprehensive Guide Through the Italian Database Research Over the Last 25 Years*, Studies in Big Data 31, DOI 10.1007/978-3-319-61893-7_21

technologies has empowered the whole discipline of data analytics, by eventually creating even new professional figures which summarize skills in computer science and databases, statistics, machine learning and economics.

However, the explosion of this new discipline is not new, and finds its path in the several contributions throughout the last 20 years. The Italian community has been particularly active in this scenario, by providing cutting-edge scientific results and prototypes in the field. As a proof of concept of this, this chapter tries to summarize some important contributions to the analysis of complex data both from the technological and the application point of view. In the following, we briefly recall the basic data types and methods that we have explored in the last years. In order to give the reader a more appealing view of the mentioned approaches, we also describe some applications that we have carried out with respect to some key important topics in the framework of the H2020 societal challenges.

2 Data

In this section, we describe some of the data types that we have widely investigated both for scientific and industrial applications in recent years.

2.1 Sets and Transaction Data

The need to deal with transactional data, i.e., tuples of variable size of categorical data, comes from many relevant applications, such as finance (where the transactions represent orders, invoices, payments), productivity (where transactions represent event plans or activity records), logistics (deliveries, storage or travel records). Thus, transactions can involve everything from a purchase order to shipping status to employee hours worked to insurance costs and claims.

High-dimensional categorical data, such as market-basket and web-usage data, is a particular facet of transactional data. Records in such datasets include a large number of attributes, typically with boolean values. Several emerging application settings require clustering techniques that provide an effective treatment of this kind of data, such as text analysis, bioinformatics, e-commerce, astronomy and insurance industry

Formally, high dimensional transactional data can be defined as follows. Let us consider a set $\mathcal{M} = \{a_1, \ldots, a_m\}$ of boolean attributes, and a data set $\mathcal{D} = \{\mathbf{x}_1, \mathbf{x}_2, \ldots, \mathbf{x}_n\}$ of tuples defined on $\mathcal{M}$. In the current literature, $a \in \mathcal{M}$ is usually denoted as an *item*, and a tuple $\mathbf{x} \in \mathcal{D}$ as a *transaction*. Usually, $\mathbf{x}$ is represented in a more compact form, as a proper subset of $\mathcal{M}$, with the meaning that all the items explicitly represented in $\mathbf{x}$ take value *true*, and the others take value *false*. For example, transaction $\mathbf{x} = \{a_1, a_5\}$ is a compact representation of a tuple defined on $\mathcal{M}$ where all attributes are *false* except for attributes a_i and a_5. Data sets composed by transactions are usually denoted as *transactional data* or *high-dimensional categorical data* (i.e., data which adhere to a schema where there are several, not necessarily boolean attributes). Notice that each data set whose attributes are categorical can be represented as transactional data (by explicitly representing each possible attribute value as a boolean attribute in $\mathcal{M}$).

The analysis of transactional data takes place in several scenarios. The most well known examples is the case of frequent itemset mining and association rules [2], but there are other interesting scenarios where clustering [19] or classification based on transactions takes place.

2.2 Sequences

Sequences are the natural evolution of transactions, and they play an important role in disparate real-life scenarios, including many relevant scientific/medical, security, and business applications. In short, sequences can be considered as transactions where the order matters, e.g., because of a temporal sequentiality. Examples of such data are indeed DNA sequences and protein sequences, and the log traces produced by various kinds of systems, networks, and business processes. As discussed in [29], a major kind of sequences are "event sequences", i.e., sequences of events (e.g., weblogs, customer purchase histories, software traces, process traces) that encode information on actions performed by some real-world entity. For example, each event in a weblog corresponds to a request of some web resource; each record in a customer purchase history provides information on a purchase event (in terms of who made the purchase, where and when the latter was made, which items were purchased, etc.); a trace of a business process represents the sequence of operations performed on one instance of the process.

Let us provide a general definition for such kinds of data, inspired to [29]. Let E be a given data type, describing the structure and semantics of the elements that may constitute each sequence. Then a *sequence* over E is an ordered list $S = s_1 \ldots s_m$, where:

1. each s_i (also denoted as $S[i]$) is a member of E, and is called an *element* of S,
2. m (also denoted as $|S|$) is the *length* of S, and
3. each number in $\{1, \ldots, |S|\}$ indicates a position in S.

According to this definition, DNA sequences are (symbolic) sequences over {`A`, `C`, `G`, `T`}, whereas both customer purchase histories and process/system traces are (event) sequences defined over complex element types.

Let us now focus on the case of business process traces, as an archetype of the challenging class of event sequences where the underlying element type encodes more information than a simple concept label/symbol.

Traces and logs. Whenever a process is enacted, a *trace* is recorded for each process instance (a.k.a. "case"), which stores the sequence of *events* occurred along the execution of the instance. Let E and T denote the universes of all possible events and traces for the process under analysis, respectively. Clearly, according to the above definition, each trace is a sequence over (the element type) E. For the sake of notation and of concreteness, let us assume that each event e in E is a tuple storing two main kinds of information: a timestamp indicating when the event occurred (denoted as $time(e)$), and a (sub-)tuple $prop(e)$ of properties capturing various aspects of the event. Some properties that appear in the events of many real logs are: the process activity/operation performed and associated parameters/results, and information about the actor involved.

Finally a (process) *log* L (over T) is a multiset of T's traces. The analysis of such data has recently gained momentum with the rapid growth of the Process Mining field of research, that addresses the "*confrontation between event data and process models*" [53, 54] through new process-aware data analysis tasks.

Different kinds of behavioral patterns and models can be extracted for a business process, based on a log gathering the traces generated by past executions of it. In particular, an important class of descriptive models that can be discovered are workflow-like process models, which summarize the typical behaviors of a process in terms of elementary activities and routing constructs (e.g., choices, parallelism, loops, synchronization). These models can be represented through simple directed graphs representing precedence relationships, or by using more expressive languages enjoying such as specific Petri net dialects. Event logs can be also exploited to discover predictive process models [28], capable to predict the outcome (in terms of a numerical performance measure, a risk indicator, or the deviation w.r.t. to some given criterion of normality) for an (a possibly unfinished) process instance, based on its associated trace.

Event/trace abstraction. In both cases, in order to discover meaningful and general models, it is often fundamental to bring the given traces to some proper level of abstraction. A common solution consists in applying some classification scheme for the events, such that any possible event will be abstracted into an event class representing some type of activity/action.

In general, an *event abstraction function* α for a given event universe E is a function mapping each event $e \in E$ onto a symbol $\alpha(e)$, which identifies a distinguished event class. Let us also denote as $\alpha(E)$ the alphabet of symbols that α would generate when applied to all possible events. Most Process Mining approaches adopt simple event abstraction functions that just project the contents of each event onto one of its attributes (usually, the task identifier).

Based on such a function, every log trace can be turned into a string over $\alpha(E)$, which represents a sequence of activity/action types. Precisely, for each trace $\tau \in T$, let $\alpha(\tau)$ be the sequence obtained by replacing each event in τ with its abstract representation: $\alpha(\tau) = \alpha(\tau[1]), \ldots, \alpha(\tau[|\tau|])$. Let also $\alpha(L)$ be multi-set that contains, for each trace τ in L, an element $\alpha(\tau)$ with the same multiplicity of the former.

2.3 Trees and Semistructured Data

In late 90's a new data type received a great deal of attention, namely semi-structured data. In particular a lot of research has been done on XML data. XML documents are represented as unordered trees, whose nodes belong to the alphabet $\mathbb{N}$ and are labeled with symbols in the alphabet Σ (alphabet Σ is assumed to be infinite, as it represents XML tags, attributes, and text).

A *tree* is a directed acyclic graph where all the nodes have indegree 1 but one (the root), which has indegree 0. A tree t is represented by the tuple $(r_t, N_t, E_t, \lambda_t)$, where $N_t \subseteq \mathbb{N}$ is the set of nodes, $E_t \subseteq N_t \times N_t$ is the set of edges, $\lambda_t : N_t \rightarrow \Sigma$ is the node labeling function and $r_t \in N_t$ is the distinguished root of t. We assume standard notions of leaf, child, parent, descendant, and ancestor node, as well as notions of tree depth and path between two nodes. Moreover, we say that a node is a *branching node* if it has at least two child nodes, and denote the set of branching nodes of a tree t as $\widehat{N}_t$.

Given a tree t, we denote its depth as $depth(t)$, and the path between its root and a node $u \in N_t$ as $\wp_t(u)$. Moreover, we say that a tree t' is a *subtree* of t if $N_{t'} \subseteq N_t$ and $E_{t'} \subseteq E_t$. Finally, we say that t is a *chain* if it has no branching node. The set of trees defined on the alphabet of node labels Σ will be denoted as T_Σ.

There is a natural way of extending the notion of pattern described for transactions of sequences, to the case of trees. A *tree pattern* p is a pair $\langle t_p, o_p \rangle$, where:

1. $t_p = (r_p, N_p, E_p, \lambda_p)$ is a tree in $T_{\Sigma'}$, where $\Sigma' = \Sigma \cup \{*\}$;
2. E_p is partitioned into the two disjoint sets C_p and D_p denoting, respectively, the *child* and *descendant* edges;
3. $o_p \in N_p$ is the distinguished output node.

Notions of child, parent, descendant, ancestor, branching, and leaf node for tree patterns, as well as the notion of subpattern, depth and path, are trivial extensions of the corresponding notions defined for trees. Given a tree pattern p and a node $u \in N_p$ we denote the set of child nodes of u as $Child_p(u)$, i.e., $Child_p(u) = \{v \mid (u, v) \in E_p\}$, and the set of ancestor nodes of u as $Ancestor_p(u)$.

2.4 Graphs and Structured Data

Graphs can be considered a generalization of trees. They are important as a data type for two main reasons. First, several domains use graphs as instances to analyse. For example, chemical data (e.g., with nodes corresponding to atoms, and links corresponding to bonds between the atoms); biological data (e.g., protein interactions, biological networks); networked infrastructures such as airport connection, ethernet, or the Web; business processes, where nodes represent activities and links represent precedence relations between such activities.

Formally, let $G = (V, E)$ be a graph where V is a set of n nodes, $E \subseteq V \times V$ is a set of m *directed* arcs, and (u, v) indicates that u is connected v. We also denote the neighborhood of a node u as $N(u) = \{v \in V : (u, v) \in E \vee (v, u) \in E\}$. Moreover let $\mathcal{F}$ denote a set of h binary features. We are given a binary $n \times h$ matrix F such that $F_{u,f} = 1$ when user u is interested in the feature f. For simplicity we denote this case also as $(u, f) \in F$. Finally, we denote all the features of the node u as $F(u) = \{f \in \mathcal{F} : (u, f) \in F\}$ and the set of all the nodes having attribute f as $V(f) = \{u \in V : (u, f) \in F\}$.

The analysis of graph data has exploded in the mid of the 2000s [23]. Graph mining usually refers to algorithms used to extract patterns, trends, classes, and clusters from graphs [37]. However, it's with the advent of social networks such as Facebook and Twitter that Network Science has come to the attention of the data mining and machine learning community, In fact, social networks build on massive graphs, and hence new specific algorithms may need to be applied to solve predictive and descriptive problems which raise from these large infrastructures.

2.5 *Streams*

Data streams are potentially infinite sources of data that flow continuously while monitoring a physical phenomenon, like temperature levels or other kind of human activities such as trajectories [43], RFID streams [44], document streams [34], and so on.

Consider an ordered set of n sources denoted by $\{s_1, \ldots, s_n\}$ producing n independent streams of data, representing the value being monitored. Each data stream can be viewed as a sequence of triplets $\langle id_s, v, ts\rangle$, where:

1. $id_s \in \{1, .., n\}$ is the source identifier;
2. v is a non negative integer value representing the measure produced by the source identified by id_s;
3. ts is a *timestamp*, i.e., a value that indicates the time when the reading v was produced by the source id_s.

3 Methods

3.1 *Supervised Learning*

Unbalanced Classification. Classification is one of the most extensively studied tasks in machine learning, pattern recognition and data mining. Given a collection of labeled training data, the aim is to learn a suitable model, referred to as a classifier, wherein the regularities in the labeled data are exploited to induce a reasonable approximation (i.e., a hypothesis) on the actual mappings between any data case from the same domain and one of multiple predefined class labels. A classifier is, hence, useful to predict the unknown class of a previously unseen case, on the basis of the other observable features of the same case. Various types of classifiers have been proposed in the literature, that meet several different requirements in a wealth of distinct applicative settings, such as decision trees, rule-based classifiers, neural networks, naïve Bayes classifiers, support vector machines and statistical classifiers [30]. In particular, rule learning is a method for inducing minimal rule-based concept descriptions, that can be used for classification. Rule-based classifiers are a mainstay of research in machine learning, because of various desirable properties such as, e.g., their expressiveness and intelligibility to humans as well as their efficiency and effectiveness in classification. Such classifiers have been empirically shown to be effective in processing (sparse) high-dimensional training data with categorical attributes [55] and are comparable in performance with other classification methods in several applicative domains [46]. Unfortunately, like most classification models, rule-based classifiers exhibit a poor classification performance in imprecise (multi-class) learning environments, which are challenging domains wherein cases and classes of primary interest for the learning task are rare. Besides, minority and majority classes can be hardly separable and the cost of misclassifying a case of a minority class as belonging to a predominant class is much higher than the cost of the dual error. Also, training data may be corrupted by noise, which further obstacles the identification of rarities.

Imprecise domains are often encountered in practical applications. Examples include fraud detection [32, 47], intrusion detection [52], manufacturing line monitoring [50], risk management, telecommunications management [31], medical diagnosis [22], text classification [56], and oil-spill detection in satellite images [42]. The peculiarities of such settings pose several challenging issues to traditional algorithms for learning rule-based classifiers, that essentially make the resulting models low sensitive to rarities.

Rarity is clearly the major obstacle. Rare classes corresponds to the well known *class imbalance* issue, i.e., an evenly distribution of classes, such that majority classes overwhelm minority ones. Instead, rare cases are very small portions of the training data, that can be viewed as exceptional sub-concepts seldom occurring within predominant or rare classes. As it is pointed out in [56], rarity actually prevents conventional algorithms for rule induction from finding and reliably generalizing the regularities within infrequent classes and exceptional cases.

Indeed, class imbalance generally leads to classification models tending to exhibit a high specificity (i.e., capability at recognizing majority classes), coupled with a low sensitivity (i.e., capability at recognizing minority classes).

Recommender Systems. An interesting application of supervised learning, particularly tailored to transactional data, is the case of recommender systems. With the increasing volume of information, products, services, resources (or, more generally, items) available on the Web, the role of *Recommender Systems* [49] and the importance of highly-accurate recommendation techniques have become a major concern both in e-commerce and academic research. The goal of a recommender is to provide users with recommendations that are non-trivial, useful, and that the user may not find on her own.

Recommendation is a form of information filtering which analyzes users' past preferences on a catalog of items to generate a personalized list of suggested items. Let $\mathcal{U} = \{u_1, \ldots, u_M\}$ be a set of users and $\mathcal{I} = \{i_1, \ldots, i_N\}$ a set of items. Users' preferences can be represented as a $M \times N$ matrix R, whose generic entry r_i^u denotes the preference value assigned by user u to item i. User preference data can be classified as *implicit* or *explicit*. Implicit data correspond to mere observations of co-occurrences of users and items, which can be recorded by analyzing clicks, users' web sessions, likes or check-in. Hence, the generic entry r_i^u of the user-item rating matrix R is a binary value: $r_i^u = 0$ means that u has not yet experienced i, whereas by $r_i^u = 1$ we denote that user u has been observed to experience item i. By contrast, explicit data record the actual ratings explicitly expressed by individual users on the experienced items.

Generally, explicit ratings can be encoded as scores in a (totally-ordered) numeric domain $\mathcal{V}$, represented as a fixed rating scale that often includes a small number of interestingness levels. In such cases, for each pair (u, i), rating values r_i^u fall within a limited range $\mathcal{V} = \{0, \ldots, V\}$, where 0 represents an unknown rating and V is the maximum degree of preference. In real-world applications, the rating matrix R is characterized by an exceptional sparseness, as individual users tend to rate a limited number of items.

Given an active user u, the goal of a recommender is to provide u with a recommendation list $L_u \subseteq \mathcal{I}$ including unexperienced items (i.e., $L_u \cap \mathcal{I}_{\mathbf{R}}(u) = \emptyset$), that are expected to be of her interest. This clearly involves predicting the interest of u towards unrated items: exploiting (implicit and/or explicit) information about users' past actions, the recommender provides a scoring function $p_i^u : \mathcal{U} \times \mathcal{I} \rightarrow \mathbb{R}$, which accurately estimates future preferences and hence can be used to predict which are the most likely products to be purchased in the future.

The idea behind recommendation is that human interests and preferences are essentially correlated, thus it is likely that a target user tends to prefer what other like-minded users have appreciated in the past. Hence, the basic approach to information filtering consists of collecting data about user interests and preferences to build reliable user profiles. These allow to predict the interests of a target user based on the known preferences of similar users and, eventually, the provision of suitable recommendations of potentially relevant items. Recommendations generated by means of *filtering* techniques can be divided into three main categories:

- *demographic filtering* approaches assume the possibility of partitioning the overall set of users into a number of classes with specific demographic features. Customization is achieved by exploiting manually-built, static rules that offer recommendations to the target user on the basis of the demographic features of the class she belongs to;
- *content-based filtering* techniques learn a model of users' interests and preferences by assuming that item and user features are collected and stored within databases information systems. The aim is to provide the target user with recommendations of unexperienced content items, that are thematically similar to those she liked in the past;

- *collaborative filtering* methods work by matching the target user with other users with similar preferences and exploits the latter to generate recommendations.

Collaborative Filtering (CF) strategies are effective with huge catalogs when information about past interactions is available. According to this assumption, several CF-based recommendation techniques have been proposed, mainly focusing on the predictive skills of the system.

It has been shown [10] that probabilistic approaches based on latent-factor models allow the most adequate degree of flexibility, as they: *(i)* allow the specification of complex yet easy to interpret latent structures; *(ii)* achieve the highest recommendation accuracy.

Typically, complex patterns can be better detected by means of co-clustering approaches [11, 41, 51]. The latter aim at partitioning data into homogeneous blocks enforcing a simultaneous clustering on both the dimensions of the preference data. This highlights the mutual relationships between users and items.

3.2 Unsupervised Learning

The Clustering Challenge. Cluster analysis represents a fundamental and widely used method of knowledge discovery, for which many approaches and algorithms have been proposed over the years [1]. An incomplete list of methods includes partition-based (e.g., *k-means*), density-based (e.g., *DBScan*), hierarchical (e.g., *BIRCH*), and grid-based (e.g., *STING*) methods. The continuing stream of clustering algorithms proposed over the years underscores the importance and difficulty of the problem, and the fact that the logical and algorithmic complexities of this many-facet problem have yet to be tamed and can still benefit from research advances. Foremost among the remaining issues is that clustering algorithms do not live up to their claim of providing *unsupervised learning*: current algorithms are effective only when the user provides several parameters that she needs to determine via a difficult exploratory process.

Clustering process traces/events. Process discovery techniques [54] have gained attention in Business Process Management applications, owing to their ability to extract a descriptive/predictive model for the behavior of a process, based on historical log data. Such models can provide actionable process knowledge, and support process analysis/design tasks, and process improvement efforts.

Two critical issues undermine the effectiveness of traditional workflow discovery methods, when applied to the logs of lowly-structured processes: *(i)* the high level of details and granularity that usually characterizes log events, which makes it difficult to provide the analyst with an easily interpretable description of the process in terms of relevant business activities, and *(ii)* the presence of various execution scenarios (a.k.a. "process variants"), exhibiting different business processing logics (and often determined by key context factors), which cannot be captured effectively with a single workflow model. When applied to such logs, indeed, most current process discovery techniques tend to yield "spaghetti-like" descriptive models (suffering from both low readability and low fitness) [17] and to overfitting predictive models. And yet, automatically discovered models would be utterly important for a lowly-structured process, seeing as no (or little) a-priori knowledge exists on its typical execution schemes.

Two kinds of approaches have been proposed in the literature to alleviate these problems: (a) partitioning the log into trace clusters [38], in order to capture homogenous execution groups, and modelling each of them with a simpler and more precise predictive/descriptive process model; (b) turning raw events into high level activities via automated *event abstraction techniques* [35], capable to automatically induce a sort of event abstraction function (cf. Sect. 2.2).

The problem of partitioning a given collection of traces into homogenous process variants has been faced by using different kinds of clustering approaches. Some approaches, in particular, looking at the (abstract) traces as symbolic sequences, exploit standard distance-based clustering schemes in combination with customized versions of string-oriented edit distances [9].

Higher scalability is typically achieved by feature-based approaches [38], which rely on projecting the traces onto a vector space, where some dimensions represent the occurrence of distinguished process activities or of more involved behavioral patterns (e.g., k-grams of activities, or variable-length "conserved" subsequences borrowed from Bioinformatics such as Tandem/Maximal Repeats) capturing frequent hidden functionalities of the process. A hierarchical clustering approach was proposed in [38], where the traces are partitioned recursively (by resorting to a special kind of sequence patterns, named "discriminant rules") based on the way they diverge from a preliminary workflow model. As a result a tree of descriptive models are discovered for the analyzed process, where each leaf represents the behavior of a set of process instances in an modular and precise way. Such a tree can be converted into a process taxonomy, as proposed in [38], by providing each non-leaf node with an abstract process model—namely the root provides the most abstract view over the executions of a process, and any other node refines this abstract model to describe a subclass of executions.

In absence of background knowledge on relevant process activities, unsupervised event-abstraction techniques [35] perform the aggregation by means of some clustering procedure over the events, so that each discovered cluster of events is viewed as an abstract activity. The automated extraction of event classes has been faced in [35] for the challenging case where a predictive model for process performances is to be discovered for a lowly-structured process log (like, e.g., those of ERP systems), where multiple data attributes are typically stored in each log event, but none of them fully characterizes what kind of activity was performed. In such a scenario, the event classes are discovered by using a conceptual clustering procedure (which specializes the general framework of predictive clustering, where the events are partitioned by way of mutually-exclusive boolean formulas (expressed over event attributes), which are easy to interpret and validate. Notably, in [35] event clusters (acting as different activity types) are discovered synergistically with trace clusters (representing different process variants), by using a (conceptual) co-clustering scheme, before eventually equipping each trace cluster with a specific predictor. Besides enjoying compelling prediction accuracy (w.r.t. current methods, combined with usual log abstractions), in real-life application cases, this approach helps recognize performance-relevant activity patterns at the right level of abstraction.

Distance-Based Outlier Detection. *Outlier detection* is a data mining task consisting in the discovery of observations which deviate substantially from the rest of the data, raising the hypothesis that they were generated by a different mechanism [40]. Outlier detection provides information which is readily usable to react in critical situations; therefore, it has many important practical applications in domains such as medical anomaly detection, sensor networks, industrial damage detection, cyber-intrusion detection, fraud detection, image processing, and textual anomaly detection [21]. Outlier detection is also quite computationally demanding, and its application to very large data sets currently requires smart algorithms and high-performance computing facilities.

Many supervised approaches to outlier mining first learn a model over example data already labelled as exceptional or not [33], and then evaluate a given input data as normal or outlier depending on how well it fits the model. Unsupervised methods, instead, are able to separate exceptional and normal data even without the help of training examples. A prominent approach to the unsupervised problem is *distance-based outlier detection*, which distinguishes an object as outlier on the base of the distance to its nearest neighbors [48]. Most of these approaches define a weight or score for every object, which summarizes its dissimilarity to its nearest neighbors by means of a function of their distances. Distance-based outlier detection is non-parametric, in that no assumption is required on the distribution of data.

In the last years our research effort focused on distance-based outlier detection. Before to highlight the main results obtained in this area, we provide some basic concepts. In the following, we assume a data set D of objects, which is a finite subset of a given metric space.

Given an object $p \in D$, the *weight* $w_k(p, D)$ of p in D is the sum of the distances from p to its k nearest neighbors in D. Let Top be a subset of D having size n; if there not exist objects $x \in Top$ and y in $(D \setminus Top)$ such that $w_k(y, D) > w_k(x, D)$, then Top is said to be the set of the *top n outliers* in D [3]. Hence *Outlier Detection Problem (ODP)* consists in finding the top n outliers in D. ODP can be solved in $\mathcal{O}(|D|^2)$ time by computing all the distances among the all objects in D, but real life applications deal with data sets of hundred thousands or millions of objects, and thus brute-force approaches are not applicable.

The first contribution has been the introduction of the concept of the *outlier detection solving set* S which is a subset of D such that, for each $y \in D \setminus S$, it holds that $w_k(y, S) \leq w^*$, where w^* is the weight of the top n-th outlier in D [3]. In practice, an outlier detection solving set S is a very small subset of D including a sufficient number of points to allow us to consider only the distances among the pairs in $S \times D$ to obtain the top n outliers.

We presented the *SolvingSet-algorithm* that computes S with sub-quadratic time requirements, providing the top n outliers in D and the weight w^*.

We also proved that, in general, the computational complexity of computing a minimum cardinality solving set is intractable.

The second contribution has been the definition of parallel data mining (PDM) and distributed data mining (DDM) algorithms to solve *ODP* [4, 5]. Indeed, we presented a distributed method to detect distance-based outliers, which is suitable to be used both in parallel and distributed scenarios. The proposed method exhibits excellent performances, in that the temporal cost in charge of each node is $O(\frac{\varrho}{\ell}T_b)$, where ϱ is the relative size of the (distributed) solving set, ℓ is the number of nodes involved in the computation, and T_b is the time required to compute all pairwise distances among data set objects, that is the time needed to determine the nearest neighbors of each object. In common settings, the term $\frac{\varrho}{\ell}$ is likely to be smaller than $\frac{1}{1000}$. Moreover, experimental results validate that the run time scales up well with respect to the number of nodes.

The third contribution draws on the recent availability of high-performance computing architectures, such as *Graphic Processing Units (GPU)*, which are receiving increasing attention justified by the fact that they characterize computers that trade-off between performance and power consumption. For this reason, our objective has been to test the effectiveness of a GPU–based solution for *ODP* [6–8]. First we developed an implementation for a GPU architecture of *SolvingSet-algorithm*, to be used as baseline in comparisons. Then we implemented several variants, both centralized and distributed, of the above algorithms, to experiment the effects of different usages of the memory hierarchy and of different implementation and optimization techniques.

The experimental results show that the approach is quite effective, with the speedup of the centralized strategies reaching two orders of magnitude in the best cases. The best results are given by the variants which address the best usage of the parallel threads and memory hierarchy rather than the variants with a reduction of memory occupancy. As for the distributed approach, experiments highlight that the speedup over the centralized GPU algorithm tends to the ideal one for data sets of increasing sizes.

4 Applications

4.1 Health, Demographic Change and Wellbeing

Nowadays, microarray experiments allow the exploration of huge amounts of gene expressions using a single chip. Moreover, the relatively moderate cost for a chip and the small sample preparation times, enable the analysis of a large number of different experimental conditions, such as points of time-series experiments or disease progression in a cohort of patients. This huge amount of data poses many challenges to the bioinformatics community such as finding the behavior of set of related genes in different conditions. This goal is often achieved by means of cluster analysis, i.e., the identification of similar patterns in different conditions. Indeed, the ability to gather genome-wide expression data has far outstripped the ability of human brains to process the raw data, thus cluster analysis can help scientists to distill the data down to a more comprehensible level by subdividing the genes into a smaller number of categories and then analyzing those. Further motivation for the exploitation of cluster analysis for biological data lies in the fact that similar patterns found by clustering may correspond to co-regulation of genes.

In [45] experimental results regarding the validity of clustering results from a biological viewpoint are presented. Indeed, clustering gene expression data is a valid support for functional annotation, tissue classification, regulatory motif identification, and other applications, but choosing the right clustering may be rather difficult. To address this issue, several proposal have been presented and in [45] the quality measure defined in [36] has been leveraged for biological data clustering evaluation since it summarizes several evaluation metric in a single measure. This analysis assures a stronger validation of the clustering results from a biological viewpoint that was confirmed by the results reported in [45].

4.2 Secure, Clean and Efficient Energy

Renewable energy is actually a strategic sector for all the European countries, due to the strategic and urgent need of reducing pollution emission. In this perspective, a nice proposal (*VI-POC*) [18], provides a framework for collecting, storing, analyzing, querying and retrieving data coming from heterogeneous renewable energy production plants (such as photovoltaic, wind, geothermal, Stirling engine, water running) distributed on a wide territory.

More in detail, it has been investigated the usage of big data technologies in order to effectively manage data coming from heterogeneous sources, possibly of different nature (i.e., photovoltaic plants generate data different from those generated by wind plants). Indeed, it also important to decouple the model and the energy source in order to make the system flexible and scalable. Moreover, *VI-POC* is intended to refine (i.e., making more efficient, effective and reliable) raw predictions usually available from national electric authorities. Vi-POC predicts real time energy needs with higher precision. This will lead to two key advantages: (1) the definition of a better offer for the energy market and (2) the definition of an accurate purchase strategy.

Furthermore, the prototype could be useful both for main players of the energy market such as the distributors and smaller companies that act between offer (trailers) and request in the supply chain in order to build better supply planning for their customers. Moreover, the synergy between modern renewable energy production sites and advanced technologies for data storage and analysis allow a continuous monitoring of the production process. Forecast is crucial in *VI-POC* and may apply to a single renewable power generation system, or refer to the aggregation of large numbers of systems spread over an extended geographic area. Accordingly, different forecasting methods are used. Forecasting methods also depend on the tools and information available to forecasters, such as data from weather stations and satellites and outputs from numerical weather prediction (NWP) models.

In this perspective, it becomes necessary to resort to data-driven solutions and, in particular, to Data Mining solutions which, on the basis of historical data of different nature, are able to forecast energy production for both small and large users. Moreover, it has been recognized that physical (e.g., wind speed and solar irradiation) properties behavior exhibit a trail called concept drift which, in the vocabulary of data stream mining, means that they change characteristics over time. To take into account concept drift and, thus, deal with non-stationarity of data, either adaptive models and time series prediction algorithms have been applied. In this respect, the adaptive models are generally considered to produce more reliable predictions, especially regarding to concept drift, but require a continuous training phase. This training phase demands for a more complex underlying architecture, such as those of Big Data, in order to guarantee the possibility of storing and processing big volumes of data which arrive at high speed.

4.3 Smart, Green and Integrated Transport

The analysis and control of transport operations is a key problem in the field of logistics networks. In a series of research projects, we developed a several techniques and tools for supporting the analysis of the activities performed in an Italian maritime container terminal. Essentially, the operational systems of the terminal supports and registers several logistic activities for each container that passes through the harbor. In our analysis, we focused on the handling of containers which arrive and depart by ship ("transhipment" flows), and on the different kinds of moves they undergo over the "yard", i.e., the main area used in the harbor for storage purposes.

According to the notation introduced in Sect. 2.2, the events stored in each of the analyzed log traces is provided with the following major event properties: *(i)* the source and destination position it was moved between, in terms of yard's blocks and (coarser grain) areas; *(ii)* the kind of operation performed, (e.g., MOVe, DRive to Bring, DRrive to Get, LOAD, DIScharge, SHuFfle, OUT); *(iii)* the type of instrument used (e.g., cranes to straddle-carriers and multi-trailers) Each trace, in its turn, is associated as a whole with "case-wide" properties, which all concern the container that has been handled, and include: he previous and next ports, and their associated countries, and several physical features (e.g., the size and height).

For each container, we also considered, as a sort of environmental variables, the hour (resp., week-day, month) when it arrived, and the total number of containers ("workload") in the port at that time.

Based on these log data, we faced to process mining tasks: *(i)* the discovery of predictive performance-oriented process models, allowing to directly forecast the value of a given performance metric (namely, the total time spent and the total distance covered by the respective containers) for an unfinished process instance, based on its associated (partial) log trace; *(ii)* the discovery of workflow-like models, allowing to gain a compact representation of the logistic process, illustrating the typical flows of work that were really carried out.

As to former task, we obtained good achievements with the co-clustering scheme of [35], which allowed us to deal with the unstructured nature of this logistic process, where no high-level process activities and no pre-defined event abstraction criteria were known. Interestingly, the clustering rules discovered with this approach, for both the events and the traces, provided us with useful descriptions of process behavior (besides supporting accurate predictions). In particular, we found that some of the discovered event clusters correspond to a subset of destination areas, some others also depend on the source area, or even further on the kind of move performed. On the other hand, the descriptions of trace clusters let us reckon the presence of different execution scenarios, linked to both context factors (e.g., the country of the previous/next port, or the line/service planned to bring the container) and to some of the discovered event clusters (hence playing as high-level activity patterns). Further details can be found in [35].

In order to analyze the usage of the different storage areas (and possibly detect overly intricate/long moving patterns), we extracted our workflow-like models from a "positional" representation of the containers histories. More specifically, for each container c, we built an abstracted trace that just encoded the sequence of positions (represented either at the level of yard sectors or at the coarser level of yard areas) that c occupied during its stay. By applying the approach in [39], we found interesting taxonomies of workflow schemas, describing quite different scenarios for the movement of containers. In particular, one of the discovered schemas gave us some insight on the path followed by a group of containers that incurred into exceptional events (witnessed by the passage of the containers through an area devoted to special checking activities). Other schemas evidenced frequent "house-keeping" moves internal to some sectors, and between a specific subset of sectors. In summary, the process taxonomies computed with this approach allowed to both recognize distinct process variants and to provide compact higher-level views over these variants, despite the very high number of event classes — which would make classical workflow discovery approaches to yield rather unreadable and imprecise process models. Moreover, the effect of using abstraction mechanisms was not merely syntactical: many of the activities that were abstracted together shared indeed some semantical affinity. Further details can be found in [39].

4.4 Europe in a Changing World - Inclusive, Innovative and Reflective Societies

Understanding the mechanisms behind adoption of new practices, ideas, beliefs, technologies and products within a population characterized by *social connections*, is a central issue for the whole of social sciences. Recently with the explosion of on-line social platforms such as `Twitter` and `Facebook`, the interest in these topics has exploded accordingly. Researchers have investigated all these issues and proposed elegant and scalable mathematical models. Taking into account the modular structure of the underlying social network provides further important insight in the phenomena described above. Hot topics in this area of research are:

- social network modeling: the study and definition of compressed models that can summarize, reproduce and predict the evolution of a social network;
- information diffusion: how information flows within the network. The objectives are the discovery of the best routes for the information transmission and of the hub and sink nodes;
- user behavior discovery: the discovery of user behaviors that are repeated over time;
- influence propagation: identify the hidden factors explaining why, on the basis of his/her ego network, a given subject is prone to experience a situation/publish specific information/perform some actions.

A particularly challenging line of research consists in addressing the above issues in an elegant and mathematically rigorous way through the use of probabilistic latent factor models. Latent factor models are a family of descriptive models assuming that collections of observations exhibit underlying hidden thematic structure. In the context of social network modeling, probabilistic latent factor models find several applications. For example, large effort has been devoted to develop community detection algorithms based on the relationships that the network peers exhibit.

Communities have a long history in social sciences, but it is with the explosion of on-line social platforms such as Twitter and Facebook, that the interest in these topics has grown accordingly and in particular has gained a computer science perspective. While most early methods are based on ad-hoc heuristics, in recent years approaches based on fitting generative latent factor models to the data have emerged.

Also, individuals tend to adopt the behavior of their social peers, so that cascades happen first locally, within close-knit communities, and become global viral phenomena only when they are able cross the boundaries of these densely connected clusters of people. Therefore, the study of social contagion is intrinsically connected to the problem of understanding the modular structure of networks (known as community detection), and together form the central core of network science. Latent factor models can be proven extremely flexible in this context, as they lend themself to a variety of predictive and descriptive tasks which can elegantly and robustly explain the behavior of an individual within the network: for example, whether a user shall adopt a given item [14], or whether two users are likely to be connected [15]. It is worth mentioning how efficient methods for the discovery of the latent factors explaining user behavior find several applications, spanning to person-to-person recommendation, homophily [12], trust [27] and other external factors of correlation [13].

4.5 Secure Societies - Protecting Freedom and Security of Europe and Its Citizens

The strong dependability between critical infrastructures and ICT has favored the design and development of new applications and services devoted to these infrastructures. For example, SCADA systems are meant to ease the monitoring and control within energy and nuclear plants. We live in a connected world where all systems communicate and interact by means of the cyberspace. These systems are particularly sensitive from a security point of view, since they control critical infrastructures and hence vulnerabilities in their components (whether hardware or software) are extremely dangerous. Throughout the years, these issues have called for the setup of dedicated services for security monitoring, early warning and incident response. These services are strongly based on data analytics and artificial intelligence, and raise important research issues.

Security metrics. How to correctly measure and quantify security and trustworthiness is still a challenging and priority task. Security metrics should be studied under three different perspectives: system-level, technology-level, organization-level. Under this perspective, they should allow to answer questions such as: how secure is my organization? Is there an improvement in the response to security threats throughout the years? Is the tool we aim at delivering secure? Are there required changes we need to accomplish by introducing new tools into our lifecycle? Is my system resilient to cyber-threats?

Situation Understanding through data mining. Collecting and analysing data by means of specific tools should allow to solve issues such as threat detection and identification, impact evaluation and role attribution. Data mining techniques are gaining an increasingly important role. A straightforward application of these techniques consists devising systems for intrusion/fraud detection/prevention, where the emphasis is aimed at automatically analysing huge data collections coming, e.g., from traffic or sensor networks. A specific class of problems concerning the analysis of user behavior is taking place, called activity monitoring. These problems require monitoring the behavior of a large population of entities, in order to detect relevant events upon which to react. Examples are: profiling a user in order to identify her typical behavioral signature, which can be used to prevent security threats; monitoring the usage of a brand throughout the web, in order to prevent misuses or steal of identities, masquerading and abuses. Finally, the analysis of RFID data is extremely relevant for devising additional security measures, especially in situation where we can prevent or predict: blocks of working machinery; access control to dangerous zones; misuse of critical tools; telemetry.

Process mining for (cyber-)security. Besides the "traditional" data mining methods, the analysis of process and system logs is especially important in the relation to security. Audit trail data characterizes the activities of several organizations which are interested in evaluating risks and in leveraging the security policies. Also, the recent adoption of standard (enforced, e.g., by the Sarbanes-Oxley Act) for security checks and monitors within the organizational processes, will likely increase the need to track processes and activities, and as a consequence to analyze such traces.

The specific focus on the research themes discussed above has triggered the setup of the Cybersecurity Technological District (https://www.distrettocybersecurity.it/), located in Cosenza and including private organizations (Poste Italiane, NTT Data, ICT Sud Consortium) and academic institutions (University of Calabria, University of Reggio Calabria) and Research Institutions (CNR). The district conducts research on big data analytics and cybersecurity, with emphasis to three main directions: protecting the end-user; protecting payment systems; secure dematerialization.

4.6 Fiscal Fraud

Fraud detection represents a challenging issue in several application scenarios, and the automatic discovery of fraudulent behavior is a very important task with great impact in many real-life situations. Examples are network traffic monitoring, where anomalous patterns could represent an attack attempt; similarly, anomalies in credit card transaction data could represent an illegal action, or detection of abnormal behavior in athlete's performance could report a drug abuse. In particular, fiscal fraud detection has witnessed an increasing interest and has become a widespread application field for data mining techniques, due both to its importance from a social point of view, and for its technical intrinsic difficulties.

In this section we describe the experience we made on the Value Added Tax (VAT) fraud detection scenario. Like any tax, the VAT is open to fraud and evasion. There are several ways in which it can be abused, e.g., by underdeclaring sales or overdeclaring purchases. However, opportunities and incentives to fraud are provided by the credit mechanism which characterizes VAT: tax charged by a seller is available to the buyer as a credit against their liability on their own sales and, if in excess of the output tax due, refunded to them. Thus, fraudulent claims for credit and refunds are an extensive and problematic issue in fiscal fraud detection. The situation is further exacerbated in Italy by current laws which allows to compensate VAT credit with other taxes, thus boosting the trend for fraudulent behavior.

In [16, 26] we tackle the VAT Fraud Detection issue raised by the credit mechanism via the adoption of data mining techniques. The objective was to design a predictive analysis tool able to identify the tax payers with the highest probability of being VAT defrauders to the aim of supporting the activity of planning and performing effective fiscal audits. The construction of the model is based on historical VAT declaration records labeled with the outcome of the audit performed by the Agency.

The context was particularly challenging both from a scientific and a practical point of view. First of all, audited data available were only 0,004% of the overall population of taxpayers who file a VAT refund request. This resource-aware restriction inevitably raises a *sample selection bias*. Indeed, auditing is the only way to produce a training set, and auditors focus only upon subjects which are particularly suspicious according to some clues. As a consequence, the number of positive subjects (individuals which are actually defrauders) is much larger than the number of negative (i.e., non-defrauders) subjects. This implies that, despite the number of fraudulent individuals is far smaller than those of non-fraudulent individuals in the overall population, this proportion is reversed in the training set.

Since auditing is resource-consuming, the number of individuals reported as possible fraudsters is of high practical impact. Hence, a scoring system should primarily suggest subjects with a high fraudulent likelihood, while minimizing false positives. Also, the scoring system should also focus on criterias such as proficiency (higher fraud amounts make defrauders more significant), equity (the amount of fraud should be related to the business volume) and efficiency (scoring and detection should be sensitive to total/partial frauds).

Classical supervised techniques typically suffer from the class-unbalance problem: Typically, only a limited number of behavioral cases is recognized as fraudulent within the training set. Also, from a socio-economic point of view, it is preferable to adopt a rule based approach to modeling. Indeed, intelligible explanations about the reason why individuals are scored as fraudulent are by far more important than the simple scores associated to them, since they allow auditors to thoroughly investigate the behavioral mechanisms behind a fraud.

The result of our study was the design of a supervised methodology capable of coping with all the above mentioned issues in a unified framework. The basic ideas consist in progressively improving the learned rules by means of (i) probabilistic smoothing techniques [26], and (ii) ensemble techniques which filter out and retrain weak rules which do not meet the profitability, efficiency and equity criteria [16].

4.7 *Risk Analysis and Entity Resolution*

Risk analysis is a key business analysis task for bank organizations and it is required to identify and assess those critical factors that may negatively affect customer successful payback or somehow lower the expected revenues.

A typical approach to risk analysis is the development of a specific data mart capable of ranking the credit risk of a customer by looking at the past insolvency history involving the customer. Besides the traditional predictive loan approval application, there are other scenarios where a business transaction identifies a creditor and a debtor. For example, a bank agency acts as a mediator in the transaction, by supporting the creditor in collecting the amount she claims. This is accomplished by providing billing services, or by anticipating the monetary flow. Here, risk analysis consists in identifying all possible critical factors which can negatively affect a given transaction. The occurrence of such critical factors are estimated and eventually employed both to identify suitable countermeasures that allow the bank organization to avoid/recover from insolvencies, partial paybacks and unexpectedly low revenues.

Besides the intrinsic difficulties in providing accurate estimations, the problem is further exacerbated by the dematerialization of all the documents involved in the transactions. Document filling is subjected to erroneous data-entry, misspelled terms, transposition oversights, inconsistent abbreviations, lack of attribute values and so forth. These further increase the heterogeneity of bank data by making the individual records of a same textual document (e.g., a collection of personal demographic information) appear with a varying structure.

The issues of dematerialization call for a data quality and consistency problem that has to be addressed in order to ensure the reliability of business-intelligence results. Large quantities of such data are stored as continuous text, such as personal demographic data, bibliographic information, phone and mailing lists. Often, the integration of such data is a problematic process, that involves handling two major issues, namely *structural* and *syntactic* heterogeneity. The former occurs when the data does not explicitly exhibit a common field structure. Schema reconciliation techniques are required to identify a common field structure for heterogeneous bank data, in order to exploit the mature relational technology for more effective information management. Further, whether or not reconciled into a common database schema, data can still be affected by *syntactic heterogeneity*. This is a fundamental issue in the context of information integration systems, that consists in discovering duplicates within the integrated data, i.e., syntactically different records that, as a matter of facts, refer to a same real-world entity. Thus, a suitable methodology for handling with data duplication is need to discover synonymies in the data. This would allow to perform tasks such as summing up the singularities (insolvencies, for instance) of seemingly distinct individuals and associating them to a unique customer, so that to mark the latter as seriously critical. Notably, both approaches were handled in a unified framework [20, 24, 25], by providing a unified architecture capably of managing large collection of textual data and incrementally performing the tasks of schema reconciliation and data deduplication.

References

1. C.C. Aggarwal, C.K. Reddy (eds.), *Data Clustering: Algorithms and Applications* (CRC Press, Boca Raton, 2014)
2. C.C. Aggarwal, M.A. Bhuiyan, M. Al Hasan, *Frequent Pattern Mining Algorithms: A Survey* (Springer International Publishing, Cham, 2014), pp. 19–64
3. F. Angiulli, S. Basta, C. Pizzuti, Distance-based detection and prediction of outliers. TKDE **18**(2), 145–160 (2006)
4. F. Angiulli, S. Basta, S. Lodi, C. Sartori, A distributed approach to detect outliers in very large data sets. Euro-Par **1**, 329–340 (2010)
5. F. Angiulli, S. Basta, S. Lodi, C. Sartori, Distributed strategies for mining outliers in large data sets. TKDE **25**(7), 1520–1532 (2013)
6. F. Angiulli, S. Basta, S. Lodi, C. Sartori, Fast outlier detection using a gpu, in *HPCS* (2013), pp. 143–150
7. F. Angiulli, S. Basta, S. Lodi, C. Sartori, Accelerating outlier detection with intra- and inter-node parallelism, in *HPCS* (IEEE, 2014), pp. 476–483
8. F. Angiulli, S. Basta, S. Lodi, C. Sartori, GPU strategies for distance-based outlier detection. IEEE TPDS **27**(11), 3256–3268 (2016)
9. T. Baier, J. Mendling, M. Weske, Bridging abstraction layers in process mining. Inf. Syst. **46**, 123–139 (2014)

10. N. Barbieri, G. Manco, An analysis of probabilistic methods for top-n recommendation in collaborative filtering, in *ECML PKDD* (2011), pp. 172–187
11. N. Barbieri, M. Guarascio, G. Manco, A block mixture model for pattern discovery in preference data, in *ICDMW* (2010), pp. 1100–1107
12. N. Barbieri, F. Bonchi, G. Manco, Cascade-based community detection, in *WSDM* (2013), pp. 33–42
13. N. Barbieri, F. Bonchi, G. Manco, Influence-based network-oblivious community detection, in *ICDM* (2013), pp. 955–960
14. N. Barbieri, F. Bonchi, G. Manco, Topic-aware social influence propagation models. Knowl. Inf. Syst. **37**(3), 555–584 (2013)
15. N. Barbieri, F. Bonchi, G. Manco, Who to follow and why: link prediction with explanations, in *KDD* (2014), pp. 1266–1275
16. S. Basta et al., High quality true-positive prediction for fiscal fraud detection, in *ICDM Workshops* (2009), pp. 7–12
17. J.C.A.M. Buijs, B.F. van Dongen, W.M.P. van der Aalst, On the role of fitness, precision, generalization and simplicity in process discovery, in *On the Move to Meaningful Internet Systems: OTM 2012*, vol. 7565 (2012), pp. 305–322
18. M. Ceci, R. Corizzo, F. Fumarola, M. Ianni, D. Malerba, G. Maria, E. Masciari, M. Oliverio, A. Rashkovska, Big data techniques for supporting accurate predictions of energy production from renewable sources, in *IDEAS* (2015), pp. 62–71
19. E. Cesario, G. Manco, R. Ortale, Top-down parameter-free clustering of high-dimensional categorical data. IEEE Trans. Knowl. Data Eng. **19**(12), 1607–1624 (2007)
20. E. Cesario, F. Folino, A. Locane, G. Manco, R. Ortale, Boosting text segmentation via progressive classification. Knowl. Inf. Syst. **15**(3), 285–320 (2008)
21. V. Chandola, A. Banerjee, V. Kumar, Anomaly detection: a survey. ACM Comput. Surv. **41**(3), 15:1–15:58 (2009)
22. N.V. Chawla, K.W. Bowyer, L.O. Hall, W.P. Kegelmeyer, Smote: synthetic minority oversampling technique. J. Artif. Intell. Res. **16**, 321–357 (2002)
23. D. Cook, L. Holder, *Mining Graph Data* (Wiley, Hoboken, 2007)
24. G. Costa, F. Folino, A. Locane, G. Manco, R. Ortale, Data mining for effective risk analysis in a bank intelligence scenario, in *ICDE Workshops* (2007), pp. 904–911
25. G. Costa, G. Manco, R. Ortale, An incremental clustering scheme for data de-duplication. Data Min. Knowl. Discov. **20**(1), 152–187 (2010)
26. G. Costa, G. Manco, R. Ortale, E. Ritacco, From global to local and viceversa: uses of associative rule learning for classification in imprecise environments. Knowl. Inf. Syst. **33**(1), 137–169 (2011)
27. G. Costa, G. Manco, R. Ortale, A generative bayesian model for item and user recommendation in social rating networks with trust relationships, in *ECML PKDD* (2014), pp. 258–273
28. A. Cuzzocrea, F. Folino, M. Guarascio, L. Pontieri, A robust and versatile multi-view learning framework for the detection of deviant business process instances. Int. J. Cooperative Inf. Syst. **25**(4), 1–56 (2016)
29. G. Dong, J. Pei, *Sequence Data Mining*, vol. 33 (Springer Science & Business Media, Boston, 2007)
30. R.O. Duda, P.E. Hart, D.G. Stork, *Pattern Classification* (Wiley, New York, 2001)
31. K. Ezawa, M. Singh, S.W. Norton, Learning goal oriented bayesian networks for telecommunications risk management, in *ICML* (1996), pp. 139–147
32. R.E. Fawcett, F. Provost, Adaptive fraud detection. Data Min. Knowl. Disc. **3**(1), 291–316 (1997)
33. T. Fawcett, F. Provost, Adaptive fraud detection. Data Min. Knowl. Disc. **1**, 291–316 (1997)
34. S. Flesca, S. Garruzzo, E. Masciari, A. Tagarelli, Wrapping PDF documents exploiting uncertain knowledge, in *CAiSE* (2006), pp. 175–189
35. F. Folino, M. Guarascio, L. Pontieri, Mining predictive process models out of low-level multidimensional logs, in *CAISE* (2014), pp. 533–547

36. I. Gat-Viks, R. Sharan, R. Shamir, Scoring clustering solutions by their biological relevance. Bioinformatics **19**(18), 2381 (2003)
37. G. Greco, A. Guzzo, G. Manco, D. Saccà, Mining and reasoning on workflows. IEEE Trans. Knowl. Data Eng. **17**(4), 519–534 (2005)
38. G. Greco, A. Guzzo, L. Pontieri, D. Saccà, Discovering expressive process models by clustering log traces. IEEE Trans. Knowl. Data Eng. **18**(8), 1010–1027 (2006)
39. G. Greco, A. Guzzo, L. Pontieri, Mining taxonomies of process models. Data Knowl. Eng. **67**(1), 74–102 (2008)
40. J. Han, M. Kamber, J. Pei, *Data Mining: Concepts and Techniques* (Morgan Kaufmann, Amsterdam, 2011)
41. R. Jin, L. Si, C. Zhai, A study of mixture models for collaborative filtering. Inf. Retr. **9**(3), 357–382 (2006)
42. M. Kubat, R.C. Holte, S. Matwin, R. Kohavi, F. Provost, Machine learning for the detection of oil spills in satellite radar images. Mach. Learn. **30**(2), 192–215 (1998)
43. E. Masciari, Trajectory clustering via effective partitioning, in *FQAS* (2009), pp. 358–370
44. E. Masciari, SMART: stream monitoring enterprise activities by RFID tags. Inf. Sci. **195**, 25–44 (2012)
45. E. Masciari, G.M. Mazzeo, C. Zaniolo, Analysing microarray expression data through effective clustering. Inf. Sci. **262**, 32–45 (2014)
46. T.M. Mitchell, *Machine Learning* (McGraw-Hill, New York, 1997)
47. C. Phua, D. Alahakoon, V. Lee, Minority report in fraud detection: classification of skewed data, in *ACM SIGKDD Explorations Newsletter*, Special issue on learning from imbalanced datasets (2004), pp. 50–59
48. S. Ramaswamy, R. Rastogi, K. Shim, Efficient algorithms for mining outliers from large data sets, in *SIGMOD* (2000), pp. 427–438
49. F. Ricci, L. Rokach, B. Shapira, P.B. Kantor (eds.), *Recommender Systems Handbook* (Springer, New York, 2011)
50. P. Riddle, R. Segal, O. Etzioni, Representation design and brute-force induction in a boeing manufacturing domain. Appl. Artif. Intell. **8**(1), 125–147 (1994)
51. H. Shan, A. Banerjee, Bayesian co-clustering, in *ICDM* (2008), pp. 530–539
52. J. Tang, Z. Chen, A. Fu, D. Cheung, Capabilities of outlier detection schemes in large datasets, framework and methodologies. Knowl. Inf. Syst. **11**, 45–84 (2007). doi:10.1007/s10115-005-0233-6
53. W. van der Aalst et al., Process mining manifesto, in *Proceedings of BPI* (2012), pp. 169–194
54. W.M.P. van der Aalst, *Process Mining: Discovery, Conformance and Enhancement of Business Processes* (Springer, Berlin, 2011)
55. J. Wang, G. Karypis, HARMONY: efficiently mining the best rules for classification, in *Proceedings of SIAM International Conference on Data Mining* (2005), pp. 205–216
56. G.M. Weiss, Mining with rarity: a unifying framework. SIGKDD Explor. **6**(1), 7–19 (2004)

Services Discovery and Recommendation for Multi-datasource Access: Exploiting Semantic and Social Technologies

Devis Bianchini, Valeria De Antonellis and Michele Melchiori

Abstract The advent of Service Oriented Architectures (SoA) in the late 90s has significantly changed the development of enterprise systems. Web application development relying on selection and reuse of services, offered as third party software components, has been proposed as a new paradigm to effectively support creativity and productivity of developers. This development paradigm strongly requires advanced discovery and recommendation techniques, able to use and combine different types of information to suggest the most suitable data services for multi-datasource access. WSDL-based, semantic-enriched service matchmaking approaches have been initially proposed to enable service discovery and composition. Subsequently, approaches for web mashup, through RESTful services and Web APIs selection based on their lightweight descriptions, have emerged to meet requirements of agile development. Recently, in this context, service discovery and recommendation techniques are being empowered by considering factors related to the social web such as the existence of developers social networks and the possibility of evaluating the experience of web application developers. According to these premises, in this chapter, we present main features of a comprehensive data service selection framework, apt to provide advanced discovery and recommendation techniques. In the framework, an experience perspective will be considered, focused on social networks of developers, where social relationships represent explicit endorsements among developers concerning their skill in Web application development and votes on data services, assigned by developers, are used to estimate developers' credibility according to a majority-based approach.

D. Bianchini · V. De Antonellis · M. Melchiori (✉)
Department of Information Engineering, University of Brescia,
Via Branze, 38, 25123 Brescia, Italy
e-mail: melchior@ing.unibs.it

D. Bianchini
e-mail: bianchin@ing.unibs.it

V. De Antonellis
e-mail: deantone@ing.unibs.it

S. Flesca et al. (eds.), *A Comprehensive Guide Through the Italian Database Research Over the Last 25 Years*, Studies in Big Data 31, DOI 10.1007/978-3-319-61893-7_22

Keywords Data service · Web API · Web application design · Developers social network · Collective knowledge · Discovery · Recommendation · Search · Ranking · Similarity

1 Introduction

The advent of Service Oriented Architectures (SoA) in the late 90s has significantly changed the development of enterprise systems. Companies have heavily invested in Web Service technologies and, as a consequence, a growing number of services is being made available over the time. Service proliferation over the Web has been initially facilitated by the development of several standards, like WSDL for description of service interface and functionalities, UDDI for service registry, SOAP for message exchange and BPEL4WS for service orchestration. Later, the REST (REpresentational State Transfer) architectural style, focusing on system resources to be accessed, has been defined and RESTful services have been developed to facilitate access to the resources. As a consequence, nowadays, more and more public and private data providers offer (semi) structured data sources on the Web that are usually exposed as Web services, which can be implemented as SOAP-based services (data that can be used are described as signature's I/O parameters in WSDL) or as RESTful services (data that can be accessed are described by Web APIs). Compared to SOAP-based services, RESTful services (often referred to as Web APIs) do not have a formal model or standard description of service capabilities (e.g., WSDL). The RESTful model is mainly based on the description of methods to be invoked, relying on HTTP protocol, as well as tags, assigned to services by those who created or used them, to classify services and enable their fast search. This lightweight description is the reason behind RESTful services widespread diffusion [1]. In fact, REST has emerged in the last few years as a predominant Web service design model and has had such a large impact on the Web that it has mostly displaced SOAP- and WSDL-based service design because it's a considerably simpler style to use.

In this context, building web applications to access and use data sources available over the web increasingly requires frameworks to support the discovery and recommendation of services that enable access to web data sources. In the course of time, WSDL-based, semantic-enriched service matchmaking approaches have been initially designed to enable service discovery and composition [2]. Subsequently, approaches for web mashup, through RESTful services and Web APIs selection based on their lightweight descriptions [3, 4], as well as based on experiences and preferences of web users [5–8], have emerged to meet requirements of agile development. Recently, service discovery and recommendation techniques are being empowered by considering factors related to the social web such as the existence of developers social networks and the possibility of evaluating the experience of web application developers.

According to these premises, in this chapter we present a data service selection framework, apt to provide advanced search and ranking techniques that take into account: (i) lightweight data service descriptions, in terms of (semantic) tags and technical aspects on which service implementations rely (e.g., protocols, formats for data exchange); (ii) previously developed aggregations of data services, to enhance selection by considering services already used in similar contexts (i.e., in applications based on similar data services); (iii) a social network of developers, where social relationships represent explicit endorsements among developers concerning their skill in Web application development starting from third party data service selection. Moreover, developers can also express votes on data service as included into existing applications, and these votes are used to estimate developers' credibility according to a majority-based approach. In the framework, service selection techniques based on single components, their descriptions and their aggregations are complemented with an experience perspective, focused on developers who used and voted components to build their own aggregations. The data service selection is based on the social network of developers and on the analysis of their social relationships. This analysis combined with a credibility evaluation of each developer determines a ranking of developers.

The chapter is organized as follows: Sect. 2 introduces a relevant application scenario that motivates the described approach, following recent trends in service discovery and recommendation; in Sect. 3 existing approaches in literature are presented; Sect. 4 describes the framework architecture, while Sect. 5 details on how the experience perspective can be engaged in service discovery and recommendation are discussed; finally, Sect. 6 closes the chapter.

2 Application Scenario

Let's consider a developer who is developing a new application for hotel booking. We envision the developer's task as organized into two main phases: firstly, the developer has to discover and select the right data services and might rely, to this purpose, on functional and non functional data service descriptive features; then, data service composition have to be addressed, in order to deploy the application required for hotel booking. Here, we focus on the former phase. Data service selection may be difficult because the developer has no control on services, and source reputation could be only partially known. These difficulties are mainly due to the high number and heterogeneity of available data services over the Web and to their often unknown origins.

Let's suppose that the developer is using one of the most popular Web API repositories (e.g., `Mashape.com`, `ProgrammableWeb.com`), and starts by specifying `hotel booking` keywords to search for services. `ProgrammableWeb` returns 79 services,[1] as shown in Fig. 1. The popularity of `Cleartrip Hotel` service (followed by 163 users) and mashups where hotel booking services have been used

[1]`https://www.programmableweb.com/search/hotel%20booking`.

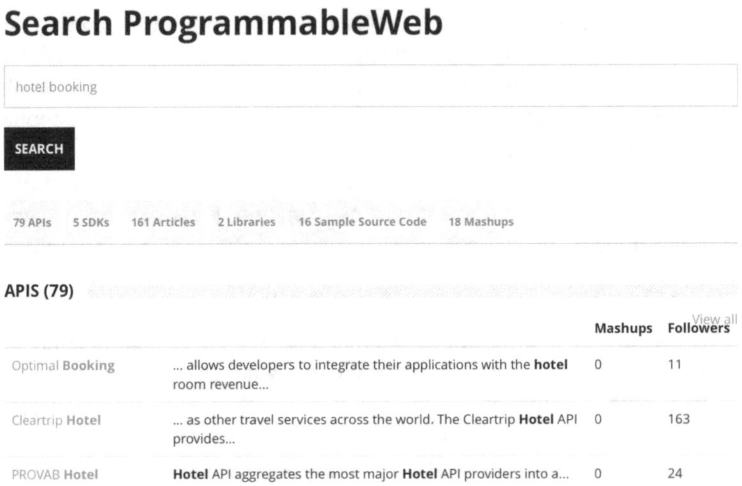

Search ProgrammableWeb

hotel booking

SEARCH

79 APIs 5 SDKs 161 Articles 2 Libraries 16 Sample Source Code 18 Mashups

APIS (79)

View all

		Mashups	Followers
Optimal **Booking**	... allows developers to integrate their applications with the **hotel** room revenue...	0	11
Cleartrip **Hotel**	... as other travel services across the world. The Cleartrip **Hotel** API provides...	0	163
PROVAB **Hotel**	**Hotel** API aggregates the most major **Hotel** API providers into a...	0	24

Fig. 1 An example of `hotel booking` service search using `ProgrammableWeb`

(highlighted in figure) are not exploited by the system and it is up to the developer to improve his/her search by considering them. On the other hand, also approaches that empower service recommendation approaches using these details provide the developer with look-up search results, without taking into account decisions and evaluations made by the developer during the development process by interacting with the search tool.

3 Related Work

In literature, data service (or data source) discovery and selection and their composition are, as we already remarked in the previous section, clearly distinguished into two distinct phases [9]. According to this viewpoint, various approaches consider as separate the activities of discovery and composition. For example, the approach described in [10], which shares with ours many preliminary definitions, focuses on data service integration and querying, considering the data service selection results as given. In this chapter we deal with data service recommendation for selection purposes, which have to be solved before beginning integration of data and metadata coming from the selected sources. In this research field, recent trends focused on collaborative filtering, users' experience and social-based service recommendation and selection.

Collaborative filtering. Other quite recent approaches are based on techniques of collaborative filtering applied to Web APIs recommendation. For example, the Web API recommender system described in [6] exploits users' ratings to refine service ranking by applying collaborative filtering techniques. In [11] tags used to annotate both RESTful data services and mashups are classified into topics through a probabilistic distribution. Topics are used to add semantics on top of traditional tagging. The API popularity is taken into account as experience dimension, computed as the number of times a Web API has been used in the past. API popularity is used during search to weight API relevance (with respect to the issued request), based on topics. With respect to these approaches, we exploit the network of social relationships between developers, integrating the credibility assessment techniques.

Users' experience for service recommendation. Recent trends on (data) service recommendation integrated users' experience aspects (ratings/votes, co-usage of the same services in several applications, service popularity, etc.) with traditional discovery techniques, mainly based on functional and non functional features. For example, matrix factorization techniques have been used in dynamic service exploration and recommendation, to identify user/topic/service-related latent factors that influence users to make service selections [7]. For service recommendation, emerging techniques based on the similarity of user's behavior and service usage can be interesting too [12]. Approaches in [13, 14] study service recommendation based on votes/ratings and introduce techniques to evaluate the reputation and credibility of raters. They are mainly focused on these aspects, compared to our approach, where credibility assessment is integrated in a comprehensive framework for data service recommendation. This enables a finer and more effective service selection.

Social-based service discovery and selection. An attempt to introduce social network exploitation for Web API/service selection is described in the SoCo (Social Composer) system [15]. In this approach, components are suggested to the user u considering other users that are similar to u in a social network. Specifically, a component is suggested to u depending on the number of times the component has been used by other users socially related to u and on the social proximity between such users and u. The notion of *social network* in [5] presents a different meaning compared to our definition. Indeed, in [5] social relationships are defined between mashups, if their tag-based similarity and the number of common services is over a pre-defined threshold, and social networks of developers are not considered to support service discovery.

4 Framework Architecture

To enable social network-based selection of data services, we discuss a framework, WISeR (Web apI Search and Ranking) [16], extended with social-based functionalities. The architecture of the framework is shown in Fig. 2. WISeR is based on a multi-layered model, developed over different perspectives:

- a *component perspective*, focused on data services, described through tags/categories, technical features (e.g., protocols, data formats);
- an *application perspective*, focused on aggregations of data services;

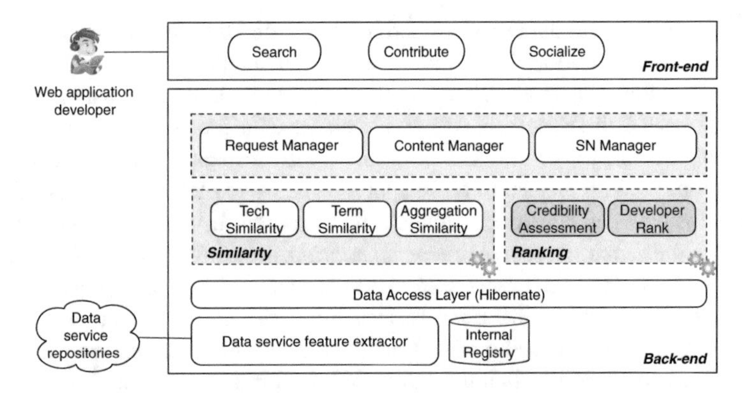

Fig. 2 The WISeR (Web apI Search and Ranking) framework architecture

- an *experience perspective*, focused on developers who used and voted data services to build their own aggregations.

Formally, data services and aggregations are defined as follows.

Definition 1 We define a <u>data service</u> s (hereafter, <u>service</u>) as an operation/method/query to access data of a web source, whose underlying data schema might be unknown to those who use the service. Within the scope of this chapter, we model a service s as $\langle n_s, \mathcal{F}_s, \{t_s\}\rangle$, where: n_s is the service name; $\mathcal{F}_s$ is an array of elements, where each element $\mathcal{F}_s^X$ represents the technical feature X (e.g., protocols, data formats, authentication mechanisms) and is modeled as a set of allowed values for that feature (e.g., XML or JSON as data formats); $\{t_s\}$ is a set of <u>tags</u>. We denote with $\mathcal{S}$ the overall set of available services.

In Fig. 3 data services taken from `ProgrammableWeb.com` for the running example are listed.

Definition 2 A <u>service aggregation</u> represents a set of services that can be composed to deploy a Web application. An aggregation g is modeled as a triple $\langle n_g, \mathcal{S}(g), d\rangle$, where: n_g is the aggregation name; $\mathcal{S}(g) = \{s_1, \ldots, s_n\}$ is the set of data services used in g; $d \in \mathcal{D}$ is the developer who designed the web application by composing

Service	Service name	Technical features	Tags
s_1	HotWire	$\mathcal{F}_{s_1}^{DataFormat} = \{\texttt{XML,JSON}\}$ $\mathcal{F}_{s_1}^{Protocol} = \{\texttt{RSS, Atom, REST}\}$	{City, Star, Hotel, Travel}
s_2	EasyToBook	$\mathcal{F}_{s_2}^{DataFormat} = \{\texttt{XML}\}$ $\mathcal{F}_{s_2}^{Protocol} = \{\texttt{SOAP}\}$	{City, Hotel, Travel}
s_3	LocalDiffusion	$\mathcal{F}_{s_4}^{DataFormat} = \{\texttt{XML,JSON}\}$ $\mathcal{F}_{s_4}^{Protocol} = \{\texttt{HTTP}\}$	{Cuisine, City, Restaurant, Tourism}

Fig. 3 Data service descriptions used in the running example

services in g. We denote with $\mathcal{G}$ the overall set of service aggregations, that is, $g \in \mathcal{G}$, and with $\mathcal{G}(s)$ the set of aggregations where s has been included.

Fictious examples of aggregations for the running example are listed in the following.
$g_1 \Rightarrow \langle \texttt{TravelPlan}, \mathcal{S}_{g_1} = \{s_1, s_3\}, d_{g_1} \rangle$
$g_2 \Rightarrow \langle \texttt{Stay\&Fun}, \mathcal{S}_{g_2} = \{s_2, s_3\}, d_{g_2} \rangle$
According to this vision, best representatives of data services are resource-oriented services (i.e., RESTful ones). For services that present a structured, operation-oriented description (e.g., WSDL), we consider only data they work on, represented through tags. We admit different ways of assigning tags to services: (i) keywords extracted from the service name, I/O names and textual description through the application of text mining techniques (such as stop words removal, stemming, camel case word decomposition, etc.) [17]; (ii) tags assigned by developers who used the service to design their own applications. Concerning application development, a modern Web application is typically implemented through different phases: firstly, the developer has to discover and select the most suitable services; furthermore, service composition has to be performed, in order to deploy the final application. Therefore, we clearly distinguish between (complete) Web applications and service aggregations, which constitute an intermediate step, as specified in the previous definition.

WISeR allows different kinds of search, either for single isolated services or for services needed to complete an existing application. Furthermore, both simple and proactive search are provided. In the *simple search*, the developer receives suggestions about relevant data services after explicitly specifying search features (e.g., tags, required values for each kind of technical features, and so on). In the *proactive search*, that is the most explorative search typology, the developer does not have in mind the services of interest, or he/she has just a partial idea of what he/she is looking for, and the framework proactively suggests candidate data services according to the aggregation that is being developed. The framework is equipped with proper wizards that guide the developer in formulating the request, depending on the specific kind of search that is being performed. Wizards have been described in [16]. Similarity metrics, based on above perspectives, have been defined in WISeR to quantify the compliance of available services compared to the specified request:

- the **tag similarity**, to denote the similarity between the request and each service based on tags, either semantically disambiguated or not; we denote tag similarity as $TagAff(\{t_{s_i}\}, \{t_{s_j}\}) \in [0, 1]$, where $\{t_{s_i}\}$ and $\{t_{s_j}\}$ are compared sets of tags;
- the **technical feature similarity**, to denote the similarity between the request and each service based on technical features; we denote similarity for a technical feature X as $TechSim^X(\{f_{s_i}\}, \{f_{s_j}\}) \in [0, 1]$, where $\{f_{s_i}\}$ and $\{f_{s_j}\}$ are compared sets of values allowed for feature X;
- the **aggregation similarity**, to denote the similarity between the request and each service based on average similarity between the aggregation that is being developed and aggregations where the service has been used in the past, respectively; we denote aggregation similarity as $AggSim(g_o, g_p) \in [0, 1]$, where g_o and g_p are compared aggregations; the rationale here is that the more similar the services used

in the two compared aggregations according to their similarity, the more similar the two aggregations.

We denote with $Sim(s_i, s_j) \in [0, 1]$ the overall similarity between two data services, computed as a combination of the above similarities. Overall similarity testing and setup of weights, to proper balance tag, technical feature and aggregation similarity, have been discussed in [16].

Data services are kept within public repositories, where they are advertised. Proper wrappers have been designed to collect relevant service features (e.g., tags already assigned within such repositories) and to maintain a reference towards original services, thus ensuring full compatibility. Wrappers extract service descriptions according to Definition 1. Service descriptions are stored within the `Internal Registry` shown in Fig. 2. Wrappers are invoked within the `Data Service Features Extractor`.

Service ratings, information about aggregations they belong to and developers' social relationships are saved within the `Internal Registry` as well. Developers interact with the framework through the front-end, that provides the following functionalities: (i) Search, to search for services; (ii) Contribute, to tag and assign votes to services and create new aggregations; (iii) Socialize, to manage social relationships between developers. Search and Contribute functionalities are performed within the WISeR framework through proper Web GUIs described in [16]. Search result is a list of candidate services that is properly sorted when displayed, where sorting is enhanced with developers' rank metric. During voting procedure, developers' credibility assessment is performed. The `Manager` modules act as controllers for the front-end functionalities. The core modules we added to the existing WISeR framework, namely `Developer rank` and `Credibility Assessment`, will be detailed in the next Sect. 5.

Concerning the Socialize functionality, the developer is redirected to the social network of developers that he/she can access using his/her credentials. Data on the developers' social relationships are stored within the `Internal Registry` through a proper wrapper, as explained above for extracting data service descriptions. Some public Web API repositories, such as `Mashape`, introduce "follower-of" social relationships between developers. Developers, who used services to design their applications, are therefore organized in a social network, formalized as follows.

4.1 Modeling the Social Network of Developers

Definition 3 The social network of developers is a pair $SN = \langle \mathcal{D}, \mathcal{E} \rangle$, where: (a) $\mathcal{D}$ is the set of developers; (b) $\mathcal{E}$ is a set of *follower-of relationships* between developers, defined as $\mathcal{E} = \{d_i \xrightarrow{f} d_j | d_i, d_j \in \mathcal{D}\}$, where $d_i \xrightarrow{f} d_j$ indicates that d_i explicitly declares to be inclined to learn from the choices made in the past by d_j for web application design purposes.

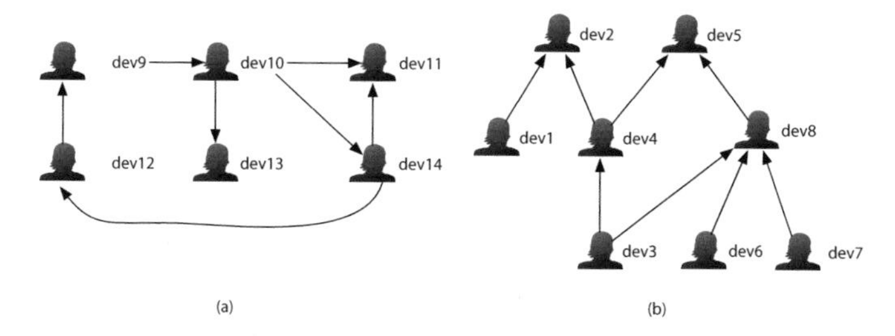

Fig. 4 Sample social networks of developers, which present peer-based (**a**) and hybrid (**b**) topologies

Each developer $d_i \in \mathcal{D}$ is modeled as $\langle \mathcal{G}(d_i), \mathcal{D}^* \rangle$, where $\mathcal{G}(d_i) \subseteq \mathcal{G}$ is the set of aggregations designed by d_i in the past, $\mathcal{D}^* \subseteq \mathcal{D}$ is the set of other developers, whom d_i declares to be inclined to learn from, in order to design web applications, that is, $\mathcal{D}^* = \{d_k | d_i \xrightarrow{f} d_k \in \mathcal{E}\}$.

The organization of the *follower-of* relationships determines the network structure as extracted from `Mashape` repository. The developers' social network can be represented as one or more directed graphs, as shown in Fig. 4, where a graph can assume different topologies. For example, it can be restricted to a hierarchy. On the other hand, a peer-based network has a topology where a hierarchy is not present; there can be pairs of developers that mutually follow each other, and this is typical of totally collaborative and open contexts. An example is the network in Fig. 4a. A third kind of topology, see Fig. 4b, represents a hybrid case, where a developer is or has been involved in different web application design projects and, maybe depending on the particular application domain, can follow different reference developers (consider, as an example, `dev3`, who declares to follow both `dev4` and `dev8`).

5 Social-Based Service Discovery and Recommendation

5.1 Service Request

A developer, who is responsible for a project based on data services, hereafter denoted as the *requester* d^r, formulates the request, denoted with s^r, that is matched against the set $\mathcal{S}$ of available data services. The aim is at finding data services to complete/expand the application that is being designed. Therefore, the request s^r is formulated as a set of desired (semantic) tags, a set of data services, that have been already included in the application, and a set of desired technical features. Formally, $s^r = \langle \{t^r\}, g^r, \{f^r\} \rangle$, where $\{t^r\}$ is the set of tags, $g^r = \{s_1, s_2, \ldots s_n\}$ is the set of already selected data service descriptions and $\{f^r\}$ is a set of technical features.

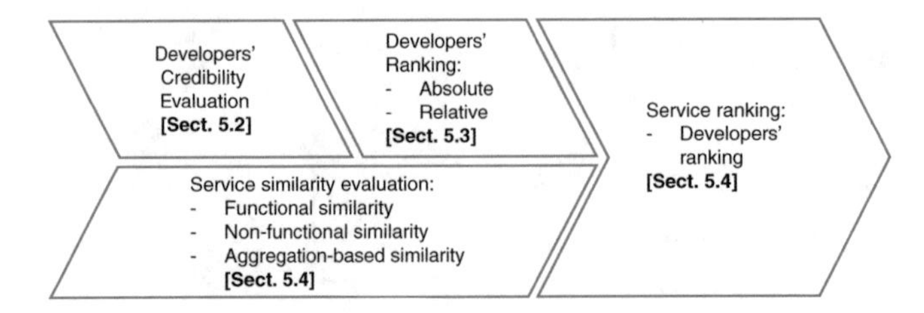

Fig. 5 Services recommendation phases in the extended version of WISeR

It worths noting that the search for a single data service (or for the first data service to be included in the new Web application that is being designed) is a particular case of the same definition, that is, $s^r = \langle\{t^r\}, g^r, \{f^r\}\rangle$, where $g^r = \emptyset$.

Answering the service request s^r is based in our approach on the following phases: (i) developers' credibility evaluation; (ii) developers' ranking; (iii) service selection; (iv) service ranking. In the following, we detail these phases. Diagram in Fig. 5 provides a short summary with references to relevant sections and where in the diagram each phase uses the outputs of the preceding ones.

5.2 Developers' Credibility Evaluation

In WISeR, a developer may assign votes to services used in the applications. In particular, since developers exchange their experiences in using services, votes become an enabling feature to this purpose. Following this perception, for example, all the most popular Web API repositories (and, among them, `ProgrammableWeb`) include a rating system. Votes are assigned by the developers with the adoption of the NIH 9-point Scoring System.[2] This scoring system has few rating options (only nine) to increase potential reliability and consistency and with enough range and appropriate anchors to encourage developers to use the full scale (from `poor`, to denote completely useless and wrong services, to `exceptional`, to denote services with very good performances and functionalities and easy to use). During the rating, the developer is provided with the set of options and corresponding meaning (such as, for example, the ones associated with `poor(9)` and `exceptional(1)` options, as mentioned above). These options are uniformly distributed over the [0, 1] interval so that the highest vote is corresponding to 1 and the lowest to 0. Our approach introduces an important distinction for service rating, compared to existing systems, because it takes into account the aggregation in which services have been evaluated according to the following definition.

[2] `http://enhancing-peer-review.nih.gov/scoring%26reviewchanges.html`.

Definition 4 Given a service $s_j \in \mathcal{S}$, we denote with $v(s_j, g_k, d_i) \in [0, 1]$ the vote assigned to the service s_j by a developer $d_i \in \mathcal{D}$ with reference to the aggregation $g_k \in \mathcal{G}$ in which s_j has been used. Votes are assigned according to the NIH 9-point Scoring System and then mapped to the [0, 1] interval.

Aggregation-contextual rating helps in properly weighting votes assigned to services. For instance, different votes might be assigned to the `Amazon` APIs, depending on the aggregations where they have been used. When a developer is looking for the average of votes assigned to a service, relevant votes to be considered are those that have been assigned with reference to aggregations that are similar to the one is being developed (according to the aggregation similarity, $AggSim()$, mentioned in Sect. 4).

It is important to estimate the credibility of a developer, who expresses votes or judgments. Therefore, we include credibility evaluation techniques with respect to which we introduce the notion of aggregation-contextual rating. Votes are used in the `Credibility Assessment` module (See Fig. 2). The basic idea is that, if the reported vote does not agree with the majority opinion, the developer's credibility is decreased, otherwise it is increased. Suppose the developer d_i assigned some votes to the service s_j with reference to the aggregations $g_1, g_2, \cdots, g_t$, respectively. For each g_m in these aggregations, we consider the set $\mathcal{A}g_m$ of aggregations $g_o \in \mathcal{G}$ that have similarity $AggSim(g_o, g_m)$ above a given threshold representing a level of minimum acceptable similarity.

A k-mean clustering algorithm is then applied to the set of votes assigned by other developers to the service s_j in the context of aggregations in $\mathcal{A}g_m$. By grouping similar ratings together we define the majority opinion in the context of specific aggregations. The rationale for clustering votes can be explained with the help of an example: if a service receives votes 1,1,1,2,9,9,8 (considered before their normalization to interval [0, 1]) and we adopt an average-based model, we obtain an overall rating of 4.4 out of 9. Actually, the average rating does not well describe the depicted situation, where the most of voters gives a very high vote (i.e., near to 1). So we choose a majority opinion approach inspired by the work [14]. The majority opinion on s_j is hence represented by the most densely populated cluster, whose centroid is taken as the majority vote:

$$M(s_j) = centroid(max_{i=1}^{k}(\mathcal{C}_i)) \tag{1}$$

where $\mathcal{C}_i$ is the i-th cluster, $k = |\{\mathcal{C}_i)\}|$ is the total number of clusters, $max()$ returns the cluster with the largest membership and $centroid()$ computes the centroid of the cluster. The number k of desired clusters is set to $\lceil \sqrt{N/2} \rceil$ where N is the number of considered votes and $\lceil x \rceil$ is the smallest integer not less than x. Therefore, let us consider a developer d_i having currently a credibility $c_n(d_i)$ after he already assigned n votes. And consider also a new vote $v(s_j, g_m, d_i)$ assigned by d_i to a service s_j when used within an aggregation g_m, then the updated credibility $c_{n+1}(d_i)$ of d_i is computed as follows:

$$c_{n+1}(d_i) = \frac{c_n(d_i) \cdot n + (1 - |M(s_j) - v(s_j, g_m, d_i)|)}{n + 1} \in [0, 1] \tag{2}$$

According to Eq. (2), if the vote $v(s_j, g_m, d_i) \in [0, 1]$ differs from the centroid $M(s_j) \in [0, 1]$, then term $1 - |M(s_j) - v(s_j, g_m, d_i)|$ tends to zero, therefore $c_{n+1}(d_i) < c_n(d_i)$ (the decrement is controlled by denominator $n + 1$, to avoid the case in which a designer looses too quickly his/her credibility for few assigned votes that are not aligned with majority opinion). Viceversa, if the vote $v(s_j, g_m, d_i)$ is close to $M(s_j)$, then term $1 - |M(s_j) - v(s_j, g_m, d_i)|$ tends to 1 and $c_{n+1}(d_i) > c_n(d_i)$ until $c_{n+1}(d_i)$ reaches 1 (max credibility). Initial values $c_0(d_i)$ are set to an intermediate value of 0.5. Note that credibility of a developer with a high value of n and that assigns a vote different from the ones expressed by the majority, is reduced only of a low amount. In fact, this type of vote is not necessarily describing an incoherent behavior of the developer and could be the result of a recent change in the service quality perceived by the voter.

5.3 Ranking of Developers

Let's suppose d^r has formulated the request s^r. Consider two candidate services s_1 and s_2, used by two developers d_1 and d_2, respectively, in aggregations that are similar to the one in s^r. If s_1 and s_2 are equally relevant with respect to s^r, then s_1 will be ranked better than s_2 in case the experience of d_1, who used s_1, is ranked better than the experience of d_2, who used s_2. The point here is at ranking the experience of developers d_1 and d_2.

The rank of a developer $d_i \in \mathcal{D}$ is computed as the product of two different rankings, according to the following formula:

$$dr(d_i) = \rho_{rel}^{d^r}(d_i) \cdot \rho_{abs}(d_i) \in [0, 1] \tag{3}$$

where: (a) a *relative ranking* $\rho_{rel}^{d}(d_i) \in [0, 1]$ ranks developer d_i based on the *follower-of* relationships between d_i and d^r (this rank is introduced to take into account the viewpoint of d^r, who explicitly declared to learn from other developers to select the right service); (b) an *absolute ranking* $\rho_{abs}(d_i)$ is based on the overall network of developers, to take into account the authority of d_i in the network independently of the developer d^r, who issued the request.

Relative ranking. In particular, the relative ranking $\rho_{rel}^{d^r}(d_i)$ is inversely proportional to the distance $\ell(d^r, d_i)$ between d^r and d_i, in terms of *follower-of* relationships, that is:

$$\rho_{rel}^{d^r}(d_i) = \frac{1}{\ell(d^r, d_i)} \in [0, 1] \tag{4}$$

If there is no a path from d^r to d_i, $\ell(d^r, d_i)$ is set to the length of the longest path of *follower-of* relationships that relates d^r to the other developers, incremented by 1, to denote that d_i is far from d^r more than all the developers within the d^r sub-network. Consider for example the network shown in Fig. 4, where the developer `dev3` is the requester and has to choose among services that have been used in the past by the developers `dev4`, `dev5`, `dev6`, `dev8` and `dev11`, whose *follower-of* relationships are depicted in the figure. In the example, ℓ(`dev3`, `dev4`) = ℓ(`dev3`, `dev8`) = 1, ℓ(`dev3`, `dev5`) = 2, and ℓ(`dev3`, `dev6`) = ℓ(`dev3`, `dev11`) = 4+1 = 5.

Absolute ranking. The absolute ranking $\rho_{abs}(d_i)\in[0, 1]$ is evaluated no matter is the viewpoint of the requester d^r. This ranking is composed of two different parts. The first one depends on the number of aggregations designed by d_i, the second one depends on the topology of the network of other developers who declared their interest for d_i past experiences, that is:

$$\rho_{abs}(d_i) = \frac{1-\alpha}{|\mathcal{D}|}\cdot|\mathcal{G}(d_i)| + \alpha\cdot\sum_{j=1}^{n}\frac{c(d_j)\cdot\rho_{abs}(d_j)}{F(d_j)} \quad (5)$$

This expression is an adaptation of the known PageRank metrics for Web search engines to the context we are considering. The original PageRank calculates an authority degree for Web pages based on the incoming links to pages. The value $\rho_{abs}(d_i)$ represents the probability that a developer will consider the example given by d_i in using a service for designing an enterprise Web application. Therefore, $\sum_i \rho_{abs}(d_i) = 1$. Initially, all developers are assigned with the same probability, that is, $\rho_{abs}(d_i) = 1/|\mathcal{D}|$. Furthermore, at each iteration of the absolute ranking computation, the absolute rank of a developer d_j, such that $d_j \xrightarrow{f} d_i$, is "transferred" to d_i according to the following criteria: (i) if d_j follows more developers, his/her rank is distributed over all these developers, properly weighted considering the credibility $c(d_j)$ of d_j(see the second term in Eq. (5), where $F(d_j)$ is the number of developers followed by d_j); (ii) a contribution to $\rho_{abs}(d_i)$ is given by the experience of d_i and is therefore proportional to the number $|\mathcal{G}(d_i)|$ of aggregations designed by d_i(see the first term in Eq. (5)). A damping factor $\alpha\in[0, 1]$ is used to balance contributions explained in (i) and in (ii). At each step, a normalization procedure is applied in order to ensure that $\sum_i \rho_{abs}(d_i) = 1$.

The algorithm actually used to compute recursively Eq. (5) is similar to the one applied for PageRank. In particular, denoting with $\rho_{abs}(d_i, \tau_N)$ the N-th iteration in computing $\rho_{abs}(d_i)$, with $\mathbf{DR}(\tau_N)$ the column vector whose elements are $\rho_{abs}(d_i, \tau_N)$, we have:

$$\mathbf{DR}(\tau_{N+1}) = \frac{1-\alpha}{|\mathcal{D}|}\cdot\begin{bmatrix}|\mathcal{G}(d_1)|\\|\mathcal{G}(d_2)|\\\vdots\\|\mathcal{G}(d_n)|\end{bmatrix} + \alpha\cdot\mathbf{M}\cdot\mathbf{DR}(\tau_N) \quad (6)$$

where $\mathbf{M}$ denotes the adjacency matrix properly modified to consider credibility, that is, $M_{ij} = \frac{c(d_j)}{F(d_j)}$ if $d_j \xrightarrow{f} d_i$, zero otherwise. As demonstrated in PageRank, computation formulated in Eq. (6) reaches a high degree of accuracy within only a few iterations.

Table 1 Details about developers' features, considered for the running example

Developer (d_i)	$\|\mathcal{G}(d_i)\|$	Credibility $c(d_i)$
dev1	5	1.0
dev2	3	0.7
dev3	2	1.0
dev4	4	0.1
dev5	3	0.7
dev6	2	0.2
dev7	2	1.0
dev8	2	0.2
dev9	2	0.7
dev10	3	0.6
dev11	3	0.7
dev12	2	0.9
dev13	1	0.5
dev14	2	0.7

For example, let's consider Table 1, that lists features (i.e., number of developed aggregations, credibility) considered for developers in the running example. In particular, we set $\alpha = 0.6$. At time τ_0 we set $\rho_{abs}(d_i) = 1/|\mathcal{D}| = 0.0714$ for all d_i. During the next iteration:

$$\rho_{abs}(\texttt{dev4}, \tau_1) = \left[\frac{1-0.6}{14} \cdot 4 + 0.6 \cdot \frac{1.0 \cdot 0.0714}{2}\right] = 0.1357$$

Similarly, $\rho_{abs}(\texttt{dev8}, \tau_1) = 0.1299$. After each iteration, normalization is applied to have $\sum_i \rho_{abs}(d_i) = 1$. In the example, after 5 iterations, the error measured as Euclidean norm of the vector $\mathbf{DR}(\tau_5) - \mathbf{DR}(\tau_4)$ is less than 0.001. And we obtain $\rho_{abs}(\texttt{dev4}) = 0.0997$ and $\rho_{abs}(\texttt{dev8}) = 0.0801$.

5.4 Service Selection and Ranking

Data service selection is performed in WISeR according to: (a) functional requirements, e.g., tags, used to classify services with respect to what services have been designed for, past use of service in aggregations that are similar to the one in the request, and (b) technical features.

Within the scope of this chapter, we remark that these requirements can be used to quantify the compliance between a service $s \in \mathcal{S}$ and the request s^r. Our aim is to combine the overall similarity value (defined in [16] and summarized in Sect. 4) with a ranking function $\rho_{serv} : \mathcal{S}^* \mapsto [0, 1]$, that is based on: (i) the ranking of developers who used $s \in \mathcal{S}^*$; (ii) the votes $v(s, g_i, d_k)$ assigned to s by each developer d_k that used s in an aggregation g_i. In particular, the better the ranking of developers who used the service s and the higher the votes assigned to s, the closer the value $\rho_{serv}(s)$ to 1.0 (maximum value). The value $\rho_{serv}(s)$ is therefore computed as follows:

$$\rho_{serv}(s) = \frac{\sum_{k=1}^{n} \sum_{i=1}^{m_k} dr(d_k) \cdot v(s, g_i, d_k)}{N} \in [0, 1] \tag{7}$$

where $d_k \in \mathcal{D}$, for each k, are the developers who used the service s in their own m_k Web application design projects, the vote $v(s, g_i, d_k)$ is weighted by $dr(d_k)$ that is the ranking of developer d_k according to Eq. (3). Moreover, N is the number of times the service s has been selected (under the hypothesis that a developer might use a data service s in $m \geq 1$ projects, then $dr(d_k)$ is considered m times), thus $N = \sum_{k=1}^{n} m_k$. The overall service similarity $Sim(s^r, s)$ and $\rho(s)_{serv}$ elements are finally combined in the following harmonic mean in order to get the ranking of service s:

$$rank(s) = \frac{2 \cdot \rho_{serv}(s) \cdot Sim(s^r, s)}{\rho_{serv}(s) + Sim(s^r, s)} \in [0, 1] \tag{8}$$

6 Conclusions

The advent of Service Oriented Architectures (SoA) in the late 90s has significantly changed the development of enterprise systems. Third party services are searched and integrated in web applications that need access to multi-datasources. WSDL-based, semantic-enriched service matchmaking approaches have been initially proposed to enable service discovery and composition. Subsequently, approaches for web mashup, through RESTful services and Web APIs selection based on their light-weight descriptions have emerged to meet requirements of agile development. In this chapter we presented a data service discovery and recommendation framework, where service selection techniques based on single components, their descriptions and their aggregations are complemented with an *experience perspective*, focused on developers who used and voted components to build their own aggregations. Metrics take into account the network of social relationships between developers, in order to exploit them for estimating the importance that a developer assigns to past experience of other developers.

Further studies are on going for extending the social-based recommendation techniques: specifically, other aspects such as the maturity of the use of services (estimated through their publishing data and the number and quality of aggregations including the services), the time sensitivity of social relationships (that is, relationships that have been established later could have a different impact for data service ranking) and specificity of the searched services (i.e., general purpose or domain-specific) may be investigated with respect to a possible influence in the search and ranking process.

References

1. C. Pedrinaci, J. Domingue, Web services are dead. Long Live Internet Services, Technical report, SOA4All White Paper (2010)
2. D. Bianchini, V. De Antonellis, M. Melchiori, Flexible semantic-based service matchmaking and discovery. World Wide Web J. **11**(2), 227–251 (2008)
3. W. Xu, J. Cao, L. Hu, J. Wang, M. Li, A social-aware service recommendation approach for mashup creation, in *IEEE International Conference on Web Services* (2013)
4. L. Yao, S. Zheng, A. Segev, J. Yu, Recommending web services via combining collaborative filtering with content-based features, in *IEEE International Conference on Web Services* (2013)
5. B. Cao, J. Liu, M. Tang, Z. Zheng, G. Wang, Mashup service recommendation based on user interest and social network, in *Proceedings of International Conference on Web Services (ICWS)* (2013)
6. B. Cao, M. Tang, X. Huang, Cscf: a mashup service recommendation approach based on content similarity and collaborative filtering. Int. J. Grid Distrib. Comput. **7**(2), 163–172 (2014)
7. X. Liu, I. Fulia, Incorporating user, topic, and service related latent factors into web service recommendation, in *Proceedings of IEEE International Conference on Web Services (ICWS 2015)* (2015), pp. 185–192
8. D. Bianchini, V. De Antonellis, M. Melchiori, Exploratory search of web data services, in *OTM Conferences*, vol. LNCS 10033 (2016), pp. 456–464
9. S. Ceri, D. Braga, F. Corcoglioniti, M. Grossniklaus, S. Vadacca, Search computing challenges and directions, in *Objects and Databases*, Lecture Notes in Computer Sciences, vol. 6348 (2010), pp. 1–5
10. S. Quarteroni, M. Brambilla, S. Ceri, A. Bottom-up, Knowledge-aware approach to integrating and querying web data services. ACM Trans. Web **7**(4), 1–33 (2013)
11. C. Li, R. Z. Z. Huai, H. Sun, A novel approach for api recommendation in mashup development, in *Proceedings of International Conference on Web Services (ICWS)* (2014), pp. 289–296
12. R. Liu, X. Xu, Z. Wang, Service recommendation using customer similarity and service usage pattern, in *Proceedings of IEEE International Conference on Web Services (ICWS 2015)* (2015), pp. 408–415. doi:10.1109/ICWS.2015.61
13. J. Al-Sharawneh, M. Williams, X. Wang, D. Goldbaum, Mitigating risk in web-based social network service selection: follow the leader, in *Proceedings of Sixth International Conference on Internet and Web Applications and Services* (2011), pp. 156–164
14. Z. Malik, A. Bouguettaya, RATEWeb: reputation assessment for trust establishment among web services. VLBD J. **18**, 885–911 (2009)
15. A. Maaradji, H. Hacid, R. Skraba, A. Lateef, J. Daigremont, N. Crespi, Social-based web services discovery and composition for step-by-step mashup completion, in *Proceedings of International Conference on Web Services (ICWS)* (2011)
16. D. Bianchini, V. De Antonellis, M. Melchiori, A multi-perspective framework for web API search in enterprise mashup design (best paper), in *Proceedings of 25th International Conference on Advanced Information Systems Engineering (CAiSE)*, vol. LNCS 7908 (2013), pp. 353–368
17. V. Gupta, G. Lehal, A survey of text mining techniques and applications. J. Emerg. Technol. Web Intell. **1**(1), 60–76 (2009)

Data Semantics Meets Knowledge Discovery in Databases

Claudia Diamantini, Domenico Potena and Emanuele Storti

Abstract In the last 30 years two important fields were born and have developed rapidly: knowledge discovery and knowledge management based on semantics. In the present chapter we provide an overview of the interlinks between them, taking the perspective of the evolution of systems and platforms supporting knowledge discovery with the help of data semantics.

1 Introduction

The term Knowledge Discovery in Databases (KDD) has been introduced for the first time at a 1989 IJCAI workshop. Reading the workshop report [42] almost 30 years later can be very instructive. One the one hand, it enlightens the need for a novel vision able to organize and structure diverse ideas converging from fields as different as Machine Learning and Databases, on the other end it layed down most of the needs and issues that paved the way of research until today. Among them, the need for "tools, tested and documented, made available to other people." Such tools, it is said, "would go far beyond the existing statistical tools and significantly enhance the human capabilities for data analysis" and "will make it easier for other people to run comparative studies". Also, a discussion on the role of domain knowledge is reported. The present chapter mainly develops along these lines, presenting the evolution of Knowledge Discovery support systems, with a standardization and domain knowledge perspective. One of the first standardization efforts has been in the KDD process

C. Diamantini · D. Potena (✉) · E. Storti
Dipartimento di Ingegneria dell'Informazione, Università Politecnica delle Marche, via Brecce Bianche, 60131 Ancona, Italy
e-mail: d.potena@univpm.it

C. Diamantini
e-mail: c.diamantini@univpm.it

E. Storti
e-mail: e.storti@univpm.it

S. Flesca et al. (eds.), *A Comprehensive Guide Through the Italian Database Research Over the Last 25 Years*, Studies in Big Data 31, DOI 10.1007/978-3-319-61893-7_23

itself. The best known definition of KDD is by Fayyad and Piatesky-Shapiro [19]: *"Knowledge Discovery in Databases is the non-trivial process of identifying valid, novel, potentially useful, and ultimately understandable patterns in data."* As a complex process, KDD involves several knowledge-intensive tasks driven by the analyst, from the choice of data to the interpretation of discovered patterns. Nowadays the two most adopted KDD process models are the one proposed by Fayyad et al. [19] and the CRISP-DM (Cross-Industry Standard Process for Data Mining) [53]. Comparing the two models, the main difference is the introduction of explicit phases of Business and Data Understanding in CRISP-DM. For the scope of this chapter, hereafter we will adopt a synthesis of the two, composed of four main phases: Domain Understanding (i.e. CRISP-DM's Business and Data Understanding), Data Preprocessing, Data Mining, and Interpretation/Evaluation (see Fig. 1).

In the *Domain Understanding* phase, as a first step the user defines the business goals on the basis of requirements and available data. Then business goals are mapped to data analysis goals, e.g. if the business goal is "avoid customers switch to the competition", a data analysis goal could be "to group customers on the basis of their behaviours and to define the best set of services for each group". At the end of this phase, the user also identifies needed resources, sketches the project plan and performs a preliminary analysis of available data. *Data Preprocessing* phase consists of the three tasks: selection, cleaning and transformation, explicitly named in the model by Fayyad et al.. The Selection phase allows the analyst to choose useful data sources and to extract relevant data from them. In order to be used as input to analysis tools, data are then cleaned filtering noise, removing outliers, correcting errors and managing missing data. Transformation is mainly devoted to find useful features to represent data depending on analysis goal. The core of the KDD process is the *Data Mining* phase, where statistical or machine learning algorithms are applied for extracting valid *patterns* or *models* for the phenomenon under study. In order to consider a model as knowledge, in the *Interpretation/Evaluation* phase, the quality of the model is evaluated and is interpreted from the point of view of business goals. Both [19, 53] highlight the iterative and interactive nature of the KDD process. If patterns are not valid, novel, useful, or understandable the analyst comes back to previous steps tuning algorithm's parameters, changing the algorithm, selecting new features or adding more data sources.

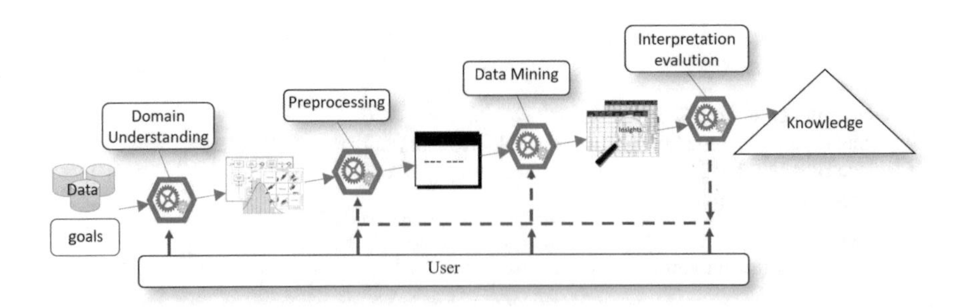

Fig. 1 The KDD process

It is noteworthy that there is no standard way to deal with a knowledge discovery problem: the design of a KDD process depends on data to be handled, business domain and goals. For each KDD phase, and in particular for Data Mining, several algorithms have been (and are) defined over time. Hence, the choice and setting of the right algorithm requires a high degree of expertise. Furthermore, analysts should be enough experienced in the (combined) use of these algorithms and have proper domain knowledge for interpreting the results. Such a scenario depicts a KDD project as a complex and collaborative work, where several users share knowledge and expertise.

In order to support the design and the execution of a KDD process, in the Literature several platforms and frameworks have been proposed, which we group into three generations. The first two generations include systems supporting a single user in local settings and platforms for distributed and parallel KDD respectively. The former moves from a scenario of several independently developed tools to suites featuring integrated DM and preprocessing algorithms. As to the latter, decentralization of users and tools raises the need for the management of distributed computations and the support to cooperative work and knowledge sharing, turning the attention to metadata and markup languages for algorithm interoperability, workflow specification and data exchange. In order to bridge the gap between technical and business view of KDD, 3rd-generation systems move from syntactic to semantic descriptions of resources to provide advanced support functionalities like algorithm selection and parameter setting. Furthermore, although tool composition and workflows received some attention in previous generations, in the 3rd generation they become a central component aimed at defining semantically correct KDD processes.

This chapter presents a survey of KDD support systems and approaches using semantics to support the KDD process. With this perspective in mind, we hasten to note that many applications of semantics that are typically grouped under the term Semantic Data Mining are out of the scope of the chapter, in particular inductive logic programming and the application of data mining techniques to analyze the Semantic Web and Linked Open Data.

In Sect. 2 we focus on the first two generations of KDD platforms and discuss the main approaches proposed in the Literature, following a historical/technological perspective. Section 3 discusses the 3rd generation focusing on how semantics has been used to represent algorithms, datasets and models. Finally, Sect. 4 provides some insights on current open issues and challenges, and sketches future trends on the topic.

2 Early Generations KDD Platforms

Early studies on KDD mainly focused on the definition of new algorithms and on scalability issues when large volumes of data are considered. Then the emphasis shifted to systems supporting the whole KDD process. In this mainstream we identify three generations of systems: stand-alone software (1st-generation), distributed architectures (2nd-generation) and semantics-based platforms (3rd-generation). Next subsections present the first two generations, while the third will be discussed in Sect. 3.

2.1 First Generation

In the 1-st generation we include systems supporting single users in local settings. Early vendor-related initiatives have led to commercial products [18, 21] whose success has been limited by their reduced flexibility and extensibility. As a matter of fact, the continuous development of new techniques made soon apparent the unfeasibility of legacy solutions. To this end, more advanced tools have been developing over time, most notably R,[1] Weka,[2] RapidMiner[3] and KNIME.[4]

The statistical toolkit R appeared for the first time in 1993. It provides a high-level programming language with several predefined functions, namely algorithms for data access, pre-processing, and mining. New functions can be added through CRAN repository.[5] These characteristics make R a very flexible tool, but it is intended for expert users with programming skills (e.g. algorithms composition is made by coding).

Weka [61], whose development began in the late '90s, it is by far the most popular and complete suite for KDD. Developed in Java, Weka can be used both through its GUIs and as a library. The latter feature leveraged its spread within other systems like RapidMiner and KNIME. New Java classes can be added in Weka guaranteeing the extensibility of the system. Furthermore it proposed the first unifying model for datasets description, the Attribute Relationship File Format (ARFF) that is still widely adopted as data interchange format.

RapidMiner's (formerly known as YALE [49]) and KNIME's (first version released in July 2006) most distinguishing feature is an advanced drag-and-drop graphic user interfaces simplifying composition. Extensions are straightforward by means of well-stocked marketplace. They are listed among market leaders together with other commercial products [31]. These systems are continuously developed to add advanced features. We like to mention for example an interesting feature of latest versions of RapidMiner, the so called "Wisdom of Crowds", a system supporting users in the choice and setting of algorithms. Suggestions come from around 250 K worldwide users of RapidMiner.

---

[1] https://www.r-project.org/.

[2] http://www.cs.waikato.ac.nz/ml/weka/.

[3] http://community.rapidminer.com/.

[4] https://www.knime.org/.

[5] https://cran.r-project.org/.

2.2 *Second Generation: Distributed Data Mining*

With the spread of network technologies, the late 90s has witnessed the birth of support systems based on distributed architectures: Client-Server, Agent-based, Service-Oriented Architectures, Grid and Cloud.

The Client-Server architecture has been adopted in [12, 38] to separate GUIs and data visualization functionalities, which are assigned to client, from computation, coordination and data management capabilities. These are managed by remote servers running on clusters of workstations and enabling parallel operations. The high volume of data to be transferred from client to server is a major limitation of this kind of architecture. Agent-based systems adopted a specular approach, based on mobile-agent paradigm: a software agent can move to the nodes where data reside to perform data analysis tasks and coordinating with others agents to achieve the KDD goal [22, 33, 46]. In this way, data are not moved and communication among agents is mainly limited to the setting of the workflow and coordination information, thus improving performances and preserving data privacy.

The decentralization of users and tools raised the need for the management of distributed computations and the support of cooperative work and knowledge sharing, turning the attention to metadata and markup languages for algorithm interoperability, workflow specification and data exchange. In [33] a language is introduced for supporting users in the composition of the workflow. In order to standardize interfaces, Papyrus [22] first proposed the Predictive Model Markup Language (PMML), a language for the exchange of predictive models whatever the data analysed and the algorithm. PADDMAS [46], Parallel And Distributed Data Mining Application Suite, supports users by means of a visual environment for workflow definition, an XML-based language for algorithm description that eases the introduction of new algorithms, and a module suggesting the most suitable algorithms based on this XML description.

Principles permeating the development of the Service Oriented computing model has been transferred to the KDD field as well, although initially mainly with a focus on performance. As for mobile-agent systems, in [29], services are sent to data sites, instead of transferring data, with the aim to help lowering consumption of bandwidth and to preserve privacy. Anteater [35] focused on improving both the communication among services and the management of parallel clustered servers in order to achieve computationally intensive processing on large amount of data.

The notion of Data Mining as a service is introduced in [51], for the first time. The approach concerns single, ready to use models that can be shared with the aim to facilitate novice users. A further development is on service composition and orchestration. For instance in [1] users can manually build a KDD process as a workflow of pre-defined services; the workflow can be then executed through Triana [32]. A similar approach is adopted in [56], where users are provided with an integrated interface for searching Data Mining services, building BPEL4WS processes and executing them. In [13], authors deal with privacy-preserving issue when a BPEL4WS workflow is shared. A contribution in the field has been also provided by the SEBD conference, where in 2003 authors of this chapter discussed the requirements of a KDD process design support system, and proposed the first version of a Service Oriented Architecture that then evolved into the KDDVM platform introduced in the next session. The original contribution was a set of core support services for wrapping legacy code, service discovery, and composition, supported by a language for DM tools description [14].

Solutions based on Grid for distributed Data Mining implement the Service Oriented Grid paradigm, where Grid resources are described and accessed as Web Services with great benefit for scalability and flexibility. Among them, [36] aims at solving computational issues by massive parallelism of Data Mining tasks, also introducing algorithm specifically suited for such architecture. Based on the Open Grid Services Architecture (OGSA) [55] standard, the system proposed in [45] focuses on a vertical and generic Grid application, which allows the execution of Data Mining workflows, mainly formed by Weka algorithms.

Recently, in the mainstream of high-performance computing, more scalable solutions based on Cloud Computing have been proposed. Among them, Googles TensorFlow,[6] the Machine Learning Service from Microsoft Azure[7] and ClowdFlows [28]. The first two are commercial products that make several KDD tools available over a huge number of computational devices. ClowdFlows is an open source platform mainly devoted to the collaborative design and execution of scientific workflows, providing users with user-friendly interfaces and supporting the integration of new tools as Web Services.

3 The 3-rd Generation: Semantics of Data and Algorithms

We identify the 3rd generation with approaches attempting to bridge the semantic gap between the technical view and the business view of the knowledge discovery process, and to promote collaboration among distributed interdisciplinary teams of experts. To this end, 3rd-generation systems move from syntactic to semantic descriptions of resources. The presentation is organized according to a classification of the kind of resources involved:

- *Computational resources*. Any application program for data manipulation useful in the context of some KDD phases. Computational resources process data with the aim to transform them in models. Computational resources can be further detailed as *computational units*, i.e. algorithms, tools and services, and *computational processes*, i.e. workflows of computational units.
- *Dataset*. The set of ground facts from an application domain.
- *Models*. The intermediate and final models produced by computational resources.

[6]https://www.tensorflow.org/.

[7]https://azure.microsoft.com/en-us/services/machine-learning/.

3.1 Computational Resources

KDD is largely a data manipulation activity, hence it relies on many automatic and semi-automatic techniques. Each technique can be described at the level of algorithm (abstract), tool or service (concrete) and execution log (instance):

- *Algorithm.* The abstract procedure that allows to solve a class of KDD problems.
- *Tool.* A particular implementation of an abstract algorithm, written in a given programming language. Many tools can implement the same algorithm, differing in the programming language, data structures, input/output interfaces, more of less efficient implementative solutions and the presence of further facilities like the capability to print intermediate results, visualization capabilities, and so forth. In distributed scenarios the same tool can be hosted by different machines and run at different sites. A *service* is a running instance of a tool.
- *Execution trace.* It is the specific executions, characterized by start/end time and possible abnormal termination, and related to instances of other resources like specific input data and actors.

The need to describe features of computational units, with more or less formalism and expressivity, is mainly motivated by the need to support users in a variety of tasks. Among them, the *discovery* of algorithms/tools on the basis of some properties (goal to achieve, user dataset, functional and non-functional characteristics), *sharing* and *composition*, i.e. the activity of linking suitable algorithms in order to build valid and useful knowledge discovery processes. As such, hereafter the discussion will cover both computational units and their composition in processes.

Early work represents knowledge about computational units by exploiting a relational meta-model [34] or the Object-Oriented paradigm [60], where models are introduced for supporting the building of a workflow of KDD tools. In the context of centralized systems, the KDDML (KDD Markup Language) [50] has been proposed as an XML-based language able to represent sequences of KDD operators forming a KDD process. These and similar approaches are suited for systems with limited heterogeneity, while richer descriptions are needed in more collaborative networks. A more generic language for tool description is KDTML (Knowledge Discovery Tool Markup Language) [26], which enables the annotation of tool characteristics, e.g. its location and execution environment, I/O interfaces and functionalities. As regards process and workflows, DPML (Discovery Process Markup Language) [2] is an XML-language for describing KDD workflows (at all abstraction levels) in the DiscoveryNet project. Such a language allows users to annotate both the whole workflow and each service for a collaborative management of remotely available tools as well as data sources. An early proposal of conceptualization of KDD tools in a hierarchy is proposed in [5] for a specific class of classification problems.

Following the mainstream of Semantic Web, more recent approaches exploit knowledge representation languages for a formal characterization of computational units at conceptual and logic level. Several ontologies have been developed in last years, mainly targeted to provide support for discovery and composition. Most

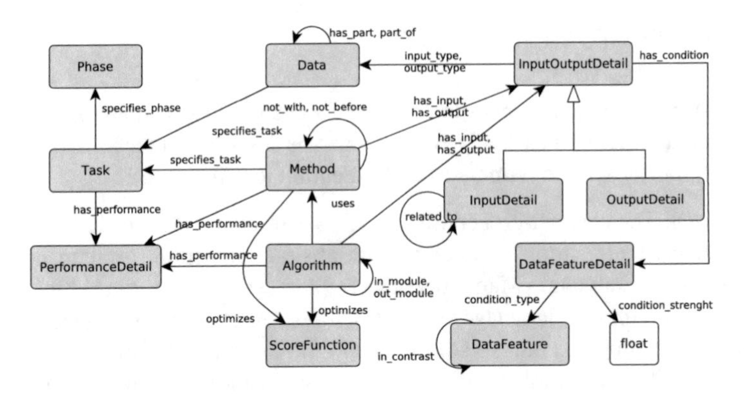

Fig. 2 KDDOnto classes and relations

ontologies focus on software and algorithms for Data Mining, without considering the whole KDD process. Among the early examples, *DAMON* (DAta Mining ONtology) [9] was built to simplify the development of distributed KDD applications on the Grid, supporting advanced discovering of tasks, methods, algorithms and software for a given goal.

Although a unified conceptualization of KDD is still an open issue, in the last decade several ontologies on KDD converged on some modeling choices. Among these, KDDONTO [15] is an ontology aimed to formally represent algorithms for KDD process. It is specifically conceived for supporting the *discovery* of algorithms and for enabling reasoning mechanisms for automatic *composition* in the KDDVM platform [16]. The core part of the ontology, shown in Fig. 2, is devoted to represent the concept of `Algorithm` and its properties in terms of `Method` (e.g. decision tree) used and the `Task` (e.g. classification) and `Phases` (e.g. data mining) it can be applied to. Algorithms interface are also described, including input/output details. The representation of pre/post conditions is a peculiar feature of KDDONTO. They provide a formalization of conditions under which the output of an algorithm can be used as input of another, and thus are the basis of process composition reasoning mechanisms. Other parts of the ontology are devoted to describe data (i.e., datasets, generic parameters, models) and performances. Similarly, the Data Mining Ontology (DMO) [27] contains information about the objects manipulated by the KDD support system (I/O Object), the needed metadata, the algorithm used by the tool and a goal description that formalizes the users desiderata. A KDD ontology [63] is proposed by Žáková et al. representing algorithms and every piece of knowledge involved in a KDD process: i.e., data, patterns, models, and tasks, according to the categorization firstly proposed by Džeroski [17]. The primary purpose of such knowledge is to enable a workflow planner to reason about which algorithms can be used for achieving a specified Data Mining task. The ontology used by the NExT system [6] is built on OWL-S and provides a relatively detailed structure of propositional Data Mining algorithms. It focuses on classical Data Mining processes, consisting of preprocessing, model induction and postprocessing. In *UKG* (Universal

Knowledge Grid) [62], the ontology is exploited for selecting algorithms and composing them in sequences, on the basis of the specific application domain they are used for. Users can browse the ontology also to search for case studies similar to the problem at hand, thus obtaining algorithms and software related to such case studies, through which solving the problem. Another support system for KDD [44] adopts OWS (Ontology Web Services) as process representation language, allowing both the execution over the Orange4WS platform and the annotation of any part of the process by means of concepts in an ontology representing tools, datasets and models. More recently, OntoDM-core [37], an extension of the previous OntoDM ontology, was proposed as a reference ontology aimed to define the most essential data mining entities. It provides a representational framework for the description of models, together with taxonomies of datasets, data mining tasks, generalizations, data mining algorithms and constraints, based on the type of data. Since it is not conceived for achieving specific goals, systems based on such an ontology can be used for a wide range of applications/use cases.

We refer the reader to a more comprehensive survey for many other intelligent data analysis systems [52], including general frameworks for workflow management, data manipulation and knowledge extraction.

3.2 Data

A huge literature exists on the application of data semantics, but approaches rarely contribute to the development of advanced features of KDD support systems. For this reason, and the lack of space, here we only briefly review the proposals that more strictly regard specific tasks of the KDD process.

Domain understanding The application of data mining techniques must be guided by an understanding of the business and data at hand, in order to properly model the knowledge discovery task, choose the proper technique and, in general, control the process. This is a non trivial step, as technical people may lack domain expertise. The adoption of domain ontologies in early stages can support newcomers to get acquainted with concepts and relations in a domain, reasoning mechanisms underlying formal ontologies may help enlightening hidden dependencies or conflicting assumptions. Despite this fact, very few systems include some support for domain understanding [7, 11]. Reference [48] provides a recent survey of general purpose tools to link structured, semi-structured and unstructured data to Linked Open Data and interpret them.

Data preprocessing First steps of data pre processing includes data selection, possibly by integration of different sources, and data cleaning. The application and advantages of semantics in these steps has been widely studies in other data base research streams, and the results obtained therein are applied in data mining as well. For instance domain knowledge expressed in the form of constraints helps correcting errors, identifying outliers, or correctly fill missed values [3, 8, 59]. An approach frequently cited in relation to data cleaning is OntoDataClean [40]. It develops a systems based on a conceptualization of operations involved in this step. Thus it is strictly related to the approaches discussed in Sect. 3.1.

One of the applications of semantics that is specific to data mining is the extraction of informative features. Most proposals adopt very simple ontologies (e.g. hierarchies) and concept matching techniques (e.g. based on string matching). The work by Phillis et al. [41] is among the first introducing the idea to exploit semantic relations between attributes to support this step. It develops a Prolog based system that takes attributes names contained in a data file and identify relations among them and new, derived, attributes by a semi-automatic procedure. Recently, approaches have been proposed to link entities of a given dataset to Linked Open Data entities, and then exploit the navigation of the LOD graph to augment the original dataset with new features. The proposals have led to extensions for RapidMiner [39, 47]. Feature extraction supported by ontologies has been intensively investigated in text analysis. In this context ontology concepts ease text interpretation and feature construction. For instance in the context of tourism information, from a text describing a city, features like tourism activities or points of interest can be discovered with the help of a domain ontology [57]. Among feature extraction approaches we include recent proposals dealing with sensor data, that apply semantics to pass from low level information to application-oriented features. Just to make few examples in the behavioral mining domain, [66] discusses the semantic lifting of location trajectories provided by GPS systems to a sequence of points of interest (hospital, park, market), [67] is a survey of methods to extract objects, poses or activities from raw image of video data. Whilst these works adopt a purely data-driven approach, in [25] probabilistic description logics integrate machine learning techniques to recognize home activities at multiple levels from sensor data. Finally, we mention the adoption of ontologies for feature selection, that allows to retain only the subset of most relevant features for the task at hand. While traditional methods typically resort to statistical measures of relevance (e.g. correlation with the class attribute, discriminative power, eigenvector analysis) ontologies may help recognizing dependencies or similarities among attributes that stems from the a priori knowledge of causal relations [20, 58].

Results interpretation Domain knowledge can be exploited also to improve the outcome of data mining algorithms and to simplify their interpretation. Early attempts appeared since the mid '90s with the use of concept taxonomies (e.g. white bread and wheat bread are two kinds of bread, bread is a kind of food) exploited to build multiple-level association rule mining techniques. Concept taxonomies provide a generalization at multiple concept levels of data items, enabling the discovery of high-level interesting relations that cannot be seen at the lowest item level because of limited support [24, 54]. WordNet is exploited in [4] to compute the semantic distance between words in the antecedent and consequent of an association rule to provide better measures of rule novelty. Similarly, the comparison of expert's hypotheses in the form of rules with discovered rules is discussed in [7] as enabled by subsumption hierarchies. Reference [10] develops a general framework combining technical interestingness with domain and business interestingness to deliver operable business rules and briefly discusses formalization of domain knowledge.

```
<ClusteringModel modelName="k-means" functionName="clustering"
                 modelClass="centerBased" numberOfClusters="4">
 ...
 <ComparisonMeasure kind="distance">
      <squaredEuclidean/>   ...
 <Cluster name="cluster_0" size="32">
      <Array n="4" type="real">
      6.9125 3.0999 5.8468 2.1312
      </Array>
 </Cluster>
```

Fig. 3 A fragment of PMML representing a clustering model

3.3 Models

Different models of knowledge are associated to different Data Mining techniques. Decision trees or neural networks can be trained to provide predictive models, association rules or clusters are examples of descriptive models. These models are characterized by a variety of structures and properties that can be represented in very different formats.

The issue of model representation and management arises within collaborative KDD and e-science environments, related to sharing and reuse, as well as verifiability and reproducibility of science.[8]

The standard solution to the problem is the adoption of markup languages to describe models characteristics. In particular, the Predictive Model Markup Language (PMML) is the de-facto standard developed by the Data Mining Group (DMG) and widely adopted by current commercial suites. Despite the name, its structure allows to describe other components of the KDD process, like input data and preprocessing operations besides models, and descriptive models like association rules and clustering models can be described as well. Figure 3 shows a short fragment of a clustering model. A detailed description of PMML can be found in [23]. Portable Format for Analytics (PFA) is an alternative JSON-based language recently proposed to improve extensibility, composition of models into workflows, easy integration in modern distributed platforms like Hadoop and Spark [43]. There is limited work in the application of formal semantics to model representation. Examples are the work by Zhu et al. [64, 65] that propose formal logical foundations of metadata languages supporting automatic reasoning, semantic conflict detection and consistency checking, and [30] that proposes an ontology for model sharing, selection and reuse in the context of self-service knowledge discovery.

---

[8]See e.g. http://www.kdd.org/kdd2017/Calls/view/kdd-2017-call-for-research-papers, http://ecmlpkdd2017.ijs.si/submission.html, sections on reproducibility.

4 Conclusions: Trends and Challenges

In this section we finally discuss some lessons learned from the survey of methods and techniques in the literature. Since the early days of customized coding and individual experimentation much work has been done. Support platforms have spread widely, a rich set of operators and tools are available and many efforts have been made in the direction of standardization and extensibility.

The trend born within the Semantic Web has also been followed in the KDD field, and the use of semantics has proven to be a successful choice in manifold directions: (1) semantics helps bridging the knowledge gap between the business and the technical view of KDD. This promises to leverage current stream towards self-service data analytics; (2) it contributes in reducing time taken in selecting, cleaning and transforming data, which are the most time-consuming steps being around the 90% of the whole process time; (3) more accurate and interpretable results can be obtained using semantics to constrain purely statistics techniques with background knowledge; (4) semantics leverages sharing and reusability of tools and reproducibility of results in the recent stream of collaborative networked systems.

Nevertheless, a number of challenges still need to be faced. The potential of ontologies has not yet fully deployed, reasoning mechanisms are often limited to subsumption on concept hierarchies. We are far from a fully fledged semantics-based system. For instance, efforts are needed to better support the domain understanding phase, in particular for the mapping of business goals to data mining tasks. The most advanced support based on semantics is provided in systems developed in specific application areas, in particular life science, while a general model of a semantics based knowledge discovery support system is still lacking. Finally, efforts are needed to standardize existing KDD ontologies and semantic representations. This will allow us to significantly step forward for taking fully advantage of proposed methodologies and available tools.

References

1. A.S. Ali, O.F. Rana, I.J. Taylor, Web services composition for distributed data mining, in *International Conference Workshops on Parallel Processing, 2005. ICPP 2005 Workshops* (2005), pp. 11–18
2. S. Alsairafi, M. Ghanem, N. Giannadakis, Y. Guo, D. Kalaitzopoulos, M. Osmond, A. Rowe, J. Syed, P. Wendel, The design of discovery net: towards open grid services for knowledge discovery. Int. J. High Perform. Comput. Appl. **17**(3), 297–315 (2003)
3. O. Arieli, A. Zamansky, A graded approach to database repair by context-aware distance semantics. Fuzzy Sets Syst. **298**, 4–21 (2016)
4. S. Basu, R.J. Mooney, K.V. Pasupuleti, J. Ghosh, Evaluating the novelty of text-mined rules using lexical knowledge, in *Proceedings of the 7th ACM SIGKDD International Conference on Knowledge Discovery and Data Mining* (2001), pp. 233–238
5. A. Bernstein, F. Provost, S. Hill, Towards intelligent assistance for a data mining process: an ontology based approach for cost-sensitive classification. IEEE Trans. Knowl. Data Eng. **17**(4), 503–518 (2005)

6. A. Bernstein, M. Dänzer, The next system: towards true dynamic adaptations of semantic web service compositions, in *Proceedings of the 4th European conference on The Semantic Web: Research and Applications, ESWC '07* (Springer, Berlin, 2007), pp. 739–748
7. L. Brisson, M. Collard, How to semantically enhance a data mining process?, in *ICEIS*, ed. by J. Filipe, J. Cordeiro, Lecture Notes in Business Information Processing, vol. 19, (Springer, Berlin, 2008), pp. 103–116
8. S. Brüggemann, H.J. Appelrath, Context-aware replacement operations for data cleaning, in *Proceedings of the 2011 ACM Symposium on Applied Computing, SAC '11* (2011), pp. 1700–1704
9. M. Cannataro, C. Comito, A data mining ontology for grid programming, in *Proceedings of the 1st International Workshop on Semantics in Peer-to-Peer and Grid Computing* (Budapest, Hungary, 2003), pp. 113–134
10. L. Cao, Y. Zhao, H. Zhang, D. Luo, C. Zhang, E.K. Park, Flexible frameworks for actionable knowledge discovery. IEEE Trans. Knowl. Data Eng. **22**(9), 1299–1312 (2010)
11. H. Cespivova, J. Rauch, V. Svatek, M. Kejkula, M. Tomeckova, Roles of medical ontologies in association mining CRISP-DM cycle, in *ECML/PKDD Workshop on Knowledge Discovery and Ontologies* (Italy, Pisa, 2004), pp. 1–12
12. J. Chattratichat, J. Darlington, Y. Guo, S. Hedvall, M. Köler, J. Syed, An architecture for distributed enterprise data mining, in *Proceedings of the 7th International Conference on High-Performance Computing and Networking, HPCN Europe '99* (Springer, London, 1999), pp. 573–582
13. W.K. Cheung, X.F. Zhang, H. fai Wong, J. Liu, Z.W. Luo, F.C.H. Tong, Service-oriented distributed data mining. IEEE Internet Comput. **10**, 44–54 (2006)
14. C. Diamantini, M. Panti, D. Potena, Una piattaforma per servizi di KDD, in *Proceedings of the 11th Italian Symposium on Advanced Database Systems* (2003), pp. 119–130
15. C. Diamantini, D. Potena, E. Storti, KDDONTO: an ontology for discovery and composition of KDD algorithms, in *Proceedings of the ECML/PKDD09 Workshop on Third Generation Data Mining: Towards Service-oriented Knowledge Discovery* (Bled, Slovenia, 2009), pp. 13–24
16. C. Diamantini, D. Potena, E. Storti, A virtual mart for knowledge discovery in databases. Inf. Syst. Front. **15**(3), 447–463 (2013)
17. S. Džeroski, Towards a general framework for data mining, in *Proceedings of the 5th International Conference on Knowledge Discovery in Inductive Databases* (Springer, Berlin, 2007), pp. 259–300
18. J. Elder, D. Abbott, A comparison of leading data mining tools, in *Proceedings of the 4th International Conference on Knowledge Discovery and Data Mining* (1998)
19. U.M. Fayyad, G. Piatetsky-shapiro, P. Smyth, *From Data Mining to Knowledge Discovery: An Overview* (American Association for Artificial Intelligence, Menlo Park, 1996), pp. 1–34
20. S. Ghosh, S. Mitra, R. Dattagupt, Fuzzy clustering with biological knowledge for gene selection. Appl. Soft Comput. **16**, 102–111 (2014)
21. M. Goebel, L. Gruenwald, A survey of data mining and knowledge discovery software tools. ACM SIGKDD Explor. **1**(1), 20–33 (1999)
22. R. Grossman, S. Bailey, A. Ramu, B. Malhi, P. Hallstrom, I. Pulleyn, X. Qin, The management and mining of multiple predictive models using the predictive modeling markup language. Inf. Softw. Technol. **41**(9), 589–595 (1999)
23. A. Guazzelli, M. Zeller, W. Lin, G. Williams, PMML: an open standard for sharing models. R J. **1**(1), 60–65 (2009)
24. J. Han, Y. Fu, Mining multiple-level association rules in large databases. IEEE Trans. Knowl. Data Eng. **11**(5), 798–805 (1999) (previously published in Proc. of the 21st VLDB Conference, Zurich, Switzerland 1995)
25. R. Helaoui, D. Riboni, H. Stuckenschmidt, A probabilistic ontological framework for the recognition of multilevel human activities, in *Proceedings of the 2013 ACM International Joint Conference on Pervasive and Ubiquitous Computing, UbiComp '13* (ACM, 2013), pp. 345–354
26. KDDVM project site, http://kdmg.dii.univpm.it/?q=KDDVM

27. J. Kiets, F. Serban, A. Bernstein, S. Fisher, Towards cooperative planning of data mining workflows, in *Proceedings of the ECML/PKDD09 Workshop on Third Generation Data Mining: Towards Service-oriented Knowledge Discovery* (Bled, Slovenia, 2009), pp. 1–12
28. J. Kranjc, R. Ora, V. Podpean, N. Lavra, M. Robnik-ikonja, Clowdflows: online workflows for distributed big data mining. Future Gener. Comput. Syst. **68**, 38–58 (2017)
29. A. Kumar, M.M. Kantardzic, P. Ramaswamy, P. Sadeghian, An extensible service oriented distributed data mining framework, in *Proceedings of the International Conference on Machine Learning and Applications* (Louisville, KY, USA, 2004), pp. 256–263
30. Y. Li, M.A. Thomas, K.M. Osei-Bryson, Ontology-based data mining model management for self-service knowledge discovery. Inf. Syst. Front. 1–19 (2016)
31. L. Kart, G. Herschel, A. Linden, J. Hare, *Magic quadrant for advanced analytics platforms* Technical report, Gartner Inc. (2016)
32. S. Majithia, M.S. Shields, I.J. Taylor, I. Wang, Triana: a graphical web service composition and execution toolkit, in *Proceedings of IEEE International Conference on Web Services* (2004), pp. 514–521
33. G.L. Martiny, A. Unruhy, S.D. Urbanz, An agent infrastructure for knowledge discovery and event detection. Technical Report MCC-INSL-003-99, Microelectronics and Computer Technology Corporation (1999)
34. K. Morik, M. Scholz, The miningmart approach to knowledge discovery in databases, in *Intelligent Technologies for Information Analysis*, ed. by N. Zhong, J. Liu (Springer, Berlin, 2004), pp. 47–65
35. D.O.G. Neto, W. Meira, R. Ferreira, Anteater: a service-oriented architecture for high-performance data mining. IEEE Internet Comput. **10**, 36–43 (2006)
36. R. Olejnik, T.F. Fortis, B. Toursel, Webservices oriented data mining in knowledge architecture. Future Gener. Comput. Syst. **25**(4), 436–443 (2009)
37. P. Panov, L. Soldatova, S. Džeroski, Ontology of core data mining entities. Data Min. Knowl. Discov. **28**(5), 1222–1265 (2014)
38. S. Parthasarathy, R. Subramonian, Facilitating data mining on a network of workstations, in *Advances in Distributed and Parallel Knowledge Discovery*, ed. by H. Kargupta, P. Chan (AAAI/MIT Press, Menlo Park, 2000), pp. 233–258
39. H. Paulheim, Exploiting linked open data as background knowledge in data mining, in *ECML/PKDD Workshop on Data Mining on Linked Data. CEUR Workshop Proceedings*, vol. 2013 (1082), pp. 345–354
40. D. Perez-Rey, A. Anguita, J. Crespo, Ontodataclean: ontology-based integration and pre-processing of distributed data, in *Biological and Medical Data Analysis: 7th International Symposium, ISBMDA 2006, Thessaloniki, Greece, December 7–8, 2006, Proceedings* (Springer, Berlin, 2006)
41. J. Phillips, B. Buchanan, Ontology-guided knowledge discovery in databases, in *1st ACM International Conference on Knowledge Capture* (Victoria, Canada, 2001), pp. 123–130
42. G. Piatetsky-Shapiro, Knowledge discovery in real databases: a report on the IJCAI-89 workshop. AI Mag. **11**(5), 68–70 (1991)
43. J. Pivarski, C. Bennett, R.L. Grossman, Deploying analytics with the portable format for analytics (PFA), in *Proceedings of the 22nd ACM SIGKDD International Conference on Knowledge Discovery and Data Mining* (2016), pp. 579–588
44. V. Podpecan, M. Zemenova, N. Lavrac, Orange4WS environment for service-oriented data mining. Comput. J. **55**(1), 82–98 (2011)
45. M.S. Přez, A. Sánchez, V. Robles, P. Herrero, J.M.P. na, Design and implementation of a data mining grid-aware architecture. Future Gener. Comput. Syst. **23**(1), 42–47 (2007)
46. O. Rana, D. Walker, M. Li, S. Lynden, M. Ward, PaDDMAS: parallel and distributed data mining application suite, in *14th International Parallel and Distributed Processing Symposium* (Cancun, 2000), pp. 387–392
47. P. Ristoski, C. Bizer, H. Paulheim, Mining the web of linked data with rapidminer. Web Semant.: Sci. Serv. Agents World Wide Web **35**(Part 3), 142–151 (2015)

48. P. Ristoski, H. Paulheim, Semantic web in data mining and knowledge discovery: a comprehensive survey. Web Semant.: Sci. Serv. Agents World Wide Web **36**, 1–22 (2016)
49. O. Ritthoff, R. Klinkenberg, S. Fischer, I. Mierswa, S. Felske, Yale: yet another learning environment, in *Proceedings of LLWA01/FGML-2001* (2001), pp. 84–92
50. A. Romei, S. Ruggieri, F. Turini, KDDML: a middleware language and system for knowledge discovery in databases. Data Knowl. Eng. **57**, 179–220 (2006)
51. S. Sarawagi, S.H. Nagaralu, Data mining models as services on the internet. SIGKDD Explor. Newsl. **2**(1), 24–28 (2000)
52. F. Serban, J. Vanschoren, J.U. Kietz, A. Bernstein, A survey of intelligent assistants for data analysis. ACM Comput. Surv. **45**(3), 31:1–31:35 (2013)
53. C. Shearer, The CRISP-DM Model: the new blueprint for data mining. J. Data Warehous. **5**(4), 13–22 (2000)
54. R. Srikant, R. Agrawal, Mining generalized association rules. Future Gener. Comput. Syst. **13**(2), 161–180 (1997) (previously published in Proceedings of the 21st VLDB Conference, Zurich, Switzerland 1995)
55. D. Talia, The open grid services architecture: where the grid meets the web. IEEE Internet Comput. **6**(6), 67–71 (2002)
56. C.Y. Tsai, M.H. Tsai, A dynamic web service based data mining process system, in *Proceedings of the 5th International Conference on Computer and Information Technology* (IEEE Computer Society, 2005), pp. 1033–1039
57. C. Vicient, D. Snchez, A. Moreno, An automatic approach for ontology-based feature extraction from heterogeneous textual resources. Eng. Appl. Artif. Intell. **26**(3), 1092–1106 (2013)
58. C. Wan, A.A. Freitas, An empirical evaluation of hierarchical feature selection methods for classification in bioinformatics datasets with gene ontology-based features. Artif. Intell. Rev. 1–40 (2017)
59. Y. Wang, S. Yang, Outlier detection from massive short documents using domain ontology, in *2010 IEEE International Conference on Intelligent Computing and Intelligent Systems*, vol. 3 (2010), pp. 558–562
60. R. Wirth, C. Shearer, U. Grimmer, T.P. Reinartz, J. Schlsser, C. Breitner, R. Engels, G. Lindner, Towards process-oriented tool support for knowledge discovery in databases, in *PKDD '97: Proceedings of the First European Symposium on Principles of Data Mining and Knowledge Discovery* (Springer, London, 1997), pp. 243–253
61. I.H. Witten, E. Frank, *Data Mining: Practical Machine Learning Tools and Techniques*, 2nd edn. (Morgan Kaufmann, San Francisco, 2005)
62. L. Yu-hua, L. Zheng-ding, S. Xiao-lin, W. Kun-mei, L. Rui-xuan, Data mining ontology development for high user usability. Wuhan Univ. J. Nat. Sci. **11**(1), 51–56 (2006)
63. M. Žáková, P. Křemen, F. Železný, N. Lavrač, Automating knowledge discovery workflow composition through ontology-based planning. IEEE Trans. Autom. Sci. Eng. **8**(2), 253–264 (2011)
64. X. Zhu, J. Yang, *An Extended Predictive Model Markup Language for Data Mining* (Springer, Berlin, 2010), pp. 218–231
65. X. Zhu, H. Wang, H. Gan, C. Gao, Construction and management of automatical reasoning supported data mining metadata, in *2011 International Conference on Business Management and Electronic Information*, vol. 5 (2011), pp. 205–210
66. L. Zhu, C. Xu, J. Guan, H. Zhang, SEM-PPA: a semantical pattern and preference-aware service mining method for personalized point of interest recommendation. J. Netw. Comput. Appl. **82**, 35–46 (2017)
67. M. Ziaeefard, R. Bergevin, Semantic human activity recognition: a literature review. Pattern Recognit. **48**(8), 2329–2345 (2015)

Analysing Trajectories of Mobile Users: From Data Warehouses to Recommender Systems

Franco Maria Nardini, Salvatore Orlando, Raffaele Perego, Alessandra Raffaetà, Chiara Renso and Claudio Silvestri

Abstract This chapter discusses a general framework for the analysis of trajectories of moving objects, designed around a Trajectory Data Warehouse (TDW). We argue that data warehouse technologies, combined with geographic visual analytics tools, can play an important role in granting very fast, accurate and understandable analysis of mobility data. We describe how in the last decade the TDW models have changed in order to provide the user with a more suitable model of the reality of interest and we also cope with the challenge of semantic trajectories. As a use case we illustrate how the framework can be instantiated for realizing a recommender system for tourists.

1 Introduction

Recent advances in mobile network devices, sensors, and positioning technologies enabled the tracking of large amounts of moving objects: vehicles, animals, vessels and in large part also humans. These technologies produce huge data streams of observations, which can be stored and used to reconstruct the original objects trajectories. These movement data represent a treasure in term of the potential applications that can benefit from their analysis. A special interesting aspect is the possibility of enriching the pure spatio-temporal data with suitable knowledge bases, to semantically annotate and transform such trajectories into more meaningful data.

In this chapter we survey a general framework for creating, analysing and exploiting (possibly semantically enriched) trajectories that we have been developing in the last decade.

The central idea is that of a *Trajectory Data Warehouse* (TDW), with spatial and temporal dimensions, which is populated, via a suitable ETL process starting from raw trajectory data (essentially, spatio-temporal points or *samples*). The TDW relies on a flexible *conceptual model* with associated *spatio-temporal dimensions*

F.M. Nardini · S. Orlando (✉) · R. Perego · C. Renso
ISTI-CNR, Pisa, Italy
e-mail: orlando@unive.it

S. Orlando · A. Raffaetà · C. Silvestri
DAIS - Università Ca' Foscari Venezia, Venezia, Italy

S. Flesca et al. (eds.), *A Comprehensive Guide Through the Italian Database Research Over the Last 25 Years*, Studies in Big Data 31, DOI 10.1007/978-3-319-61893-7_24

and *hierarchies*. More specifically, the spatial domain can be structured according to the application requirements, by exploiting hierarchies of regular grids (like in [7, 8]) or of regions with ad-hoc shapes [6]. While a hierarchy of regular grids can be used to analyse objects that can move freely in the space, hierarchies with ad-hoc shapes are useful for objects whose movements are constrained, such as objects that can only move along a road network (e.g., cars).

The TDW is provided with an interface that allows for *visual* OLAP operations for the analysis of aggregate trajectory data, by integrating OLAP tools with visual analytics [2]. This permits to overcome the limits of the usual OLAP user interfaces. In fact, the table based representation commonly adopted by OLAP tools makes it very difficult for the user to grasp the relationships between areas in the same neighbourhood, the evolution of spatial measures in time, or the correlations of different measures. Visualisation is crucial: it can be seen simultaneously as the output and end-product of a knowledge discovery cycle and the starting point for further, interactive and visual, analysis.

The TDW, as described above, suffices to study several quantitative properties of trajectories, such as speed, traveled distance, or presence. However, in order to analyze information concerning semantic aspects such as the kind of places visited, the goals of the trajectories, the performed activity, transportation means, a semantic enrichment of the trajectory data is necessary. We discuss how semantic trajectories can be constructed from the original collected samples by properly combining movement data with suitable knowledge bases. We describe how the conceptual model of the TDW has to be modified to implement a *Semantic Trajectory Data Warehouse* allowing us to analyse semantic trajectories according to the above mentioned semantic dimensions.

As a use case for semantic analysis of trajectories we discuss a touristic recommendation system. We illustrate how trajectories, which have been semantically enriched, can be used to recommend personalized tours. Specifically, we outline the whole process, starting from the selection of a set of tourist trajectories, the enrichment step for properly transforming and enriching the trajectories for our purposes, and finally how the obtained semantic trajectories are exploited to suggest personalized sightseeing tours, by modeling and maximizing user interest and visiting time-budget.

Figure 1 summarizes the overall (semantic) trajectory analysis framework discussed in this chapter. On the top of the figure we show the source data, i.e., a set of trajectory samples represented as small colored circles. Samples for the same trajectory are filled with the same color and connected by a gray dashed line that represents the actual movement of the object. These samples are fed to a module in charge of reconstructing the trajectories followed by the moving objects, possibly using a map of the visited geographic region. Once reconstructed, the trajectories can be processed directly by an ETL module to populate a TDW, which allows us to perform visual OLAP analyses (Sect. 3). Alternatively, we can exploit some knowledge bases, for example a set of categorized geographic Points of Interest (PoIs), to semantically enrich and eventually transform trajectories. In this last case, the ETL module is specialized to populate a semantic TDW, with a suitably extended concep-

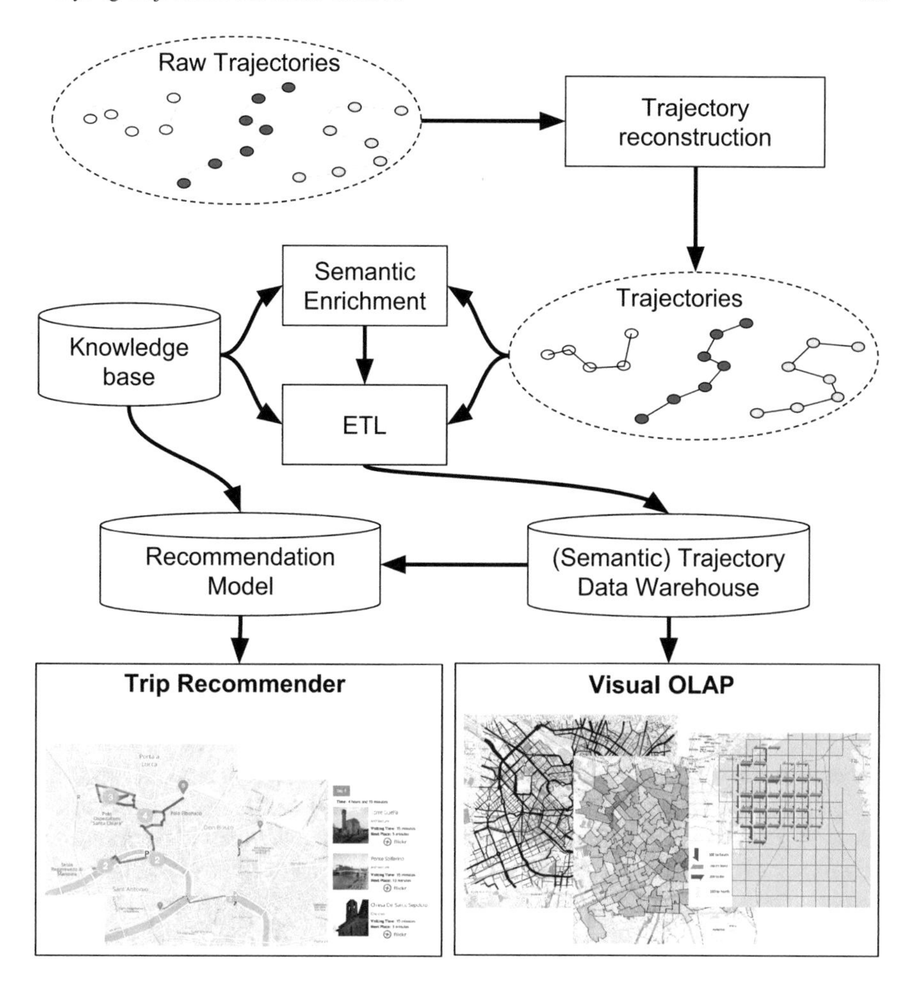

Fig. 1 Overview of the (semantic) trajectory analysis framework

tual model (Sect. 4). The overall framework of Fig. 1 can be instantiated for particular applications. In Sect. 5 we illustrate how it can be used for building a recommender system for tourists.

2 Preliminaries

Several works in the literature address the analysis of trajectory data [15]. Even for the definition of a trajectory several variants exist, formalizing the general idea of a trajectory as a representation of the spatio-temporal evolution of a moving object. Since trajectories are usually collected by means of position-enabled devices, the notion of trajectory has to deal with the concept of *sampling* that is the action of the device to detect spatial and temporal points at given temporal intervals. Here we call *raw trajectory* the discrete representation of a trajectory as a sequence of spatio-temporal points or *samples* as collected by the device.

Definition 1 (*Raw Trajectory*) A trajectory T is an ordered list of spatio-temporal points or *samples* $p_1, p_2, p_3, \ldots, p_n$. Each p_i is a tuple (id, x_i, y_i, t_i) where id is the identifier of a trajectory, x_i, y_i are the geographical coordinates of the sampled point, and t_i is the timestamp in which the point has been collected, with $t_1 < t_2 < t_3 < \ldots < t_n$.

From these sampled data, according to the application requirements, we need to reconstruct the approximation of the real trajectory, modeled as a continuous function from time to geographic coordinates.

The possible methods for reconstruction are different, and depend on the scenarios on which we focus on. Objects can move almost freely in the space (e.g., vessels on the sea), or object movements can be constrained (e.g., cars moving along a road network). In the first case, in order to reconstruct the whole trajectory, *local interpolation* can be used. According to this method, objects are assumed to move between the observed points following some rule. For instance, a *linear* interpolation function models a straight movement with constant speed, while other polynomial interpolations can represent smooth changes of direction. If we consider the alternative scenario of cars moving along a road network, in turn modeled as a graph embedded in the Euclidean 2D-space, we have that the movements of objects are completely constrained, since cars are supposed to stay on the network. So reconstruction must take into account the topology of the road network to determine the path followed by each object between two consecutive sampled positions in the raw data [4]. The reconstruction phase produces a sequence of lines in a spatio-temporal space, each representing the continuous "development" of the moving object during a time interval. Notice that the spatial projection of these lines are segments of the road network or portions of these segments.

3 Trajectory Data Warehouses

The motivation behind a TDW is to transform trajectories into valuable knowledge that can be used for decision making purposes in ubiquitous applications, such as Location-Based Services (LBS) or traffic control management. Intuitively, the high volume of raw data produced by sensing and positioning technologies, the complex nature of data stored in trajectory databases and the specialized query processing demands, they all make extracting valuable information from such spatio-temporal data a challenging task. For this reason, the idea is to develop specific aggregation techniques to produce summarized trajectory information and provide *visual* OLAP style analyses.

3.1 The Conceptual Model

Our first proposal [8] of TDW consists of a fact table containing keys to dimension tables and a number of measures expressing properties about sets of trajectories. The dimensions of analysis are the spatial dimensions X, Y ranging over spatial intervals, and the temporal dimension T ranging over temporal intervals. A regular three-dimensional grid obtained by discretizing the corresponding values of the dimensions is defined and a set-grouping hierarchy is associated with each dimension. The measures of interest are the number of trajectories inside each cell of the grids, their average, maximum/minimum speed, the covered distance and the time spent inside the cell. Then in [11] we add a new dimension *OBJECT_PROFILE_DIM* in order to take into account demographical information, such as gender, age, job, of moving objects. However, these approaches suffer from a main limitation: they are restricted to freely moving objects. Thus, they do not allow to explicitly account for constrained movements, for example due to the presence of a road network. Moreover, they support only spatio-temporal hierarchies consisting of regular grids.

In [6] we define a framework relying on a more flexible conceptual model with associated spatio-temporal dimensions and hierarchies. More specifically, the spatial domain can be structured according to the application requirements, by exploiting hierarchies of regular grids (like in [8, 11]) or of regions with ad-hoc shapes. While a hierarchy of regular grids can be used to analyse objects that can move freely in the space, hierarchies with ad-hoc shapes are useful for objects whose movements are constrained, such as objects that can only move along a road network (e.g., cars). Furthermore, Voronoi tessellation can be employed in order to build hierarchies of regions based on the actual distribution of the points forming the trajectories. This kind of partitioning turns out to be particularly suited for highlighting the directions of the trajectory movement.

The resulting model is presented in Fig. 2. We distinguish two classes of facts, namely INTRA- GRANULE and INTER- GRANULE facts. Intuitively, intra-granule facts express properties related to trajectories inside a *single* granule whereas inter-granule facts describe properties concerning the movement of trajectories between two granules. We recall that a *base granule* is obtained by partitioning both the spatial and temporal dimensions and this partition is the finest one. From this base granularity other *coarser* partitions can be defined by merging together spatial regions and temporal intervals, respectively. Informally, a granule can be defined as a contiguous spatial region during a given time interval.

More specifically, the INTRA-GRANULE facts model events that are related to a *single base granule* concerning a certain object group. For a given object group U and a granule g, the measures are:

- *visits*: the number of trajectories belonging to group U which start from or enter into granule g;
- *start/end*: the number of trajectories belonging to group U starting/ending in granule g;

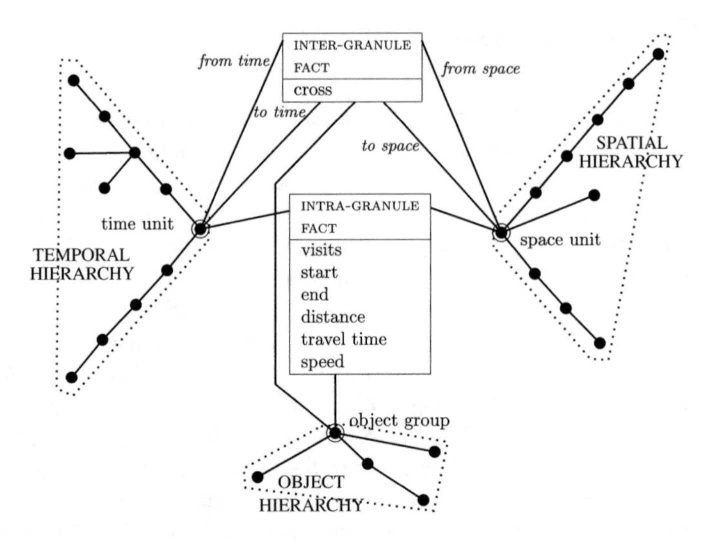

Fig. 2 TDW Conceptual model [6]

- *travel time/distance*: the time spent/distance travelled by all trajectories belonging to group U while moving inside granule g;
- *speed*: the average speed of trajectories belonging to group U traversing granule g.

The INTER-GRANULE facts model events that are related to *pairs of granules* and are concerned with a specific object group. For a given group U and pair of granules g and g', a measure of interest is

- *cross*: number of times the border from g to g' has been traversed by trajectories belonging to group U.

Note that the measure *cross* is interesting only for adjacent granules (for non-adjacent granules it is invariably 0). However, in general, inter-granule facts can model events which are meaningful for all pairs of granules. An example could be the *origin-destination* measure, which, for any pair of granules, represents the number of trajectories starting from the first and ending into the second granule.

Clearly, the presented measures are not an exhaustive collection, but they correspond to a set of common measures which we found interesting and useful in different scenarios.

The TDW provides efficient OLAP roll-up operations since for all the defined measures, values at a coarser granularity can be computed by using values at a finer granularity. In particular, for the measures *start*, *end*, *travel time*, *distance* and *cross*, we use the *distributive* function *sum* as aggregate function whereas for *visits* and *speed* we use *algebraic* aggregate functions. The aggregate function for *speed* is computed as the ratio between the measures *distance* and *travel time*, as expected. On the other hand, for the measure *visits* we use the auxiliary measure *cross*. To give an intuition, let us consider a granule g composed by two finer granules g_1 and g_2.

Hence the number of visits in the granule g is obtained by summing up the visits in g_1 and g_2, subtracting the number of trajectories crossing the border between g_1 and g_2. This is motivated by the fact that the border between two finer granules, g_1 and g_2 composing g, is completely inside g. Hence trajectories moving from g_1 to g_2 (or vice versa) increase the number of visits in g_1 (or g_2) but they should not be counted as visits in the coarser granule g because the movement is completely inside g, i.e., they do not enter g. We refer the reader to [6] for a formal definition of such aggregate functions.

It is worth noting that measure *visits* can provide an accurate approximation of measure *presence*, which counts the number of *distinct* trajectories occurring in a spatio-temporal granule. The aggregate function for *presence* is *holistic*: the raw data are needed to compute the exact result at all granularities. This is due to the fact that trajectories might span multiple granules. Hence in the aggregation phase we have to cope with the so called *distinct count problem* [17]: if an object remains in the query region for several timestamps during the query interval, one should avoid to count it multiple times in the result. *Holistic* functions represent a big issue for data warehouse technology. In [6] we discussed about the computation of measure *presence* and we showed that the proposed solution, i.e., the use of measure *visits*, is a more precise approximation with respect to some common approaches [9, 17] facing the same problem.

3.2 *Visual OLAP*

In the analysis of spatial and spatio-temporal data, the use of suitable, interactive, visualization tools is of paramount importance to help the analytic user in effectively grasping the information hidden in those complex data. For this reason, we have provided the TDW with an interface that allows for OLAP visual operations, based on V-Analytics [1, 2], an interactive visual analytics system running on the Java® Virtual Machine. This system permits a user to view georeferenced data over a map and run analyses on them, for example to find clusters or to tessellate the space. It also offers functionalities to handle temporal data, by using graphs or animations, according to the type of data to analyse.

In the following we report some examples that highlight several kinds of OLAP analyses on different scenarios. A first one, illustrated in Fig. 3a, shows the fishing effort index in the Northern Adriatic Sea during 2007, at the base spatial granularity. The fishing effort index is a value indicating how much a given area has been exploited by the boats fishing in it. The space is partitioned into a regular grid and granules in darker colours are the most exploited. By using a drill-down operation on the temporal dimension, we can inspect the situation at a higher level of detail. For instance, Fig. 3b shows the fishing effort in the trimester July-September of 2007. The fact that it is sensibly reduced with respect to effort in the whole period is somehow expected due to a law which prevents most fishing activities during August.

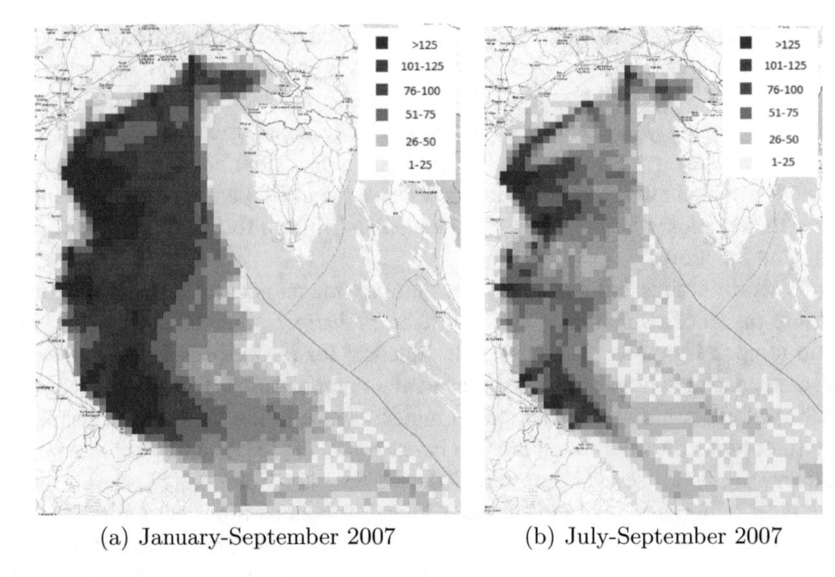

(a) January-September 2007 (b) July-September 2007

Fig. 3 Fishing effort distribution [6]

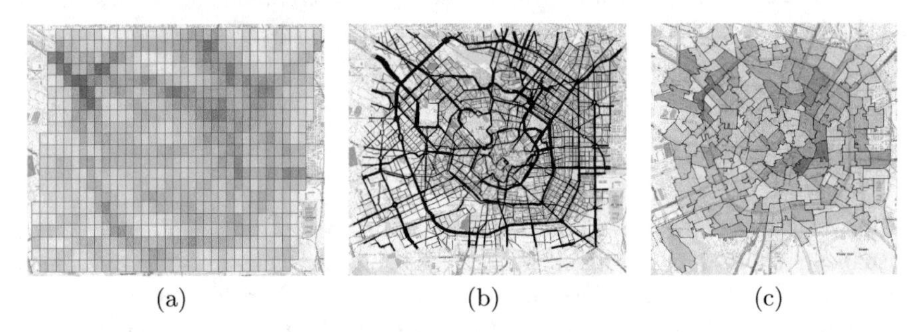

(a) (b) (c)

Fig. 4 **a** Grid based spatial dimension, and **b** street segment based spatial dimension with **c** dimensional attribute having polygon spatial type. [6]

The flexibility in the definition of the spatio-temporal hierarchy offered by the presented TDW model allows the user to adopt a suitable model of the reality, thus obtaining a much more meaningful visual representation of the information contained in the TDW. The images in Fig. 4 are relative to a different example that concerns trajectories of cars moving in the city of Milan. Specifically, they visually represent the number of visits to spatial granules during the time interval corresponding to a particular temporal granule. Each image corresponds to a different spatial granularity: in Fig. 4a granules are cells of a regular grid, whereas in Fig. 4b, c granules are respectively street segments and city districts. The results obtained with a regular grid may be suited for getting an initial overview of the data. However, a more detailed exploration is complicated since the cells do not bear any semantics and do not correspond to the real geographic and topographic properties of the data. This is why it is important to have also streets and district for the analyses. For example, by using the streets we can detect which are the most busy roads and how the traffic flows.

4 Semantic Trajectory Data Warehouse

The concept of semantic trajectory has been proposed as a way to overcome the lack of semantics characterizing raw trajectories. A well known definition of semantic trajectory relies on the "stop and move" approach: a trajectory is segmented into parts where the object is stopped (the "stop") and the parts where the object is changing his/her position (the "move") [16]. This approach evolved to the more general definition of *episodes* to represent segments of a trajectory complying to some predicate representing the semantics of that segment, like the transportation mean, the goal or activity [10]. A further evolution towards this direction brought to the definition of a conceptual model for semantic trajectories as proposed in [3] where several contextual aspects contribute to create the concept of semantic trajectory.

Definition 2 (*Semantic Trajectory*) A semantic trajectory is a trajectory that has been enhanced with annotations and/or one or several complementary segmentations.

Note that, according to the specific requirements of applications, such semantic trajectories can be transformed and abstracted so to adhere to a model, e.g., the "stop and move" one. Therefore, while semantics enrichment can add meaningful information to trajectories, the obtained semantic trajectory can actually lose some of the information contained in the original one, by keeping only that useful for the specific application goal.

4.1 The Semantic TDW Conceptual Model

This section introduces the Mob-Warehouse model [18] which is organized around the notion of semantic trajectory where different aspects contribute to describe the context. The model is based on the so called *5W1H* (Who, Where, When, What, Why, How) framework [19], recurrently used by journalists as a guide for narrating a fact.

To semantically enrich a trajectory, each narrative question of the 5W1H model is mapped to a specific trajectory feature. In this way, we describe the moving object (*Who*) moving by a transportation means and/or having a certain behavior (*How*), performing an activity (*What*), for a certain reason (*Why*), at a given time (*When*) and place (*Where*). The increased level of semantic information into our model trajectory allows us to perform more meaningful queries about moving object habits.

The resulting semantic TDW conceptual model consists of six dimensions, as illustrated in Fig. 5. It reflects the *semantic* structure of trajectories and in the fact

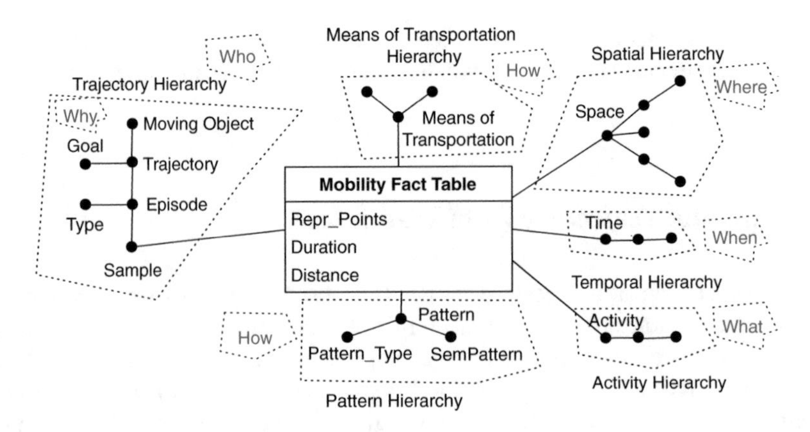

Fig. 5 Semantic TDW Conceptual model [18]

table we store detailed information, no more aggregate properties about trajectories as in the model presented in Sect. 3. Specifically, the dimensions *Space* and *Time* essentially correspond to the previous model dimensions and they represent respectively the *Where* and *When* questions of the 5W1H Model. It is worth noting that the spatial domain can be structured according to the application requirements, providing the user with a great flexibility. The third dimension named *object group* in Fig. 2 representing features of the objects under analysis, here it is called dimension *Trajectory*, as it becomes a central component of our model and it is used to represent the trajectory of the moving objects. At the base granularity it represents a single sample (id, x, y, t) belonging to the trajectory identified by *id*. The hierarchy having *Sample* as a root mixes together semantic and geometric features. A sample belongs to an *episode*, which can be classified according to its *Type* (e.g., a *stop* or a *move*) and it is grouped into a *Trajectory*. Each *Trajectory* is associated not only with the *Moving object* but also to a *Goal*, which is the main objective of such a trajectory. This dimension allows one to model *Who* is performing the action (the moving object) and the attribute *goal* answers the question *Why*. A fourth dimension, called *Activity*, states the activity the object is doing in a certain sample. This allows one to describe in a very detailed manner *What* is going on at the different samples of a trajectory. We can build a hierarchy of activities which classifies properly the variety of tasks an object can perform. Usually this hierarchy is application dependent hence in the general model it is not specified and should be instantiated case by case depending on the application requirements. Then the dimension *Means of Transportation* represents which transportation means the object is using for the movement. The last dimension, called *Pattern*, collects the patterns mined from the data under analysis. In this way we can directly relate trajectories to the patterns they belong to. The latter two dimensions express the concept of *How* the movement is performed.

The fact table stores measures about the samples of the trajectories. Unlike [6, 7, 11] where at the minimum granularity data were already aggregated, in this model we record the most detailed information. This gives the user the ability to analyze the behavior according to various points of view: at the minimum granularity we store information related to a single sample of a trajectory, specifying the kind of activity is doing, the means of transportation is using, the patterns, the space and time it belongs to. Then by aggregating according to the described hierarchies we can recover also properties concerning the whole trajectory or groups of trajectories satisfying certain conditions.

In the fact table we store measures related to a given sample $s = \langle id, x, y, t \rangle$

- *Repr_Points* is a spatio-temporal measure containing the spatial and temporal component of the sample, i.e., (id, x, y, t);
- *Duration* is the time spent to reach the sample from the previous point of the same trajectory in the same granule. It is set to 0, if this is the first point of the trajectory in such a granule;
- *Distance* is the traveled distance from the previous point to the sample of the trajectory in the same granule. It is set to 0, if this is the first point of the trajectory in such a granule.

As far as the aggregate functions are concerned, for the measures *Duration* and *Distance*, we use the *distributive* function *sum*: super-aggregates are computed by summing up the sub-aggregates at finer granularities. On the other hand, the aggregate function for the measure *Repr_points* can be defined in different ways according to the application requirements. The simplest way is to use the *union* operator to join together the points satisfying given conditions. Differently one can return a bounding box enclosing all the points or compress the points removing the ones which are spatio-temporally similar.

5 Use Case: Trip Planning Recommender for Tourists

Planning a travel itinerary is a difficult and time-consuming task for tourists approaching their destination for the first time. Different sources of information such as travel guides, maps, on-line institutional sites and travel blogs are consulted in order to devise the right blend of Points of Interest (PoIs) that best covers the subjectively interesting attractions and can be visited within the limited time planned for the travel. However, the user still need to guess how much time is needed to visit each single attraction, and to devise a *smart* strategy to schedule them moving from one attraction to the next one. Furthermore, tourist guides, and even blogs, reflect the point of view of their authors, and they may result to be not authoritative sources of information when the tourist preferences diverge from the most popular flow.

We show how, relying on our framework, we can build a personalized plan of visit by exploiting the wisdom-of-the-crowds by past tourists. First of all we have to select and/or create the *Knowledge bases* (see Fig. 1) that can be used both for the Semantic Enrichment and during the ETL phase. In order to suggest interesting itineraries, we have to identify the set of PoIs in the geographical region that tourists would like to visit. Given the bounding box BB_{city} containing the city of interest, we download all the geo-referenced Wikipedia pages falling within this region. We assume each geo-referenced Wikipedia named entity, whose geographical coordinates falls into BB_{city}, to be a fine-grained Point of Interest. For each PoI, we retrieve its descriptive label, its geographic coordinates as reported in the Wikipedia page, and the set of categories which the PoI belongs to. Categories are reported at the bottom of the Wikipedia page, and are used to link articles under a common topic. They form a hierarchy, although sub-categories may be a member of more than one category. By considering the set C of categories associated with all the PoIs, we generate the normalized relevance vector of each PoI.

We then perform a density-based clustering to group in a single PoI sightseeing entities which are very close one to each other.[1] Clustering very close PoIs is important since a tourist in a given place can enjoy all the attractions in the surroundings even if she does not take photos to all of them. Moreover, it aims at reducing the sparsity that might affect trajectory data. Finally, we obtain the relevance vector for the clustered PoIs by considering the occurrences of each category in the members of the clusters and by normalizing the resulting vector. The final result is a knowledge base consisting of a set of PoIs $\mathcal{P} = \{p_1, \ldots, p_N\}$ and each POI is associated with the *relevance vector* $\boldsymbol{v_p} \in [0, 1]^{|C|}$.

Now we need a method for collecting users $\mathcal{U}$ and the long-term itineraries crossing the discovered PoIs. We query Flickr to retrieve the metadata (user id, timestamp, tags, geographic coordinates, etc.) of the photos taken in the given area BB_{city}. The assumption we are making is that photo albums made by Flickr users implicitly represent sightseeing itineraries within the city. To strengthen the accuracy of our method, we retrieve only the photos having the highest geo-referenced accuracy given by Flickr.[2] This process thus collects a large set of geo-tagged photo albums taken by different users within BB_{city}. We preliminary discard photo albums containing only one photo and the resulting set represents the set of *Raw Trajectories* in Fig. 1.

Then, we apply a *Semantic Enrichment* step. We spatially match the photos in the raw trajectories against the set of PoIs previously collected. We associate a photo with a PoI when *it has been taken within a circular buffer of a given radius having the PoI as its center*. Note that in order to deal with clustered PoIs, we consider the distance of the photo from all constituent members: in the case the photo falls within the circular region of at least one of the members, it is assigned to the clustered PoI. Moreover, since several photos by the same user are usually taken close to the same PoI, we consider the timestamps associated with the first and last of these photos as the starting and ending time of the user visit to the PoI. At the end of this step we have a set of *semantic* trajectories consisting of sequences of PoIs belonging to P and each PoI is annotated also with the time the user is assumed to enter and to exit from such a PoI.

---

[1] E.g., the beautiful marble statues in the *Loggia dei Lanzi* in Florence are only a few meters far one from each other but have a distinct dedicated page in Wikipedia.

[2] http://www.flickr.com/services/api/flickr.photos.search.html.

It is worth noting that in this case semantic trajectories are sequences of *stops* since the selected dataset does not provide any information about the movements of the user from one stop to the following. For the purpose of our application, in the ETL phase the *moves* are computed as the shortest path between two consecutive stops by using Googlemaps. Moreover, the set of PoIs is further annotated with the *visiting time* and the *popularity* index. The visiting time for a PoI p is the time spent by users in p and it is computed as the average of the durations of the visits to p. The popularity of each PoI is computed as the number of distinct users that take at least one photo in its circular region. The set of PoIs is used to build the spatial dimension of the Semantic TDW.

Finally, it is possible to associate a *preference vector* with each user by summing up and normalizing the relevance vectors of all the PoIs occurring in the semantic trajectories of such a user.

The general Semantic TDW model of Fig. 5 when instantiated to this use case includes only the dimensions *Space*, *Time* and *Trajectory*, since the raw data do not provide any specific information on means of transportation, activities and patterns. It is important to highlight that this data warehouse can support analyses at different levels of abstraction: from very detailed data involving samples to semantic trajectories modeled as sequences of stops and moves.

The *Trip Planning Recommendation* [12–14] is an example of analysis that can be performed on top of the Semantic TDW. The aim is to generate visiting plans made up of actual touristic itineraries that are the most tailored to the specific preferences and the temporal constraints of the tourist. The Trip Planning Recommendation is defined as a set cover problem, formulated as an instance of the Generalised Maximum Coverage (GMC) problem [5]. We model each visiting pattern by means of the PoIs and the associated Wikipedia categories, and the GMC profit function by considering PoIs popularity and the actual user preferences over the same Wikipedia categories. The cost function is instead built by considering the average visiting time for the PoIs in the patterns plus the time needed to move from one PoI to the next one.

Given a tourist, the Trip Planning Recommendation problem can be thus solved by looking at the set of semantic trajectories fitting the available time budget and covering the PoIs, that maximises the user interests. Determining an exact solution for the Trip Planning Recommendation problem is NP-hard. We solve it by employing the efficient greedy approximation algorithm proposed in [5]. Trajectories are then scheduled and provided to the user as an agenda of activities to be performed in the city. An example of the recommendation produced for the city of Pisa is shown in Fig. 6.

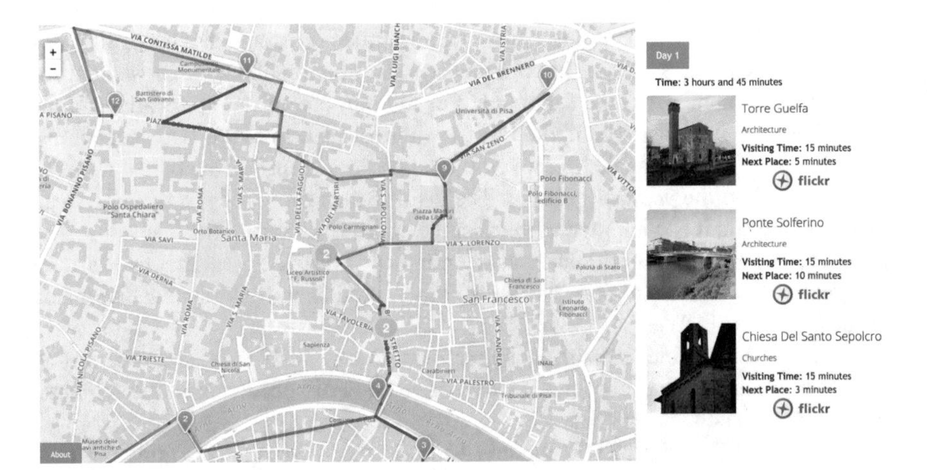

Fig. 6 An example of trip plan recommendation [12–14]

6 Conclusion

This chapter surveyed some research results obtained by the authors in the field of trajectory data analysis. Our achievements are discussed by referring to a general framework that encompasses many steps, from semantic enrichment of trajectories, in turn reconstructed from sequences of samples, to Data Warehousing.

The central part of this chapter refers to a general conceptual model for TDW presented in [6], with associated spatio-temporal dimensions, where the spatial domain can be structured according to the application requirements and it is no longer restricted to consist of simple regular grids as in previous works. Moreover, the TDW is provided with a set of spatial and temporal visualisation techniques, supporting OLAP analysis of movement data, which permits a user to view geo-referenced data over a map and run insightful analyses on them.

An extension of the TDW model for semantically enriched trajectories is also discussed, by presenting the Mob-Warehouse model [18]. In this case, the most notable contribution is the semantic conceptual TDW model based on the so called *5W1H* (Who, Where, When, What, Why, How) framework. Each narrative question of the 5W1H model is mapped to a specific trajectory feature. In this way, we describe an object (Who) moving by a transportation means and/or having a certain behavior (How), performing an activity (What), for a certain reason (Why), at a given time (When) and place (Where).

We finally illustrated a trip planning recommender in the context of a tourism scenario [12–14]. We analyzed its various steps to eventually recommend a trip plan to a user, where her profile and time budget is known, on the basis of a recommendation model extracted from past trajectories of tourists.

References

1. G. Andrienko, N. Andrienko, S. Wrobel, Visual analytics tools for analysis of movement data. ACM SIGKDD Explor. **9**(2), 28–46 (2007)
2. N. Andrienko, G. Andrienko, Visual analytics of movement: an overview of methods, tools and procedures. Inf. Vis. **12**(1), 3–24 (2013)
3. V. Bogorny, C. Renso, A.R. de Aquino, F. de Lucca Siqueira, L.O. Alvares, CONSTAnT - a conceptual data model for semantic trajectories of moving objects. Trans. GIS **18**(1), 66–88 (2014)
4. S. Brakatsoulas, D. Pfoser, R. Salas, C. Wenk, On map-matching vehicle tracking data, in *Proceedings of VLDB* (2005), pp. 853–864
5. R. Cohen, L. Katzir, The generalized maximum coverage problem. Inf. Process. Lett. **108**(1), 15–22 (2008)
6. L. Leonardi, S. Orlando, A. Raffaetà, A. Roncato, C. Silvestri, G.L. Andrienko, N.V. Andrienko, A general framework for trajectory data warehousing and visual OLAP. GeoInformatica **18**(2), 273–312 (2014)
7. G. Marketos, E. Frentzos, I. Ntoutsi, N. Pelekis, A. Raffaetà, Y. Theodoridis, Building real world trajectory warehouses, in *Proceedings of MobiDE* (2008), pp. 8–15
8. S. Orlando, R. Orsini, A. Raffaetà, A. Roncato, C. Silvestri, Trajectory data warehouses: design and implementation issues. J. Comput. Sci. Eng. **1**(2), 240–261 (2007)
9. D. Papadias, Y. Tao, P. Kalnis, J. Zhang, Indexing spatio-temporal data warehouses, in *Proceedings of ICDE* (2002), pp. 166–175
10. C. Parent, S. Spaccapietra, C. Renso, G. Andrienko, N. Andrienko, V. Bogorny, M.L. Damiani, A. Gkoulalas-Divanis, J.A. Macedo, N. Pelekis, Y. Theodoridis, Z. Yan, Semantic trajectories modeling and analysis. ACM Comput. Surv. **45**(4) (2013)
11. A. Raffaetà, L. Leonardi, G. Marketos, G. Andrienko, N. Andrienko, E. Frentzos, N. Giatrakos, S. Orlando, N. Pelekis, A. Roncato, C. Silvestri, Visual mobility analysis using T-warehouse. Int. J. Data Warehous. Min. **7**(1), 1–23 (2011)
12. I. Ramalho Brilhante, J.A.F. de Macêdo, F.M. Nardini, R. Perego, C. Renso, Where shall we go today? planning touristic tours with TripBuilder, in *Proceedings of CIKM*, eds. by Q. He, A. Iyengar, W. Nejdl, J. Pei, R. Rastogi (ACM, 2013), pp. 757–762
13. I. Ramalho Brilhante, J.A.F. de Macêdo, F.M. Nardini, R. Perego, C. Renso, TripBuilder: a tool for recommending sightseeing tours, in *Proceedings of ECIR* (2014), pp. 771–774
14. I. Ramalho Brilhante, J.A.F. de Macêdo, F.M. Nardini, R. Peregob, C. Renso, On planning sightseeing tours with TripBuilder. Inf. Process. Manag. **51**(2), 1–15 (2015)
15. C. Renso, S. Spaccapietra, E. Zimányi (eds.), *Mobility Data: Modeling, Management, and Understanding* (Cambridge University Press, Cambridge, 2013)
16. S. Spaccapietra, C. Parent, M.L. Damiani, J.A. de Macedo, F. Porto, C. Vangenot, A conceptual view on trajectories. Data Knowl. Eng. **65**(1), 126–146 (2008)
17. Y. Tao, G. Kollios, J. Considine, F. Li, D. Papadias, Spatio-temporal aggregation using sketches, in *Proceedings of ICDE* (2004), pp. 214–225
18. R. Wagner, J.A.F. de Macêdo, A. Raffaetà, C. Renso, A. Roncato, R. Trasarti, Mob-warehouse: a semantic approach for mobility analysis with a trajectory data warehouse, in *Proceedings of SeCoGIS*, LNCS, vol. 8697 (2013), pp. 127–136
19. L. Yang, Z. Hu, J. Long, T. Guo, 5W1H-based conceptual modeling framework for domain ontology and its application on STPO, in *Proceedings of SKG* (2011), pp. 203–206

Part V
Security, Privacy and Health Systems

Big Data Security and Privacy

Elisa Bertino and Elena Ferrari

Abstract Recent technologies, such as IoT, social networks, cloud computing, and data analytics, make today possible to collect huge amounts of data. However, for data to be used to their full power, data security and privacy are critical. Data security and privacy have been widely investigated over the past thirty years. However, today we face new issues in securing and protecting data, that result in new challenging research directions. Some of those challenges arise from increasing privacy concerns with respect to the use of such huge amount of data, and from the need of reconciling privacy with the use of data. Other challenges arise because the deployments of new data collection and processing devices, such as those used in IoT systems, increase the attack potential. In this paper, we discuss relevant concepts and approaches for Big Data security and privacy, and identify research challenges to be addressed to achieve comprehensive solutions to data security and privacy in the Big Data scenario.

1 Introduction

Technological advances and novel applications, such as sensors, cyber-physical systems, smart mobile devices, cloud systems, data analytics, social networks, Internet of Things (IoT), are making possible to collect, store, and process huge amounts of data, referred to as Big Data, about everything from everywhere and at any time.[1] The Big Data term denotes a data management and analytics paradigm featuring 5V: huge data Volume, high Velocity (i.e., timely response requirements), high Variety of data formats, low Veracity (i.e., uncertainties in the data), and high Value.

[1] Data, data everywhere. The Economist, 25 February 2010, available at http://www.economist.com/node/15557443.

E. Bertino
Computer Science Department, Purdue University, West Lafayette, IN, USA
e-mail: bertino@purdue.edu

E. Ferrari (✉)
Department of Theoretical and Applied Sciences, University of Insubria,
Via Mazzini 5, Varese, Italy
e-mail: elena.ferrari@uninsubria.it

S. Flesca et al. (eds.), *A Comprehensive Guide Through the Italian Database Research Over the Last 25 Years*, Studies in Big Data 31, DOI 10.1007/978-3-319-61893-7_25

Recent advances in sensors, actuators, and embedded computing devices in the physical environment and into physical objects - referred to as Internet of Things (IoT) - further multiply the ability to collect data and act on the physical environment [7]. Gartner forecasts predict that by the year 2020 20.8 billions of IoT devices will be deployed. As IoT grows, so do the volumes of data it generates. CISCO estimates that IoT devices will generate 507.5 zettabytes of data per year by 2019. Moreover, not only today we have technology, such as cloud and high-performance computing systems, for storing and processing huge data sets, we also have advanced data analytics tools that allow one to extract useful knowledge from data and predict trends and events. This will open a number of opportunities for new data-intensive applications in a number of different fields, such as manufacturing and energy management, healthcare management and urban life, just to mention few of them. However, such a scenario increases the threats to the security and privacy of the managed data. Damage and misuse of data affect not only single individuals or organizations, but may have negative impacts on entire social sectors and critical infrastructures. Moreover, smart IoT objects as well as end users are today interconnected by different software platforms. For instance, Online Social Networks (OSNs), represent today the huge example of this trend, with an estimation of around 2.67 billion social media users around the globe by 2018.[2] Such connections multiply the possible threats to security and privacy because they increase the paths on which data may flow. As a result, increasing numbers of attacks have been reported that aim at stealing data through sophisticated attacks, including insider attacks [6].

The problem of data security and privacy is not a new problem; research addressing this problem dates back from the early 70's [23] (see for instance [8] for a short history of research efforts on data security). However, early security and privacy techniques were designed for data stored in corporate database systems and therefore today we need to complement and adapt such early techniques in order to provide full spectrum data protection for Big Data.

In this paper, we first briefly discuss key data security and privacy requirements. We then focus on Big Data and identify key research challenges related to their protection. We then focus on two crucial application domains, namely IoT and OSNs. We finally outline a few concluding remarks. The remainder of the paper is organized as follows. Next section provides an overview of the main data protection requirements. Section 3 illustrates the main research issues in the field of Big Data security and privacy, whereas Sects. 4 and 5 discuss security and privacy issues in the Iot and OSN scenario, respectively. Finally, Sect. 8 concludes the paper.

2 Data Protection Requirements

Traditionally, protecting data requires to ensure three main security properties, that is, data confidentiality, integrity and availability [8], also known as the *CIA triad*. Confidentiality refers to data protection from unauthorized read accesses, whereas

[2]https://www.statista.com.

integrity deals with data protection from unauthorized modifications. Data integrity has been further generalized to data trustworthiness, which refers to making sure not only that data are not modified by unauthorized subjects, but also that data are free from errors, up to date, and originating from reputable sources. Assuring data trustworthiness is thus a difficult problem which often depends on the application domain. Its solution requires combining different techniques, ranging from cryptographic techniques for digitally signing the data, to access control, for checking that only authorized parties modify the data, to data quality techniques, for automatically detecting and fixing data errors [4], provenance techniques [48], for determining from which sources data originate, and reputation techniques, for assessing the reputation of data sources. Finally, availability is the property of assuring that data are available to authorized users. These three requirements are still very critical today, and meeting them is today much more challenging because data attacks are more sophisticated and the data attack surface has expanded, due to increasing data collection activities from many different sources and to data sharing.

In addition to the CIA properties, privacy has emerged as a new critical requirement. Many definitions of data privacy have been so far proposed, and the concept of privacy has evolved over time as a result of the evolution of the means to acquire personal information. One of the first systematic written discussion on the concept of privacy was made by Samuel Warren and Louis Brandeis in their 1890 essay titled "The Right to Privacy" [51], where they define privacy as "the right to be let alone". Warren and Brandeis focused mostly on the press and on the publicity effects produced by the new emerging technological inventions of the time, such as photography and widely distributed newspapers. With the development of more advanced technological products that enabled the acquisition, the carriage, the dissemination, and the persistence of information, such as the video-camera, video-tape, telephone, fax, etc., information privacy continued to attract valuable interest. The appearance and the spread of Internet and of the World Wide Web have made possible to collect massive records of information about individuals (e.g., financial and credit history, medical records, purchase history, telephone calls) that may not exactly know what information is stored about them, by whom, and who has access to it [44]. Today, one of the most used definition of data privacy is due to Allan Westin that defined data privacy as "the claim of individuals, groups, or institutions to determine for themselves when, how, and to what extent information about them is communicated to others" [52].

Very often data privacy is seen as the same requirement as data confidentiality, but referring to personal data. There are however relevant differences between the two requirements. It is true that data privacy requires ensuring data confidentiality, because if data are not protected against unauthorized accesses, privacy cannot be ensured. However, privacy has additional issues deriving from the need of taking into account requirements from legal privacy regulations, as well as individual privacy preferences. For example, the concept of purpose is fundamental when dealing with privacy, in that an individual may be fine with sharing his/her own data for research purposes, whereas another individual may not be. Therefore, systems managing privacy-sensitive data may have to collect and record the privacy preferences concerning the individuals to whom the data refer to. Also data subjects may change their privacy preferences over time. Addressing privacy thus requires, among other things, systems able to enforce not only the access control policies that an organization may have in place to govern accesses to the data, but also data subject preferences and legal regulations. This might require to manage additional dimensions related to the access control decision, such as obligations and user consent. Recently, research has started to address this issue by proposing privacy aware access control systems for relational DBMSs [10, 20, 21] and, more recently for NoSQL databases [17, 18, 35, 37]. However, most of the proposed privacy-enhancing techniques only focus on privacy and do not address the key problem of reconciling data privacy with an effective use of data, especially when the use is for security applications, including cyber security, homeland protection, and health security.

In what follows, we focus on two of the most important data protection requirements, that is, confidentiality and privacy. We then consider two of the most relevant application domains, that is, IoT and social networks.

3 Research Issues in Big Data Confidentiality and Privacy

Several techniques to assure data confidentiality and preserve privacy have been proposed over the last fifteen years, ranging from cryptographic techniques, such as oblivious data structures [50] that hide data access patterns, to data anonymization techniques, that transform data to make more difficult to link specific data records to specific individuals [12], to advanced access control models [19]. However, many research challenges still remain to be addressed. In what follows, we discuss some of them. The presentation in this section is partially based on the discussion in [7].

3.1 Data Confidentiality

Several data confidentiality techniques and mechanisms exist - the most notable being access control and encryption. Both have been widely investigated. However, with respect to access control systems for Big Data we need approaches for:

- **Access control policies merging and integration**. In many cases, Big Data analysis entails integrating data sets originating from multiple, possible heterogeneous, sources; these data sets may be associated with their own access control policies, and these policies must be enforced even when a data set is integrated with other data sets. Therefore, policies need to be integrated and conflicts solved, possibly by using some automated or semi-automated policy integration system [36]. Policy integration and conflict resolution are, however, much more complex when dealing with privacy-aware access control models, as these models allow one to specify policies that include the purpose for which the access to a protected data item is allowed, obligations arising from the use of data, and special privacy-related conditions that must be meet in order to access the data. Automatically integrating such type of policies and solving conflicts is a major challenge.

- **Authorizations management**. If fine-grained access control is required, manual administration of authorizations on large data sets is not feasible. We need techniques by which authorizations can be automatically granted, possibly based on the user digital identity, profile, and context, and on the data contents and metadata. A first step towards the development of machine learning techniques to support automatic permission assignments to users is by Ni et al. [45]. However, more advanced approaches are needed to deal with dynamically changing contexts and situations.
- **Enforcing access control on Big Data platforms**. Some of the recent Big Data systems allow their users to submit arbitrary jobs encoded in general programming languages. For example, in Hadoop, users can submit arbitrary MapReduce jobs written in Java. This creates significant challenges in order to efficiently enforce fine grained access control for different users. Although there is some initial work [49] that tries to inject access control policies into submitted jobs, more research is needed on how to efficiently enforce such policies in recently developed Big Data stores, especially if access control policies are enforced though the use of fine-grained encryption. Additionally, the variety of data models and query languages adopted by the existing NoSQL datastores make the definition of a general purpose access control mechanism a challenging task. However, some research efforts have been recently started towards the definition of a unifying query language for NoSQL datastores (see e.g., JSONiq [25] and SQL++ [46]) that can be exploited for that purpose. For instance, [17] relies on SQL++ to provide a general approach to support fine grained ABAC (Attribute-based Access Control) within NoSQL platforms. However, more research is needed to define techniques for enforcing fine-grained access control with a reasonable overhead for any query type and policy coverage.

3.2 Privacy

Although solutions to protect data confidentiality represent the core modules for privacy preservation, protecting privacy for Big Data requires to investigate further relevant issues, which include:

- **Techniques to check that data are used for the intended purpose**. The issue here is how to verify that data returned to a user are used for the data owner intended purpose. An initial pioneering approach was proposed in [11] that associates with each data item a set of possible purposes, from an ontology of purposes, for which the data can be used. When a user accesses some data items, the user indicates in the access request the purpose(s) for which the data items are being accessed. The query purposes are then matched against the purposes associated with the data items to verify that the query purposes comply with the intended use associated with the requested data items. Such an approach needs to be complemented with techniques for automatically and securely identifying the data access purposes, instead of relying on indications given by users as part of their access requests.

- **Support for both personal privacy and population privacy**. In the case of population privacy, it is important to understand what is extracted from the data as this may lead to discrimination [27]. Also when dealing with security with privacy, it is important to understand the tradeoff of personal privacy and collective security.
- **Usability of data privacy policies and user preferences**. Usability is a big issue when dealing with big data privacy and security. Privacy and access control policies must be easily understood by users. We need tools for the average users that help them in specifying their preferences and understand their effect in terms of privacy risks they incur and possible benefits they can get in sharing the data. One direction towards this goal is to empower the user with a secured logical space, a Personal Data Store (PDS) [22] acting as a centralized repository of his/her data. The PDS can then be equipped with a set of analytical tools to reason about the collected data and their sharing with third parties.
- **Privacy-aware access control**. As mentioned before dealing with privacy requires to address further issues wrt data confidentiality, such as obligations, user preferences, and user consent [19]. Although some preliminary work in this direction have been done in the context of Big Data, they mainly focus on specific Big Data platforms (e.g., MongoDB [16, 18]).
- **Risk models**. Different relationships exist between privacy risks and Big Data. One the one hand, Big Data can increase privacy risks, in that they multiply data analysis opportunities; on the other hand, the availability of Big Data sets can reduce security risks in many domains (e.g., national security). The development of models for these two types of risk is critical in order to identify suitable tradeoff and privacy-enhancing techniques to be used.
- **Privacy-aware data lifecycle framework**. A comprehensive approach to privacy for Big Data needs to be based on a systematic data lifecycle approach. Phases in the lifecycle need to be defined and their privacy requirements and implications need to be identified. This is also required by new privacy regulations. For instance, the General Data Protection EU Regulation (GDPR)[3] which has been approved in April 2016 and will enter into application on May 2018 has introduced the privacy by design principle [1]. This will mandatory require that when designing a new system or service that manage personal data, data protection considerations are taken into account starting from the early stages of the design process. Furthermore, the GDPR introduces the Data Protection Impact Assessments (DPIA) which should start prior to the start of processing the personal data, with the goal of identify high risks to the privacy rights of individuals when processing their personal data and possible countermeasures to address them.
- **Data ownership**. The question about who is the owner of a piece of data is often a difficult question. For instance, a notable example is that of photo management

[3] http://europa.eu/rapid/press-release_MEMO-15-6385_en.htm.

in Facebook as users are able to avoid being tagged in a photo,[4] in order to prevent it from being accessible through their profile, but they cannot state how this photo has to be shared in the network. It is perhaps better to replace this concept with the concept of stakeholder. Multiple stakeholders can be associated with each data item. The concept of stakeholder ties well with risks. Each stakeholder would have different (possibly conflicting) objectives and this can be modeled according to multi-objective optimization. In some cases, a stakeholder may not be aware of the others. For example, a user to whom a data item pertains (and thus a stakeholder for the data item) may not be aware that a law enforcement agency is using this data item. Although some preliminary work on this issue has been done in the context of OSNs (see e.g., [30] [31]), technological solutions for Big Data platforms still need to be investigated on support of multiple stakeholder policies.
- **Privacy versus security tradeoff**. The problem of how to reconcile privacy and security is today a major challenge. However, to date very few approaches have been proposed that are suitable for large scale datasets. An example of an initial approach along such direction is the scalable protocol for privacy-preserving data matching by Cao et al. [13] which combines secure multiparty computation (SMC) techniques and differential privacy [24] to address scalability issues.

4 IoT Data Security and Privacy

IoT represents an important emerging trend that, according to various forecasts, will have a major economic impact. IoT applications are changing and improving our every-day lives in a variety of forms, such as with wearable devices that track our sport activities and health status, or with smart home technologies supporting home automation services. However, while on one side, IoT will make many novel applications possible, on the other side, it increases the risks of cyber security attacks to data. In addition, because of its fine-grained, continuous, and pervasive capabilities for data acquisition and control/actuation capabilities, IoT raises concerns about privacy and safety. The OWASP Internet of Things Project[5] has shown that many IoT vulnerabilities arise because of the lack of adoption of well-known security techniques, such as encryption, authentication, access control. This is due to a variety of reasons, such as the cost of deploying privacy and security solutions or the security and privacy unawareness by IT companies involved in the IoT space. But one fundamental reason is because existing security and privacy techniques, tools, and products may not be easily deployed to IoT devices and systems, for reasons such as the variety of hardware platforms and limited computing resources of many types of IoT devices, as well as the underlying decentralized architecture. Therefore, addressing IoT data security and privacy requires extending or re-engineering existing solutions as well as to develop new solutions to fit the specific requirements of

[4]Facebook Help Center - Tag Review. https://www.facebook.com/help/247746261926036/.

[5]https://www.owasp.org/index.php/OWASP_Internet_of_Things_Project.

IoT. Such solutions must ensure protection while data are transmitted and processed at the devices. In addition, in many cases, data availability is critical and therefore solutions minimizing data losses must be devised. In what follows, we survey some projects that cover different aspects of Iot data security and privacy [5].

4.1 Cryptographic Protocols

When dealing with very large IoT systems, one of the key research challenges is to devise efficient and scalable encryption scheme, able to cope with smart objects with very limited processing capabilities. An example of research project in this direction is the certificate-less sign and encryption protocol proposed in [47], that is, a protocol not requiring key certificates and supporting both message encryption and authentication. The protocol works for many different devices, including Raspberry Pi2, and Android. As this protocol does not use expensive pairing operations, it is highly efficient compared to other similar protocols. Other projects along this line have been proposed for specific Iot scenarios, such as for instance the protocol proposed in [43] for efficient authentication operations for networked vehicles. The protocol is able to manage multiple concurrent authentication operations with real-time response time. Response time is critical in that, if a vehicle has to stop suddenly, information about this event has to reach the other vehicles in a very short time so that these vehicles have enough time to break. Therefore, it is crucial that authentication operations both at the sender and the receivers have minimal overhead. To address such requirement, the implementation of the authentication operations takes advantage of the GPU usually present in systems-on-chips today used in vehicles. Another interesting project focuses on encryption protocols for networks consisting of small sensors and drones [53]. In such networks, sensors are on the ground and acquire data of interest from the environment, whereas drones fly over the sensors to collect and aggregate data from them. The main issue here is to save energy and to make sure that drones do not have to wait too long for sensors to start generating encryption keys. To address such requirements, the approach is to use low power listening (LPL) techniques at the sensors and dual radio channels at the drones. In this way, the sensors can timely start generating the cryptographic keys when drones approach.

Results from those projects show that a careful engineering is critical to the effective deployment of cryptographic protocols in IoT. In particular, it is critical to analyze in details the protocols in order to determine the expensive operations so to replace or optimize them, and to understand how to take advantage of specific hardware features of the devices in order to enhance the implementation of the different steps of the protocols.

4.2 *Network Security*

Security techniques at network level are critical in order to minimize data losses. Such minimization is crucial for many applications, such as monitoring applications and control systems. In order to minimize data losses, it is critical to be able to quickly diagnose the cause of data packet losses so to repair the network as soon as possible. A recent project [40] has addressed this requirement by developing a Fine-Grained Analysis (FGA) tool that investigates packet losses and reports their most likely cause. Such FGA tool is based on profiling the wireless links between the nodes as well as their neighborhood, by leveraging resident parameters, such as RSSI and LQI, available within every received packet. By using those profiles, the tool is able to determine whether the cause of a packet loss is a link that has been jammed or a sensor that has been compromised. In the former case, the FGA tool is able to quite reliably detect the source of interference. The design of the system is fully distributed and event-driven, and its low overhead makes it suitable for resource-constrained entities such as wireless motes. This project is however just an initial approach. Research is needed to develop more advanced FGA tools able to deal with mobile systems and heterogeneous communication technologies which may require using different profiling parameters.

4.3 *Application Security*

Protecting applications is crucial for data security as attacks to steal data often use application vulnerabilities as stepping stones. It is important to notice that even though today we have several techniques for program analysis and hardening, such techniques need substantial extensions to fit IoT devices. The research in this direction is still in its infancy. A first example of research projects in this area is represented by techniques to protect programs against code injection attacks and code reuse attacks [26]. A well-known approach to protect against those attacks is to instrument the application binary code by inserting a static check statement before any instruction that modifies the program counter. Such check verifies that the target address, to which the program execution has to move, is the correct address, that is, that the next instruction to be executed is the expected one and not an instruction to which the attacker is trying to redirect the execution. Such techniques have been shown to be quite efficient. As the run-time overhead ranges between 0.51 and 12.22%, based on the benchmarked applications [26]. However, the instructions that can modify the program counter are different for different platforms; such variations thus require devising specific instrumentation techniques for specific platforms. Other relevant attacks are those exploiting memory vulnerabilities. An approach for applications written in a variant of the C language specific for TinyOS applications has been proposed in [41]. Such an approach statically analyzes an application to identify memory vulnerabilities. As in some cases it is not possible to statically determine if a certain piece of code will lead to a vulnerability at run-time, the approach adds some code to check at run-time whether a vulnerability occurs. Also in this project the main issue is to minimize the run-time overhead as this is critical for devices with limited capabilities. Both those projects show that significant work is required to modify existing application program security techniques for use in IoT systems.

4.4 Privacy and Access Control

IoT is more and more evolving into a loosely coupled, decentralized system of cooperating *smart objects*, where high-speed data processing, analytics and shorter response times are becoming more necessary than ever. Such decentralization has a great impact on the way personal and sensitive information generated and consumed by smart objects should be protected, because, without a centralized data management entity, it is more difficult to control how data generated by smart objects are combined and used, even to infer new information. In this scenario, there is the need of defining new enforcement mechanisms for both access control policies and privacy preferences. In this respect, some proposals have recently emerged. For instance, [29] proposed a distributed capability-based access control mechanism exploiting public key cryptography to share information among smart objects. In [39], a two layered architecture is proposed for protecting users' privacy in smart city applications: a first trusted layer, where information is stored and processed by the platform's components, and an open and untrusted second layer, where only generic and unidentifiable information is made available to the external applications. [14] proposed a system for specifying and enforcing privacy preferences in the IoT scenario. The framework provides an expressive language to specify privacy preferences and a mechanism to automatically generate preferences when new information is generated as a result of the data processing. However, the proposal presented in [14] considered a centralized architecture, that is, a scenario where IoT devices have only the capability to sense the data and send them to a data center for being analyzed. In particular, in the framework proposed in [14] sensed data are forwarded by a message broker to a Complex Event Processing system (CEP) as append-only streams of tuples, where registered queries analyze, combine and aggregate them generating new output data. The enforcement monitor statically analyzes every data consumer query and decides if privacy policies of the consumer satisfies the privacy preferences specified by owners of devices generating the data. A challenging issue to be addressed is thus that of designing a fully decentralized privacy enforcement mechanism, where compliance check of data owner privacy preferences is performed directly by smart objects. This has to cope with non negligible overhead that may arise and with the reduced processing capabilities of many smart objects (e.g., sensors). Also in a decentralized setting, the enforcement mechanism should be robust against malicious and colluding smart objects.

5 OSNs Data Security and Privacy

OSNs are one of the most relevant phenomenon in the Big Data area, with billion of users worldwide. OSNs have introduced substantial changes to the way people communicate and socialize within and out of their communities. As a matter of fact, they represent today the biggest available repository of personal information, However, despite all their benefits they also create serious privacy and confidentiality concerns given the nature of information users share over them on almost a daily basis. Users publish their personal stories and updates, and they might also express their opinion by interacting on information shared by others, but, in most cases, they are not fully aware of the size of the audience that gets access to their information. Current commercial OSNs provide very basic form of data protection [15]. In what follows, we survey some of the main security and privacy issues in the realm of OSNs, by covering related relevant projects.

6 Privacy-Aware Access Control

In OSNs, data protection has been mainly approached with Relationship-Based Access Control (ReBAC). According to ReBAC access control decisions are taken by tracking the interpersonal relationships established between users in the network and allowing the formulation of access policies based on them [15]. Privacy settings currently available in commercial OSNs operate under a limited form of ReBAC, but remain both complicated to use, and not flexible enough to model all the privacy preferences that users may require [38]. Most of the mechanisms and techniques that have been suggested for achieving ReBAC in OSNs fall under one of two categories: trust-based or encryption-based. The trust-based approach has mainly been explored under the centralized design of OSNs, where a central entity has full knowledge of the network graph including its nodes, edges, and data ownership, and it is in charge of performing access control. On the other hand, the encryption-based approach has been mostly investigated to address the access control problem under the emerging Decentralized OSNs (DOSNs) scenario. A DOSN is a system that offers OSN services in a peer to peer manner. The concept of DOSNs aims at bringing back control to OSN users and freeing them from the observance of the central service providers. Deploying data encryption to manage access control in DOSNs means that anyone could retrieve the encrypted content but only those who have the corresponding keys can interpret it. This implies that one of the requirements is to offer a mechanism for the distribution, management, and revocation of the corresponding keys, which can introduce a significant overhead due to the huge OSNs population (see e.g., [9, 32]). Whilst such solutions might ensure high data security levels, they have scalability problems and are not flexible enough to support the fine granularity and complex access scenarios required for data dissemination in OSNs. Therefore, what is needed is an investigation of alternative paradigms to perform access control in DOSNs, wrt the preventive one commonly adopted. A step in this direction is represented by [3], where the authors propose an audit-based mechanism to perform a posteriori access control in DOSNs.

7 Identity Validation

All the access control models and methods discussed thus far assume a mechanism in the system by which subjects have been identified and authenticated. However, identities in OSNs are very loose. To facilitate their adoption and encourage people to join them, only a valid email address is required for a user to create an identity in the OSN. The problem of fake accounts and identity related attacks in OSNs has attracted considerable interest from the research body. An example of research project in this area is SybilLimit [54], that leverages on the fast mixing principle, by which honest nodes should converge to having high connectivity to the rest of the network, to detect Sybil attacks. L. Jin et al. suggest in [34] a framework for the detection of cloning attacks, based on attribute and friends'network similarities. All such approaches, and others following the same approach, aim at detecting malicious nodes that follow identified and formalized attack trends. Another important issue is how to validate identity across multiple social networks. Along this line is the work in [28]. However, despite the many methods so far proposed, real case OSN scenarios demonstrate that malicious activities are still taking a huge share [42]. This is due to the fact that almost all detection mechanisms could catch fake nodes only after they have demonstrated some malicious activity or abnormal behavior [33]. Moreover, this detection tends to fail when fake nodes succeed in establishing enough links with good profiles and imitating normal features and behavior. Therefore, the development of effective methods to detect fake accounts is still an open issue. A complementary promising approach to increase the immunity of OSNs to such threats is to empower their honest users with tools that provide them with guarantees or indications on the trustworthiness of the other peers they want to start interacting with. Along this direction, [2] proposes to exploit the OSN crowd to collaboratively estimate the validity of OSN user identities based only on the information they provide on their profiles.

8 Conclusion

While there is no doubt that the Big Data revolution has created substantial benefits to businesses as well as end users, there are commensurate risks that go along with using Big Data. The need to secure data, to protect private information, being at the same time able to ensure data quality, exists whether data sets are big or small. However, the specific properties of Big Data (volume, variety, velocity, veracity, and value) create new types of risks that necessitate to be addressed. In this paper, we have highlighted some of them, by also focusing on two key Big Data scenarios, namely Iot and Social Networks. As a final remark, we would like to point out that addressing the today and tomorrow challenges in data security and privacy require multidisciplinary research drawing from many different areas, including computer science and engineering, information systems, statistics, economics, social sciences, political sciences, psychology. We believe that all these perspectives are needed to achieve effective solutions to the problem of security and privacy in the era of Big Data and pervasive data acquisition and use, and especially, to the problem of reconciling security with privacy.

Acknowledgements The work reported in this paper is partially supported by NSF under the grant ACI-1547358.

References

1. T. Antignac, D. Le Metayer, Privacy by design: from technologies to architectures, in *Bart Preneel and Demosthenes* ed. by Ikonomou, Privacy Technologies and Policy, LNCS, vol. 8450 (Springer, Berlin, 2014)
2. L. Bahri, B. Carminati, E. Ferrari, COIP - continuous, operable, impartial, and privacy-aware identity validity estimation for OSN profiles. ACM Trans. Web **10**(4), 23:1–23:41 (2016)
3. L. Bahri, B. Carminati, E. Ferrari, CARDS - collaborative audit and report data sharing for a-posteriori access control in DOSNs, in *Proceedings of the 1st IEEE Conference on Collaboration and Internet Computing (CIC 2015)* (2015)
4. C. Batini, M. Scannapieco, *Data and Information Quality Dimensions, Principles and Techniques* (Springer, Berlin, 2016)
5. E. Bertino, Data privacy for IoT systems: concepts, approaches, and research directions, in *Proceedings of the IEEE International Conference on Big Data (BigData 2016)* (2016)
6. E. Bertino, *Data Protection from Insider Threats*. Synthesis Lectures on Data Management (Morgan & Claypool Publishers, 2012)
7. E. Bertino, Data security and privacy: concepts, approaches, and research directions, in *Proceedings of the 40th IEEE Computer Software and Applications Conference (COMPSAC 2016)* (2016)
8. E. Bertino, R. Sandhu, Database security: concepts, approaches, and challenges. IEEE Trans. Dependable Sec. Comput. **2**(1), 2–19 (2005)
9. O. Bodriagov, G. Kreitz, S. Buchegger, Access control in decentralized online social networks: applying a policy-hiding cryptographic scheme and evaluating its performance, in *Pervasive Computing and Communications Workshops (PERCOM Workshops)* (2014)
10. J.W. Byun, N. Li, Purpose based access control for privacy protection in relational database systems. The VLDB J. **17**(4) (2008)
11. J.W. Byun, E. Bertino, N. Li, Purpose based access control of complex data for privacy protection, in *Proceedings of the 10th ACM Symposium on Access Control Models and Technologies (SACMAT 2005)* (2005)
12. J.W. Byun, A. Kamra, E. Bertino, N. Li, Efficient k-anonymization using clustering techniques, in *Proceedings of the 12th Conference on Database Systems for Advanced Applications (DASFAA 2007)* (2007)
13. J. Cao, E.-Y. Rao, E. Bertino, M. Kantarcioglu, A hybrid private record linkage scheme: separating differentially private synopses from matching records, in *Proceedings of the 31st Conference on Data Engineering (ICDE 2015)* (2015)

14. B. Carminati, P. Colombo, E. Ferrari, G. Sagirlar, Enhancing user control on personal data usage in internet of things ecosystems, in *Proceedings of the IEEE International Conference on Services Computing (SCC 2016)* (2016)
15. B. Carminati, E. Ferrari, M. Viviani, *Security and trust in online social networks*. Synthesis Lectures on Information Security, Privacy, and Trust (Morgan & Claypool Publishers, 2013)
16. P. Colombo, E. Ferrari, Enhancing MongoDB with purpose based access control. IEEE Trans. Dependable Sec. Comput. to appear
17. P. Colombo, E. Ferrari, Towards a unifying attribute based access control approach for NoSQL datastores, in *Proceedings of the 33rd IEEE Conference on Data Engineering (ICDE 2017)*, to appear
18. P. Colombo, E. Ferrari, Towards virtual private NoSQL datastores, in *Proceedings of the 32nd IEEE Conference on Data Engineering (ICDE 2016)* (2016)
19. P. Colombo, E. Ferrari, Privacy aware access control for big data: a research roadmap. Big Data Res. **2**(4), 145–154 (2015)
20. P. Colombo, E. Ferrari, Enforcing obligations within relational database management systems. IEEE Trans. Dependable Sec. Comput. **11**(4), 318–331 (2014)
21. P. Colombo, E. Ferrari, Enforcement of purpose based access control within relational database management systems. IEEE Trans. Knowl. Data Eng. **26**(11), 2703–2716 (2014)
22. Y.A. De Montjoye, E. Shmueli, S.S. Wang, A.S. Pentlan, openPDS: protecting the privacy of metadata through safe answers, in *PLoS One* (2014)
23. D.E. Denning, P.J. Denning, Data security. ACM Comput. Surv. **11**(3), 227–249 (1979)
24. C. Dwork, A. Roth, The algorithmic foundations of differential privacy. Found. Trends Theor. Comput. Sci. **9**(3–4), 211–407 (2014)
25. D. Florescu, G. Fourny, JSONiq: the history of a query language. IEEE Int. Comput. **17**(5) (2013)
26. J. Habibi, A. Panicker, A. Gupta, E. Bertino, DisARM: mitigating buffer overflow attacks on embedded devices, in *Proceedings of the 9th Conference on Network and System Security (NSS 2015)* (2015)
27. S. Hajian, J. Domingo-Ferrer, A. Monreale, D. Pedreschi, F. Giannotti, Discrimination- and privacy-aware patterns. Data Min. Knowl. Discov. **29**(6), 1733–1782 (2015)
28. B.-Z. He, C.-M. Chen, Y.-P. Su, H.-M. Sun, A defense scheme against identity theft attack based on multiple social networks. Expert Syst. Appl. **41**(5), 2345–2352 (2014)
29. J.L. Hernandez-Ramos, D.G. Carrillo, R. Marin-Lopez, A.F. Skarmeta, Dynamic security credentials pana-based provisioning for IoT smart objects, in *Proceedings of the IEEE 2nd World Forum on Internet of Things (WF-IoT'15)* (2015)
30. H. Hu, G.J. Ahn, J. Jorgensen, Multiparty access control for online social networks: model and mechanisms. IEEE Trans. Knowl. Data Eng. **25**, 1614–1627 (2013)
31. P. Ilia, B. Carminati, E. Ferrari, P. Fragopoulou, S. Ioannidis, SAMPAC: socially-aware collaborative multi-party access control, in *Proceedings of the 7th ACM Conference on Data and Applications Security and Privacy (CODASPY 2017)* (2017)
32. S. Jahid, S. Nilizadeh, P. Mittal, N. Borisov, A. Kapadia, Decent: a decentralized architecture for enforcing privacy in online social networks, in *Pervasive Computing and Communications Workshops (PERCOM Workshops)* (2012)
33. J. Jiang, Z.F. Shan, X. Wang, L. Zhang, Y.F. Dai, Understanding sybil groups in the wild. J. Comput. Sci. Technol. **30**(6), 1344–1357 (2015)
34. L. Jin, H. Takabi, J.B. Joshi, Towards active detection of identity clone attacks on online social networks, in *Proceedings of the 1st ACM Conference on Data and Application Security and Privacy* (2011)
35. D. Kulkarni, A fine-grained access control model for key-value systems, in *Proceedings of the 3rd ACM Conference on Data and Application Security and Privacy (CODASPY 2013)* (2013)
36. D. Lin, P. Rao, E. Bertino, N. Li, J. Lobo, EXAM: a comprehensive environment for the analysis of access control policies. Int. J. Inf. Sec. (IJIS) **9**(4), 253–273 (2010)
37. J. Longstaff, J. Noble, Attribute based access control for big data applications by query modification, in *Proceedings of IEEE BigDataService* (2016)

38. M. Madejski, M.L. Johnson, S.M. Bellovin, The failure of online social network privacy settings. Columbia University Academic Commons (2011)
39. O. Mazhelis et al., Towards enabling privacy preserving smart city apps, in *Proceedings of the IEEE Smart Cities Conference* (2016)
40. D. Midi, E. Bertino, Node or Link? Fine-Grained Analysis of Packet Loss Attacks in Wireless Sensor Networks. ACM Trans. Sens. Netw. **12**(2) (2016). Accepted for publication
41. D. Midi, T. Payer, E. Bertino, nesCheck: Memory Safety for Embedded Devices, submitted for publication
42. S. Mitter, C. Wagner, M. Strohmaier, Understanding the impact of socialbot attacks in online social networks. arXiv preprint arXiv:1402.6289 (2014)
43. A.A. Mudgerikar, A. Singla, I. Papapanagiotou, A.A. Yavuz, HAA: hardware-accelerated authentication for internet of things in mission critical vehicular networks, in *Proceedings of the 34th Conference for Military Communications (IEEE MILCOM 2015)* (2015)
44. A. Narayanan, V. Toubiana, S. Barocas, H. Nissenbaum, D. Boneh, A critical look at decentralized personal data architectures. arXiv preprint arXiv:1202.4503 (2012)
45. Q. Ni, J. Lobo, S.B. Calo, P. Rohatgi, E. Bertino, Automating role-based provisioning by learning from examples, in *Proceedings of the 14th ACM Symposium on Access Control Models and Technologies (SACMAT 2009)* (2009)
46. K.W. Ong, Y. Papakonstantinou, R. Vernoux, The SQL++ unifying semi-structured query language, and an expressiveness benchmark of SQL-on-Hadoop, NoSQL and NewSQL databases. CoRR, arXiv:1405.3631 (2014)
47. S.H. Seo, J. Won, E. Bertino, pCLSC-TKEM: a Pairing-free Certificateless Signcryption-tag key encapsulation mechanism for a privacy-preserving IoT. Trans. Data Priv. (2016)
48. S. Sultana, E. Bertino, A Distributed system for the management of fine-grained provenance. J. Database Manag. **26**(2), 32–47 (2015)
49. H. Ulusoy, P. Colombo, E. Ferrari, M. Kantarcioglu, E. Pattuk, GuardMR: fine-grained security policy enforcement for MapReduce systems, in *Proceedings of the 10th ACM Symposium on Information, Computer and Communications Security (ASIACCS'15)* (2015)
50. H.X. Wang, K. Nayak, C. Liu, E. Shi, E. Stefanov, Y. Huang, Oblivious data structures. IACR Cryptology ePrint Archive (2014)
51. S.D. Warren, L.D. Brandeis, The Right to Privacy. Harvard Law Review (1890), pp. 193–220
52. A. Westin, *Privacy And Freedom* (Atheneum, New York, 1967), p. 7
53. J. Won, S.H. Seo, E. Bertino, A secure communication protocol for drones and smart objects, in *Proceedings of the 10th ACM Symposium on Information, Computer and Communications Security (ASIACCS '15)* (2015)
54. H. Yu, P.B. Gibbons, M. Kaminsky, F. Xiao, Sybillimit: a near-optimal social network defense against sybil attacks, in *Proceedings of the IEEE Symposium on Security and Privacy* (2008)

Not Only Databases: Social Data and Cybersecurity Perspective

Francesco Buccafurri, Gianluca Lax, Serena Nicolazzo and Antonino Nocera

Abstract The title of this chapter describes the common denominator of the contributions given by the research group of the University Mediterranea of Reggio Calabria to the SEBD community. These contributions are mainly devoted to study non-core problems of databases, by moving towards border line issues to look for cross-fertilization coming from different related domains. This chapter offers a view over these contributions given in three different domains: efficient structures for approximate OLAP, social data and social networks, and the cybersecurity perspective.

Keywords Compression for OLAP · Histograms · Data streams · Social network data · Data security.

1 Introduction

In this chapter, we try to depict the long story of work done by the research group of the University Mediterranea of Reggio Calabria in the SEBD trace, how was their view about database issues, in which way they contributed to the growth of the conference and to enrich the discussion in the SEBD community. There was a common denominator in this research: to study non-classical problems of databases, by moving from core problems towards border line ones (w.r.t. the database domain) to establish a sort of bridge to different related domains and to favor cross-fertilization.

F. Buccafurri (✉) · G. Lax · S. Nicolazzo · A. Nocera
DIIES University Mediterranea of Reggio Calabria, Via Graziella,
Località Feo di Vito, 89122 Reggio Calabria, Italy
e-mail: bucca@unirc.it

G. Lax
e-mail: lax@unirc.it

S. Nicolazzo
e-mail: nicolazzo@unirc.it

A. Nocera
e-mail: a.nocera@unirc.it

S. Flesca et al. (eds.), *A Comprehensive Guide Through the Italian Database Research Over the Last 25 Years*, Studies in Big Data 31, DOI 10.1007/978-3-319-61893-7_26

This explains the title of this chapter, but maybe this is also coherent with a general evolution of the database community as, since the birth of databases, the value and the role of data in the information society is progressively changed.

We may identify three main streams of research: (1) efficient structures for approximate OLAP, (2) social data and social networks, and (3) the cybersecurity perspective.

In the first stream, the effort was to find efficient hierarchical structures to allow fast online analytical processing, by paying a little price in terms of accuracy. In [6], they studied the problem in the one-dimension case, by understanding how to use a small additional storage space, in order to improve the estimation inside histogram buckets. In [7], they applied the tree-like approach above to data streams, by supporting efficiently sliding windows and range queries. Finally, in [15], they proposed the use of an improved 64-bit index, called *5LT*, which takes into account the practical topic of 64-bit computer architectures.

Concerning the stream (2), they addressed two different problems, both falling into the multiple-social-network perspective. As a matter of fact, this a truly data-oriented view, because the attempt is to abstract concepts from single social networks to a higher layer. Specifically, in [9], they defined a multi-social-network model abstracting the main concepts occurring in social networks and a set of meta-API to support multi-social-network programming. By means of this model, in [13], they studied how to match different profiles of the same digital identity spread out over different social networks.

From the cybersecurity perspective, whose role is more and more important also in databases, they analyzed in [10] how to manage the problem of cloud query integrity in case of range query and high-frequency append operations, and in [5], how to exploit the information-sharing power of social networks to implement a lightweight electronic signature protocol.

2 Efficient Structures for Approximate OLAP

The first main stream of research concerns efficient hierarchical structures to allow fast online analytical processing by paying a little price in terms of accuracy. The next section is devoted to this problem in the one-dimension case, by understanding how to use a small additional storage space to improve the estimation inside histogram buckets. In Sect. 2.2, a tree-like approach is defined to support sliding windows of data streams and efficient querying.

2.1 Data Compression for OLAP

In many application contexts, such as datawarehouses, transaction recording systems, OLAP applications, data mining activities, intrusion detection systems, scientific databases, there is the need to handle a huge number of detailed records. However, useful knowledge can extracted by exploiting only condensed information and often histograms are the structure used to summarize data: for example, histograms have been used for approximating probability distributions in statistical databases [24], for improving the join processing in temporal databases [28], and for approximating query answering in order to reduce the query response time in on-line decision support systems and OLAP [23].

For a given storage space reduction, the problem of determining the best histogram is crucial. Indeed, different partitions lead to dramatically different errors in reconstructing the original data distribution. To better explain the problem, consider a typical case of recovering original data from a histogram: the evaluation of range queries. Think of a histogram defined on the attribute X of a relation R as a set of non-overlapping intervals of X covering all values of X in R. To each of these intervals, say B, the number of occurrences (called *frequency*) in R, having the value of X belonging to the interval B, is associated (and included into a data structure called *bucket*). A *range query*, defined on an interval Q of X, evaluates the number of occurrences in R with value of X in Q. Thus, buckets embed a set of pre-computed disjoint range queries covering the whole active domain of X in R. As a consequence, the histogram does not give, in general, the possibility of evaluating exactly a range query not corresponding to one of the pre-computed embedded queries. It turns out that it is convenient to define the boundaries of buckets in such a way that the estimation error of the non-precomputed range queries is minimized. This issue is being investigated since some decades, and a large number of techniques for arranging histograms have been proposed [21, 22, 26].

In [6], the authors studied how to use a small additional storage space, in order to improve the estimation inside histogram buckets. In particular, they proposed to use an additional 32-bit memory word for each bucket, called *4LT* because this 32-bit memory word is organized as a 4-Level Tree, in order to store partial sums internal to the bucket in a hierarchical fashion, using a tree-like index inducing a further partition of the bucket into a number of equal-size sub-buckets (corresponding to the leaves of the tree).

In Fig. 1, an example of 4LT built over a bucket containing 16 data is shown. The root stores the exact value of the bucket sum, say c. The value of gray nodes is stored, whereas that of white nodes is computed as difference between the parent node and the left-sibling node. The value of the leaf nodes is encoded by 4 bits and the approximate value of each of the 4 partial sums is denoted by $\delta_{1/8}$, $\delta_{3/8}$, $\delta_{5/8}$ and $\delta_{7/8}$ — let $L_{i/8}$, $i = 1, 3, 5, 7$, denote such 4-bit strings. The remaining 16 bits are used as follows: the partial sums $\delta_{1/4}$ and $\delta_{3/4}$ are approximated by the 5-bit strings $L_{1/4}$ and $L_{3/4}$, respectively, while the partial sum $\delta_{1/2}$ with a 6-bit string $L_{1/2}$. As a result, the larger the intervals, the higher is the number of bits used. The 8 L strings are constructed as follows ($\langle x \rangle$ stands for $round(x)$):

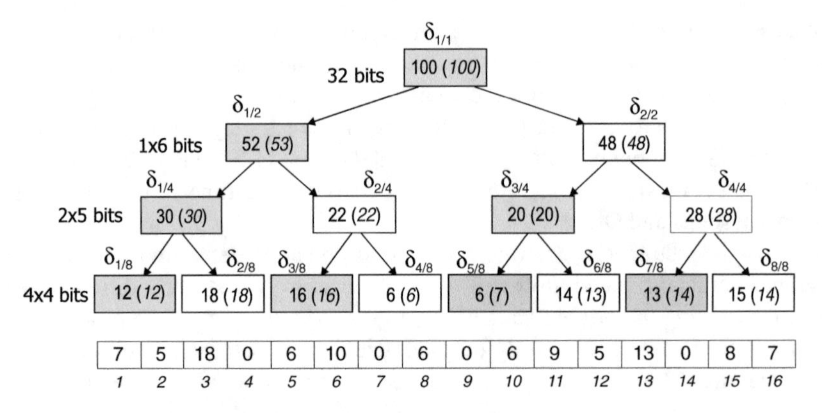

Fig. 1 An example of 4-level tree

$$
\begin{aligned}
L_{1/2} &= \langle \tfrac{\delta_{1/2}}{\delta_{1/1}} \cdot (2^6 - 1) \rangle \\
L_{i/4} &= \langle \tfrac{\delta_{i/4}}{\delta_{j/2}} \cdot (2^5 - 1) \rangle \quad (i = 1 \wedge j = 1), (i = 3 \wedge j = 2) \\
L_{i/8} &= \langle \tfrac{\delta_{i/8}}{\delta_{j/4}} \cdot (2^4 - 1) \rangle \quad (i = 1 \wedge j = 1), (i = 3 \wedge j = 2), \\
&\qquad\qquad\qquad\qquad\quad (i = 5 \wedge j = 3), (i = 7 \wedge j = 4)
\end{aligned}
$$

The approximate values for the partial sums are eventually computed as:

$$
\begin{aligned}
&\widetilde{\delta}_{1/1} = \delta_{1/1} = c \qquad \widetilde{\delta}_{1/2} = \tfrac{L_{1/2}}{2^6 - 1} \times \widetilde{\delta}_{1/1} \; \widetilde{\delta}_{2/2} = \widetilde{\delta}_{1/1} - \widetilde{\delta}_{1/2} \\
&\widetilde{\delta}_{i/4} = \tfrac{L_{i/4}}{2^5 - 1} \times \widetilde{\delta}_{j/2} \quad (i = 1 \wedge j = 1), (i = 3 \wedge j = 2) \\
&\widetilde{\delta}_{i/4} = \widetilde{\delta}_{j/2} - \widetilde{\delta}_{i-1/4} \; (i = 2 \wedge j = 1), (i = 4 \wedge j = 2) \\
&\widetilde{\delta}_{i/8} = \tfrac{L_{i/8}}{2^4 - 1} \times \widetilde{\delta}_{j/4} \quad (i = 1 \wedge j = 1), (i = 3 \wedge j = 2) \\
&\qquad\qquad\qquad\qquad\quad (i = 5 \wedge j = 3), (i = 7 \wedge j = 4) \\
&\widetilde{\delta}_{i/8} = \widetilde{\delta}_{j/4} - \widetilde{\delta}_{i-1/8} \; (i = 2 \wedge j = 1), (i = 4 \wedge j = 2) \\
&\qquad\qquad\qquad\qquad\quad (i = 6 \wedge j = 3), (i = 8 \wedge j = 4)
\end{aligned}
$$

Example 1 Consider the 4-level tree in Fig. 1. The 32 bits store the following approximate cumulative frequencies: $L_{1/2} = 33, L_{1/4} = 18, L_{3/4} = 13, L_{1/8} = 6, L_{3/8} = 11, L_{5/8} = 5, L_{7/8} = 7$.

Once 4LT as been built, the authors of [6] studied if the use of 4LT, which has the drawback of reducing the overall number of buckets, can improve estimation accuracy. A rich experimental campaign to evaluate the effects of the combination of the 4LT index with the existing methods for building histograms, showed that 4LT improves significantly the accuracy of the state-of-the-art histograms (see [16] for details). Finally, the use of an improved index of 64 bits, called *5LT*, has been described in [15], which takes into account the practical topic of 64-bit computer architectures.

2.2 Data Streams

Data streams are indefinite sequences of data which continuously vary in time, often very quickly [3]. There are many application contexts characterized by the presence of data streams, such as sensor networks, financial applications, telecommunication data management. Often, the analysis of the most recent part of a data stream is enough to give meaningful information. From a semantic point of view, this means giving more importance to recent knowledge w.r.t. past one, assuming that recent information, called *sliding window*, is more reliable and significant than old information [18].

Also in the context of data streams, *range queries* are one of the most common types of queries (for example, "Return the total amount of traffic crossing the router in a given time interval"). The approach seen in Sect. 2.1 based on a tree index has been suitably improved to be exploited also in this context: indeed, the 4-Level Tree approach used for persistent data is not applicable to data streams because it does not support the continuous updating of data. In [7], the index *c-Tree* is proposed: this structure is based on a hierarchical structure, a tree, and its nodes contain, in an aggregation hierarchy, pre-computed range-sum queries, stored by a bit-saving encoding. For this reason, the structure directly supports the estimation of arbitrary range queries (in particular, range queries of type *sum*). Indeed, range queries are either embedded in the histogram or derivable by linear interpolation. Reduction derives from both the aggregation implemented by leaves of the tree (discretization), and the saving of bits obtained by representing range queries with less than 32 bits. This structure is efficiently dynamic, because both updates and query answers can be executed in logarithmic time (w.r.t. the window size).

In the initial configuration, C-TREE consists of:

1. A full binary tree with n levels, where $2 \leq n \leq log_2 w$ is a parameter set according to the amount of memory allowed for the synopsis and w is the sliding window size. Each node N stores an integer value denoted by $val(N)$. In the initial state, all nodes contain the value 0.
2. A buffer $B = \langle e, s \rangle$, where e (number of elements) and s (sum) are non-negative integer values. This buffer is used to accumulate $d = \frac{w}{2^{n-1}}$ elements of the data stream. In the initial state, $e = 0$ and $s = 0$.
3. A positive integer value P, with $1 \leq P \leq 2^{n-1}$, used as a pointer to a leaf node. In the initial state, P is set to 1.

After the arrival of a new data D, C-TREE is updated as follows. First, D is stored in the buffer. This is done by adding 1 to the number of elements in the buffer and by updating their sum. Then, if the number of elements in the buffer is less than $d = \frac{w}{2^{n-1}}$ (i.e., the buffer is not full), then the procedure ends. Otherwise (i.e., when the number of elements in B is equal to d), the value s of B, which is the sum of the last d points, is moved to the binary tree as follows. Let N be the leaf node pointed by P and δ the difference between s and $val(N)$ (i.e., the current value of N). Then, δ is added to all the nodes belonging to the path from N to the root of the binary tree. Finally, the buffer is emptied and P is updated to point to the next leaf. By the use of the modulus operator, leaf nodes are managed as a cyclic array. P points to the

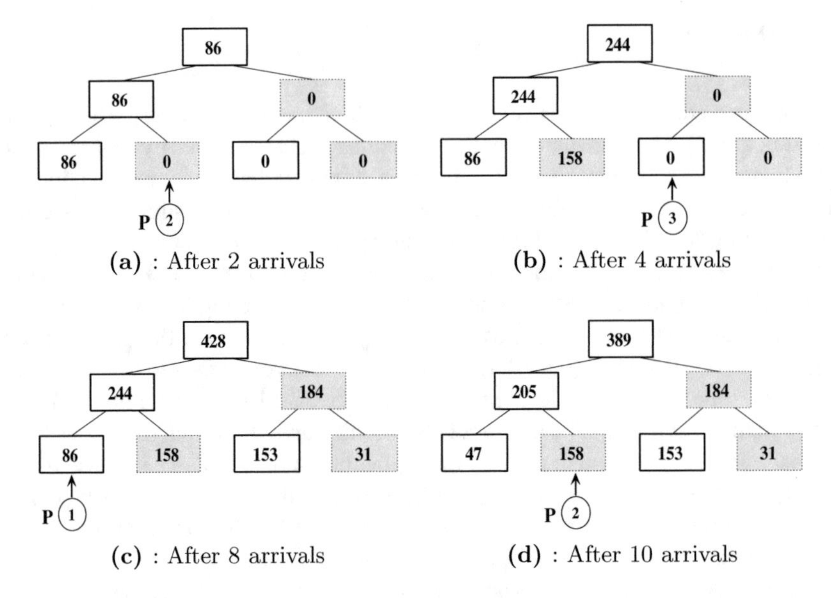

Fig. 2 Building of C-TREE

leaf node containing the least recent data, which will be replaced by the next data coming from the buffer.

Observe that the value of each right-hand child node is not stored because it can be derived as a difference between the value of the parent node and the value of the sibling node. As a consequence, given a C-TREE with n levels, only 2^{n-1} nodes (instead of $2^n - 1$) are saved.

An example of C-TREE building is shown in Fig. 2. Let $\langle 35, 51, 40, 118, 132, 21, 15, 16,$ be a data stream ordered by arrival time increasing and assume that the sliding window size is 8. Initially, $e = 0$, $s = 0$, $P = 1$ and the value of all nodes is 0. The first data coming from the stream is 35, thus $e = 1$ and $s = 35$. Since $e < d = w/2^{n-1} = 2$ no other operation has to be done at this step. The next data coming from the stream is 51, thus $e = 2$ and $s = 86$. Since $e = d$, the leaf node pointed by P (that is, the first one), is set to the value s, and the nodes belonging to the path from such a leaf node to the root are increased by $\delta = 86$. Finally, $P = 2$, and e and s are set to zero. In Fig. 2a, the resulting C-TREE is reported. Therein (as well as in the other figures of the example), they omit the values of the buffer since they are zero. Moreover, the authors represent those nodes that are not stored (i.e., the right-hand child nodes) by gray boxes, since they are derived from the other ones. For the first 8 data arrivals, the updates proceed as before. In Fig. 2b and c, they report the snapshots after 4 and 8 updates, respectively. Then, the pointer P is updated to the value 1. Now, the value coming from the data stream is 18. Therefore, $e = 1$ and $s = 18$. At the next arrival, $e = 2$ and $s = 47$. Since $e = 2$, $\delta = 47 - 86 = -39$ is added to the leaf node pointed by P (that is, the first leaf) which takes the new value 47. Moreover, also the nodes belonging to the path from such a leaf node to the root are increased by δ. Finally, P is updated to the value 2. The final C-TREE is shown in Fig. 2d.

In [8], the authors by an extensive experimental campaign showed that C-TREE has better performances than the state of the art [20].

3 Social Network Data

The second main stream of research addresses two problems falling into the multiple-social-network perspective. In the next section, a model abstracting the main concepts occurring in social networks is formalized to support multiple-social-network programming. In Sect. 3.2, the problem of matching different profiles of the same digital identity spread out over different social networks is dealt with.

3.1 A Model for Multiple Social Networks

Over the past years, online social networks have became part of people's live. Nowadays, most people have a profile in one or more online social networks. This leads to a new phenomenon known with the name of *multiple social network*. Figure 3 shows a graphical representation of a possible scenario involving three social networks, namely `Twitter`, `Facebook`, and `LinkedIn`, to highlight similarities and differences among social networks. In this example, nodes from 1 to 6 are `Twitter` accounts, nodes from 7 to 12 are `Facebook` accounts and, finally, nodes from 13 to 18 are `LinkedIn` accounts. As for edges, they represent friendship relationships

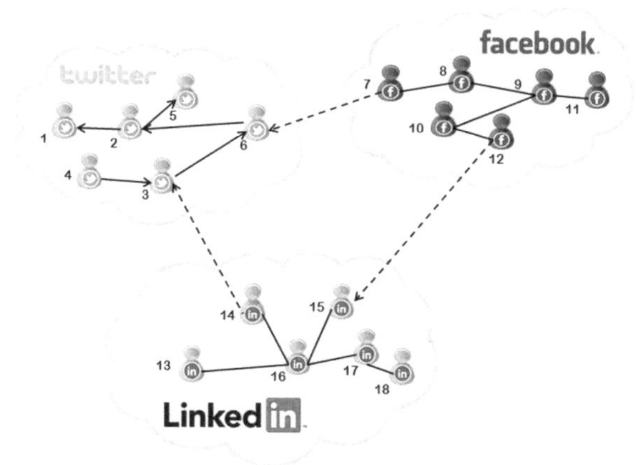

Fig. 3 An example of a multiple social network scenario

among users. However, while edges among `Facebook` and `LinkedIn` actors are bidirectional, those among `Twitter` users are directed, according to the typology of relationship allowed by the social network. Finally, edges (14, 3), (12, 15) and (7, 6) connect accounts of the same user on different OSNs.

The authors of [9] focused their attention on this topic and observed that despite the conceptual uniformity of the social-network universe in terms of structure, basic mechanisms, main features, etc., each social network has in practice its own terms, resources, actions. For instance, contacts are represented by *friends* in Facebook and the relationship is symmetric, while they are represented by *followers* and *followings* in Twitter and the corresponding relationship is not symmetric. Again, the concept of appreciation becomes +1 in Google+ and *endorsement* in about.me. Importantly, similar concepts can be mapped to each other but they have in general different features. This heterogeneity is a strong handicap for the design and implementation of applications enabling internetworking functions among multiple social networks. On the other hand, the power of the social-network substrate can be fully exploited only by moving from a single-social-network to a multiple-social-network perspective, still keeping the user-centered vision [25, 29].

Consider, for example, the possibility of building the complete profile of users by merging all the information they spread out over the joined social networks. This could give a considerable added value to market analysis and job recruitment strategies, as membership overlap among social networks is often an expression of different traits of users personality (sometimes almost different identities). From the above observations, it clearly follows that even though each single social network is an extraordinary source of knowledge, the information power of the social network can be considerable increased if a huge global social network is considered, composed of autonomous components with strong correlation and interaction. Thus, social-network-based programming should work at this abstraction level.

In [9], a model aimed at generalizing concepts, actions and relationships of existing social networks has been proposed. Specifically, the following key concepts have been identified. *Profile*, a form of individual (or group) homepage, which provides a description of each registered user. Profiles are constructed by filling out forms on the site. *Links to external social networks*, an important feature provided by all social networks to add accounts in another social site or external website. By this concept, it is possible to see different social sites as members of a multiple-social-network environment. *Friendship*, by means of which a user may receive automatic notifications about the activity of another user. *Resources*, a Web asset such as a status update, a photo, a Web link or a video created and loaded by a user in his profile. *Actions on resources*, after a resource is published, several actions can be performed on this resource: other users can appreciate it, or re-share it, or it can be associated with a user through a mention on his profile.

Starting from these concepts, the authors have identified eleven technical entities, of which three concepts and eight relationships. The abstract multiple-social-network model is defined by a direct graph $G = \langle N, E \rangle$, in which nodes represent the concepts and edges encode the relationships. The set of nodes is partitioned into three disjoint sets P, R, and B, which correspond to the set of social profiles, the set of resources, and the set of bundles (which are resource containers), respectively. Further, the set of edges is partitioned into eight disjoint sets F, M, Pu, S, T, Re, L, and Co, each corresponding to one of the eight relationships identified above.

After defining the conceptual model, the authors showed in [11] how to practically map real-life data from social networks to each component of the model, in such a way to build a data structure that can be used at application level. Moreover, they described how this model has been profitably applied to two applications very relevant in the context of social network analysis. The first application regards the extraction of information from a multiple-social-network scenario, the second one concerns a particular analysis done on social network data.

3.2 Identity Matching

In a scenario of multiple social networks, two users can interact with each other even though they joined two different OSNs and did not know each other: it is sufficient that their communication passes through a *bridge* user, i.e. a user who created an account in both the OSNs. As a consequence, bridge users represent a key aspect of a multiple social network. The links connecting the different accounts of the same bridge user in different OSNs are called `me` edges and, clearly, play a key role in a multiple-social-network scenario. Several users explicitly specify their different accounts in the different OSNs they joined, and, consequently, their `me` edges. Even though OSNs facilitate this activity by providing users with suitable support tools, many users, due to disparate reasons, do not perform this specification. However, the knowledge of these *undeclared* edges could be extremely useful not only to allow users of different OSNs to communicate with each other but also to better construct the complete profile of a user. This feature could be very relevant in various application fields: for instance, in e-commerce, it could help to better identify users needs and desires and to perform personalized offers; in e-recruitment, it could help to better know the complete profile of a candidate taking into account not only the information officially declared in his curriculum vitae but also the information he informally expressed in the joined OSNs.

The detection of `me` edges is related to several problems well studied in the literature: identifying users on the Web, link prediction (i.e., the task of link mining aimed to predict the even future existence of a link between two objects), and re-identification of anonymized social network data. However, a solution for this specific problem has been proposed in [13]. Here, the approach used to discover a hidden `me` edge between two accounts of two different OSNs examines both the similarity of the account screen-name corresponding to the two nodes and the node neighbors. Concerning the first element, several string similarity functions (e.g., Jaro-Winkler, Levenshtein, QGrams, Monge-Elkan, Soundex [17]) have been experimentally used and compared to verify which is the best one (i.e., the one leading our approach to obtain the best performances). Concerning the second element, to determine the neighborhood of a node, the information about the corresponding user has been used. It is declared by means of XFN (XHTML Friends Network) [30] and FOAF (Friend-Of-A-Friend) [4], two standards specifically conceived for encoding human relationships. As the number of node pairs to consider for the possible presence of a `me` edge is enormous, a mechanism to consider only a reasonable number of very promising pairs is mandatory. More specifically, from the examination of the explicitly declared `me` edges, it results that with a high probability some of the nodes belonging to the neighbors of two nodes linked by a `me` edge are, in their turn, linked by a `me` edge. As a consequence, this approach starts from a set of already known `me` edges and examines only the neighbors of the nodes involved in these edges.

A rich experiment campaign aimed at computing the correctness of the approach has been carried out on a sample consisted of 93169 nodes and 146325 edges, 745 of which were `me` edges. The adopted account name similarity function was QGrams because it proved to assure the best precision. The response given by the technique about the presence (yes or not) of a hidden `me` edge between two nodes has been verified by a human expert, who really visited the pages corresponding to the nodes of each edge. For each edge, possible answers of the expert were *true*, *false* and *unknown* (given when he was not able to access the page associated with a node or to give an answer with an absolute certainty). As for performance metrics, the correctness has been used, computed as

$$corr = \frac{t' + f'}{t' + f' + t'' + f''} \cdot \left(\frac{t'}{t' + f'}\right) + \frac{t'' + f''}{t' + f' + t'' + f''} \cdot \left(\frac{t''}{t'' + f''}\right)$$

where $\frac{t'+f'}{t'+f'+t''+f''}$ and $\frac{t''+f''}{t'+f'+t''+f''}$ represent the weights to be associated with `me` and not `me` edges, respectively (they depend on the number of edges evaluated as *true* or *false* by the human expert with an absolute certainty), $\frac{t'}{t'+f'}$ represents the correctness for `me` edges and $\frac{t''}{t''+f''}$ denotes the correctness for not `me` edges. At the end of the experiment, the measured correctness has been 0.85, showing that this technique has good performances [14].

4 Data Security

This stream of research concerns the cybersecurity perspective, focusing on how to guarantee integrity of range query in cloud (in the next section) and how to implement a lightweight electronic signature protocol relying on the information-sharing power of social networks (in Sect. 4.2).

4.1 Integrity in Cloud

Cloud computing allows users to move their computation and storage to servers transparently distributed over the Internet. Among others, services supporting the storage of huge quantity of data and the querying on them, is a very common application. More and more often, both private and public organizations rely on cloud-based services supplied by third parties. This happens in many cases, for those companies that do not choose to implement private clouds and for those governments in which national clouds are not adopted. In these cases, the trustworthiness of the cloud provider becomes a critical issue. Among other threats, the possibility that the cloud does not return intact responses to queries has to be considered. Moreover, a user may have the (legal) necessity of proving that query results are not compromised.

This research topic is well known in the literature, under the name of *query integrity*, and the increasing attention towards the cloud has also renewed the interest in this issue [27]). Solving this problem means to allow users to verify that query results are *complete* (i.e., no qualifying tuples are omitted), *fresh* (i.e., the newest version of the results are returned), and *correct* (i.e., the result values are not corrupted).

In this general framework, [10] identifies the specific and very up-to-date scenario of data streams in which append operations and range queries are dominant, and the efficiency is a critical factor. The considered scenario is about video surveillance setting, even though the approach is more general since other possible application settings could be identified (i.e., all the cases in which sensors produce time series, in e-health, SCADA systems, the Internet of Things, smart environments, etc.). The referred application context is composed of a number of *(battery-powered) cameras* (such as, drones, micro-drones, insect spy drones, etc.) forming a network able to monitor a high-size area and store images into the cloud. Beside allowing data storage, this server provides an interface to access data and to perform query processing on behalf of the data owner, who administrates and analyzes query results in accordance to specific application-related requirements.

In a similar scenario, classical general-purpose deterministic techniques for query integrity appear little suitable, as the computation required to update the extra data structures could become a bottleneck. To overcome this drawback, the authors of [10] propose a new deterministic technique, which is proved to be more efficient on insertions than the state-of-the-art techniques, and also efficiently supporting range queries, which are the relevant queries in this setting (a user typically looks for video temporal intervals).

The proposed approach works by organizing database tuples associated with F^i in a chain. Given a tuple $f^i_{t_j} = \langle a_1, \ldots, a_p \rangle$, an attribute encoding a link towards the next tuple in the database is added, according to the timestamp value.

Therefore, it holds that: $\hat{f}^i_{t_j} = \langle t_j, attr_2, \ldots, attr_p, MAC \rangle$ where:

- MAC is a message authentication code and is computed by means of the function $HMAC(v, K_i)$ implementing the $HMAC$ protocol with SHA-256 as cryptographic hash function;
- $v = (f^i_{t_j} || e)$, K_i is a secret shared by the s_i camera and the data owner;

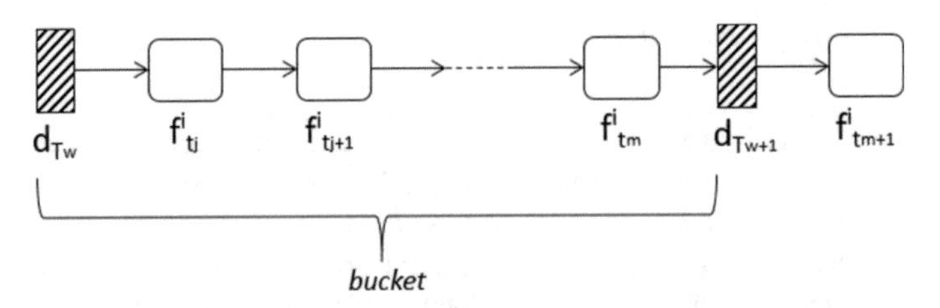

Fig. 4 The structure of the chain, where dashed elements $d_{T_w} \ldots d_{T_{w+1}}$ represent *markers*, whereas the grey ones are normal tuples

- e can be either the next tuple $f^i_{t_{j+1}}$ of s_i, or a special element, called *marker* defined hereafter.

Indeed, the chain is completed with the insertion of dummy entries representing *markers* that are used to both validate the head of the chain and to reduce the integrity verification costs by splitting F^i in time buckets. These elements are pre-added to the database and are known to all the actors involved in the scenario (i.e., they are part of the public scheme of the protocol) (Fig. 4).

Concerning the *markers*, they have the following basic structure: $\hat{d}_{T_w} = \langle T_w, ID_b, MAC_1, \ldots, MAC_n \rangle$ where:

- T_w is the *marker* pre-fixed time (chosen by the owner during the system initialization phase),
- ID_b is the bucket identifier.
- MAC_i is a message authentication code associated with the device s_i and is computed by means of the function $HMAC(v^i, K_i)$
- $v^i = T_w||ID_b||e$, K_i is a secret shared by the s_i camera and the data owner;
- e can be either the next s_i tuple $f^i_{t_{j+1}}$ (the first tuple in the corresponding bucket) or the next *marker* $d_{T_{w+1}}$.

This *flat* structures support insertions in constant time (if a single-block hash computation as unitary cost is considered) opposite to the logarithmic cost of the hash-tree based solutions. This is a relevant aspect, if the devices responsible of insertions (as in our application setting) could perform these operations autonomously, but they could be strictly bound by power consumption and computational resource constraints. All details on the algorithms for the insertion of a new tuple and the verification of a query are described in [10]. Moreover, [12] provides a detailed performance analysis showing the computational advantages of this solution w.r.t. classical ones based on Merkle tree data structures, and a complete security analysis proving the effectiveness of this solution.

4.2 Security via Social Network

In this section, we discuss again about data integrity. Usually, it is obtained by means of digital signature, which allows the user to be aware of the identity of the person who created an electronic document (a text file, an image, a video, etc.) and to ensure that this document has not been altered since its creation.

In [5], the authors propose an alternative to digital signature called *social signature*, consisting in a lightweight procedure to guarantee integrity and authenticity of documents. Differently from digital signature, social signature does not rely on a certification authority, asymmetric cryptography, or signature device such as smart card or USB key. This solution is suitable for closed domains, in which according to legislation [1, 2], electronic signatures are applied to document exchange, for example, between municipal public offices and registered citizens, university and its students, or private company and employees.

Social signature is based on the use of the famous social network Twitter, so that each entity involved in the procedure of generating or verifying a social signature has a Twitter profile.

There are two other entities involved in social signature: the first is called *company* and overviews the whole signature procedure; the second entity is formed by the signers, called *employees* (because it includes all the employees of the company), who use social signature to provide integrity and authenticity of the documents they create, and to verify integrity and authenticity of documents created by other employees of the same domain.

To use social signature, all actors have to carry out the *Registration* procedure, which works as follows.

First, the company creates an account on Twitter. Clearly, this is done by a person who is authorized to act on behalf of the company. Let `@Company` be the username chosen for the account on Twitter. Next, the signers also create an account on Twitter. Suppose that an employee chooses `@Company_Name _Surname` as username.

Each time an employee completes his registration on Twitter, declares a *following* relationship towards `@Company` and vice versa (i.e., `@Company` becomes a follower of the employee account). In this phase, the company is responsible of the verification of the employee identity. Moreover, `@Company` *tweets* the message `#X is Y`, where X (which is hashtagged) is the username of the registered employee and Y is an information identifying the employee. Y is typically the name and surname of the employee; however, further information, such as the employee id, is added to manage cases of homonymy. The employee completes this phase by tweeting the message `I am an employee of #Company` (i.e., the username of the company hashtagged). Finally, the employee receives the software that is used to generate social signatures, which is installed on the computer and/or notebook used by the employee. This software accesses public data contained in Twitter by exploiting Twitter APIs.

Once an employee has completed the registration procedure, he is enabled to create a social signature on a document with scope and validity relevant to the working domain. This is done by the procedure described below. First, the employee runs the social signature software and selects the file to sign. Thereafter, the signer is prompted to enter his Twitter username and password. The signature software computes the hash of the file by the cryptographic hash function SHA-256. Let H be the hexadecimal representation of the resulting digest. Now, the software allows the user to post the tweet `I have signed the document #H`, which is shorter than 140 characters (i.e., the maximum tweet length). In this phase, `@Company` receives the tweet and then tweets the message `@X has signed the document #H`. This message is called *confirmation* tweet.

Any employee or the company itself can verify the social signature generated by an employee through the procedure described below. First, the user selects the file whose signature has to be validated. Then, the software computes the hexadecimal representation H of the digest of the selected file by means of SHA-256 and search for the tweets with hashtag `#H`. If no confirmation tweet from the company with the message `@X has signed the document #H` is found, then the signature verification fails. Otherwise, the signature is considered valid and the identity of the employee who posted this tweet is returned. Observe that, more than one tweet (from different accounts) with this message can be found, this means that more employees have signed this document.

In [5], the authors prove also that social signature provides the basic features typically required to an advanced electronic signature and discuss several attacks that could be done and how they are disarmed.

5 Conclusion

In this chapter, we presented some important milestones of the route followed during the SEBD story by the research group of the University Mediterranea of Reggio Calabria. As depicted throughout the chapter, they studied non-core problems of databases, by focusing their attention towards topics bridging different domains, like social data and cybersecurity. This is a proof that the SEBD community has, since its birth, a very open view of research in the field of databases, thus believing, as the authors of this chapter, that weak ties can be beneficial for every community (as Granovetter taught us [19]). This is the spirit that will drive the future work of the research group of the University Mediterranea of Reggio Calabria during the long and successful future story of SEBD.

References

1. DPCM 22 Febbraio 2005, http://www.agid.gov.it/sites/default/files/leggi_decreti_direttive/dpcm_22_febbraio_2013_-_nuove_regole_tecniche.pdf (2013)
2. Decreto Legislativo 7 Marzo 2005, n. 82, http://www.funzionepubblica.gov.it/media/672080/dlgs-822005-aggiornato.pdf
3. B. Babcock, S. Babu, M. Datar, R. Motwani, J. Widom, Models and issues in data stream system, in *PODS* (2002), pp. 1–16
4. D. Brickley, L. Miller, The Friend of a Friend (FOAF) project, http://www.foaf-project.org/ (2017)
5. F. Buccafurri, L. Fotia, G. Lax, S. Nicolazzo, A. Nocera, A lightweight electronic signature scheme using twitter, in *SEBD* (2015), pp. 160–167
6. F. Buccafurri, G. Lax, L'istogramma nlt: una codifica approssimata di range query gerarchiche, in *SEBD* (2003), pp. 81–92
7. F. Buccafurri, G. Lax, Rappresentazioni compresse di data stream attraverso istogrammi, in *SEBD* (2004), pp. 350–361
8. F. Buccafurri, G. Lax, Approximating sliding windows by cyclic tree-like histograms for efficient range queries. Data Knowl. Eng. **69**(9), 979–997 (2010)
9. F. Buccafurri, G. Lax, S. Nicolazzo, A. Nocera, A Model for Handling Multiple Social Networks and its Implementation. Submitted to the Italian Symposium on Advanced Database Systems (SEBD 2017)
10. F. Buccafurri, G. Lax, S. Nicolazzo, A. Nocera, Completeness, correctness and freshness of cloud-managed data streams, in *Proceedings of the Italian Symposium on Advanced Database Systems (SEBD 2016)* (Lecce, IT, 2016), pp. 134–141
11. F. Buccafurri, G. Lax, S. Nicolazzo, A. Nocera, A model to support design and development of multiple-social-network applications. Inf. Sci. **331**, 99–119 (2016)
12. F. Buccafurri, G. Lax, S. Nicolazzo, A. Nocera, Range query integrity in cloud data streams with efficient insertion, in *Proceedings of the 15th International Conference on Cryptology and Network Security (CANS 2016)*, Milan, Italy (Springer, Berlin, 2016), pp. 719–724
13. F. Buccafurri, G. Lax, A. Nocera, D. Ursino, Discovering hidden me edges in a social internetworking scenario, in *SEBD* (Citeseer, 2012), pp. 15–26
14. F. Buccafurri, G. Lax, A. Nocera, D. Ursino, Discovering missing me edges across social networks. Inf. Sci. **319**, 18–37 (2015)
15. F. Buccafurri, G. Lax, D. Sacca, Progresses on tree-based approaches to improving histogram accuracy, in *SEBD* (2006), pp. 147–158
16. F. Buccafurri, G. Lax, D. Saccà, L. Pontieri, D. Rosaci, Enhancing histograms by tree-like bucket indices. Int. J. Very Large Data Bases **17**(5), 1041–1061 (2008)
17. A. Elmagarmid, P. Ipeirotis, V. Verykios, Duplicate record detection: a survey. IEEE Trans. Knowl. Data Eng. **19**(1), 1–16 (2007)
18. M.M. Gaber, A. Zaslavsky, S. Krishnaswamy, Mining data streams: a review. SIGMOD Rec. **34**(2), 18–26 (2005)
19. M.S. Granovetter, The strength of weak ties. Am. J. Sociol. **78**(6), 1360–1380 (1973)
20. S. Guha, On the space–time of optimal, approximate and streaming algorithms for synopsis construction problems. The VLDB J. **17**(6), 1509–1535 (2008)
21. S. Guha, K. Shim, J. Woo, Rehist: relative error histogram construction algorithms, in *VLDB* (2004), pp. 300–311
22. H.V. Jagadish, N. Koudas, S. Muthukrishnan, V. Poosala, K.C. Sevcik, T. Suel, Optimal histograms with quality guarantees, in *Proceedings 24th International Conference Very Large Data Bases, VLDB* (1998), pp. 275–286
23. N. Koudas, S. Muthukrishnan, D. Srivastava, Optimal histograms for hierarchical range queries (extended abstract), in *Proceedings of the Nineteenth ACM SIGMOD-SIGACT-SIGART Symposium on Principles of Database Systems* (ACM Press, 2000), pp. 196–204
24. F.M. Malvestuto, A universal-scheme approach to statistical databases containing homogeneous summary tables. ACM Trans. Database Syst. (TODS) **18**(4), 678–708 (1993)

25. V.S.A. Menezes, G. Zimbrão, J.M. Souza, Group and link analysis of multi-relational scientific social networks. J. Syst. Softw. **86**(7), 1819–1830 (2013)
26. V. Poosala, P.J. Haas, Y.E. Ioannidis, E.J. Shekita, Improved histograms for selectivity estimation of range predicates, in *Proceedings of the 1996 ACM SIGMOD international conference on Management of data* (ACM Press, New York, 1996), pp. 294–305
27. P. Samarati, Data security and privacy in the cloud, in *ISPEC* (Springer, Berlin, 2014), pp. 28–41
28. I. Sitzmann, P.J. Stuckey, Improving temporal joins using histograms, in *Database and Expert Systems Applications* (2000), pp. 488–498
29. Z. Sun, L. Han, W. Huang, X. Wang, X. Zeng, M. Wang, H. Yan, Recommender systems based on social networks. J. Syst. Softw. **99**, 109–119 (2015)
30. XFN. XHTML Friends Network, http://gmpg.org/xfn (2017)

Confidentiality Protection in Large Databases

Sabrina De Capitani di Vimercati, Sara Foresti, Giovanni Livraga, Stefano Paraboschi and Pierangela Samarati

Abstract A growing trend in today's society is outsourcing large databases to the cloud. This permits to move the management burden from the data owner to external providers, which can make vast and scalable infrastructures available at competitive prices. Since large databases can include sensitive information, effective protection of data confidentiality is a key issue to fully enable data owners to enjoy the benefits of cloud-based solutions. Data encryption and data fragmentation have been proposed as two natural solutions for protecting data confidentiality. However, their adoption does not permit to completely delegate query evaluation at the provider. In this chapter, we illustrate some encryption-based and fragmentation-based solutions for protecting data confidentiality, discussing also how they support query execution.

1 Introduction

Starting from the pioneering Database-as-a-Service (DaaS) paradigm [20], recent years have seen an ever-growing trend towards the outsourcing of large data collections to external providers. By delegating data storage and management to external third parties, data owners can enjoy the immediate benefits of a reduced overhead at their side. The rapid advancements in cloud computing have accelerated this trend: the availability of a rich cloud market allows data owners to store and maintain huge

S. De Capitani di Vimercati · S. Foresti · G. Livraga · P. Samarati (✉)
Università degli Studi di Milano, via Bramante 65, 26013 Crema, Italy
e-mail: Pierangela.Samarati@unimi.it

S. De Capitani di Vimercati
e-mail: Sabrina.decapitani@unimi.it

S. Foresti
e-mail: Sara.Foresti@unimi.it

G. Livraga
e-mail: Giovanni.Livraga@unimi.it

S. Paraboschi
Università degli Studi di Bergamo, via Marconi 5, 24044 Dalmine, Italy
e-mail: parabosc@unibg.it

S. Flesca et al. (eds.), *A Comprehensive Guide Through the Italian Database Research Over the Last 25 Years*, Studies in Big Data 31, DOI 10.1007/978-3-319-61893-7_27

data collections in the cloud at competitive (and typically pay-as-you-go) prices. The benefits of data outsourcing are multiple and not confined to economic factors, ranging from high service availability, to improved scalability and elasticity. However, no lunch comes for free, and one of the major issues arising when resorting to the cloud for data storage and management is the inevitable loss of control by the owner over her own data, and consequent risks to data protection in this context. Cloud providers are typically considered *honest-but-curious*, that is, trustworthy for correctly managing data but not trusted for accessing their content. Besides the intuitive need for protecting data confidentiality against external unauthorized subjects, it is then essential to protect it also against the provider itself.

Encryption represents a natural means for providing data confidentiality: by wrapping data with a layer of encryption, data are made unintelligible to unauthorized parties (which do not know the encryption key). If data are encrypted by their owner before being outsourced to the cloud, encryption also effectively provides data confidentiality against the storage provider itself [26]. Encrypting an entire data collection however can represent an overdo in many scenarios, where what is sensitive is the association among data items, rather than the data themselves. For instance, while knowing that a hospitalized patient is named Alice, and that a patient at the same hospital has flu might not be sensitive, the fact that Alice suffers from flu can be sensitive. The sensitive association between patients' names and diseases can be protected by simply storing names and diseases in two unlinkable data chunks, reducing the need for encryption. Following this intuition, the research community has proposed to combine encryption with data fragmentation [15]. In a nutshell, fragmentation-based techniques protect sensitive associations among data by splitting them into different unlinkable fragments (e.g., [1, 5, 7]).

Both encryption and fragmentation, while proved effective for confidentiality protection, cam impair query execution or make it more complex. Indeed, the cloud provider neither knows the encryption keys used to protect sensitive data nor can join fragments. Hence, it cannot evaluate user queries formulated on the original (plaintext and non-fragmented) relation. A naive solution would require the provider to return the entire encrypted or fragmented relation to the requesting user who (being authorized to issue queries) can decrypt or recombine it, to evaluate the query locally. This solution is however not viable, as it would nullify the benefits of resorting to the cloud. We will see how the use of indexes (metadata) associated with the encrypted relation or specific strategies allow the (partial) query evaluation directly at the provider, without requiring to decrypt data or join fragments (e.g., [1, 5, 7, 13, 20]).

In this chapter, we illustrate encryption-based and fragmentation-based techniques for protecting the confidentiality of large data collections outsourced at cloud providers, discussing also their support for query execution. Clearly, data confidentiality is only one aspect of the more general problem of ensuring proper protection to data. Many other problems have been investigated (e.g., [16, 22, 23]) but they are outside the scope of this chapter. The remainder of this chapter is organized as follow. Section 2 describes encryption-based solutions, while Sect. 3 presents fragmentation-based techniques, describing both the protection model and query evaluation approaches. Section 4 finally concludes the chapter.

2 Encryption-Based Approaches

A natural solution for protecting data confidentiality consists in wrapping data with an *encryption* layer. Not knowing the encryption key, unauthorized subjects cannot decrypt the data and access their plaintext content. In this section, we discuss the use of encryption to protect the confidentiality of outsourced data, and illustrate different techniques that can be used to (partially) delegate query evaluation on encrypted data to the cloud provider. We first focus on the index-based approach (Sect. 2.1), and then illustrate solutions that support the execution of queries directly over encrypted data (Sect. 2.2). For simplicity, we focus on outsourced data organized as a relation r defined over a relational schema $R(a_1, \ldots, a_n)$, with the note that the discussed protection techniques can also be applied to different (semi-)structured data models. We further note that the encryption schemas supporting keyword-based searches over generic encrypted data are outside the scope of this chapter and therefore we do not discuss them.

2.1 Encryption and Indexes

Data can be encrypted with symmetric as well as asymmetric encryption algorithms but, since symmetric encryption is typically cheaper, many proposals adopt symmetric encryption [26]. The outsourced relation can be encrypted at different granularity levels: cell level, attribute level, tuple level, or relation level. The chosen granularity level impacts on the query evaluation process, with consequences on its performance. For instance, relation-level encryption would require to return to the requesting user the entire encrypted relation. In general, finer granularity levels enable users to be more precise in downloading the encrypted content of interest but, on the other hand, cause a high overhead due to encryption/decryption operations. Viceversa, coarser granularity levels imply a lower overhead for encryption/decryption operations, but require to download larger encrypted chunks than necessary for query evaluation. Tuple-level encryption represents a good tradeoff between the overhead caused by encryption/decryption operations and query execution efficiency [26].

With tuple-level encryption, a relation r defined over relation schema $R(a_1, \ldots, a_n)$ is outsourced at the cloud provider as an encrypted relation r^e defined over schema R^e(<u>`tid`</u>, *enc*), with `tid` the primary key added to the encrypted relation and *enc* the encrypted tuple. Each tuple t in r is represented as an encrypted tuple t^e in r^e, where t^e[`tid`] is a random identifier and $t^e[enc] = E_k(t)$ is the encrypted tuple content, with E a symmetric encryption function with key k. To illustrate, consider

FINANCIALDATA

SSN	Name	Race	Job	Salary	Ins
123-45-6789	Alice	white	teacher	40K	160
234-56-7890	Bob	while	farmer	25K	100
345-67-8901	Carol	asian	nurse	20K	100
456-78-9012	David	black	lawyer	50K	200
567-89-0123	Eric	black	secretary	20K	100
678-90-1234	Fred	asian	lawyer	40K	180

(a)

c_1={SSN}
c_2={Name, Salary}
c_3={Name, Ins}
c_4={Salary, Ins}
c_5={Race, Job, Ins}

(b)

Fig. 1 An example of a relation (**a**) and of confidentiality constraints over it (**b**)

FINANCIALDATA^e

tid	enc
1	4tBf
2	lkG7
3	wF4t
4	m;Oi
5	n:8u
6	xF-g

(a)

FINANCIALDATA$_\iota^e$

tid	enc	$\mathbf{I}_r$	$\mathbf{I}_j$	$\mathbf{I}_s$	$\mathbf{I}_i$
1	4tBf	α	δ	ζ	ξ
2	lkG7	α	δ	η	ν
3	wF4t	β	ϵ	θ	ν
4	m;Oi	γ	ϵ	κ	ξ
5	n:8u	γ	δ	λ	ν
6	xF-g	β	ϵ	μ	ξ

(b)

Fig. 2 An example of encrypted (**a**) and encrypted and indexed (**b**) versions of relation FINANCIALDATA in Fig. 1a

relation FINANCIALDATA in Fig. 1a, reporting financial information about a set of individuals. Figure 2a illustrates an example of an encrypted version of it.

Query evaluation. To enable query evaluation over encrypted data at the cloud provider, without the need of decryption, index-based solutions complement the encrypted relation with indexes. Indexes are metadata that preserve some of the properties of the attributes on which they have been defined, and therefore can be used for query evaluation. Indexes are represented as additional attributes in the encrypted (and indexed) relation r_ι^e, which is then defined over schema R_ι^e(tid, *enc*, $I_{i_1}, \ldots, I_{i_j}$), with $I_{i_l}, l = 1, \ldots, j$ the index defined over attribute a_{i_l} in R. Note that not all attributes must be associated with an index – on the contrary, only those that are expected to be involved in conditions in query evaluation need to be indexed. Figure 2b illustrates an example of an encrypted and indexed version of relation FINANCIALDATA in Fig. 1, with indexes over attributes Race, Job, Salary, and Ins. For simplicity, indexes are represented in the figure with Greek letters.

Different indexing techniques have been proposed, depending on the mapping between plaintext and index values and on the supported conditions [17, 18]. Equality conditions of the form $a = v$, with v a value in the domain of a, are supported by many indexing techniques, such as *encryption-based* [13], *bucket-based* [20], and *hash-based* [13] indexes. Encryption-based indexes associate index value $E_k(t[a])$ with plaintext value $t[a]$, where E is a symmetric encryption scheme and k is the encryption key. Bucket-based indexes partition the domain of an attribute a in disjoint subsets of contiguous values each of which is associated with a *label*. A plaintext value $t[a]$ is represented in the index with the label l of the partition to which $t[a]$ belongs. Hash-based indexes instead adopt a secure hash function h that generates collisions, and the index value associated with plaintext value $t[a]$ is computed as $h(t[a])$. For instance, consider the encrypted and indexed relation in Fig. 2: index I_r is an encryption-based index over attribute Race of relation FINANCIALDATA in Fig. 1a; index I_i is a partition-based index over attribute Ins, where the domain has been partitioned in two intervals: [100, 150] with label ν, and [151,200] with label ξ; and index I_j is a hash-based index over attribute Job, where the hash function is defined as follows: $h(\text{teacher}) = h(\text{farmer}) = h(\text{secretary}) = \delta$ and $h(\text{nurse}) = h(\text{lawyer}) = \epsilon$.

Range conditions of the form a IN $[v_i, v_j]$, with a an attribute and $[v_i, v_j]$ a range in the domain of a are supported by bucket-based indexes, if the labels associated with them are defined so to preserve the order among attribute values (leaking however the order of attribute values to the provider). An indexing technique specifically designed to support range queries is based on a B+-tree index defined over the plaintext attribute to be indexed [13]. The B+-tree index is represented at the provider through an additional encrypted relation. Alternative solutions to support range conditions rely on *Order Preserving Encryption Schemas* (OPES [2, 28]).

Aggregate operators [19, 21] such as SUM and AVG are supported by indexes defined using homomorphic encryption [4, 11, 12, 19, 25], which allows the evaluation of arithmetic operators directly on encrypted data. The downside of these indexes is represented by the high computational overhead caused by homomorphic encryption schemes.

Given an encrypted and indexed relation r_ι^e over schema R^e(<u>tid</u>, *enc*), a query q formulated by the user on the original relation schema $R(a_1, \ldots, a_n)$ is translated into two queries q_p and q_u, operating at the provider and at the user sides, respectively. The original conditions in q are translated into equivalent conditions on the indexes in R^e to define the query q_p operating at the provider. Query q_u operates on the result of q_p to evaluate conditions that cannot be delegated to the provider (e.g., conditions over attributes that are not associated with indexes), and to filter the spurious tuples. For instance, with reference to the encrypted and indexed relation in Figs. 2b, 3 illustrates an example of query execution over the encrypted and indexed relation in Fig. 2b.

q_u: SELECT Name	q_p: SELECT tid, enc
FROM *Decrypt*(R_p.*enc*,k)	FROM FinancialData^e
WHERE Job='teacher'	WHERE $I_r = \alpha$ AND $I_j = \delta$

Fig. 3 Execution of query "SELECT Name FROM FinancialData WHERE Race='white' AND Job='teacher'" over the encrypted and indexed relation of Fig. 2b as subqueries at the provider side (q_p) and at the user side (q_u)

2.2 Encrypted Data Processing

An alternative to indexes to partially delegate query execution to the provider consists in adopting encryption techniques that support the execution of operations or the evaluation of conditions directly over encrypted data. For instance, deterministic encryption supports the evaluation of equality conditions, Order Preserving Encryption (OPE) supports the evaluation of range conditions (e.g., [2, 28]), and fully homomorphic encryption supports the evaluation of any function (e.g., [19]). Taking these encryption techniques as basic building blocks, some encrypted database systems have been developed (e.g., CryptDB [24], Monomi [27], and Cipherbase [3]), which support query processing over encrypted data. Among these systems, we describe CryptDB since it specifically focuses on query processing. Monomi focuses more on the physical design of the encrypted database with respect to a given workload and Cipherbase, being based on a trusted hardware on the untrusted cloud provider (which can perform arbitrary computation) focuses on how and where (client, server, trusted hardware) execute a computation.

CryptDB chooses a different encryption schema (depending on the conditions to be supported) for each attribute, and applies encryption at the cell level. To support different operations/conditions on the same attribute, multiple encryption layers are wrapped around a cell, forming an onion-like structure [24]. Note that the encryption layers are the same for all the cells in the same column, but they may vary from an attribute to another (depending on the kinds of queries to be supported). The outermost level features the strongest encryption (i.e., *randomized encryption*, a probabilistic scheme where two equal values are mapped onto different ciphertexts with non-negligible probability [24]), while the innermost level represents plaintext data. Proceeding from the outermost to the innermost layers, the adopted encryption scheme provides weaker security guarantees but supports more computations over encrypted data. For instance, attribute `Salary` in relation FINANCIALDATA in Fig. 1 can be encrypted with randomized encryption (to maximize protection in storage), deterministic encryption (to support equality conditions), and OPE (to support aggregates and range conditions), as illustrated in Fig. 4.

Encryption is dynamically regulated depending on the operations to be evaluated, by removing encryption layers. For instance, consider a query q : SELECT AVG(`Salary`) FROM `FinancialData`, and assume that the encryption layers are as in Fig. 4. To enable the cloud provider to compute the average salary, the randomized and deterministic encryption layers are removed. Note that once a layer

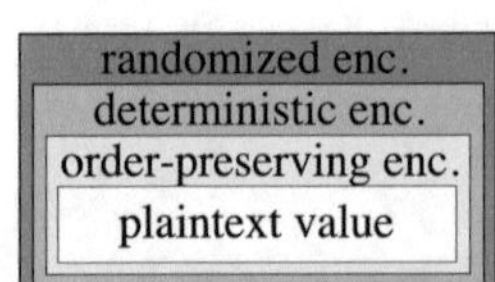

Fig. 4 An example of encryption layers adopted by CryptDB [24]

of encryption is removed from an attribute, it cannot be restored as data have been exposed to the provider, which cannot be considered oblivious.

Query evaluation. Query execution with CryptDB assumes a trusted proxy mediating the communications between the users and the cloud provider. The proxy stores a secret master key k, the database schema, and keeps track of the current encryption layers protecting each attribute in the relation. Given a query q issued by a user, the proxy rewrites it into an equivalent query $\hat{q}$ operating over the encrypted relation choosing, for each attribute on which the query operates, the most appropriate encryption layer in its onion structure. If one (or more) encryption layer(s) should be removed from one (or more) attribute(s), the proxy issues an UPDATE query removing it (them). The proxy forwards $\hat{q}$ to the cloud provider, which executes it. The provider then returns the encrypted result of $\hat{q}$ to the proxy, which finally decrypts it and sends the plaintext result to the user.

3 Fragmentation-Based Approaches

Besides complicating query execution, encryption could also be an overdo since, in many scenarios, what is sensitive is the *association* among data (e.g., the identity of a patient and her salary), rather than the data items singly taken. In these cases, confidentiality of associations can be more conveniently protected by storing different portions of data in different, non-linkable fragments. In the context of fragmentation, sensitive attributes and sensitive associations among attributes can be formulated as *confidentiality constraints*, that is, sets of attributes whose joint visibility should be protected [1]. Singleton sets (*singleton constraints*) correspond to sensitive attributes; non-singleton sets (*association constraints*) correspond to sensitive associations. Figure 1b illustrates a set $\mathcal{C}$ of confidentiality constraints over relation FINANCIALDATA in Fig. 1a. In particular, c_1 is a singleton constraint stating that the SSN of the individuals is sensitive per se; $c_2, \ldots, c_5$ are association constraints stating that: the salary and insurance of an individual are sensitive (c_2 and c_3), the associations between a certain salary and the insurance (c_4), and among the race, job, and insurance of an individual (c_5) are sensitive. In this section, we illustrate three solutions that rely on fragmentation for protecting data confidentiality. We focus on solutions that assume attributes to be independent, while noting that fragmentation can also consider dependencies and correlations among attributes, which could introduce inferences channels [14].

3.1 *Two Can Keep a Secret*

The first approach protecting data confidentiality through fragmentation guarantees data protection splitting the original relation into two fragments that do not include sensitive attributes or associations. The two fragments are guaranteed to be unlinkable as they are stored at two non-communicating providers [1].

Confidentiality constraints are satisfied by properly combining fragmentation and *encoding* [1]. Encoding an attribute a consists in splitting it into two attributes a^i and a^j, both necessary to reconstruct the values of a (i.e., $a = a^i \diamond a^j$, with $\diamond$ a non-invertible composition operator). Encryption represents a possible encoding technique, placing the ciphertext in a^i and the encryption key in a^j, with $\diamond$ the encryption algorithm. For simplicity, we will assume encryption as the adopted encoding technique.

A relation r over a schema R fragmented according to the proposal in [1] produces a fragmentation $\mathcal{F} = \{F_1, F_2, E\}$, with $F_1, F_2, E \subseteq R$. F_1 and F_2 are the sets of attributes represented in plaintext in the two fragments, while E is the set of encrypted attributes. Singleton confidentiality constraints are satisfied by encrypting the sensitive attributes (i.e., by placing them into E). Association constraints are typically satisfied by splitting the attributes they include between F_1 and F_2. Being the number of fragments fixed to two, it might happen that an attribute cannot be stored at any of the two fragments without violating some constraints. In such a situation, the attribute is encrypted. To illustrate, consider constraints c_2, c_3, and c_4 in Fig. 1b: it is not possible to split attributes Name, Salary, and Ins between two fragments without violating a constraint. A fragmentation $\mathcal{F} = \{F_1, F_2, E\}$ is *correct* iff the following two conditions hold: *(i)* $\forall c \in \mathcal{C} : c \not\subseteq F_1, c \not\subseteq F_2$ (*confidentiality*); and *(ii)* $F_1 \cup F_2 \cup E \ = R$ (*completeness*). Condition *(i)* ensures that neither F_1 nor F_2 store all the attributes in a confidentiality constraint in plaintext, while condition *(ii)* ensures that all attributes of the original relation are included in the fragmentation (i.e., no information is lost due to fragmentation). A correct fragmentation guarantees that sensitive values and associations are not accessible to non-authorized users (since they do not know the encryption key, and the two providers do not communicate), and that the original relation r can be reconstructed by authorized users from $\mathcal{F}$. Considering the relation and the confidentiality constraints in Fig. 1, an example of a correct fragmentation is $\mathcal{F} = \{F_1, F_2, E\}$, with F_1 = {Name, Race, Job}, F_2 = {Salary}, E = {SSN, Ins}.

At the physical level, fragments F_1 and F_2 are represented by two *physical fragments* F_1^e and F_2^e, where each physical fragment F_i^e stores the attributes in F_i in plaintext, a tuple identifier tid, and the encrypted attribute values (or the corresponding keys). The tuple identifier is needed for authorized users to recombine the content of the fragments, to reconstruct the original relation. Each tuple t in r is represented by a tuple t_1^e in F_1^e and a tuple t_2^e in F_2^e sharing the tuple identifier tid. Tuples t_1^e and t_2^e include in plaintext the values of the attributes in F_1 and F_2, respectively. Also, tuple t_1^e includes the encrypted values of the attributes in E, while t_2^e includes the corresponding encryption keys (or viceversa). Figure 5 illustrates the physical fragments F_1^e and F_2^e representing a correct fragmentation of relation FINANCIALDATA in Fig. 1.

Given a relation schema R and a set $\mathcal{C}$ of confidentiality constraints over it, there might exist different correct fragmentations satisfying $\mathcal{C}$. For instance, a fragmentation where fragments F_1 and F_2 are empty, and $E = R$ is clearly correct, but is

F_1^e

tid	Name	Race	Job	SSN[1]	Ins[1]
1	Alice	white	teacher	$E(123\text{-}45\text{-}6789, k_{\mathtt{SSN}}1)$	$E(150, k_{\mathtt{Ins}}1)$
2	Bob	while	farmer	$E(234\text{-}56\text{-}7890, k_{\mathtt{SSN}}2)$	$E(100, k_{\mathtt{Ins}}2)$
3	Carol	asian	nurse	$E(345\text{-}67\text{-}8901, k_{\mathtt{SSN}}3)$	$E(100, k_{\mathtt{Ins}}3)$
4	David	black	lawyer	$E(456\text{-}78\text{-}9012, k_{\mathtt{SSN}}4)$	$E(200, k_{\mathtt{Ins}}4)$
5	Eric	black	secretary	$E(567\text{-}89\text{-}0123, k_{\mathtt{SSN}}5)$	$E(100, k_{\mathtt{Ins}}5)$
6	Fred	asian	lawyer	$E(678\text{-}90\text{-}1234, k_{\mathtt{SSN}}6)$	$E(180, k_{\mathtt{Ins}}6)$

F_2^e

tid	Salary	SSN[2]	Ins[2]
1	40K	$k_{\mathtt{SSN}}1$	$k_{\mathtt{Ins}}1$
2	25K	$k_{\mathtt{SSN}}2$	$k_{\mathtt{Ins}}2$
3	20K	$k_{\mathtt{SSN}}3$	$k_{\mathtt{Ins}}3$
4	50K	$k_{\mathtt{SSN}}4$	$k_{\mathtt{Ins}}4$
5	20K	$k_{\mathtt{SSN}}5$	$k_{\mathtt{Ins}}5$
6	40K	$k_{\mathtt{SSN}}6$	$k_{\mathtt{Ins}}6$

Fig. 5 Two can keep a secret: an example of a correct fragmentation of relation FINANCIALDATA in Fig. 1a

typically undesirable since queries can be evaluated at the user side only. The authors of [1] then propose a metric to evaluate the quality of a fragmentation (where an optimal fragmentation minimizes the cost of evaluating a sample query workload on the user-side), and a heuristic algorithm to solve the (NP-hard) problem of computing such an optimal fragmentation.

Query evaluation. Since fragmentation should be transparent to final users, queries are formulated over the original relation r, and then are translated into a set of equivalent queries operating on the fragmentation $\mathcal{F}$. The intuition behind such query translation process is that all conditions operating on attributes stored plaintext at F_1^e (F_2^e, resp.) can be easily evaluated by the provider storing F_1^e (F_2^e, resp.). All conditions over attributes in E or over pairs of attributes split between F_1 and F_2 cannot be evaluated by the providers; these conditions are evaluated at the user-side. Given a query q, it is then translated into three queries: q_1, which can be evaluated by the provider storing F_1^e; q_2, which can be evaluated by the provider storing F_2^e; and q_u, which must be evaluated by the user on the results of q_1 and q_2. Naturally, the translation must guarantee equivalence (i.e., the evaluation of q_1, q_2, and q_u must produce the same result as q), and it should push as much computation as possible to the providers to limit the burden at the user side. The query evaluation can follow two different strategies: the *parallel* strategy, where each provider evaluates its query and returns the result to the user, who joins them and evaluates query q_u to finally obtain the query result; and the *sequential* strategy where one of the two providers goes first and returns the result of its query to the user, who then passes the tuple identifiers in the query result to the second provider, and the user finally combines the two results of q_1 and q_2 and evaluates q_u. Both the parallel and the sequential strategies permit to correctly evaluate a query q on a fragmentation $\mathcal{F} = \{F_1, F_2, E\}$. The parallel strategy, while reducing response time (the providers execute their queries at the same time), is likely to cause a higher communication costs. The sequential strategy may instead cause a delay in obtaining the final query result since the results from a provider are needed before a query is sent to the other provider.

Figure 6 illustrates an example of query execution over the fragments in Fig. 5. Here, condition `tid` IN $\{4, 6\}$ in q_2 of the sequential strategy is needed to consider in the execution of q_2 only the tuples returned by q_1.

Parallel strategy	Sequential strategy
q_1: SELECT tid, Name, $\texttt{Ins}^1$ FROM F_1^e WHERE Job='lawyer'	q_1: SELECT tid, Name, $\texttt{Ins}^1$ FROM F_1^e WHERE Job='lawyer'
q_2: SELECT tid, $\texttt{Ins}^2$ FROM F_2^e WHERE Salary=40K	q_2: SELECT tid, $\texttt{Ins}^2$ FROM F_2^e WHERE tid IN {4,6} AND Salary=40K
q_u: SELECT Name FROM R_1 JOIN R_2 ON R_1.tid=R_2.tid WHERE *Decrypt*($\texttt{Ins}^1$, $\texttt{Ins}^2$)=180	q_u: SELECT Name FROM R_1 JOIN R_2 ON R_1.tid=R_2.tid WHERE *Decrypt*($\texttt{Ins}^1$, $\texttt{Ins}^2$)=180

Fig. 6 Execution of query "SELECT Name FROM FinancialData WHERE Job='lawyer' AND Salary = 40K AND Ins = 180" over the fragments of Fig. 5 as subqueries at the providers side (q_1 and q_2) and at the user side (q_u) with parallel and sequential strategies

3.2 Multiple Fragments

Two can keep a secret approach guarantees the unlinkability of fragments, needed to ensure satisfaction of constraints, by assuming the two providers to neither communicate nor collude. However, this assumption is difficult to enforce in practice, and thus reduces the applicability of this model in real world scenarios. The *multiple fragments* approach overcomes this assumption by defining an arbitrary number of uninkable fragments [5].

Unlinkability among fragments is guaranteed by requiring that no plaintext attribute is included in more than one fragment. If each fragment singly taken satisfies the constraints, then all fragments of a fragmentation can be stored at the same provider without risks for confidentiality.

Confidentiality constraints are satisfied (like in the proposal in [1]) by properly combining fragmentation and encryption. More precisely, singleton constraints are satisfied by encrypting the attribute they involve. Association constraints can instead be satisfied by either encrypting at least one of the attributes they include (satisfaction through encryption), or storing these attributes in different fragments (satisfaction through fragmentation). A fragmentation $\mathcal{F} = \{F_1, \ldots, F_m\}$ is *correct* iff the following two conditions hold: *(i)* $\forall c \in \mathcal{C}$, $\forall F \in \mathcal{F}$: $c \not\subseteq F$ (*confidentiality*); *(ii)* $\forall F_i, F_j \in \mathcal{F}, i \neq j$: $F_i \cap F_j = \emptyset$ (*unlinkability*). Condition *(i)* ensures that no fragment in $\mathcal{F}$ can contain all the attributes included in a confidentiality constraint. Condition *(ii)* ensures instead that all fragments in the fragmentation are disjoint. For instance, $\mathcal{F}$={{Name,Job}, {Salary}, {Race,Ins}} is a correct fragmentation of relation FINANCIALDATA in Fig. 1a with respect to the constraints in Fig. 1b.

The multiple fragments approach has two immediate advantages over the solution in [1]. First, the entire fragmentation can be stored at a single cloud provider (as fragments $F_1, \ldots, F_n$ of $\mathcal{F}$ are disjoint), thus removing the need for having multiple non-communicating providers. Second, being the fragmentation not limited to two

F_1^e

salt	enc	Name	Job
s_{11}	xTb:	Alice	teacher
s_{12}	o;!G	Bob	farmer
s_{13}	Ap'L	Carol	nurse
s_{14}	.u7t	David	lawyer
s_{15}	y"e3	Eric	secretary
s_{16}	(l1!	Fred	lawyer

F_2^e

salt	enc	Salary
s_{21}	hg5=	40K
s_{22}	mB71	25K
s_{23}	:k?2	20K
s_{24}	Q14,	50K
s_{25}	-kGd	20K
s_{26}	p[Mz	40K

F_3^e

salt	enc	Race	Ins
s_{31}	bP5	white	160
s_{32}	*Cx	white	100
s_{33}	1Bny	asian	100
s_{34}	Oj)6	black	200
s_{35}	vT7/	black	100
s_{36}	l1fY	asian	180

Fig. 7 Multiple fragments: an example of a correct fragmentation of relation FINANCIALDATA in Fig. 1a

fragments, it is always possible to satisfy an association constraint through fragmentation, hence limiting encryption to the satisfaction of singleton constraints only. Such an approach increases the *visibility* over the data, improving the performance in accessing data (e.g., for query evaluation): the plaintext inclusion of an attribute a in a fragment F allows for the evaluation of conditions over a directly at the cloud provider storing F. The multiple fragments approach aims therefore at computing fragmentations that *maximize plaintext visibility*, that is, where each attribute not included in a singleton constraint is plaintext represented in *at least* one fragment. Note that, combining this requirement with the unlinkability condition, each attribute not involved in a singleton constraint is plaintext included in *exactly* one fragment. For instance, $\mathcal{F}$={{Name,Job}, {Salary}, {Race,Ins}} is a correct fragmentation of relation FINANCIALDATA in Fig. 1 that maximizes visibility.

At the physical level, each fragment F_i is represented by a *physical fragment* F_i^e storing: the attributes in F_i in plaintext; the attributes in $R \setminus F_i$ in encrypted form; and a primary key salt containing random values. Each tuple t in r is represented by a tuple t_i^e in each physical fragment F_i^e, where $t_i^e[a]$=$t[a]$ for all $a \in F_i$; t_i^e[salt] is a random nonce; and $t_i^e[\texttt{encr}] = E_k(t[a_j, \ldots, a_k] \oplus t^e[\texttt{salt}])$, with $\{a_j, \ldots, a_k\} = R \setminus F_i$. Since each physical fragment stores, either plaintext or encrypted, all the attributes in R, every query can be evaluated on a single fragment. Figure 7 illustrates the physical fragments of a correct fragmentation of the relation in Fig. 1a with respect to the constraints in Fig. 1b.

Given a relation schema R and a set $\mathcal{C}$ of confidentiality constraints over it, there might exist different correct fragmentations (i.e., satisfying $\mathcal{C}$) that maximize visibility. For instance, a fragmentation where all attributes not involved in singleton constraints are stored at a different fragment would be correct but likely undesirable, complicating the execution of queries involving more than one attribute. Different metrics have therefore been proposed to evaluate the quality of a fragmentation $\mathcal{F}$, aimed at minimizing the number of fragments in $\mathcal{F}$ (e.g., [5, 10, 14]), or the cost needed to execute a query workload (e.g. [6, 8]).

Query evaluation. Since all physical fragments of a fragmentation include, either plaintext or encrypted, all the attributes of R, a query q can be executed over any fragment. However, from a user's point of view, it is clearly more convenient to use the fragment that permits to delegate to the cloud provider most of the query

q_u: SELECT Name	q_p: SELECT salt, enc, Name
FROM *Decrypt*(R_p.enc⊕salt,k)	FROM F_1^e
WHERE Salary=40K	WHERE Job='teacher'

Fig. 8 Execution of query "SELECT Name FROM FinancialData WHERE Salary = 40K AND Job='teacher'" over fragment F_1^e of Fig. 7 as subqueries at the provider's side (q_p) and at the user's side (q_u)

evaluation (i.e., the fragment storing in plaintext most of the attributes in which the conditions of q operate). The original query is then translated into an equivalent pair of queries q_p and q_u, operating at the provider and at the user sides, respectively. Query q_p, which can include conditions over the attributes plaintext represented in the fragment, is sent to the provider, which returns the result to the user. The user decrypts the encrypted attributes in the result of q_p (if any), and evaluates q_u, evaluating conditions over the attributes not plaintext included in the fragment. With reference to the relation in Fig. 1a and the fragmentation in Figs. 7, 8 illustrates an example of query execution, assuming to choose fragment F_1^e.

3.3 Keep a Few

While both the two can keep a secret and the multiple fragments approaches build upon a combination of fragmentation and encryption, the *keep a few* approach [7, 9] completely departs from encryption. Such an approach can be adopted when the data owner (or a trusted third party) is available for storing a limited portion of the data. Confidentiality constraints are then satisfied by combining fragmentation with owner-side storage. More precisely, singleton constraints are satisfied by storing at the owner side the attribute they involve. Similarly, association constraints are satisfied by storing at least one of the attributes they include at the owner side. A fragmentation $\mathcal{F}$ of a relation r defined over relation schema R is then a pair $\mathcal{F} = \langle F_o, F_p \rangle$ of fragments, with $F_o, F_p \subseteq R$, and where F_o is stored at the data owner and F_p is outsourced at a cloud provider. A fragmentation $\mathcal{F} = \langle F_o, F_p \rangle$ is correct iff the following two conditions hold: *(i)* $\forall c \in \mathcal{C}, c \not\subseteq F_p$ (*confidentiality*); *(ii)* $F_o \cup F_p = R$ (*losslessness*). Condition *(i)* ensures that F_p does not contain all the attributes involved in a confidentiality constraint. Note that the confidentiality condition is not needed to hold on F_o, since this fragment is stored at the owner and hence is not accessible to non-authorized users. Condition *(ii)* ensures that the fragmentation includes all the attributes in R, guaranteeing that no information is lost due to fragmentation. Note that, since including the same attribute in both F_o and F_p would be redundant, the two fragments are typically required to be disjoint (i.e., $F_p \cap F_o = \emptyset$). For instance, $\mathcal{F} = \langle F_o, F_p \rangle$ with $F_o = \{$SSN, Name, Ins$\}$ and $F_p = \{$Race, Job, Salary$\}$ is an example of a correct and non-redundant fragmentation

F_o^e

tid	SSN	Name	Ins
1	123-45-6789	Alice	160
2	234-56-7890	Bob	100
3	345-67-8901	Carol	100
4	456-7 8-9012	David	200
5	567-89-0123	Eric	100
6	678-90-1234	Fred	180

F_p^e

tid	Race	Job	Salary
1	white	teacher	40K
2	while	farmer	25K
3	asian	nurse	20K
4	black	lawyer	50K
5	black	secretary	20K
6	asian	lawyer	40K

Fig. 9 Keep a few: an example of a correct fragmentation of relation FINANCIALDATA in Fig. 1a

of relation FINANCIALDATA in Fig. 1a, with respect to the confidentiality constraints in Fig. 1b.

At the physical level, fragments F_p and F_o are translated into two *physical fragments* F_p^e and F_o^e, including all attributes (plaintext, as no encryption is adopted with this approach) in F_p and F_o, respectively. Both physical fragments are also enriched with a common tuple identifier `tid` that authorized users can use to correctly reconstruct the original relation. Figure 9 illustrates the physical fragments of a correct fragmentation of relation FINANCIALDATA in Fig. 1a.

Given a relation schema R and a set $\mathcal{C}$ of confidentiality constraints over it, there can exist different correct and non-redundant fragmentations. For instance, a fragmentation $\mathcal{F} = \langle F_o, F_p \rangle$ where $F_p = \emptyset$ and $F_o = R$ is correct and non-redundant, but likely to be undesirable since no attribute is outsourced to the cloud. Several metrics have been proposed to evaluate the quality of a fragmentation, aimed at minimizing the burden for the data owner, which can be measured in terms of the number/size of the attributes stored owner-side at F_o, or of the computational overhead left at the data owner based on a query workload [7, 9].

Query evaluation. A query q formulated over the original relation schema R is transformed into two equivalent queries q_p and q_o, operating at the provider and at the owner sides, respectively: q_p includes conditions operating only on attributes in F_p; q_o includes conditions operating on attributes in F_o (or comparing attributes in the two fragments). The evaluation of q can follow two strategies, depending on the order in which q_p and q_o are evaluated. In the *provider-owner* strategy, the provider evaluates q_p and returns the result to the owner, who joins it with its fragment and evaluates q_o to obtain the result of q. In the *owner-provider* strategy, the owner evaluates conditions in q that involve only attributes in F_o, and sends then the identifier of the tuples satisfying such conditions to the provider. The provider evaluates q_p on these tuples, and returns the result to the owner. The data owner joins her fragment with the result computed by the provider and evaluates the conditions that involve attributes in both fragments, obtaining the result of q. Figure 10 illustrates an example of query execution according to these two strategies.

While both the provider-owner strategy and the owner-provider strategy correctly compute the result of a query q, the latter can leak sensitive information. In fact, if the provider knows the original query q, it can learn which are the tuples in F_p that satisfy the conditions evaluated by the data owner (the individuals with `Ins` equal

Provider-Owner strategy	Owner-Provider strategy
q_p: SELECT tid, Name FROM F_p^e WHERE Salary=20K	q_o: SELECT tid FROM F_o^e WHERE Ins=100
q_o: SELECT Name FROM F_p^e JOIN R_p ON F_p^e.tid=R_p.tid WHERE Ins=100	q_p: SELECT tid, Name FROM F_p^e WHERE (tid IN {2,3,5}) AND Salary=20K
	q_{po}: SELECT Name FROM F_o^e JOIN R_p ON F_o^e.tid=R_p.tid

Fig. 10 Execution of query "SELECT Name FROM FinancialData WHERE Salary=20K AND Ins=180" over the fragments of Fig. 9 as subqueries at the provider side (q_p) and at the owner side (q_o and q_{po}), with provider-owner and owner-provider strategies

to 100 in our example), even if the provider is not authorized to see the content of the attributes in F_o.

4 Conclusions

We have illustrated encryption-based and fragmentation-based solutions for protecting confidentiality in large databases when they are outsourced to the cloud. Since both encryption and fragmentation complicate or even prevent query execution at the provider, we have also illustrated some of the existing approaches that enable (partial) query execution directly at the provider, without the need for decrypting encrypted data, or of joining fragments.

Acknowledgements This work was supported in part by the EC within the H2020 under grant agreement 644579 (ESCUDO-CLOUD), and within the FP7 under grant agreement 312797 (ABC4EU).

References

1. G. Aggarwal, M. Bawa, P. Ganesan, H. Garcia-Molina, K. Kenthapadi, R. Motwani, U. Srivastava, D. Thomas, Y. Xu, Two can keep a secret: a distributed architecture for secure database services, in *Proceedings of CIDR* (Asilomar, CA, USA, 2005)
2. R. Agrawal, J. Kierman, R. Srikant, Y. Xu, Order preserving encryption for numeric data, in *Proceedings of SIGMOD* (Paris, France, 2004)
3. A. Arasu, S. Blanas, K. Eguro, M. Joglekar, R. Kaushik, D. Kossmann, R. Ramamurthy, P. Upadhyaya, R. Venkatesan, Secure database-as-a-service with cipherbase, in *Proceedings of SIGMOD 2013* (New York, USA, 2013)
4. Z. Brakerski, V. Vaikuntanathan, Efficient fully homomorphic encryption from (standard) $textsfLWE$. SIAM J. Comput. **43**(2), 831–871 (2014)

5. V. Ciriani, S. De Capitani di Vimercati, S. Foresti, S. Jajodia, S. Paraboschi, P. Samarati, Fragmentation and encryption to enforce privacy in data storage, in *Proceedings of ESORICS* (Dresden, Germany, 2007)
6. V. Ciriani, S. De Capitani di Vimercati, S. Foresti, S. Jajodia, S. Paraboschi, P. Samarati, Fragmentation design for efficient query execution over sensitive distributed databases, in *Proceedings of ICDCS* (Montreal, Canada, 2009)
7. V. Ciriani, S. De Capitani di Vimercati, S. Foresti, S. Jajodia, S. Paraboschi, P. Samarati, Keep a few: outsourcing data while maintaining confidentiality, in *Proceedings of ESORICS* (Saint Malo, France, 2009)
8. V. Ciriani, S. De Capitani di Vimercati, S. Foresti, S. Jajodia, S. Paraboschi, P. Samarati, Combining fragmentation and encryption to protect privacy in data storage. ACM TISSEC **13**(3), 22:1–22:33 (2010)
9. V. Ciriani, S. De Capitani di Vimercati, S. Foresti, S. Jajodia, S. Paraboschi, P. Samarati, Selective data outsourcing for enforcing privacy. JCS **19**(3), 531–566 (2011)
10. V. Ciriani, S. De Capitani di Vimercati, S. Foresti, G. Livraga, P. Samarati, An OBDD approach to enforce confidentiality and visibility constraints in data publishing. JCS **20**(5), 463–508 (2012)
11. J.S. Coron, A. Mandal, D. Naccache, M. Tibouchi, Fully homomorphic encryption over the integers with shorter public keys, in *Proceedings of CRYPTO* (Santa Barbara, CA, USA, 2011)
12. J.S. Coron, D. Naccache, M. Tibouchi, Public key compression and modulus switching for fully homomorphic encryption over the integers, in *Proceedings of EUROCRYPT* (Cambridge, UK, 2012)
13. E. Damiani, S. De Capitani di Vimercati, S. Jajodia, S. Paraboschi, P. Samarati, Balancing confidentiality and efficiency in untrusted relational DBMSs, in *Proceedings of CCS* (Washington, DC, USA, 2003)
14. S. De Capitani di Vimercati, S. Foresti, S. Jajodia, G. Livraga, S. Paraboschi, P. Samarati, Fragmentation in presence of data dependencies. IEEE TDSC (2014), to appear
15. S. De Capitani di Vimercati, S. Foresti, G. Livraga, P. Samarati, Practical techniques building on encryption for protecting and managing data in the cloud, in *Festschrift for David Kahn*, ed. by P. Ryan, D. Naccache, J.J. Quisquater (Springer, Berlin, 2016)
16. S. De Capitani di Vimercati, S. Foresti, P. Samarati, Managing and accessing data in the cloud: privacy risks and approaches, in *Proceedings of CRiSIS* (Cork, Ireland, 2012)
17. S. De Capitani di Vimercati, S. Foresti, P. Samarati, Protecting data in outsourcing scenarios, in *Handbook on Securing Cyber-Physical Critical Infrastructure*, ed by S. Das, K. Kant, N. Zhang (Morgan Kaufmann, 2012)
18. De Capitani di Vimercati, S., Foresti, S., Samarati, P.: Selective and fine-grained access to data in the cloud. In: Jajodia, S., Kant, K., Samarati, P., Swarup, V., Wang, C. (eds.) Secure Cloud Computing. Springer (2014)
19. C. Gentry, Fully homomorphic encryption using ideal lattices, in *Proceedings of STOC* (Bethesda, MA, USA, 2009)
20. H. Hacigümüs, B. Iyer, S. Mehrotra, Providing database as a service, in *Proceedings of ICDE* (San Jose, CA, USA, 2002)
21. H. Hacigümüs, B. Iyer, S. Mehrotra, Efficient execution of aggregation queries over encrypted relational databases, in *Proceedings of DASFAA* (Jeju Island, Korea, 2004)
22. R. Jhawar, V. Piuri, Fault tolerance and resilience in cloud computing environments, in *Computer and Information Security Handbook*, ed by J. Vacca, 2nd edn (Morgan Kaufmann, Burlington, 2013), pp. 125–141
23. R. Jhawar, V. Piuri, P. Samarati, Supporting security requirements for resource management in cloud computing, in *Proceedings of CSE* (Paphos, Cyprus, 2012)
24. R. Popa, C. Redfield, N. Zeldovich, H. Balakrishnan, CryptDB: protecting confidentiality with encrypted query processing, in *Proceedings of SOSP* (Cascais, Portugal, 2011)
25. R. Rivest, L. Adleman, M. Dertouzos, On data banks and privacy homomorphisms, in *Foundation of Secure Computations*, ed by R. DeMillo, R. Lipton, A. Jones (Academic Press, Cambridge, 1978)

26. P. Samarati, S. De Capitani di Vimercati, Cloud security: issues and concerns, in *Encyclopedia on Cloud Computing*, ed by S. Murugesan, I. Bojanova (Wiley, New Jersey, 2016)
27. S. Tu, M.F. Kaashoek, S. Madden, N. Zeldovich, Processing analytical queries over encrypted data. Proc. VLDB Endowment **6**(5), 289–300 (2013)
28. H. Wang, L. Lakshmanan, Efficient secure query evaluation over encrypted XML databases, in *Proceedings of VLDB* (Seoul, Korea, 2006)

Database Community and Health Related Data: Experiences Through the Last Decade

Pietro H. Guzzi, Giuseppe Tradigo and Pierangelo Veltri

Abstract Database community has been involved in topics related to improve data-related techniques or to solve data access efficiency. Health domain has been attracting the interest of database community as an application domain for many database research topics, including: (i) health data heterogeneity (e.g., different health bioimages protocols), (ii) data size (e.g., patient health related data), (iii) biomedical signals (e.g., electrocardiography data, ECG), (iv) geographical data (e.g., epidemiological one), and more recently (v) genomic and proteomic data as well as NGS data. In this chapter we present experiences from the last decade, made in a medical school, where we used database experiences to manage and analyse clinical, biological and health related data. The methodology is problem oriented and shows how to start from a problem defined in the medical domain and choose and apply techniques often known by the database community. In this chapter interesting results, in terms of applications to the clinical and medical domains, are reported.

1 Introduction

The recent evolutions of healthcare technology made available a large number of data to healthcare community. Such data span a broad range in terms of format and dimensions. Despite the large heterogeneity the correct management of such data may improve considerably research and help clinicians and researcher as well as healthcare decision makers. More recently, medicine is moving to a data-science approach in order to face the so-called precision medicine approach [10]. Technologies for data productions enable the investigation of a large number of biological aspects consid-

P.H. Guzzi
Department of Medical and Surgical Sciences, University of Catanzaro, Catanzaro, Italy
e-mail: hguzzi@unicz.it

G. Tradigo · P. Veltri (✉)
DIMES, University of Calabria, Rende, Italy
e-mail: veltri@unicz.it

G. Tradigo
e-mail: gtradigo@dimes.unical.it

S. Flesca et al. (eds.), *A Comprehensive Guide Through the Italian Database Research Over the Last 25 Years*, Studies in Big Data 31, DOI 10.1007/978-3-319-61893-7_28

ering different points of view, from molecular aspects to the analysis of bio-signals as well as the integration of geographically related data. Clinical and health related fields are producing constantly growing datasets, containing data from many different actors, for instance: clinicians, biologists, biomedical engineers, bioinformaticians, patients, citizens. Latter for instance are improved in terms of information coming from health related applications (e.g., app devices for runners or bikers). Interactions among these actors are based on the exchange of information such as disease protocols, treatments and rules for early disease detection [22]. This huge quantity of data can be extremely valuable not only for medical activity assessment or performance verification, but also useful to researchers and social managers.

Recently, data is used for information prediction and not only for a posteriori statistical analyses [5]. Health structures are indeed managed by calculating the amount of money worth the clinical services provided by the health structure in the previous quarter, but no prediction model is used to decide where and how to intervene in the clinical process in order to optimize costs. Data is traditionally used to create statistical models, while the new trend is to use data as a fundamental tool for analysis, prediction and decision support for business intelligence (e.g., analytics, OLAP).

Moreover the availability of health related knowledge derived from this data is an important aim for many governments, that may decide their strategic politics on the basis of updated reports. Similarly, open data protocols are increasing data sharing and interchange among structure to allow the extraction of information useful to study chronic diseases, as well as reducing costs [7].

Consequently, database community has found a large field of research in order to design systems to help in managing huge quantity of information. Such systems are designed to support researchers, physicians and healthcare operators while managing health-related information [6]. For instance, managing information flow regarding patients in cardiology emergency intervention unit, is necessary to: (i) optimize processes while scaling in terms of number of patients versus number of operators and (ii) report data regarding research interesting topics.

In this chapter we report about experiences through last decade of applying database knowledge at University of Catanzaro, by following clinical different division cases and, for each one, presenting application scenarios. The chapter presents examples of different cases in the fields of: cardiology and hemodynamic, electrophysiology, magnetic resonance analysis, patient data sharing and computational epidemiology, genome-wide association studies and proteomics, early detection and remote data analysis.

2 Database Techniques for 'Omics' Data

Advances in genomic and proteomic research make possible to study new methodologies and frameworks linking genomic information (i.e. origins of diseases or cellular malfunctions) with clinical data. To enrich knowledge about diagnosis and therapies and treatments (to be included in EHR software tools), studies have been doing to verify biological hypotheses by means of extracting information from large clinical databases.

We studied many techniques for analysing proteomics data as well as genomic data by starting from raw data. Starting from previous studies in 2004–2008 [9, 24], passing through protein structure coordinates acquisition and prediction [16, 19], arriving recently in studying genomics and information extraction from genes data (both in NGS, next generating sequencing and microarray) [4, 13].

2.1 Managing Mass Spectrometry Data

Mass spectrometry (MS) is a well-known technique among physics researchers, which has been attracting biologists. By using MS, a spectrum containing couples of values [intensity, m/z] can be extracted by a biological sample. Biological sample of a biomolecule with a certain mass to charge ratio (m/z) is represented. Specialists may extract from spectra information on biological samples. MALDI-TOF data is produced by a Matrix-Assisted Laser Desorption/Ionization (MALDI) mass spectrometer where [intensity, m/z] couples are identified by a time of flight (TOF) detector. We developed a first experience by using MS data and by storing it in an ad hoc defined database management system.

Mass spectrometer is able to detect and characterize biomolecules, such as proteins/peptides, oligosaccharides and oligonucleotides, with molecular masses between 400 and 350,000 Dalton (Da). Intensity values in an m/z range (also called peaks), can be associated to peptide sequences contained in the original sample. Such an association is performed by querying publicly available databases containing theoretical peaks values associated to known proteins/peptides.

Each spectrum is usually stored in a flat file managed by the file system such that no meta information is associated to experiments thus limiting results sharing and reusing. Indeed, each mass spectrometer experiment is associated to such meta information: such as operator, spectrometer type, ionization technique, parameters, and so on. Even if in some scenario it is possible to speed up spectra file management operations (e.g., by using compression techniques), the main focus in using spectrometers is to simplify information extraction from spectra data. Each spectrum stored as flat file, may be as large as hundreds of Kilobytes to some Gigabytes. Large volume of data may occur while increasing the instruments resolution or by using a more precise technique. Retrieving biological information from peaks contained in each spectrum, adding annotations to it about biological intuition, and cross validating experimental results require high performance platforms and ad hoc data management strategies.

We experienced in defining a database system for managing spectra data, called SpecDB. It allows to manage data coming from different nodes, each one associated to a biological laboratory [24]. Spectra data produced in the laboratory are stored and queried against nodes. Data mining techniques and preprocessing have been used for proteins identification. SpecDB was designed to work with grid environment, that can be considered for the tome of publication (2005) could be considered pre-cloud global environment.

2.2 *Enhancing Information from Spectra Data*

The result of an experiment is a MS/MS spectrum, i.e. a collection of spectra signals. Each spectra is a large collection of pairs (mass-to-charge ratio, intensity). In such field we used database technology for improving the identification of peptides (protein portions) in mass spectrometry analysis, such as in MS/MS ICAT analysis. The target is to increase the number of peptides identified and quantified in input samples.

Figure 1 shows an example of a MS spectrum where *[m/z, intensity]* pairs are related to the presence of a biomolecule in the input sample, with mass-to-charge ratio *m/z* and abundance expressed by the *intensity* value [17]. Typically, these biomolecules are peptides. In fact they are more suitable for MS/MS sequencing than a whole protein. The MS/MS process performs multiple MS analysis steps by generating a mass spectrum for fragments related to a subset of selected peaks identified in a previous step.

Protein/peptide identification from MS/MS spectra consists in the computation of *qualitative* information and is performed by querying publicly available databases. Proteomics literature presents a large number of highly specialized repositories and tools for storing and handling large scale MS/MS proteomics datasets.

The protocol of the experiment marks two input protein mixtures (sample S_1 and sample S_2) with H and L labels, having identical chemical properties but different masses. Then, the labelled peptides are selectively captured by affinity chromatography. Identical peptides belonging to the same protein, but originated from different samples are detected at different *m/z* values because of the difference in mass, which corresponds to different flight times. After a database search, peptide sequence identification is performed to produce tables of proteins/peptides with their relative expression levels.

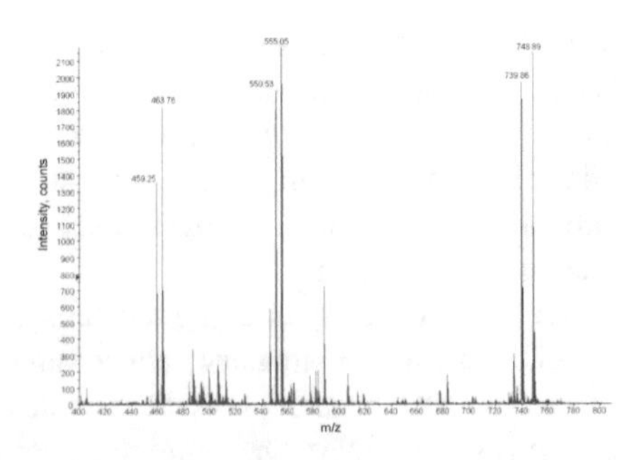

Fig. 1 Mass spectrum of a biological sample; X axis contains mass/charge ratio, Y axis indicates intensities at the detector

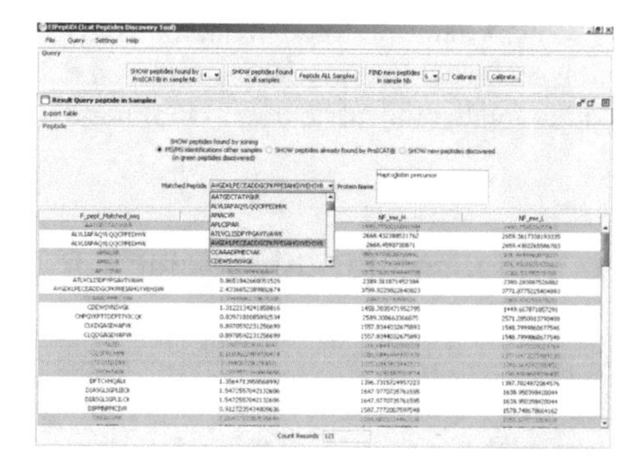

Fig. 2 The graphical user interface of eipeptidi showing the results of the analysis of data stored into the internal database

Database enrichment protocols have been used to implement a system, (see Fig. 2), that is used to discover new peptides [2].

2.3 Managing Next Generation Data for Population Wide Studies

Recently, possible applications of using Precision Medicine (PM) have been studied. Personalized-based Medicine revisits current workflows in patients treatment and healthcare [18]. The successful sequencing of the human genome [12], together with subsequent studies about the discovery of single nucleotide polymorphism (SNP) and genotyping efforts [8] showed that the human genome has a structure which is much more complex than the sequence of its nucleotides (i.e., primary sequence). One of the main discoveries is the linkage disequilibrium (LD) blocks which are separated by points of recombination throughout the entire genome. It has also been observed that it is possible to examine the DNA of an individual just by using a fraction of the million SNPs existing in the population, which has given a huge boost in genomics [20]. Genome-Wide Association Studies (GWAS) are a mature tool towards to implement PM, being able to link genetic features (e.g., particular gene variations and mutations) to phenotypical ones (e.g., diseases, analytes, life expectancy, BMI index, sex, race, age).

A major goal is to support physicians in finding the right drug for the right patient when they need it. This could also help patients affected by rare diseases for which it has been economically impracticable to find a cure so far. GWAS can be considered as the final step of a health workflow model in next generation clinical workflows. GWAS studies often involve hundreds of thousands of genotyped markers for several thousands individuals. Novel genetic data acquisition instruments can also acquire entire genomes, allowing for genome-wide analyses, for groups of patients or even populations, feasible. A GWAS tool, apart from the usual data management functionalities, can perform: (*i*) summary statistics, (*ii*) population stratification, (*iii*) association analyses, (*iv*) identity-by-descent estimation.

GWAS data are stored in flat files and partially structured. We experienced in mapping such data into partially structured databases to perform information extraction and to relate clinical data with familiar data obtained by genomics data.

3 Database Techniques for DICOM Images

In this section we report about our experiences in applying database techniques in acquiring, storing and managing radiological based images. Data is produced by devices that respect the DICOM protocol [15] and stored in well defined informatics systems called RIS PACS. Nevertheless, we treated problems related to manage DICOM images due to physicians requirements which were not able to use such sources of information adapted to their own protocols or to the department experiences.

We report about an experiences (from 2006 until today) regarding acquiring and managing biomedical images. We report about two experiences. The former regards DICOM images acquired during emodynamic surgery intervention. The latter regards the acquisition of DICOM images and their fusion in a single environment for annotation and statistical analysis.

3.1 Acquiring and Annotating Cardiovascular Images

Vessels in human are often subject to specific diseases that make difficult the flow of blood inside them. Such pathologies are often lethal without surgical intervention. The most used surgical intervention is known as angioplasty. Coronary stents are placed during a percutaneous coronary intervention, also known as angioplasty. We experiences in developing a system able to acquire and manage DICOM images stored in a database and annotated. Currently, optimization techniques used by physicians to measure coronary stenoses, are included in software tools coming with angiographic equipment. Cartesio [11] is an innovative software tool used by physicians working in emodynamics surgery rooms. It helps in making a pre-implant analysis for the estimation of the dimensions of the stent to be implanted. The tool interacts with virtually any angiographic equipment by acquiring its high-resolution video signal and offering a set of functions to play with images and to draw a virtual stent over the acquired video frames. It allows the operator to calibrate the stent before implanting it. Measurements help physicians to evaluate the exact dimension of the stenosis and to define the physical parameters for the virtual stent in such a way that it will be compatible with the vessel structure.

The tool interacts with any angiographic equipment by acquiring its high-resolution video signal and offering a set of functions to zoom, pan, playback, measure and draw a virtual stent over the acquired video frames. Each rendered measurement or stent pre-implant analysis can be exported as a bitmap image on the

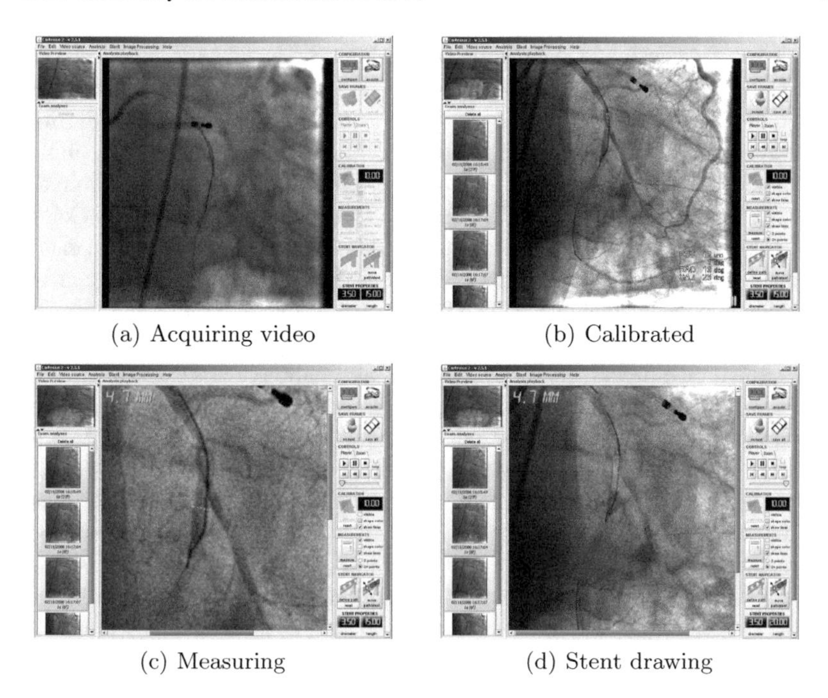

(a) Acquiring video (b) Calibrated

(c) Measuring (d) Stent drawing

Fig. 3 Four interaction phases with the *Cartesio* tool

file system or saved in an experiment repository on a relational database for future reference. The software uses a balloon catheter with radio-opaque iridium markers positioned at 10 mm from each other. A calibration tool supports such phases and it automatizes all the steps of image acquisitions, the reconstruction of the stents, and storing data acquired during analysis in the DICOM standard image format on a local repository, data can be reused for case studies [14]. Figure 3 reports screenshots related to the Acquisition, Calibration, Measuring and Stent drawing and positioning simulation phases are depicted.

The informatics system based on Cartesio allows to manage experiences and to support students and clinical doctors while simulating cases by performing measurements on real cases.

3.2 *Annotating DICOM Images*

Systems have been developed to manage medical data and information for clinical, therapeutic and administrative purposes in the oncological domain. A Tag Management has been implemented to connect and correlate clinical data with DICOM images. A module has been developed to integrate data provided by nuclear medicine and radiology for oncology domain. A novel DICOM-CDA XML integration module has been designed to integrate DICOM images and CDA (Clinical Document Architecture). Synthetically, a set of DICOM images are managed by a server PACS and grouped related to acquisition mode and image orientation; then, information are extracted by these images and converted in XML/HL7 format by using CDA document. PACS system collects and manages DICOM images coming from difference diagnostic sources (e.g. PET - Positron Emission Tomography, CT - Computer Tomography, or MR - Magnetic Resonance) that are saved into a DICOM database. The images can be visualized by using a specific DICOM viewer tool. A conversion of images in other format as jpeg can be performed; moreover, in order to interface the two different HL7 and DICOM format, HL7 CDA has been used. By DICOM images and their annotations, tags or relevant information can be extracted to support diagnosis, to acquire general knowledge or to search into the database. In particular, a first set of tags are extracted automatically by header of DICOM file, while a second set of tags are extracted manually respect to the tags used by physician to study a single case. These tags and information are converted in standard format and they are saved in a XML database. In this way, they can be used by other modules and guarantee the interoperability between physician users, clinical departments and health structures. To execute this module, libraries have been used and integrated into the system. Moreover, import and export functions have been provided to manage CDA HL7 documents. Import function allows the physician to load oncology DICOM images of a patient that represents the results of a diagnostic examination.

The system allows to annotate DICOM images using clinical heterogeneous data used for diagnosis as: (i) personal data for patient, (ii) clinical and oncology data, (iii) instrumental data from radiology or nuclear medicine departments, and (iv) data and information related to acquisition methods, codify and communication protocol that have been used. Users insert these annotations by using a graphical user interface. System provides the automatic translation of the inserted data into XML-CDA format.

4 Databases For: Clinical Records, Audiological Signals and Voice Signals

In this section we present our experiences on using databases to support clinicians in screenings clinical signals for disease early detection. The idea is to use on-line systems to acquire health related signals to support patients and clinicians while monitoring health related parameters. Web based systems can be used to monitor patients' health status, e.g., as a follow-up of surgical procedures. Analysis results, integrated with patients clinical information, can be used for early detection and parameters monitoring.

4.1 Sharing Electronic Patient Records

Sharing Electronic Patient Records (EPRs) among different hospitals is a key technology enabling both the improvement of document and information sharing as well as the improvement of patient status. Peer-to-peer (P2P) infrastructures may be used to exchange information about specific pathologies or treatments among different health structures. The sharing of EPRs, as the one reported in the SIGMCC [3] system, enable the easy collection and sharing of EPR data among different hospital. The system is based on the use of an XML-based distributed EPR repository enables nodes to share information among different health structures managing data by using personalized databases and systems. A P2P network might map the exchange for clinical data processing in a network of health structures, where physicians may perform queries and obtain results about patients or treatments. An XML repository can be used to store metadata extracted from heterogeneous EPRs. Health structure operators formulate queries against the meta-EPR schema from one node of a network (i.e., a health structure). Queries are distributed to the connected hospitals through a P2P infrastructure hosting the meta-EPR instances [23].

By using our database experiences we implemented a system composed by four modules: (*i*) a data wrapper, which is in charge of collecting data from distributed EPR data sources in clinical structures; (*ii*) a module able to map information into an metadata base (e.g., XML repository) storing subsets of EPRs; (*iii*) a query engine able to compose and distribute queries among nodes of a peer-to-peer network; (*iv*) a security and update management module able to guarantee privacy and data updates (i.e., keeping updated information in the XML database).

An XML based Metadata has been defined and used to share data among healthcare structures and manage updates.

4.2 Managing Audiology and Vestibular Data

The vestibular human apparatus is the inner part of the ear (also called labyrinth), constituted by bones and soft tissues. It is the sensory system that allows to detect the position and motion of the head, i.e., rotation and motion, allowing balance and spatial orientation. While head is moving, the vestibular apparatus stabilizes the eyes and adjusts neck and body muscle tone during movements 15 to send information useful to balance movements. The brain uses such information to analyse movements and to guide muscles to dynamically balance body cinematic.

We present experiences while acquiring data obtained by the audiology clinical unit, and stored in a EPR for audiology unit, which aims to support audiology physicians for: (i) managing patients and (ii) extracting hidden information from diagnosis as well as instruments data.

There usually are many instrumental tests carried out on the patients in audiology. We used data obtained by one of the most important, i.e. the Vestibular Evoked

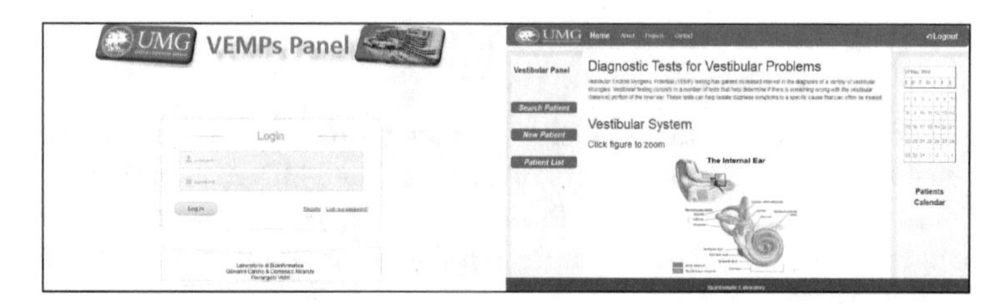

Fig. 4 Vestibular data acquisition

Myogenic Potentials (VEMPs) instrument. The VEMPs is a neurophysiological technique used to assess the function of the otolithic organs of the inner ear. When a test is performed, the instrumentation gives to the operator a set of numerical data printed in a graph which represents evoked potentials caused by stimulus. The operator observes the trend of the peaks of the evoked potentials and to associates such a trend to possible pathological conditions. This evaluation is typically carried out by taking into account the delay time of formation of the first positive peak (P_1) and the first negative peak (N_1) as well as the peak-to-peak value between P_1 and P_2.

The audiometric test is the measurement of hearing acuity at varying sound intensities and frequencies. There are two main types of tests: (i) the vocal test, used to measure the capability of hearing and understanding recorded speech signals and (ii) the tonal test, used to estimate the capability of hearing pure tone recorded signals. A system has been developed to acquire data and to perform experiments. Figure 4 shows the web based system implemented and used for vestibular data acquisition and analysis.

Algorithms (such as data mining, machine learning based ones, decision support systems) have been used to support correct diagnosis of specific pathology related to a patient. Tests have been performed for predictive purpose to find interesting patterns in the data, as well as clusters and subgroups of data. The system has been used to elaborate data regarding almost two thousands patients data. Relation between patients and diagnosis has been analyzed to extract possible features in vestibular data. It has been release and it is in use by the clinical unit at University Medical School, audiology unit.

4.3 *Voice Signal Analysis*

Similarly, voice analysis has been realised for supporting patients in an on line analysis for acquisition and monitoring of patient signals. Offline analysis is performed to monitor patient status. Database management systems can be used to store voice signals in order to give indication to medical doctors for tuning cures and strategy. Patients record voice signals and give feedbacks acquired and measured by the system. The main modules for a system for managing data related to the analysis are:

1. Acquisition component, responsible for collecting clinical data from patients;
2. Screening component, implementing the hearing/voice health test;
3. Data analysis component;
4. Data management component;
5. Database component, which stores all the data and signals.

By using remote analysis it is possible to extract functional parameters from vocal signals and to measure the voice quality, e.g., the frequency of fundamental tone or the pitch, and to associate data features to the presence of anomalies.

Audio files have been used to check for anomalies. Among the others, the following phases can be performed server-side: (*i*) pro-actively notify specialist physicians with acquisitions requiring attention, (*ii*) analyse novel acquired voice signals by extracting fundamental features from signal data (e.g., fundamental frequency or pitch, jitter, shimmer, noise level), (*iii*) store finished analyses in the database. We experienced the use of online and offline remote systems for signal analysis as the possibility for the physicians to create statistics and applying data mining models to perform population studies on vocal related pathologies. The developed systems have been presented in [1].

5 Geographical Database Systems and Clinical Data

A common and frequent problem for a better understanding of clinical datasets is associating clinical features (e.g., diseases) with geographical and environmental data. Geographical software modules are able to associate information about disease diffusion with geographical layers containing land information. GIS technologies can be applied to analyze clinical data containing health information about large populations. Often, clinical datasets already contain geographical data (i.e., zip codes, addresses) which can be used to geolocate clinical events (e.g., patients, diseases). For instance, Geomedica [21] geocodes existing clinical data and extracts geographical coordinates positions which can then be geographically mapped and analyzed/queried using both SQL-like languages and web-based graphical user interfaces.

5.1 GeoBlood: A System for Managing Biological Data

We designed and developed Geoblood, a system able to store and analyze biological data. It is based on a web infrastructure and on the use of geographical information. We tested the system in a collaboration with the university magna graecia hospital that shared clinical data about patients. After the anonymization of data, we provided geographical information as well as habits for each patient. We show that by using

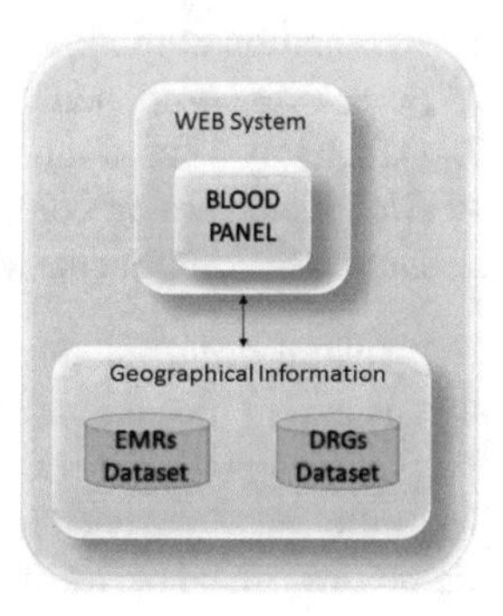

Fig. 5 Functional architecture

geographic data as common layer among different data sources, it is possible to extract and read useful information.

Figure 5 reports the web based architecture of GeoBlood. Blood Panel is the module deputed to the interface with the user. Blood Panel communicates with the back end server through a secure socket. The back end is constituted by two cloud databases storing clinical information (EMRs datasets), hospital information (DRGs Datasets) and geographic information.

Thanks to a web interface, users can access and analyze biological analytes data, navigate through them by using statistics and export relevant data. By using the system biologists and physicians can have a more precise idea, for instance of: (*i*) how analytes change over time in a single patient, and (*ii*) what are the statistics of blood values by sex and age. We are using this system to monitor biological data mapped on a geographic readable format.

5.2 Administrative Analysis by Using GIS

Database have been used to support a methodology which is able to extract and integrate information related to: (i) biological analytes; (ii) diagnosis codes (DRGs); (iii) geographical information; and (iv) quality of drinking waters. The methodology has been implemented in a framework based on a multi-step process: (i) we cross reference data by using a semantics-based clustering procedure, (ii) we extract information from EMRs and then, (iii) we cluster them by looking for similar patterns of diseases. Biological records in each disease cluster are analyzed to evaluate intra-cluster similarity by selecting near analytes typologies and values. Finally, biological data is related to diagnosis codes and mapped into geographical areas to visualize outlier patients. The DRGs have been obtained from a hospital database containing both medical and costs information. They also contain medical information organized in MDC. To enhance the geographical information extracted from EMRs, the framework presents a geographical layer containing street and administrative features. Administrative and geopolitical layers are loaded from open access available

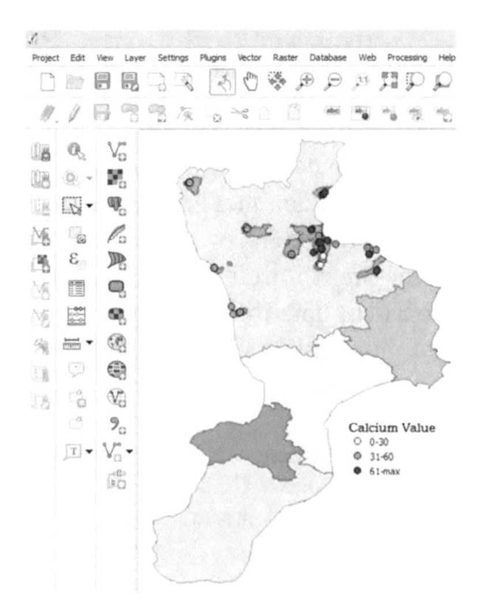

Fig. 6 Cartographic map reporting result datapoints from a spatial query

data sources and used to map patient information. The integration module is able to import and merge additional information about environmental facts or geometries which can be related to pathologies (e.g. water sources). Geographical data and blood tests are extracted and used in the disease prediction models. Finally, a hydric drinking water geographic network layer is included. The framework has been tested proving it is possible to: (*i*) associate data information extracted from EMRs (health and administrative) with geographical data and (*ii*) correlate diseases (as well as bio-analytes outliers) with drinking water quality. The goal is to show that heterogeneous information extracted from health data sources (e.g., EMRs and bio-analytes data sets), as well as from administrative information, can be used to: (*i*) cluster patients by looking at similar diseases, (*ii*) analyze bio-analytes (e.g. glycemia) in order to identify subgroups (i.e., patients showing high intra-cluster similarities), outliers or extreme values (i.e. patients showing high intra-cluster differences), (*iii*) generate geographical heat-maps for the outliers patients, and (*iv*) relate chronic diseases with drinking water and water sources.

The proposed methodology has been tested and validated in two case studies. For the first case study, we performed experiments starting from the groups of MDC and identified the outlier patients by considering the intracluster similarities among analytes. We built a set of heat maps by spatially querying the number of outliers per areas of interest (i.e. provinces and municipalities). For the second case study, we performed experiments analyzing water quality and diseases in an integrated way. Figure 6 shows the graphical user interface of the realized system showing datapoints from a spatial query performed on the clinical database. We evaluated the correlation between levels of calcium and magnesium ions present in drinking water supplied by municipalities of Calabria with large number of patients affected by heart-related diseases.

6 Summary and Future Perspectives

We are still making research in a medical and biological department making experiences with colleagues that ask for new techniques for analysing and managing data. Database techniques and research topics find interesting applications in health related data. Analytic and prediction techniques may be currently considered the more interesting areas for clinical and biological data. Future perspectives can be considered the use of data for the purpose of predicting and supporting in planning investment and intervention on health management and also to support studies related to cure chronic diseases.

Acknowledgements We are grateful to all colleagues working with us during latter decades. Many of them had roles in studying and defining problems and topics that have then been studied and that brings to the development of tools.

References

1. F. Amato, M. Cannataro, C. Cosentino, A. Garozzo, N. Lombardo, C. Manfredi, F. Montefusco, G. Tradigo, P. Veltri, Early detection of voice diseases via a web-based system. Biomed. Signal Proc. Control **4**(3), 206–211 (2009)
2. M. Cannataro, G. Cuda, M. Gaspari, S. Greco, G. Tradigo, P. Veltri, The eipeptidi tool: enhancing peptide discovery in ICAT-based LC MS/MS experiments. BMC Bioinform. **8** (2007)
3. M. Cannataro, D. Talia, G. Tradigo, P. Trunfio, P. Veltri, SIGMCC: a system for sharing meta patient records in a peer-to-peer environment. Future Gener. Comput. Syst. **24**(3), 222–234 (2008)
4. F. Cristiano, P. Veltri, M. Prosperi, G. Tradigo, On the identification of long non-coding rnas from RNA-Seq, in *2016 IEEE International Conference on Bioinformatics and Biomedicine (BIBM)* (IEEE, New York, 2016), pp. 1103–1106
5. L. Federico, P. Franco, A. Minelli, A. Perri, L. Caroprese, R. Picarelli, G. Tradigo, E. Vocaturo, F. Dattola, A. Fortunato, P. Lambardi, S. Laurita, I. Pellegrino, A. Garro, A. Pugliese, A. Tagarelli, P. Veltri, E. Zumpano, SINSE+: a software for the acquisition and analysis of open data in health and social area, in *24th Italian Symposium on Advanced Database Systems, SEBD 2016, Ugento, Lecce, Italy, June 19-22, 2016, Ugento, Lecce, Italia, June 19–22* (2016), pp. 310–317
6. J.D. Fernández, M. Lenzerini, M. Masseroli, F. Venco, S. Ceri, Ontology-based search of genomic metadata. IEEE/ACM Trans. Comput. Biology Bioinform. **13**(2), 233–247 (2016)
7. D.B. Fridsma, Health informatics: our domain, our challenge. J. Am. Med. Inf. Assoc. **23**(6), 1202–1202 (2016)
8. S.B. Gabriel, S.F. Schaffner, H. Nguyen, J.M. Moore, J. Roy, B. Blumenstiel, J. Higgins, M. DeFelice, A. Lochner, M. Faggart et al., The structure of haplotype blocks in the human genome. Science **296**(5576), 2225–2229 (2002)
9. F. Gullo, G. Ponti, A. Tagarelli, G. Tradigo, P. Veltri, A time series approach for clustering mass spectrometry data. J. Comput. Sci. **3**(5), 344–355 (2012)
10. L. Hood, S.H. Friend, Predictive, personalized, preventive, participatory (P4) cancer medicine. Nat. Rev. Clin. Oncol. **8**(3), 184–187 (2011)
11. C. Indolfi, M. Cannataro, P. Veltri, G. Tradigo, Cartesio: a software tool for pre-implant stent analyses, in *9th International Conference Computational Science - ICCS 2009, Baton Rouge, LA, USA, May 25–27, 2009, Proceedings, Part I* (2009), pp. 810–818

12. E.S. Lander, L.M. Linton, B. Birren, C. Nusbaum, M.C. Zody, J. Baldwin, K. Devon, K. Dewar, M. Doyle, W. FitzHugh et al., Initial sequencing and analysis of the human genome. Nature **409**(6822), 860–921 (2001)
13. H.N. Manners, M. Jha, P.H. Guzzi, P. Veltri, S. Roy, Computational methods for detecting functional modules from gene regulatory network, in *Proceedings of the Second International Conference on Information and Communication Technology for Competitive Strategies* (ACM, New York, 2016), p. 3
14. S. Matl, R. Brosig, M. Baust, N. Navab, S. Demirci, Vascular image registration techniques: a living review. Med. Image Anal. **35**, 1–17 (2017)
15. P. Mildenberger, M. Eichelberg, E. Martin, Introduction to the dicom standard. Eur. Radiol. **12**(4), 920–927 (2002)
16. L. Palopoli, S.E. Rombo, G. Terracina, G. Tradigo, P. Veltri, Improving protein secondary structure predictions by prediction fusion. Inf. Fus. **10**(3), 217–232 (2009)
17. E.F. Petricoin, A.M. Ardekani, B.A. Hitt, P.J. Levine, V.A. Fusaro, S.M. Steinberg, G.B. Mills, C. Simone, D.A. Fishman, E.C. Kohn et al., Use of proteomic patterns in serum to identify ovarian cancer. The lancet **359**(9306), 572–577 (2002)
18. G. Pio, D. Malerba, D. D'Elia, M. Ceci, Integrating microrna target predictions for the discovery of gene regulatory networks: a semi-supervised ensemble learning approach. BMC Bioinf. **15**(S-1), S4 (2014)
19. K. Predrag, C. Mirabello, G. Tradigo, I. Walsh, P. Veltri, G. Pollastri, Toward an accurate prediction of inter-residue distances in proteins using 2d recursive neural networks. BMC Bioinf. **15**(1), 6 (2014)
20. R. Sachidanandam, D. Weissman, S.C. Schmidt, J.M. Kakol, L.D. Stein, G. Marth, S. Sherry, J.C. Mullikin, B.J. Mortimore, D.L. Willey et al., A map of human genome sequence variation containing 1.42 million single nucleotide polymorphisms. Nature **409**(6822), 928–933 (2001)
21. G. Tradigo, P. Veltri, S. Greco, Geomedica: managing and querying clinical data distributions on geographical database systems. Procedia Comput. Sci. **1**(1), 979–986 (2010)
22. L. Vaira, M.A. Bochicchio, S.B. Navathe, Perspectives in healthcare data management with application to maternal and fetal wellbeing, in *24th Italian Symposium on Advanced Database Systems, SEBD 2016, Ugento, Lecce, Italy, June 19–22, 2016* (2016), pp. 31–40
23. S. Van de Velde, P. Roshanov, T. Kortteisto, I. Kunnamo, B. Aertgeerts, P.O. Vandvik, S. Flottorp, Tailoring implementation strategies for evidence-based recommendations using computerised clinical decision support systems: protocol for the development of the guides tools. Implement. Sci. **11**(1), 1 (2016)
24. P. Veltri, M. Cannataro, G. Tradigo, Sharing mass spectrometry data in a grid-based distributed proteomics laboratory. Inf. Process. Manag. **43**(3), 577–591 (2007)

From Rome to Milan, from Appennines to Alps, from Data to Big Data

Carlo Batini

Abstract In the last 25 years my life changed a lot. In the paper I describe the evolution of my research activities, and to some extent the change in my favorite hobby, that is climbing mountains, and in my life, from Rome to Milan. Moving from Rome to Milan was initially a bit traumatic, but at end I made it. In the story, I also mention the influence in my research activity of the ten years period I spent at the Authority for Informatics in Public Administration, where the size of projects I faced has been several orders of magnitude larger than toy examples conceived in my previous papers. I also mention the wonderful scholars that joined me in these 25 years.

Keywords Conceptual database design · Data quality · Service science

1 Change of Life

In 1993 my professional life changed suddenly. My research activity was formerly in the area of applications of the Entity Relationship model to data base design, schema integration, visual query languages and data dictionaries; in those days I published with Pino Di Battista and Giuseppe Santucci the paper [?], where we tried to solve what we believed was a relevant methodological problem in the area of conceptual design. Assume you have, say, one hundred conceptual schemes (see Fig. 1) and you need to integrate them. A one-shot integration is unfeasible, since the resulting integrated schema is too big to be reproduced on a screen or else on a paper sheet. So we have to find some mechanism to reduce complexity, while achieving the integration goal. The idea has been to adopt, besides integration, a new kind of transformation, the *abstraction transformation*, that, when applied to a schema, produces a new schema more compact than the previous one, while representing the same real world. In this way, using iteratively the integration and abstraction transformations

C. Batini (✉)
Dipartimento di Informatica, Sistemistica E Comunicazione, Università degli Studi di Milano Bicocca, Viale Sarca 336, Milano, Italy
e-mail: carlo.batini@unimib.it; batini@disco.unimib.it

S. Flesca et al. (eds.), *A Comprehensive Guide Through the Italian Database Research Over the Last 25 Years*, Studies in Big Data 31, DOI 10.1007/978-3-319-61893-7_29

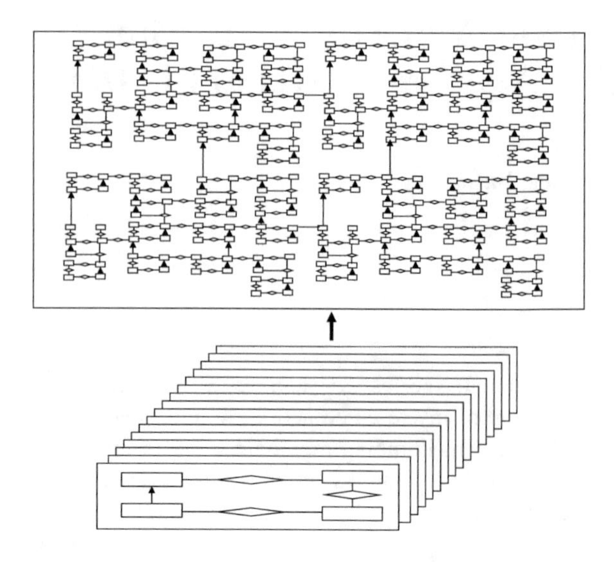

Fig. 1 Big concerns in the integration of big amount of schemas

together, we may produce a pyramid of schemes that we call *repository of schemes*. The advantages are:

- the whole conceptual content associated to the one hundred schemes is represented in an integrated fashion;
- the comprehension of the conceptual content is much richer due to the possibility of looking around among the abstraction levels of the pyramid.

In 1993 I decided to leave university to go to work in a new institution, the Authority for Informatics in Public Administration (Autorità per l'Informatica nella Pubblica Amministrazione, Aipa in the following). The main goal of Aipa was to boost in Italian central public administration the use of information technologies to improve services provided to citizens and companies.

The problems I faced in AIPA were immense; fortunately I survived, and as a first activity I was the responsible of a survey on the adoption of ICT technologies in Italian central public administration; part of the survey was focused on the most relevant 400 databases managed in the Ministries. Suddenly, we were forced to move from previous toy examples to a big project of adoption of the repository in the abstraction-integration of 400 schemes, corresponding to about 5.000 entities and as many relationships. In producing the repository I was aided by Guglielmo Longobardi. The pyramid of schemes was seven layers deep; in Fig. 2 the fragment of the repository representing the schemes of interest of the Land agency is shown.

During my period at Aipa I was almost overwhelmed, as I said, by the extent of problems I had to face, and by the imperative of using my research achievements for useful practical objectives. For instance: once produced the repository, how we could use it to improve the overall database architecture of Italian Public Administration? More importantly: how could we use the repository to fit the original goal of Aipa,

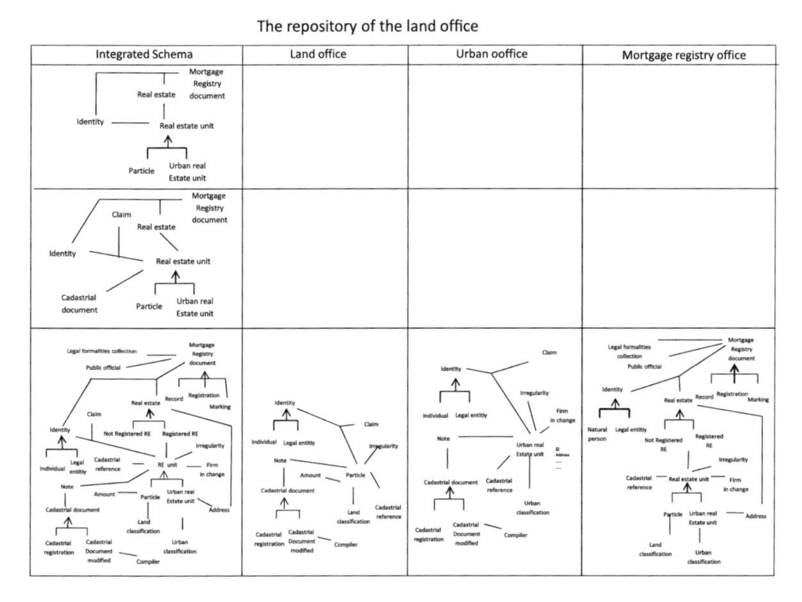

Fig. 2 The small dictionary of schemas of the land agency

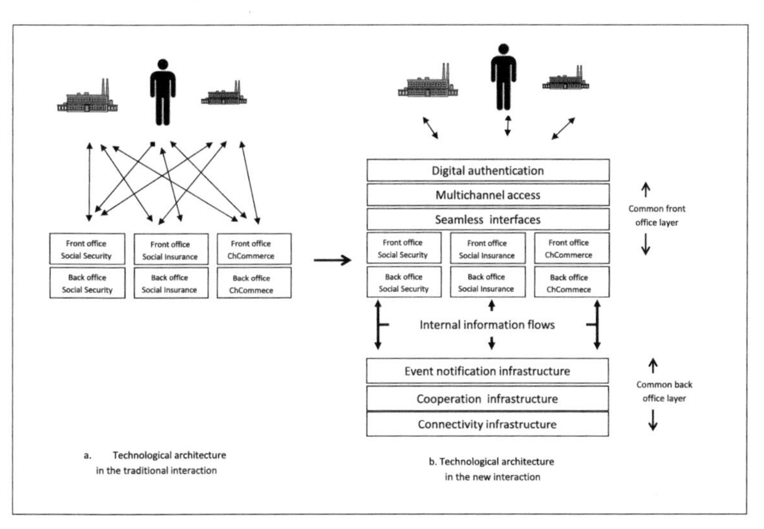

Fig. 3 The small dictionary of schemas of the land agency

namely to improve services offered to citizens and businesses? We needed to communicate with administrations; better, we needed to convince the administrations to share data among them. To enable a cooperative approach, we conceived at Aipa a *cooperative application architecture*, shown in Fig. 3, where the evolution from the traditional "stovepipe" interaction between users and administrations to the new cooperative approach is highlighted.

Agency	Identifier	Name	Type of activity	Address	City
Agency 1	CNCBTB765SDV	Meat production of John Ngombo	Retail of bovine and ovine meats	35 Niagara Street	New York
Agency 2	0111232223	John Ngombo canned meat production	Grocer's shop, beverages	9 Rome Street	Albany
Agency 3	CND8TB76SSDV	Meat production in New York state of John Ngombo	Butcher	4, Garibaldi Square	Long Island

Fig. 4 Three records representing the same company in different registries

In 1999 I was involved in a new project that leveraged the cooperative architecture. The goal of the project "Services to businesses" was to drastically improve services provided to businesses by agencies such as Social Security, Social Insurance and Chambers of Commerce. The project focused initially on simplifying the huge amount of transactions required for a business in order to register itself with agencies, as well as to update their existing registry entry.

Complicating this project was the fact that similar data about one business was likely to appear in multiple databases, each autonomously managed by different agencies that were not able to share the company data. The problem was aggravated by the many errors contained in those databases, that commonly result in mismatches among different records that refer to the same business (see in Fig. 4 a figurative example of the three records referring to the same business in the three registries owned by Social Security, Social Insurance and Chambers of Commerce).

One major consequence of having multiple disconnected views representing the same data, was that businesses experienced severe service degradation in their interaction with the agencies. Because of these complications, the comprehensive approach chosen for the project followed two main strategies, aimed at improving the state of existing business data and at maintaining correct record alignment for all future data, respectively:

1. Extensive record matching and data cleaning was performed on the existing business information, resulting in the reconciliation of a large amount of business registry entries;
2. A "one stop shop" approach was followed to simplify the life of a business and to ensure the correct propagation of its data. In this approach, one single agency is selected as a front-end for all communications with businesses. Once the information received by a business is certified, it is made available to other interested agencies through a publish/subscribe mechanism. The enabling technology was a new database that provided a thin layer of integration, consisting in the three keys of each business in agency registries linked together.

Another project that exploited the cooperative architecture has been "Normeinrete" (Laws in the network), see [17], where I was aided by Caterina Lupo.

The two above mentioned projects (and others) unveiled me another crucial issue in information systems of Public Administration, namely *data quality*. In a collaboration with the Italian National Bureau of Census (Istat in the following) we conceived a methodology characterized by a rich spectrum of strategies and techniques that allow for its adaptation to many contexts and domains. The principal reason for this is the complexity of the structure of the public administration, which in most countries is characterized by at least three tiers of agencies:

- central agencies, located close to each other, usually in the capital city of the country;
- peripheral agencies, corresponding to organizational structures distributed thorough the territory, hierarchically dependent on central agencies;
- local agencies, that are usually autonomous from central agencies, corresponding to districts, regions, provinces, municipalities, and other smaller administrative units.

The above is an example of the organizational structure of a public administration sector, that has many variants in different countries. However some common aspects include:

- its complexity, in terms of interrelations, processes, and services in which they are involved, due to the fragmentation of competencies among agencies. This frequently involves information flows exchanged between several agencies at the central and local level;
- agencies' autonomy, which makes it difficult to enforce common rules;
- the high heterogeneity of meanings and representations that characterize databases and data flows, and the high overlapping of usually heterogeneous records and objects.

Improving data quality in such a complex structure results usually in a broad and costly project, needing an activity that may last several years. In order to solve the most relevant issues related to data quality, in the Istat methodology (see in Fig. 5 its organization in terms of phases) attention is primarily focused on the most common type of data exchanged between agencies, namely, address data.

2 From Rome to Milan, from Apennines to Alps

When I left Rome and moved to Milan, in 2002, I joined my family but I left behind me, besides my relatives and friends, a huge amount of hikes and climbings in the Apennines. Let me mention my dear friend and companion of so many hikes and climbings, Mario Lucertini, that knew and managed very well my idiosyncrasies in climbings; Mario has passed in 2002.

In Milan I had to reorganize from scratch my research activity and to find new collaborators. Furthermore, my aim was to preserve the legacy of Aipa; I had dedicated ten years of my life to Aipa, and, conversely, Aipa enriched me with great

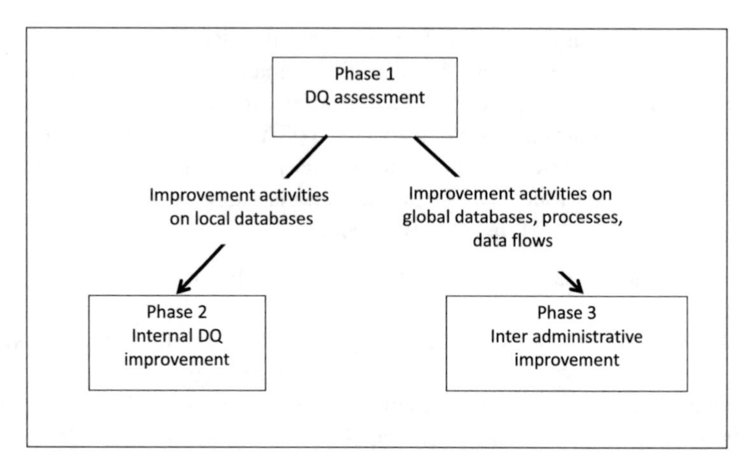

Fig. 5 The Istat methodology

experiences and challenging innovation projects. Let me say that I will be always indebted with Guido Rey that offered me such opportunity.

In Milan I found excellent scholars: they have been in the course of years Andrea Barone, Federico Cabitza, Marco Comerio, Angela Locoro, Andrea Maurino, Anisa Rula, Matteo Palmonari, Gianluigi Viscusi. With their help, I focused my subsequent research activity on (at least) five topics that I had experienced in practice in Aipa: methodologies for the eGovernment life cycle, further applications of repositories of conceptual schemes, data and information quality, service science, and social and economic value of data. I will address such topics in the rest of the paper.

2.1 eGovernment

The main contributions in this area are [13] and the book [23]. The book aims to describe a comprehensive methodology for service oriented information systems planning. The concept of service is at the center of the book; the methodology is focused on quality of service as a key factor for eGovernment initiatives. Furthermore, the methodology aims at encompassing the relationships existing between ICT technologies and social contexts of service provision, organizational issues, and juridical framework, looking at ICT technologies more as a means than an end.

The available approaches to eGovernment usually provide only one perspective to public managers and local authorities on the domain of intervention, either technological, organizational, legal, economic or social. Our aim has been to provide a methodology which structurally supports the choice of the optimal eGovernment development plans, considering all the above mentioned perspectives together. The quality driven construction of the eGovernment plan is initially influenced by the social, legal, and organizational perspective, while it subsequently achieves its final shape when considering the economic and technological perspectives.

Another contribution of the book is in the data centric approach to eGovernment. Data, information and knowledge are the most relevant strategic resources used by public administrations in eGovernment processes and services delivered. As a consequence, the methodology pays specific attention to data and to the quality of data managed in administrative processes and services, whose inner quality is strictly related to the quality of the data they use.

2.2 *Repositories of Conceptual Schemes*

In [7, 8, 12] we describe the main results of the applied research activity in the area of repositories of conceptual schemes. A first manual methodology is provided for the production of the repository of schemas of central public administrations in Italy. Then an heuristic semi-automated methodology is described that exploits the central repository for the production of the repository of an Italian region. The former exact methodology and the heuristic one are compared according to their correctness, completeness, and efficiency. Finally, it is shown how such repositories can be used in eGovernment initiatives for planning activities and in the identification and ranking of projects.

The growth in the availability of data requests public administrations for an effective control over their information asset. Furthermore, having a global representation of the core concepts of such an asset in a repository implies to manage large set of conceptual schemas, and, consequently to adopt quality processes in their management. In [6] several quality properties of repositories are investigated, analyzing them within a real, large scale experience.

2.3 *Data and Information Quality*

In the investigation of issues related to data quality I was very lucky in encountering Monica Scannapieco, initially as a student and subsequently as scholar. With Monica we have investigated a huge amount of issues in the area of data and information quality.

Due to our common interest in cooperative architectures, shared also with Massimo Mecella (I have been lucky with him too) we started to investigate cooperative information systems, that as we have seen in the discussion on the cooperative architecture, are based on services to be offered and broad casted, and on the opportunity of building coordinators and brokers on top of such services. The quality of data exchanged and provided by different services hampers such approach, as data of low quality can spread all over the cooperative system. In [18] a service-based framework for managing data quality in cooperative information systems is presented. An XML-based model for data and quality data is proposed, and the design of a broker, which selects the best available data from different services, is presented. Such a broker also supports the improvement of data based on feedbacks to source services.

The experience in Aipa on data quality and an extensive literature review have been distilled in the book [21], whose main contributions in the area of relational data are:

- a detailed description of several data quality dimensions and metrics,
- a discussion on techniques and methodologies for data quality assessment and improvement
- a focus on record linkage.

Inter-organization business processes involve the exchange of structured data across information systems. In [1] we assume that data are exchanged under given conditions of quality (offered or required) and prices. We describe a brokering algorithm for obtaining data from peers, by minimizing the overall cost under quality requirements constraints. The algorithm extends query processing techniques over multiple database schemes to automatically derive an integer linear programming problem that returns an optimal matching of data providers to data consumers under realistic economic cost models.

The paper [3] is a first attempt to extend previous data quality methodologies for relational data to other types of semi-structured or loosely structured data. Reference [4] is a systematic and comparative description of methodologies for the selection, customization, and application of data quality assessment and improvement techniques. Methodologies are compared along several dimensions, including the methodological phases and steps, the strategies and techniques, the data quality dimensions, the types of data, and, finally, the types of information systems addressed by each methodology.

The contribution in [2] is a conceptual framework for the automatic discovery of dependencies between data quality dimensions. Dependency discovery consists in recovering the dependency structure for a set of data quality dimensions measured on attributes of a database. This task is accomplished through a data mining methodology, by learning a Bayesian Network from a database. The bayesian network is used to learn dependencies between data quality dimensions. The proposed framework is instantiated on a real world database. The network model shows how data quality can be improved while satisfying budget constraints.

The contribution in [9] identifies and discusses key topics characterizing recent data quality research and their impact on future research perspectives in a context where information is increasingly diverse. The investigation considers basic issues related to data quality definitions, dimensions, and factors referring to information systems, information representation, influence of the observer and of the task. The paper, written by a computer science scholar and two scholars graduated in philosophy, concludes by discussing how philosophical studies can contribute to a better understanding of some key foundational problems that emerged in our analysis.

In the last years, the investigation has concerned the evolution from data to big data, especially referring to the *variety* dimension. Reference [10] investigates the

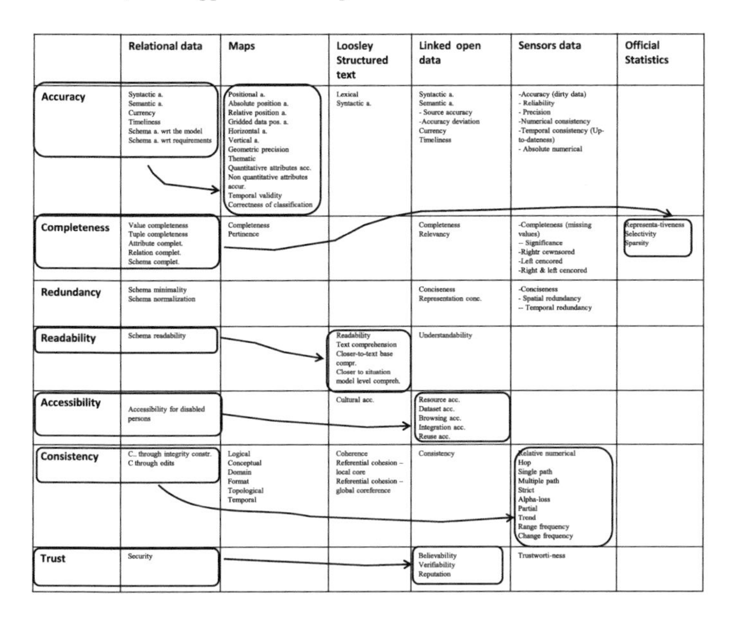

	Relational data	Maps	Loosley Structured text	Linked open data	Sensors data	Official Statistics
Accuracy	Syntactic a. Semantic a. Currency Timeliness Schema a. wrt the model Schema a. wrt requirements	Positional a. Absolute position a. Relative position a. Gridded data pos. a. Horizontal a. Vertical a. Geometric precision Thematic Quantitativre attributes acc. Non quantitative attributes accur. Temporal validity Correctness of classification	Lexical Syntactic a.	Syntactic a. Semantic a. - Source accuracy -Accuracy deviation Currency Timeliness	-Accuracy (dirty data) - Reliability - Precision -Numerical consistency -Temporal consistency (Up-to-dateness) - Absolute numerical	
Completeness	Value completeness Tuple completeness Attribute complet. Relation complet. Schema complet.	Completeness Pertinence		Completeness Relevancy	-Completeness (missing values) -- Significance -Rightr cewnsored -Left cencored -Right & left cencored	Representa-tiveness Selectivity Sparsity
Redundancy	Schema minimality Schema normalization			Conciseness Representation conc.	-Conciseness - Spatial redundancy -- Temporal redundancy	
Readability	Schema readability		Readability Text comprehension Closer-to-text base compr. Closer to situation model level compreh.	Understandability		
Accessibility	Accessibility for disabled persons		Cultural acc.	Resource acc. Dataset acc. Browsing acc. Integration acc. Reuse acc.		
Consistency	C.. through integrity constr. C through edits	Logical Conceptual Domain Format Topological Temporal	Coherence Referential cohesion – local core Referential cohesion – global coreference	Consistency	Relative numerical Hop Single path Multiple path Strict Alpha-loss Partial Trend Range frequency Change frequency	
Trust	Security			Believability Verifiability Reputation	Trustworti-ness	

Fig. 6 Quality dimensions and their evolution for different phenomena

evolution of data quality issues from traditional structured data according to three coordinates that are significant in the context of the *big data* phenomenon: the data type considered, the source of data, and the application domain (see in Fig. 6 quality dimensions, phenomena and evolutions considered in the paper). The framework allows ascertaining the relevant changes in data quality emerging in big data, with special focus on the analysis of intrinsic data characteristics that drive the change in quality dimensions and metrics.

After ten years from the Data Quality book, in 2016 Monica Scannapieco and me decided that it was time to extend our overall investigation on data quality issues to new topics; our effort produced a new book [11], whose main contributions are:

- an extension of quality dimensions from relational data to linked open data, images, maps, loosely structured texts, the domains of law texts, sensor networks and official statistics.
- a thorough comparative description of the wide literature in the last ten years on record linkage and object identification.
- an extension of assessment and improvement methodologies to heterogeneous types of data in input to the process.

2.4 Service Science

I mention here my research activity in service science just to put in evidence how previous research in data bases has deeply influenced the investigation on services.

Since 2008 we have proposed a semantic repository to represent in an integrated fashion services offered by a public administration [20]. In the paper and in subsequent papers such as [15] we adopt for a service repository the two basic *generalization* and *aggregation abstractions*, and show how the integrated view provided by the repository can positively influence the efficiency and effectiveness of the production and delivery processes of a service provider. Finally in [19] we adapt similarity measures that are used in database schema integration methodologies to the integration of service repositories.

2.5 Economic and Social Value of Data

Digital data transform markets and are a powerful driver for boosting business economic value. At the same time, they can provide social utility, in terms of improvement of quality of life. For instance, in 2011 the Economist published the results of a research made by the University of Stockholm on open data available in Uganda referring to the quality of care in hospitals; the availability of such indicators enabled households to choose hospitals for children pathologies, resulting in a cut of the death rate of under fives by a third.

In [5] we investigate economic value. The paper discusses two concepts that have been associated with various approaches to data and information, namely *capacity* and *value*, focusing on data base architectures, and on two types of technologies diffusely used in integration projects, namely *data integration*, in the area of enterprise information integration, and *publish and subscribe*, in the area of enterprise application integration. Furthermore, the paper proposes and discusses a unifying model for information capacity and value, that considers also quality constraints and run time costs of the data base architecture.

Social vaLue of data is investigated in [14, 22, 24]. In such papers an original methodology is proposed by which to assess the construct of the *situated social value* of open data; such methodology is applied to the health care domain in regard to information by which hospitals can be ranked to compare service providers. Our methodology encompasses a questionnaire-based user study and a method by which to rank information items by their perceived social value in situated scenarios. While the social component of the construct is addressed traditionally, the main assets of our contribution are:

- to ground the assessment of information value on a multidimensional space of potential situations where that information can be perceived as valuable
- to inquiry the respondents of the user study about their empathized perceptions in daily life scenarios, which are defined to cover the dimensional space properly.

The findings of the user study are reported and the implications on the construct assessment discussed in the aim to enable IS benchmarking and more focused interventions of data quality improvement.

Fig. 7 Lake Antermoia, the most wonderful place in the world

Finally, in [16] we extend to *info-graphics* previously described methods to assess information quality on different dimensions, to take into account both formal and substantial aspects; interaction quality along dimensions like usability and ease of use. The goal is to measure whether the quality of infographics affects the perception of information and the users' interaction. The overall results suggest that, although interactive infographics are perceived as more complex, the experience with them is better. From our observations, we derived a model to assess the overall quality of static and interactive infographics, based on information, interaction and design quality dimensions.

3 Conclusion

These 25 years have passed very fast. As I said, I consider myself a lucky person. Let me conclude that one reason of my satisfaction has been to move in my hikes from the bare while fascinating landscapes of the mountains in the Apennines to the wonderful panoramas of the Alps (thank you Barbara for accompanying me!) and Dolomites. Believe me, Lake Antermoia in the Dolomites (see Fig. 7) is the most wonderful place in the world.

References

1. A. Avenali, C. Batini, P. Bertolazzi, P. Missier, Brokering infrastructure for minimum cost data procurement based on quality-quantity models. Decis. Support Syst. **45**(1), 95–109 (2008)
2. D. Barone, F. Stella, C. Batini, Dependency discovery in data quality, in *International Conference on Advanced Information Systems Engineering* (Springer, 2010), pp. 53–67

3. C. Batini, D. Barone, F. Cabitza, G. Ciocca, F. Marini, G. Pasi, R. Schettini, Toward a unified model for information quality. QDB/MUD **2008**, 113–122 (2008)
4. C. Batini, C. Cappiello, C. Francalanci, A. Maurino, Methodologies for data quality assessment and improvement. ACM Comput. Surv. (CSUR) **41**(3), 16 (2009)
5. C. Batini, C. Cappiello, C. Francalanci, A. Maurino, G. Viscusi, A capacity and value based model for data architectures adopting integration technologies, in *A Renaissance of Information Technology for Sustainability and Global Competitiveness. 17th Americas Conference on Information Systems, AMCIS 2011, Detroit, Michigan, USA, 4–8 Aug 2011* (2011)
6. C. Batini, M. Comerio, G. Viscusi, Managing quality of large set of conceptual schemas in public administration: methods and experiences, in *International Conference on Model and Data Engineering* (Springer, 2012), pp. 31–42
7. C. Batini, R. Grosso, G. Longobardi, Design of repositories of conceptual schemas for large-scale e-government projects. Electron. Gov. Int. J. **3**(3), 306–328 (2006)
8. C. Batini, R. Grosso, C. Piemonte-Riccardo, Reuse of a repository of conceptual schemas in a large scale project. Adv. Topics Database Res. **5**(5), 170 (2006)
9. C. Batini, M. Palmonari, G. Viscusi, Opening the closed world: A survey of information quality research in the wild, in *The Philosophy of Information Quality* (Springer, 2014), pp. 43–73
10. C. Batini, A. Rula, M. Scannapieco, G. Viscusi, From data quality to big data quality. J. Database Manag. **26**(1), 60–82 (2015)
11. C. Batini, M. Scannapieco, *Data and Information Quality: Dimensions, Principles and Techniques* (Springer, Berlin, 2016)
12. C. Batini, G. Viscusi, D. Barone, A repository based approach to data reverse engineering. in *Proceedings of the First International Working Session on Reverse Engineering Techniques for Application Portfolio Management,(RE4APM 2007), co-located with the 23rd IEEE International Conference on Software Maintenance (ICSM 2007), Paris*, vol. 5 (2007)
13. C. Batini, G. Viscusi, D. Cherubini, Govqual: a quality driven methodology for e-government project planning. Gov. Inf. Q. **26**(1), 106–117 (2009)
14. F. Cabitza, A. Locoro, C. Batini, A user study to assess the situated social value of open data in healthcare. Procedia Comput. Sci. **64**, 306–313 (2015)
15. M. Comerio, C. Batini, M. Castelli, S. Grega, M. Rossetti, G. Viscusi, Service portfolio management: a repository-based framework. J. Syst. Softw. **104**, 112–125 (2015)
16. A. Locoro, F. Cabitza, R. Actis-Grosso, C. Batini, Static and interactive infographics in daily tasks: a value-in-use and quality of interaction user study. Comput. Hum. Behav. **71**, 240–257 (2017)
17. C. Lupo, C. Batini, A federative approach to laws access by citizens: the normeinrete system, in *EGOV* (2003)
18. Mecella, M., Scannapieco, M., Virgillito, A., Baldoni, R., Catarci, T., Batini, C.: Managing data quality in cooperative information systems. On the Move to Meaningful Internet Systems 2002: CoopIS, DOA, and ODBASE pp. 486–502 (2002)
19. F. Narducci, M. Comerio, C. Batini, M. Castelli, A similarity-based framework for service repository integration. Data Knowl. Eng. **106**, 18–35 (2016)
20. M. Palmonari, G. Viscusi, C. Batini, A semantic repository approach to improve the government to business relationship. Data Knowl. Eng. **65**(3), 485–511 (2008)
21. D. Quality, *Concepts, methodologies and techniques* (Batini, Scannapieco, 2006)
22. G. Viscusi, C. Batini, Information production and social value for public policy: a conceptual modeling perspective. Policy Internet **8**(3), 334–353 (2016)
23. G. Viscusi, C. Batini, M. Mecella, *Information Systems for eGovernment: a Quality of Service Perspective* (Springer, 2010)
24. G. Viscusi, M. Castelli, C. Batini, Assessing social value in open data initiatives: a framework. Future Internet **6**(3), 498–517 (2014)

Author Index

S. Flesca et al. (eds.), *A Comprehensive Guide Through the Italian Database Research Over the Last 25 Years*, Studies in Big Data 31, DOI 10.1007/978-3-319-61893-7

Printed in the United States
By Bookmasters